AF704167

WITHDRAWN

Introduction to Integral Equations with Applications

Introduction to Integral Equations with Applications

Second Edition

ABDUL J. JERRI
Clarkson University

A Wiley-Interscience Publication
JOHN WILEY & SONS, INC.
New York • Chichester • Weinheim • Brisbane • Singapore • Toronto

QA
431
.J47
1999

This text is printed on acid-free paper. ♾

Copyright © 1999 by John Wiley & Sons, Inc. All rights reserved.

Published simultaneously in Canada.

No part of this publication may be reproduced, stored in a retrieval system or transmitted in any form or by any means, electronic, mechanical, photocopying, recording, scanning or otherwise, except as permitted under Sections 107 or 108 of the 1976 United States Copyright Act, without either the prior written permission of the Publisher, or authorization through payment of the appropriate per-copy fee to the Copyright Clearance Center, 222 Rosewood Drive, Danvers, MA 01923, (978) 750-8400, fax (978) 750-4744. Requests to the Publisher for permission should be addressed to the Permissions Department, John Wiley & Sons, Inc., 605 Third Avenue, New York, NY 10158-0012, (212) 850-6011, fax (212) 850-6008, E-Mail: PERMREQ @ WILEY.COM.

For ordering and customer service, call 1-800-CALL-WILEY.

Library of Congress Cataloging-in-Publication Data:

Jerri, Abdul J., 1932–.
Introduction to integral equations with applications / Abdul J. Jerri. — 2nd ed.
p. cm.
"A Wiley-Interscience publication."
Includes bibliographical references and index.
ISBN 0-471-31734-9 (alk. paper)
1. Integral equations—Numerical solutions. I. Title.
QA431.J47 1999
515'.45—dc21 99-14638

Printed in the United States of America

10 9 8 7 6 5 4 3 2 1

In memory of my father and mother

Contents

Preface

The goal of this present second edition is still the same as that of the first edition. It is to present the subject of integral equations, their varied applications, and basic methods of solutions on a level close to that of a first (sophomore) course in ordinary differential equations. This is not such an easy task, especially when we assume only the basic calculus and differential equations as prerequisites. The main thrust here is that a variety of applied problems have their natural mathematical setting as integral equations, thus they have the advantage of usually simpler methods of solution. In addition, a large class of initial and boundary value problems, associated with differential equations, can be reduced to integral equations, whence enjoy the advantage of the above integral representation. Such topics also bring to light the unity of differentiation and integration. It may be said that such a basic integral equations course would complement the elementary differential equations course, especially when the actual coverage in the latter is (most often) limited to initial value problems, and for obvious historical reasons. This being that differential equations began following the work of Leibnitz and Newton, with the flavor of applications in dynamics, which had occurred a long time before integral equations started to get attention at the very beginning of this century.

We should point out here that for this elementary presentation of integral equations – assuming only calculus and differential equations preparation – the treatment in all chapters, except for (the optional) Chapter 6, is *formal*. This is in the sense that clear procedures and steps for arriving at the solution or some basic results are emphasized, without necessarily stopping to give their complete mathematical justification. The latter most often requires more advanced mathematics preparation. Thus we shall be limited to give those justifications that would not require us to go beyond the level of this basic applicable undergraduate text.

In this second edition all comments, suggestions, and corrections relayed by students, colleagues from around the world, and the expert reviewers of the journals of mathematics and other concerned professions, were addressed. They all deserve my sincere thanks and appreciation. Such suggestions, it is hoped, will help this edition in attaining even more the same goal set in the first edition for an undergraduate focusing integral equations text to serve the students of science, engineering, and mathematics. To stay with this important goal, and keep the required text material to a comparable size to that of the first edition, we decided to have a new (optional) Chapter 7 for the detailed numerical methods. This includes using higher quadrature rules for the numerical approximation of the integrals. The main changes made for this second edition, in light of the suggestions received, are:

1. Discussion of the basic theory and illustration of solutions to Fredholm integral equations of the first kind as done in Section 5.4.

2. Detailed discussion and illustration of numerical integration with the basic higher quadrature rules, and numerical methods of solving Fredholm and Volterra integral equations. The use of the high quadrature rules is covered in a new (optional) Chapter 7.

3. More exercises for each section including some challenging ones.

4. More emphasis on clear statements of the basic theorems for the existence and uniqueness of the solutions of integral equations. The brief introduction of basic theory in Chapter 6 can be considered optional, when seen in light of the main goal of this elementary text.

5. Conditions for the existence of integral transforms, their inverses, and important operations are spelled out. A more detailed treatment is found in the author's undergraduate-graduate book on the subject (Marcel Dekker, 1992) entitled "Integral and Discrete Transforms with Applications and Error Analysis."

6. Very clear examples of singular integral equations with general discussions of their solutions. Such discussions must be taken in light of the (undergraduate) level set for this book, in which no complex analysis preparation is assumed for the reader.

7. More applications to update, replace, and complement the already ample variety of applied problems as recognized by all reviews of the first edition. These now include some relevant problems in higher dimensions.

8. More emphasis are placed on the interrelation between the integral equations and the differential equations representations of boundary and initial value problems. This is also to emphasize that differentiation and integration are inseparable.

9. All detected and reported typographical as well as other errors are corrected, and some examples are deleted and replaced by more appropriate ones. Almost all the suggestions made by the expert reviewers of the journals of our and other concerned professions have been very seriously addressed, keeping in mind the main goal of an undergraduate book for scientists, engineers, as well as mathematicians. This includes the reviews of three critical experts for the first draft of this new edition, and another three reviewers of the final draft.

10. For this edition we now have a "Student's Solution Manual" to accompany this book. It contains very detailed solutions to all the odd numbered problems in the text (see the end of the preface for details).

With these changes and additions, the first chapter still starts with the statements of a number of problems from different subjects, to illustrate their integral equation representation. Although the reader is warned against expecting a full understanding of some of these problems from such a brief presentation, a very detailed formulation of them is given in Chapter 2. This is followed by the usual classification of integral equations and a clear derivation and illustration of some important integral and differential identities needed for the formulations in Chapter 2 and later chapters. Such identities are essential for showing how we can go from the integral equations representation to the differential equations representation and vice versa. We have also improved upon the self-contained (short but simple) presentation of the Laplace and Fourier transforms with clear statements for the existence of the transforms. Chapter 1 is concluded by a section on simple elements of numerical integration which represents only the essentials necessary for the numerical solutions of Fredholm and Volterra integral equations that are discussed in Chapters 5 and 3, respectively. The higher quadrature numerical integration rules along with their needed tables are covered in a new (optional) Chapter 7. They are well illustrated for the numerical integration, setting up the numerical approximation of Volterra and Fredholm equations, and the numerical solution of these integral equations. Chapter 2 involves very detailed modeling of problems as integral equations with a new section on integral equations in higher dimensions illustrated with the Schrödinger equation integral representation in the momentum space. This includes population dynamics, control, mechanics, radiation transport, and boundary and initial value problems. Chapter 3 deals with methods of solving Volterra integral equations, including approximate and numerical methods, which are presented in detail. Chapter 4 is devoted to the construction and properties of Green's functions, which is very important for reducing boundary value problems to Fredholm integral equations. Chapter 5 deals with basic theory and detailed methods of solving Fredholm integral equations including the use of the Green's functions, and a detailed presentation of the familiar approximate and numerical methods of solutions. Methods of estimating the eigenvalues of homogeneous Fredholm integral equations are also presented. In this edition a new special section (Section 5.4) is added for a very elementary theory and a method of solving Fredholm integral equations of the first kind.

Also, more varied numerical methods are used in the new Chapter 7 for solving the different integral equations, compared to the very basic ones in Chapters 3 and 5 as it was the case in the first edition. In Chapter 6 we have a brief and descriptive discussion of the theory regarding the convergence of the methods of solving linear as well as nonlinear integral equations. For the basic introductory undergraduate course, this chapter is clearly optional.

In each chapter we have attempted to present many clear examples in every section, followed by a good number of related exercises at the end of each section with hints to (almost) all exercises and answers to all the exercises.

Suggestions for Course Adoption

To use this text for an elementary one-semester or one-quarter course in integral equations, we suggest that from Section 1.4 of Chapter 1 only the very essential elements, that are necessary for the convolution theorems, of the Laplace and the Fourier transforms be included, a selected number of mathematical modeling problems from Chapter 2 be covered, depending on the students' interest, and some selected subsections of the relatively long Chapter 5 be omitted. An exposure to the very basic numerical methods of solution in Chapters 3 and 5, with their exercises and detailed answers, is very beneficial. Another possibility is to present the most basic material for a one-semester course in integral equations with boundary value problems as part of the senior or first-year graduate course in methods of applied mathematics for scientists and engineers. For this purpose, we have included the added Chapter 7 on the numerical methods using higher quadrature rules.

The Student's Solutions Manual

A Student's Solutions Manual to accompany this book with Additional Solved Problems is now available from the author directly. Student's Solutions Manual to Accompany "Introduction to Integral Equations with Applications – Second Edition, Wiley & Sons, Inc., 1999 by Abdul J. Jerri" with Additional Solved Problems, Sampling Publishing, 1999 by Abdul J. Jerri (ISBN 0-9673301-0-6).

To order: Telephone (315)265-2755 and (315)265-1005, Fax (315)265-2755, e-mail: solnman@ hotmail.com and jerria@clarkson.edu. Also see website: http:// www.clarkson.edu/~jerria/solnman. Send $29.95 plus $2.95 for shipping and handling in the United States and Canada, and $4.95 abroad (all in US currency) to: Attention S.A. Jerri, 69 Leroy Street, Potsdam, NY 13676, USA.

Acknowledgments

I would like to acknowledge many helpful suggestions from colleagues and students during the preparation and use of the first edition of this book as well as using the manuscript of the present second edition. I would also like to thank all of those colleagues and students who read the first draft of this second edition, and made valuable remarks and corrections. Professors C.A. Roberts and A. Aluthge read the prefinal draft of this edition and made very valuable suggestions that helped a great deal in steering this book towards its main stated goal of serving as the very first introduction to the subject for undergraduate students in science and engineering and I owe them my deep gratitude. Professor A. Bastys made the most thorough reading of the prefinal draft, attending to the very details of the text with very candid suggestions and corrections, and I owe him my deepest gratitude.

I would like to thank especially the reviewers here and abroad who made very constructive criticisms and detailed suggestions, which I have attempted to address very seriously, and which I hope will contribute to the desired quality and purpose of this book. In particular, Prof. J. Chochran made the most detailed critical evaluation with constructive suggestions. Indeed I have also requested him to review this manuscript, which he did with suggestions that have contributed to the further focusing of this book toward being the first introductory book on the subject for undergraduates in the applied fields. Professor Chochran deserves my gratitude. Also Professors I. Fenyô and M. Putinar made very detailed and constructive reviews of the first edition, and I owe them my sincere thanks.

All along the process of developing this second edition, Prof. M.Z. Nashed has been very generous in his constructive suggestions and valuable criticisms, thus I am deeply indebted to him. Mr. J. Craparo read the first draft of the manuscript, made suggestions and supplied detailed numerical solutions, and he deserves my thanks.

The staff of Wiley and Sons, especially Ms. J. Downey, Ms. A. Loredo, and Ms. S. Liu, deserve my thanks for their effective cooperation. I am grateful to Ms. C. Smith for typing the prefinal draft of the first edition and the final camera-ready form of this edition, and for typing the changes and additions to this new edition. Mr. J. Hruska, Jr. deserves thanks for making the drawings with patience and care.

I owe my deepest thanks to my wife Suad and my daughter Huda for their continued support and patience during the long hours of preparing this edition.

A.J.J.

1

Integral Equations, Origin, and Basic Tools

An integral equation is an equation in which the unknown function $u(x)$ appears under an integral sign. A general example of an integral equation in $u(x)$ is

$$u(x) = f(x) + \int K(x,t)u(t)dt \tag{1.1}$$

where $K(x,t)$ is a function of two variables called the *kernel* or *nucleus* of the integral equation. According to Bôcher [1914], the name *integral equations* was suggested in 1888 by du Bois-Reymond, although the first appearance of integral equations is accredited to Abel for his thesis work on the Tautochrone, which was published in 1823 and 1826, and which we shall present shortly. There is also the opinion that such first appearance was in Laplace's work in 1782 as it shall make sense when we speak of the inverse Laplace transform in Section 1.4.1. For example, the Laplace transform of the given (known) function $f(t)$, $0 < t < \infty$ is

$$\mathcal{L}\{f(t)\} = F(s) = \int_0^\infty e^{-st} f(t)dt, \qquad s > a \tag{1.2}$$

provided that the integral converges for $s > a$. So, if we are now given $F(s)$, say $F(s) = \frac{1}{s^2}$, $s > 0$, and we are to find the original function (now as unknown) $f(t)$, or the inverse Laplace transform of $F(s)$, i.e., $f(t) = \mathcal{L}^{-1}\{F(s)\}$,

$$\frac{1}{s^2} = \int_0^\infty e^{-st} f(t)dt, \tag{1.2a}$$

then we are against solving an integral equation (1.2) in (the unknown) $f(t)$. So it does make sense that integral equations started with Laplace, since he was, in the final analysis, after recovering the original function $f(t)$ from knowing $F(s)$ in (1.2). In our above example, $f(t) = t$.

In the same vein, Fourier in 1820 solved for the inverse $f(x)$ of the following Fourier transform $F(\lambda)$ of $f(x)$, $-\infty < x < \infty$

$$\mathcal{F}\{f\} = F(\lambda) = \int_{-\infty}^{\infty} e^{-i\lambda x} f(x)dx \tag{1.3}$$

as

$$f(x) = \mathcal{F}^{-1}\{F\} = \frac{1}{2\pi}\int_{-\infty}^{\infty} e^{i\lambda x} F(\lambda)d\lambda. \tag{1.4}$$

Hence in finding the (unknown) $f(x)$ in equation (1.3), he solved an integral equation in $f(x)$ with the solution given in (1.4). With such an explicit solution $f(x)$ for (1.3), it is not surprising that some historians consider this Fourier (inverse transform) result of (1.4) as the first very clear and reachable solution of an integral equation [Bell, 1945 p. 525]. We note that the formula for the inverse Laplace transform [as shown in (1.65)] is not as accessible as the above (1.4) of the inverse Fourier transform, where the mere statement of the former in (1.65) needs familiarity with complex contour integration (that we shall not pursue at the level of this introductory text).

Some problems have their mathematical representation appear directly, and in a very natural way, in terms of integral equations. Other problems, whose direct representation is in terms of differential equations and their auxiliary conditions, may also be reduced to integral equations. Problems of a "hereditary" nature fall under the first category, since the state of the system $u(t)$ at any time t depends by definition on all the previous states $u(t-\tau)$ at the previous times $t-\tau$, which means that we must sum over them, hence involve them under the integral sign in an integral equation. We may then say that such problems, among others, have integral equations as their natural mathematical representation. The examples, which we illustrate next, are from population dynamics, the surge in birth rates, the mortality of equipment and their rate of replacement, biological species living together, population of fish and game, the torsion of a wire or rod, the control of a rotating shaft, the shape of a hanging chain, the deflection of a rotating rod, and the shape of a wire that allows a bead to descend on it in a predetermined time (Abel's problem). More problems are included in this edition, and in particular, those in higher dimensions. This includes the potential distribution in a disc and Schrödinger equation in the three-dimensional momentum space. We will present almost all of these problems with their basic clear statements as they are represented by integral equations, leaving the detailed derivation of a good selection of them for Chapter 2. The rest of the examples are problems that are formulated in terms of ordinary or partial differential equations with initial and/or boundary conditions that are reduced to an integral equation or equations. The advantage here is that the auxiliary conditions are automatically satisfied, since they are incorporated in the process of formulating the resulting integral equation. The other advantage of the integral equation form is in the case when both differential equations as well as integral equations forms do not have exact, closed-form solutions in terms of elementary known functions. In this case we must resort to numerical or approximate computations, where the integral representation is more suitable.

As mentioned above the detailed formulations of many of the problems presented in the next section are given in Chapter 2, but first we need to familiarize ourselves with the various types of integral equations and acquire some basic mathematical tools necessary for facilitating such formulations.

For future reference the integral equation,

$$u(x) = f(x) + \int K(x,t)u(t)dt \tag{1.1}$$

may be written in the *operational* (abbreviated) form or notation as

$$u(x) = f(x) + (\mathcal{K}u)(x) \tag{1.5}$$

or

$$u = f + \mathcal{K}u$$

where $\mathcal{K}$ is an *integral operator* for the integral in (1.1) that maps the *function u*, as an input, to an output $(\mathcal{K}u)(x) \equiv \int K(x,t)u(t)dt$ in the range of the integral operator $\mathcal{K}$. As seen in (1.1), the image of the function u under such mapping becomes $u - f = (\mathcal{K}u)$. Such mapping $(\mathcal{K}u)$ of the function u by the (integral) operator $\mathcal{K}$ is similar to the usual mapping done by function $F(x)$ on the *variable x*. The difference here is that the domain of the operator $\mathcal{K}$ is a class of functions $u(x)$ instead of the numbers x for the operation of the function $F(x)$. This operator notation in (1.2) does help describing, in a very brief and elegant way, the more general transformation of functions instead of just numbers, and thus the topic of analysis of functions as functional analysis, a subject that we shall not pursue further in this introductory book, except for, possibly, a brief mention in the (optional) Chapter 6.

We may mention that some situations may also involve *a rate of change*, besides the cumulative nature, where a *derivative* is used. Hence the unknown $u(x)$ is involved inside an integration as well as a differentiation operation. The result of such mathematical model is a *hybrid* equation termed *integro-differential equation*. For example,

$$a_0(x)u(x) + a_1(x)\frac{du}{dx} = f(x) + \int K(x,t)u(t)dt \tag{1.6}$$

is an integro-differential equation in (the unknown) $u(x)$. See also equations (1.13) and (1.14) for the two species living together.

1.1 VARIOUS PROBLEMS AS INTEGRAL EQUATIONS

In this section we present statements of a number of problems from many different fields which will be classified, primarily, according to whether they are formulated directly in terms of integral equations, or are represented in terms of differential equations with auxiliary conditions that can be reduced to integral equations. We must emphasize again that what we are about to present are clear statements of the problems and their integral equation representations, which is for the main purpose of familiarizing the reader with the various applications of integral equations. We must caution against the temptation of expecting the formulation of all these problems on the spot! A good number are simple enough to avail themselves to such expectations. Others, which are placed here because of their interesting and representative nature, need more detailed derivation, which we shall cover in Chapter 2. Our purpose here is to convey a feeling for modeling in terms of integral equations. For example, the integral equation in $u(t)$,

$$u(t) = \int_0^t K(t,\tau)u(\tau)d\tau \tag{1.7}$$

relates the present state $u(t)$ to the accumulation (integral) of what had happened to all its previous values $u(\tau)$ from $\tau = 0$ to the present time $\tau = t$. The formulation of most of these problems will be given in detail in Chapter 2, others are left for exercises, and for the remainder, which are not as suitable for solution here, we refer the interested reader elsewhere for detailed treatment. For such problems we will give the appropriate references, which are included in the bibliography at the end of the text.

This section consists of four parts: part A covers applied problems of hereditary nature that have integral equations as their natural settings, part B addresses finding the inverse of integral transforms, such as the Laplace transform for an example, as solving integral equations (of the first kind), part C deals with ordinary differential equations associated with initial or boundary conditions that reduce to (Volterra or Fredholm) integral equations, and part D presents an example of an integral equation in two dimensions, which is associated with a partial differential equation and boundary conditions that reduces to a (Fredholm) integral equation in two dimensions.

A. Integral Equations as the Problems' Natural Setting

Human population

The problem of forecasting human population may be one of the clearest examples formulated as an integral equation since the population $n(t)$ at time t depends on the number of the initial population $n(0) = n_0$ surviving to time t, and, more importantly, all children born during the time interval $0 < \tau < t$ who survive to time t. The dependency of the population $n(t)$ on the initial population n_0 and the previous populations $n(\tau)$, $0 < \tau < t$, is represented by the integral equation

$$n(t) = n_0 f(t) + k\int_0^t f(t-\tau)n(\tau)d\tau \tag{1.8}$$

where n_0 is the number of people present at time $t = 0$ and $f(t)$ is the *survival function* (Figure 1.1), which is the fraction of the number of people that survive to age t. With regard to the integral in (1.8) we may remark that $kf(t-\tau)n(\tau)\Delta\tau$

represents the number of children born in the time interval $\Delta\tau$ around time τ that survive to time t. It is clear that their number is proportional to $n(\tau)$, the population present at time τ, and that their survival function at time t is $f(t-\tau)$ since they are then of age $t-\tau$. The detailed formulation of (1.8) is presented in (2.3) to (2.8) of Section 2.1.

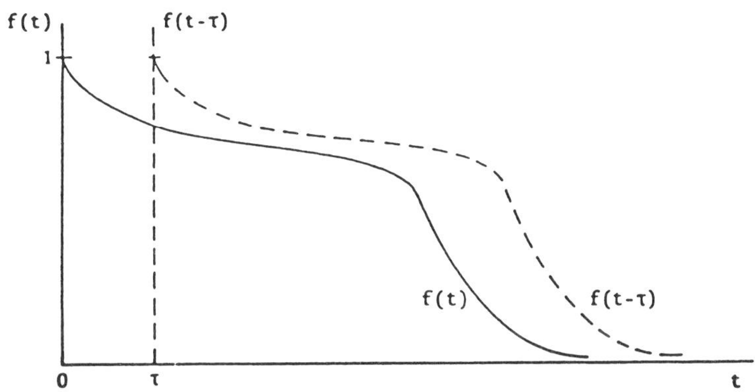

Fig. 1.1 The survival function. From Jerri [1982, 1986], courtesy of COMAP, Inc.

In Section 1.2 we will discuss the classification of integral equations with their two main classes, namely, *Volterra* and *Fredholm integral equations*. These two (different) equations are characterized by having a *variable* and a *fixed* limit of integration, respectively. Hence, equation (1.8) is a Volterra integral equation, since the (upper) limit of integration is the variable $\tau = t$. As we shall see soon, equation (1.19) that describes the small deflection $y(x)$ of a rotating shaft is with fixed integration limits $\xi = 0, l$ hence it is a Fredholm integral equation.

Periodicity in the surge of birthrates

The study of population dynamics includes determination of the surge in the birthrate $b(t)$ at any time t to allow for future necessary planning. The dependence of the birthrate $b(t)$ on previous birthrates $b(t-\tau)$, for women in the childbearing age range $\alpha < \tau < \beta$, is given by *Lotka's integral equation*,

$$b(t) = g(t) + \int_{\alpha}^{\beta} b(t-\tau)\rho(\tau)m(\tau)d\tau \tag{1.9}$$

where $\rho(\tau)$ is the probability that a female lives to age τ and $m(\tau)\Delta\tau$ is the probability that she will give birth to a female in the time interval $\Delta\tau$. $g(t)$ is a term added to allow for girls already born before the oldest childbearing woman (of age $\tau = \beta$) was born. The formulation of (1.9) is the subject of an exercise in Section 2.1 (Exercises 4), which is supported by detailed leading hints.

Mortality of equipment and its rate of replacement

Equipment wears out, so to maintain a fixed number of items $f(t)$ at any time t, it must be replaced at a certain (unknown) *rate* $r(t)$, which requires knowing (from past experience) the survival function $s(t-\tau)$ of all equipment bought at time τ previous to time t. This problem is represented by the integral equation

$$f(t) = f(0)s(t) + \int_0^t r(\tau)s(t-\tau)d\tau \tag{1.10}$$

where $s(0) = 1$ to make all newly bought equipment present at $t = 0$ [i.e., $f(0) = f(0)s(0)$] and $r(t)$ is the rate at which the equipment must be replaced. We may note here the common "hereditary" nature of both (1.8) and (1.10). We may add that $r(\tau)\Delta\tau$ in the integral of (1.10) represents the number of new machines added during the time interval $\Delta\tau$ about time τ. These will be of age $t-\tau$ at time t; hence their survival function is $s(t-\tau)$ and their surviving number at time t is about $r(\tau)s(t-\tau)\Delta\tau$. So we have to integrate these increments of the new added machines to find their total from the initial time $\tau = 0$ to time $\tau = t$ as the integral of the second term in (1.10). The first term $f(0)s(t)$ in (1.10) represents the number $f(0)s(t)$ of the initial equipment $f(0)$ that stayed operational until time t and the ones that did not need replacement.

Propagation of stocked fish in a new lake

A very closely related problem[1] to the above problem of the equipment rate of replacement is the problem of the propagation rate of fish in a new lake. New artificial lakes are stocked with fish for recreation purposes. This is the same situation with game in a protected park. From known data of similar lakes, the fish is stocked at a given (*known*) rate of supply $s(t)$ per year. Add to this that the fish multiply (or propagate) at an unknown *rate* of $r(t)$ per year. This means that in the time interval $\Delta\tau$ we have $(s(\tau)+r(\tau))\Delta\tau$ fish added in the lake. But such a population, naturally, declines, where we assume the simple exponential decay $e^{-\lambda t}$. For the future time t, such added fish during the time interval $\Delta\tau$ around time τ are of age $t-\tau$. So their survival function is $e^{-\lambda(t-\tau)}$. The result is that at future time t we will have only $(s(\tau)+r(\tau))\Delta\tau e^{-\lambda(t-\tau)}$, of the total accumulated fish in the time interval $\Delta\tau$ around $t = \tau$, survive to time t. So, to find the total number of the fish $N_{s,r}(t)$ due to the (known) supply rate $s(t)$ and the natural multiplying (propagation) rate $r(t)$ we integrate from the initial time $\tau = 0$ to $\tau = t$ to have

$$N_{s,r}(t) = \int_0^t e^{-\lambda(t-\tau)}[s(\tau)+r(\tau)]d\tau. \tag{1.11}$$

It is, of course, desired to keep the number of fish in the lake at a certain (given) level $N(t)$. This level is kept and watched by sampling the fish in the lake by selective netting. Before we supply the stocked fish, which will multiply to give the

[1] Wing [1991], courtesy of SIAM.

total number of fish $N_{s,r}(t)$ in the integral of (1.11), we assume that the lake had initial number of fish $N(0) = N_0$. But this fish will decline to $N(0)e^{-\lambda t}$ at time t. So if we add this number to the supplied and propagated number of the integral in (1.11), we have the total number of fish $N(t)$ which we would like to keep it at this (given-known) level,

$$N(t) = N_0 e^{-\lambda t} + \int_0^t [s(\tau) + r(\tau)]e^{-\lambda(t-\tau)} d\tau. \tag{1.12}$$

This is a (Volterra) integral equation in the unknown rate of propagation function $r(\tau)$, where $s(\tau)$ and $N(t)$ are assumed as known functions.

Biological species living together

Consider two separate species with numbers $n_1(t)$ and $n_2(t)$ at time t, where the first species increases and the second decreases. If they are put together, assuming that the second species will feed on the first, there will be an increase in the rate of the second species dn_2/dt which depends not only on the present population $n_1(t)$ but also on all previous values of the first species. When a steady-state condition or equilibrium is reached between these two species, it is described by the following pair of *integro-differential equations* (as n_1 and n_2 appear under both integration and differentiation operations):

$$\frac{dn_1}{dt} = n_1(t)\left[k_1 - \gamma_1 n_2(t) - \int_{t-T_0}^{t} f_1(t-\tau)n_2(\tau)d\tau\right], \qquad k_1 > 0 \tag{1.13}$$

$$\frac{dn_2}{dt} = n_2(t)\left[-k_2 + \gamma_2 n_1(t) + \int_{t-T_0}^{t} f_2(t-\tau)n_1(\tau)d\tau\right], \qquad k_2 > 0 \tag{1.14}$$

where k_1 and $-k_2$ are the coefficients of increase and decrease of the first and second species (had they stayed separate), respectively. The parameters γ_1, f_1 and γ_2, f_2 are dependent on the respective species. T_0 is assumed to be the finite heredity duration of both species. The detailed formulation of (1.13) and (1.14) is given in Section 2.1.2.

There are many other problems that are modeled as integral equations that include some examples, which we covered briefly in this section of the first edition, but shall only mention here. We give their original references, and leave for the interested reader their, somewhat, lengthy derivation. These include the propagation of nervous impulse,[2] the smoke filtration in a cigarette,[3] and the chance to find a time gap T in order to cross a dense traffic.[4] We have also attempted in this second edition to add a few new examples with emphasis on their more simple and clear derivation. The other new ones were placed as exercises for Chapter 2, which is designated for the

[2] Rashevsky [1960, p. 426].
[3] Noble [1967, p. 153].
[4] Green [1969, p. 139].

detailed mathematical modeling of problems as integral equations. Such exercises are supported with detailed hints.

Torsion of a wire

Many physical problems are also of a *hereditary* nature; for example, if we apply a torque $m(t)$ to twist a wire or bar, the torsion $\omega(t)$ will depend on this present torque in the form $m(t) = h\omega(t)$ as well as on all torques applied in times $(-\infty, t)$ previous to t. Such accumulation of twists changes the physical properties of the wire, thus introducing a hereditary (cumulative) effect. We will assume that we have some data that tells us at time t how the proportionality factor $\phi(t,\tau)$, instead of the usual constant proportionality factor h, had been affected by the continuous previous torques $m(\tau)$, $-\infty < \tau < t$. If we add the first torque $h\omega(t)$ to the accumulation of previous torques as the integral $\int_{-\infty}^{t} \phi(t,\tau)\omega(\tau)d\tau$, we have, for static equilibrium, this problem represented by the integral equation

$$m(t) = h\omega(t) + \int_{-\infty}^{t} \phi(t,\tau)\omega(\tau)d\tau. \tag{1.15}$$

As mentioned above, h is a constant and $\phi(t,\tau)$ is a function that takes care of how previous torques had affected the physical properties of the wire, and hence the present torsion. We may note that the function ϕ depends on t and the previous times τ in a more general way $\phi(t,\tau)$ than the dependence on their difference $\phi(t-\tau)$ in most of the foregoing problems. We may note how the (positive) accumulation integral in (1.15) effectively reduces the proportionality constant h of the first term $h\omega(t)$.

We may also note that this integral equation in (1.15) has a new feature, which is that one of the limits of the integration is unbounded, namely, the lower limit of the integral is $-\infty$. As we shall see when we classify integral equations in the next section (Section 1.2), such equations are called *singular* integral equations. Another property that lends an integral equation singular is when the kernel [see $K(x,t)$ in (1.1)] becomes unbounded at a point or points in the domain of the integral equation (or, more precisely, is when the kernel is not integrable on the domain of the integral). We must note here that the methods of solving such singular integral equation are more involved for the purpose of this book, where they require basic knowledge of complex analysis. We shall not pursue a discussion of such methods in this book, but we will illustrate solving a number of singular integral equations that are tractable via some basic tools that we shall present in Section 1.4.

Automatic control of a rotating shaft

The problem of correcting for the deviation $\phi(t)$ between $\theta_s(t)$, the angle of steering (or rotating) a shaft, and $\theta_i(t)$, the angle of the direction indicator to be followed, is similar to that of the above twisted wire problem.

To correct for this deviation $\phi(t)$, a torque proportional to such deviation, $a\phi(t)$, and in the opposite direction, must be applied to overcome the instantaneous deviation at time t. Another torque, $b(d\phi/dt)$, proportional to the rate of change $d\phi/dt$ of the deviation, must also be applied to dampen such a change. To do even better, a torque

$b\int_0^t \phi(\tau)d\tau$, proportional to the accumulation of the deviation from the starting time $t = 0$ to present time t, must also be applied to take care of all previous deviations. If we let I be the moment of inertia of the rotating shaft, then according to Newton's second law of motion, the torque $m_s(t)$ applied by the shaft to rotate with angle $\theta_s(t)$ is

$$m_s(t) = I\frac{d^2\theta_s}{dt^2}.$$

This torque must be equal in magnitude, but opposite in direction, to the sum of the three correction torques,

$$I\frac{d^2\theta_s}{dt^2} = -a\phi(t) - b\frac{d\phi}{dt} - c\int_0^t \phi(\tau)d\tau. \tag{1.16}$$

We note here that the unknown $\phi(t)$ of (1.16) is being differentiated in the term $-b\frac{d\phi}{dt}$ as well as integrated in the last term $-c\int_0^t \phi(\tau)d\tau$. Such equation is called an *integro-differential* equation in $\phi(t)$.

Shape of an elastic thread (The hanging chain)

An example of a physical problem that results naturally in an integral equation is to find how a variable density $\rho(x)$ must be distributed along an elastic thread in order that the thread assumes a given shape $f(x)$. For an elastic thread of length l under a horizontal tension T_0, the resulting integral equation in $\rho(x)$ is

$$f(x) = g\int_0^l G(x,\xi)\rho(\xi)d\xi, \tag{1.17}$$

where

$$G(x,\xi) = \begin{cases} \dfrac{x(l-\xi)}{T_0 l}, & 0 \le x \le \xi \\ \dfrac{\xi(l-x)}{T_0 l}, & \xi \le x \le l, \end{cases} \tag{1.18}$$

and g is the acceleration of gravity. The formulation of this problem is rather simple, but a bit more detailed. The least we need is a figure of the above function $G(x,\xi)$ with its "two branches" on $(0,\xi)$ and (ξ,l), as shown in Figure 1.2, which we shall derive in Section 2.3.1. In Figure 1.2 we note that $y(x)$ is taken to be positive in the (downward) direction of gravity.

We will show then that this $G(x,\xi)$ is the resulting shape of the elastic string, as a function of x, that is due to a vertical (point) force of unit magnitude located at $x = \xi$. So, the vertical force $\Delta F(\xi) = g\rho(\xi)\Delta\xi$, due to the weight of the increment $\Delta\xi$ of the length of the string, would cause the displacement of the string in the form $\Delta y(x,\xi) = G(x,\xi)g\rho(\xi)\Delta\xi$. What remains is to find the total displacement $y(x)$, $0 < x < l$ due to the gravity force along the whole string, which is obtained by superimposing all these displacements $\Delta y(x,\xi)$ of the elements of the string. This is accomplished by integrating over $dy(x,\xi)$ from $\xi = 0$ to $\xi = l$ to obtain the final shape of the elastic string $f(x)$, which is what we have in (1.17). We note that (1.17) is a Fredholm integral equation in the (unknown) density distribution function $\rho(x)$,

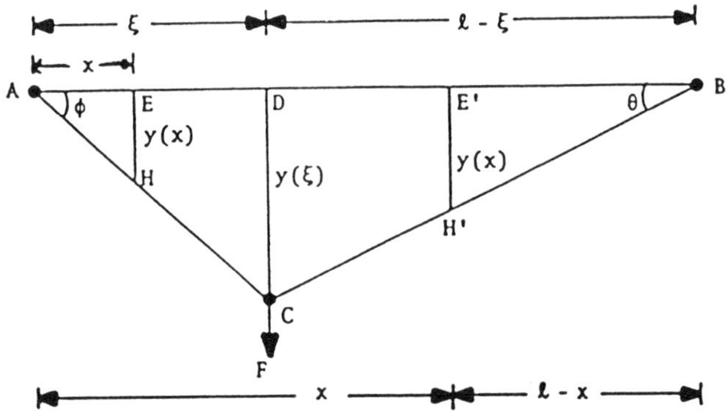

Fig. 1.2 Displacement due to a single vertical force F at ξ. From Jerri [1986], courtesy of COMAP, Inc.

since the limits of integration are fixed as $\xi = 0$ and $\xi = l$. Also, as we shall discuss the classification of integral equations in Section 1.2, when the unknown $y(x)$ is not present in a term outside the integral of an integral equation, the equation is called of *the first kind.* Hence in (1.17) we have a Fredholm integral equation of the first kind. The very detailed derivation of (1.17) and (1.18) is presented in Section 2.3.1.

Small deflection of a rotating shaft

Consider a shaft of length l rotating with angular velocity about the x axis. When it is disturbed a little, there results a deflection from the original position along the x axis, as shown in Figure 1.3. To formulate the problem for the deflection $y(x)$ at x of the bar from its original rotating position along the x axis, we will assume that we know the function $F(x, \xi)$, which gives the displacement in the y direction at the point x due to a unit force applied at another point $x = \xi$. The details of deriving such a function are similar to those of the above hanging chain problem, (which shall be derived in Section 2.3.1). To find how the force of the segment of length $\Delta\xi$ of the bar affects the displacement $y(x)$, we must find the centrifugal force of this rotating segment whose mass is $\Delta m = \rho(\xi)\Delta\xi$, radius $r = y(\xi)$, and angular velocity ω. Hence the centrifugal force is $\Delta m\omega^2 r = \rho\Delta\xi\omega^2 y(\xi)$, where $\rho(\xi)$ is the linear density. According to the definition of $F(x, \xi)$, this centrifugal force of $\Delta\xi$ around ξ will affect a displacement $\Delta y(x) = F(x, \xi)\rho(\xi)\Delta\xi\omega^2 y(\xi)$. Now, if we sum up all the contributions to the $y(x)$ displacement at x from all the segments $\Delta\xi$ along the length of the bar $(0, l)$, we obtain

$$y(x) = \omega^2 \int_0^l F(x, \xi)\rho(\xi)y(\xi)d\xi \tag{1.19}$$

which is a Fredholm integral equation in $y(x)$, the deflection of the bar at x. Since there is no external term outside the integral of (1.19) that is independent of the unknown function $y(x)$, this equation is termed *homogeneous*.

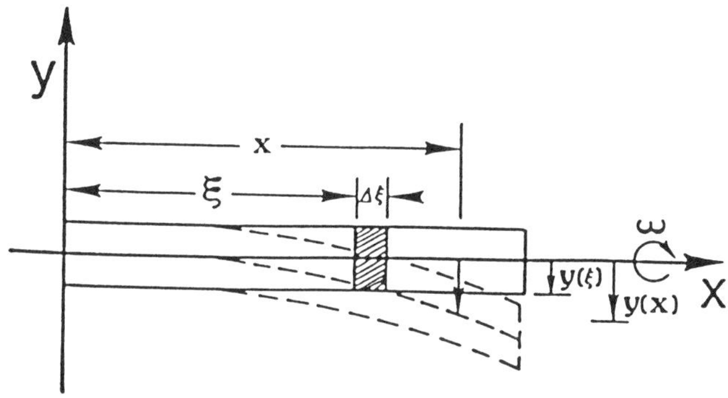

Fig. 1.3 Small deflection of a rotating bar.

Sliding a bead along a wire: Abel's problem

One of the earliest problems formulated as an integral equation was *Abel's problem.* It describes the shape of a wire $\phi(y)$ in a vertical plane (Figure 1.4)

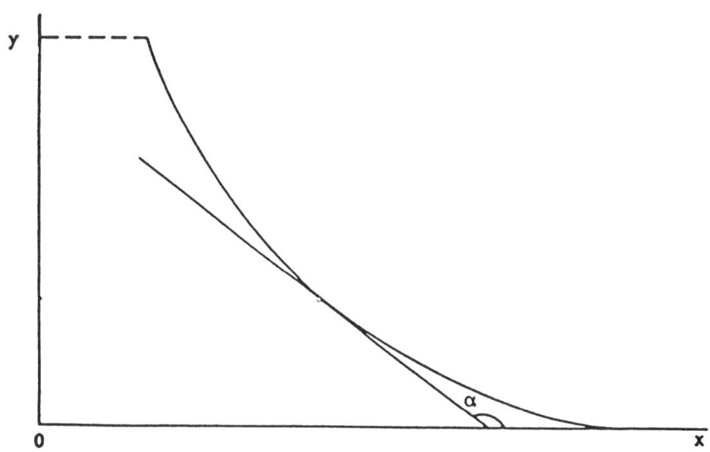

Fig. 1.4 The sliding bead on a wire—Abel's problem.

along which a bead must descend (under the influence of gravity) a distance y in a predetermined time $f(y)$. This is represented by *Abel's integral equation* in $\phi(y)$,

$$-\sqrt{2g}f(y) = \int_0^y \frac{\phi(\eta)d\eta}{\sqrt{y-\eta}} \tag{1.20}$$

where $\phi(y) = 1/sin\alpha$, the angle α is shown in Figure 1.4, and g is the acceleration of gravity. The detailed formulation of this problem is presented in Section 2.3.2. There we shall see that for $s(y)$ the length of the path as a function of the vertical distance y, we have $\frac{dy}{ds} = -\sin\alpha$. The unknown $\phi(y)$ in (1.20) is defined as $\phi(y) = \frac{1}{\sin\alpha}$.

Here, we have resorted to using η for y of $\phi(y)$ as the dummi variable of integration so that we can write the upper limit of integration as y. Most references use y for the variable of integration, and designate the upper limit of integration as y_0, which may be confused with a constant limit y_0. We had to stay with the variables y and η, since y is the vertical distance traveled. Abel in 1823-1826 formulated and solved this and more general problems. This was followed, independently, by Liouville's work in 1832-1839.

Example 1

Verify that $\phi(y) = \frac{1}{2}$ is a solution of the following special case of Abel's problem:

$$y^{\frac{1}{2}} = \int_0^y \frac{\phi(\eta)d\eta}{\sqrt{y-\eta}}. \tag{E.1}$$

We substitute $\phi(\eta) = \frac{1}{2}$ in the integral of (E.1) to obtain

$$\frac{1}{2}\int_0^y \frac{dy}{\sqrt{y-\eta}} = -\left(\frac{1}{2}\right) 2(y-\eta)^{\frac{1}{2}}\Big|_{\eta=0}^{y}$$

$$= -\left(0 - y^{\frac{1}{2}}\right) = y^{\frac{1}{2}} \tag{E.2}$$

which is the left side of (E.1); hence $\phi(y) = \frac{1}{2}$ is a solution of Abel's integral equation (E.1).

We note that this case of Example 1 corresponds to $\phi(y) = 1$ in (1.20) for a body falling a—not so interesting path!—of direct vertical fall of distance y. This is the case since for such a fall, we have $y = \frac{1}{2}gt^2$; $t = \sqrt{\frac{2}{g}y}$, where $y = 0$ corresponds to $t = 0$, so in our case we write $t = f(y) = -\sqrt{\frac{2}{g}y}$. If we substitute this value in the left side of (1.20), we have $\sqrt{2g}\cdot\sqrt{\frac{2y}{g}} = 2\sqrt{y}$. With this factor of 2, the solution to (E.1) is $\phi(y) = (2)(\frac{1}{2}) = 1 = \frac{1}{\sin\alpha}$, which results in $\alpha = \frac{\pi}{2}$, the direct vertical fall!

This is not such an interesting, if not dangerous, special case of path of descent for (1.20). The following *Tautochrone* problem is much more interesting special case of (1.20), and is also the first integral equation problem that started Abel's interest in the subject.

The Tautochrone

This is Abel's original problem that he later generalized to the integral equation (1.20). As a special case of (1.20) it deals with finding the path where the time required for descent along such path as shown in Figure 1.5 from any point (x, y) to the origin is a constant T, i.e., $f(y) = T$ is *independent* of the starting point. For

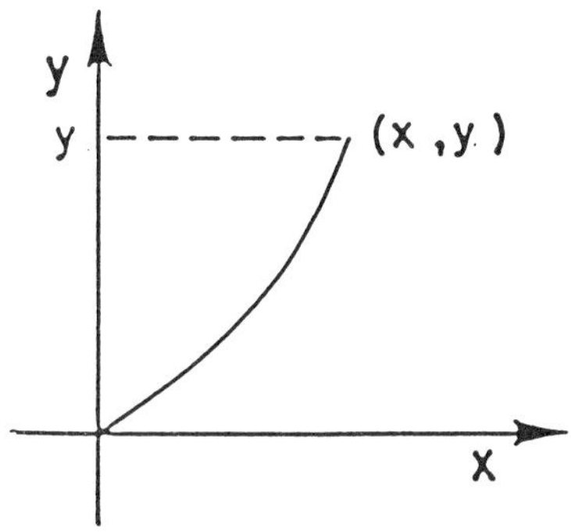

Fig. 1.5 Abel's Tautochrone problem.

this $f(y) = T$, (1.20) becomes (noting here that y corresponds to $t = 0$, and $y = 0$ corresponds to $t = T$)

$$\sqrt{2g}T = \int_0^y \frac{F'(\eta)d\eta}{\sqrt{y-\eta}}, \quad F'(\eta) = -\phi(\eta) \tag{1.21}$$

Here $F(y) = s(y)$, the length of the path as a function of the vertical distance y, and $\frac{dF}{dy} = \frac{ds}{dy} = -\phi(y)$ in (1.20), $\left(\frac{ds}{dy}\right)^2 = \left(\frac{dF}{dy}\right)^2 = 1 + \left(\frac{dx}{dy}\right)^2$. Historically, Abel is credited with the first conscientious effort in stating problems as integral equations as in (1.20) and (1.21) and offering their solutions. For example, he gave the solution to the Tautochrone problem (1.21) as

$$F(y) = s(y) = \left(\frac{2T}{\pi}\right)\sqrt{2gy},$$

Which is the equation of a *cycloid*. He also solved (1.20) and its generalization, which we shall return to in Section 1.4.1 when we cover the Laplace transform, that will facilitate obtaining all such solutions to Abel's problems.

We may note that in the above Abel problems, we see first that the kernel $K(y,\eta) = \frac{1}{\sqrt{y-\eta}}$ is unbounded at the point $\eta = y$, hence they are singular (Volterra) integral equations. Second, that the unknown term is absent outside the integral, when compared with the general integral equation in (1.1),

$$u(x) = f(x) + \int K(x,\xi)u(\xi)d\xi. \tag{1.1}$$

As we mentioned earlier, such integral equations are termed of the first kind as opposed to the those of the second kind when the unknown term is present outside the integral as in (1.1). So Abel's integral equations (1.20) and (1.21) are singular Volterra equations of the first kind.

As we shall see in Section 5.4 for the Fredholm integral equations of the first kind, such equations, very often, represent some major difficulties. For Abel problems, this is complicated even more where the problem is also singular, which is another difficult situation for integral equations. These difficulties were documented theoretically long after Abel's time. So, we may say that Abel not knowing of such, often, formidable difficulties, he took such problems by stride, as they turned out to be among the few without major apparent difficulties! We shall return in Section 2.3.2 to derive Abel's integral equation (1.20), with its solution accomplished by the use of the Laplace transform (see Example 8 in Section 3.2.1 and Exercise 5 of Section 1.4.)

The Tautochrone problem will be the subject of Exercise 5 in Section 2.3, where the derivation is supplied with detailed leading hints.

Example 2 Bernoulli's Problem

One of the simplest integral equations arises from a problem in geometry called *Bernoulli's problem*, which deals with finding the shape of a curve $y = f(x)$ for which the area A under the curve $\int_0^x f(\xi)d\xi$ on the interval $(0, x)$ is only a fraction k of the area of the rectangle circumscribing it which is $kxf(x)$ (Figure 1.6). Thus, this problem is easily represented by the integral equation

$$kxf(x) = \int_0^x f(\xi)d\xi. \qquad (E.1)$$

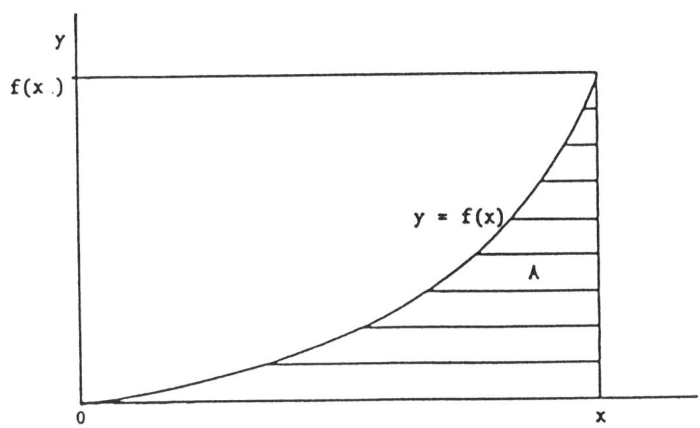

Fig. 1.6 The Bernoulli problem.

For a simple demonstration of (E.1) we can easily verify that the area under the parabola $y = x^2$ is one-third of the rectangle circumscribing it. So, if we substitute $k = \frac{1}{3}$ and $f(x) = x^2$ in the Bernoulli equation (E.1), we have

$$\frac{1}{3}x \cdot x^2 = \frac{1}{3}x^3 = \int_0^x \xi^2 d\xi \qquad (E.2)$$

which is the case when we perform the integration of (E.1):

$$\int_0^x \xi^2 d\xi = \frac{1}{3}\xi^3 \Big|_{\xi=0}^{x} = \frac{1}{3}x^3.$$

Radiation transport – Determining the energy spectrum of neutrons

We shall present here a simple example of the absorption of radiation[5] (say neutrons) in a slab with fixed thickness as illustrated in Figure 1.7. The measured

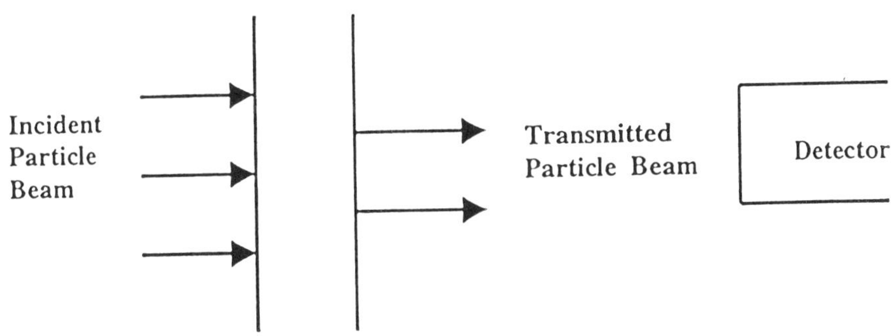

Fig. 1.7 Simple experiment for determining the energy spectrum of particles.

neutrons $g(x)$ on the other side of the slab (after the *absorption*) for *finite* number of different thicknesses x of the same material can be used to determine the neutron spectrum $f(E)$. Here $f(E)$ is the number of neutrons at the energy level E. The result is a Fredholm integral equation of the first kind in the neutron spectrum function $f(E)$. Before we start the derivation of such an equation, we need to define what we mean by the *cross section* σ of the nuclei of the material to the incoming radiation (or neutrons) with energy E. Simply speaking, it is "the effective area" that the neutrons see of the nucleus as a target. Of course, the cross section σ depends on the material, and, principally, on the energy distribution (spectrum) $f(E)$ of the colliding neutrons. Also when the particles (or neutrons) collide with the nucleus, they may be absorbed, scattered, or create new neutrons (by fission). So it is important, first, to know the probability of the collision. It can be shown easily (see Exercise 14) that

[5]Wing [1991, p. 9], courtesy of SIAM.

the probability of a neutron traveling a distance x with no collision is

$$p(x) = e^{-\sigma x}. \tag{1.22}$$

Now let us consider a beam of neutrons transmitted through a slab of uniform thickness x as shown in Figure 1.7. We shall assume in such experiment that the neutrons are only absorbed (and none are scattered). Hence the cross section is that of absorption, and, of course, it depends on the neutrons' energy spectrum $f(E)$, $E_{min} \leq E \leq E_{max}$, which we shall denote by $\sigma(E)$. We are to determine this spectrum $f(E)$ from knowing the "measured" neutrons $g(x)$ that cross the slab of thickness x. It is understood that doing such measurement for a slab gives only one sample value of the function $g(x)$. So the thickness is varied for a finite number of different thicknesses to have a good idea about "the measured" output $g(x)$.

Now in the small range of energy ΔE around E, we have $f(E)\Delta E$ neutrons which have x distance to cross without being absorbed. The probability for this to happen, according to (1.22) is $e^{-x\sigma(E)}$, hence the number of these neutrons that have crossed the slab is $e^{-x\sigma(E)}f(E)\Delta E$. If we sum this contribution over the whole energy spectrum of the neutrons from $E = E_{min}$ to $E = E_{max}$, we have the total number of the escaped (and to be measured) neutrons as

$$g(x) = \int_{E_{min}}^{E_{max}} e^{-x\sigma(E)} f(E) dE. \tag{1.23}$$

This is a Fredholm integral equation of the first kind in the neutron spectrum $f(E)$. In Exercise 15 we will present discussion concerning the difficulty in solving such Fredholm integral equations of the first kind, especially when the known function $g(x)$ in (1.23) has the inaccuracy of the measurement.

Also, very much related problem to this neutron transport one is the subject of (a detailed) Exercise 4 in Section 2.2. It deals with determining the strength of the neutrons source in a uniform rod, where it results in Fredholm integral equation of the first kind in the unknown function of the neutrons source strength.[6]

Electric potential on the rim of a unit disc

Consider the potential $u(r, \theta)$ in a unit disc, which is due to the given potential $u(1, \theta) = g(\theta)$, $-\pi < \theta < \pi$ on the rim of the unit disc. The solution $u(r, \theta)$ to this Dirichlet boundary value problem due to the given $g(\theta)$ is the well known *Poisson integral formula*,

$$u(r, \theta) = \frac{1}{2\pi} \int_0^{2\pi} \frac{1 - r^2}{1 - 2r\cos(\theta - \phi) + r^2} g(\phi) d\phi. \tag{1.24}$$

However, if we ask for the solution of *the inverse problem*, namely, to find a potential distribution $g(\theta)$ on the rim of the unit disc that would result in a given desired

[6]For this problem and other interesting problems, modeled as Fredholm integral equations of the first kind, see Wing [1991, p. 18], courtesy of SIAM.

potential distribution $u(r,\theta)$ inside the disc, then we face the above equation (1.24) as an integral equation in the unknown function $g(\theta)$. We note here that ϕ in the integral of (1.24) is a dummy variable, and just like θ, it is the polar angle, $-\pi < \phi < \pi$. Further discussion of this problem, including the derivation of the result in (1.24) is found in Section 4.1.4 [with the above result as (4.69)].

B. Inverse Problems

The Laplace and other integral transforms

As we have indicated at the beginning of this section, the most familiar example of an integral equation in $u(x)$ comes from searching for a function $u(x)$ whose *Laplace transform* $U(s)$ is known:

$$U(s) = \int_0^\infty e^{-sx} u(x)dx. \tag{1.25}$$

Another example of an integral equation arises when we have $U(\lambda)$ the *Fourier transform* of $u(x)$,

$$U(\lambda) = \int_{-\infty}^\infty e^{-i\lambda x} u(x)dx. \tag{1.26}$$

Here the solution $u(x)$ of the integral equation (1.26) is given in Section 1.4.2 as

$$u(x) = \frac{1}{2\pi} \int_{-\infty}^\infty e^{i\lambda x} U(\lambda)d\lambda. \tag{1.27}$$

This Fourier integral inversion (1.27) of (1.26) is considered historically to be the earliest in the direction of solving integral equations such as (1.26) in $u(x)$. Indeed, one of the most general integral transforms of $f(x)$,

$$F(\lambda) = \int_a^b \rho(x)K(\lambda,x)f(x)dx \tag{1.28}$$

can be considered as an integral equation in $f(x)$. Here $\rho(x)$ in (1.28) is a (known) weight function.

In Section 1.4 we present the Laplace, Fourier, and few other transforms and discuss some of their basic properties, which will be used to solve certain types of integral equations. In Appendix A we present the Hankel transform with Bessel function kernel $K(x,t) = J_n(xt)$[7], [and weight function $\rho(x) = x$ in (1.28)].

[7]Here $J_n(x)$ is the Bessel function, of the first kind of order n, which is one of the two solutions of Bessel differential equation $x^2u'' + xu' + (x^2 - n^2)u = 0$ which is bounded at $x = 0$,

$$J_n(x) = \sum_{k=1}^\infty \frac{(-1)^k (\frac{\pi}{2})^{n+2k}}{k!(n+k)!}.$$

It is needed in Chapter 2 for formulating and solving the *dual* integral equations representation of the electrified disc problem. In Section 2.6 we will derive the simpler problem of the electrified plate, where the Fourier transform of Section 1.4.2 is employed. The electrified disc dual integral equations setting and their solution are done in Example 1 of Appendix A.

Example 3

Verify that

$$u(x) = p_a(x) = \begin{cases} A, & |x| \leq 1 \\ 0, & |x| > a \end{cases} \tag{E.1}$$

is a solution of the integral equation

$$\frac{2A \sin a\lambda}{\lambda} = \int_{-\infty}^{\infty} e^{-i\lambda x} u(x) dx \tag{E.2}$$

which is a special case of (1.26).

We substitute $u(x)$ from (E.1) in the integral of (E.2) to obtain

$$\begin{aligned} \int_{-\infty}^{\infty} e^{-i\lambda x} u(x) dx &= \int_{-\infty}^{-a} e^{-i\lambda x}(0) dx + \int_{-a}^{a} e^{-i\lambda x}(A) dx + \int_{a}^{\infty} e^{-i\lambda x}(0) dx \\ &= A \int_{-a}^{a} e^{-i\lambda x} dx = A \frac{e^{-i\lambda} x}{-i\lambda} \bigg|_{-a}^{a} \\ &= \frac{A}{i\lambda}\left(e^{i\lambda x} - e^{-i\lambda x}\right) = \frac{2A \sin \lambda a}{\lambda} \end{aligned} \tag{E.3}$$

which is the left side of (E.2), after using the identity

$$\sin \lambda a = \frac{e^{i\lambda a} - e^{-i\lambda a}}{2i}.$$

C. Differential Equations with Auxiliary Conditions Reduced to Integral Equations

The above variety of problems in part A represent varied applications, where integral equations is the natural setting for the mathematical model of such problems. They are characterized by their "hereditary" or accumulative nature. In this section we shall present examples that illustrate the reduction of differential equations with auxiliary (initial or boundary) conditions to (Volterra or Fredholm) integral equations.

Initial value problems

In Section 2.4 we will use repeated integration and some integral identities from Section 1.3 to show that, for example, the following initial value problem associated with a second-order differential equation,

$$\frac{d^2u}{dx^2} = \lambda u(x) + g(x), \quad x > 0 \tag{1.29}$$

$$u(0) = 1 \tag{1.30}$$

$$u'(0) = 0 \tag{1.31}$$

reduces to an integral equation in $u(x)$,

$$u(x) = 1 + \lambda \int_0^x (x - \xi)u(\xi)d\xi + \int_0^x (x - \xi)g(\xi)d\xi \tag{1.32}$$

which represents the first main class of integral equations, the *Volterra integral equations*.

Boundary value problems

In Section 2.5 we will show, for example, that the following two-point or boundary value problem,

$$\frac{d^2u}{dx^2} = \lambda u, \qquad a < x < b \tag{1.33}$$

$$u(a) = 0 \tag{1.34}$$

$$u(b) = 0 \tag{1.35}$$

reduces to an integral equation in $u(x)$,

$$u(x) = \lambda \int_a^b K(x, \xi)u(\xi)d\xi \tag{1.36}$$

where

$$K(x, \xi) = \begin{cases} \dfrac{(x - b)(\xi - a)}{b - a}, & a \le \xi \le x \le b \\ \dfrac{(x - a)(\xi - b)}{b - a}, & a \le x \le \xi \le b. \end{cases} \tag{1.37}$$

This represents the other main class of integral equations, the *Fredholm integral equations*.

We may mention again that the integral equation (1.32) for the initial value problem has a variable limit of integration x, while (1.36) for the boundary value problem has fixed limits a and b. This points out a main distinction in classifying integral equations and hence the nature of their different methods of solution. This leads us to classify integral equations along these two lines in the following Section 1.2, where, as indicated above, (1.32) and (1.36) are special cases of the two main classes, Volterra and Fredholm integral equations, respectively.

D. Integral Equations in Higher Dimensions

All our examples of integral equations up till now are done for functions of one variable, or in one dimension. The following problem of the *electric potential* on a unit disc is a representative of integral equations in two dimensions. Examples of problems in three dimensions are left for Chapter 2. They include the *electrified plate* and *disc problems* in Section 2.6, and *Schrödinger equation* in the three-dimensional momentum space in Section 2.7.

Charge density for a potential on a unit disc

The general problem of solving for the potential in a disc is discussed in details in Section 4.1.4, and where partial differential equations are used along with some boundary conditions. For example, the potential $u(r, \theta)$ inside a unit disc with grounded rim, which is charged with charge density distribution $f(r, \theta)$, is given as

$$u(r, \theta) = \int_0^1 d\rho \int_0^{2\pi} G(r, \theta; \rho, \phi) f(\rho, \phi) d\phi. \tag{1.38}$$

What concerns us here is the *inverse problem*, i.e., to find the charge distribution $f(r, \theta)$, which results in a desired potential distribution inside (and outside) the disc. This means that the above equation (1.38) represents an integral equation in the two-dimensional (unknown) function $f(r, \theta)$.

We may conclude that in this section we presented a good variety of basic applied problems as integral equations. For the reader who is interested in the more detailed and realistic, though somewhat involved!, applications of integral equations, we refer to the specialized references (and the references therein) at the end of the book. In particular, Kanwal [1971, 2nd ed., 1997] covers applications in fluid dynamics, Wing [1991] has applications for radiation transport among other applications, Porter and Stirling [1980] covers spectral theory, while Pogorzelski [1966] covers the detailed theory and applications.

Exercises 1.1

1. Verify that $u(x) = x$ is a solution of the integral equation (a Laplace transform of $u(x)$)

$$\frac{1}{s^2} = \int_0^\infty e^{-sx} u(x) dx, \quad s > 0.$$

2. **(a)** Reduce the integral equation (E.1) of the Bernoulli problem in Example 2 to a differential equation. *Hint:* Differentiate both sides of (E.1) using the fundamental theorem of calculus: $\frac{d}{dx} \int_a^x f(\xi) d\xi = f(x)$.

   (b) Solve the resulting differential equation in part (a) for $f(x)$, the solution of the Bernoulli problem (E.1).

   For the following Exercises 3 to 5 verify that the given function $u(x)$ is a solution to the indicated (Volterra) integral equation.

3. $u(x) = 3$,

$$x^3 = \int_0^x (x - t)^2 u(t) dt$$

4. $u(x) = 1 - x$,
   $x = \int_0^x e^{x-t} u(t) dt$

Hint: Take the factor e^x outside the integral, then there is a simple integration, part of which involves integration by parts.

5. $u(x) = e^x$,

$$u(x) = 1 - \lambda \int_0^1 \sin xtu(t)dt$$

6. $u(x) = x - x^3/6$,

$$u(x) = x - \int_0^x \sinh(x-t)u(t)dt$$

For the following Exercises 6 to 10 show whether or not the given function $u(x)$ is a solution to the indicated (Fredholm) integral equation of that particular Exercise.

7. $u(x) = \sin(\pi x/2)$,

$$u(x) = \frac{x}{2} + (\pi^2/4)\int_0^1 K(x,t)u(t)dt, \tag{E.1}$$

where the kernel $K(x,t)$ is defined (with two branches) on the interval $(0,1)$ as

$$K(x,t) = \begin{cases} \dfrac{x(2-t)}{2}, & 0 \le x \le t \le 1 \\ \dfrac{t(2-x)}{2}, & 0 \le t \le x \le 1. \end{cases} \tag{E.2}$$

Hint: In substituting the kernel $K(x,t)$, with its two different branches, in the integral of (E.1), you should write the integral as the sum of two integrals on $0 \le t \le x$ and $x \le t \le 1$, where the second and the first branches of $K(x,t)$ in (E.2) are used for these two subintervals, respectively.

8. $u(x) = 1$,

$$u(x) = e^x - x - \int_0^1 x(e^{xt} - 1)u(t)dt$$

9. Show that $u(x) = (\sin ax)/\pi x$ is a solution of the integral equation

$$\int_{-a}^{a} e^{-i\lambda x}u(x)dx = p_a(\lambda) = \begin{cases} 1, & |\lambda| < a \\ 0, & |\lambda| > a \end{cases}.$$

Hint: Use (1.27) with $U(\lambda) = p_a(\lambda)$, or consult Example 3.

10. Verify that $u(x) = e^{-x}$, $x \ge 0$, is a solution of the integral equation (Fourier cosine transform)

$$\frac{1}{1+\lambda^2} = \int_0^\infty \cos\lambda x u(x)dx.$$

11. Verify that

$$u(x) = \begin{cases} 1 - x^2, & |x| < 1 \\ 0, & |x| > 1 \end{cases}$$

is a solution of the integral equation (Fourier transform)

$$\frac{4}{\lambda^3}(\lambda \cos \lambda - \sin \lambda) = \int_{-\infty}^{\infty} e^{-i\lambda x} u(x) dx.$$

Hint: Use the identity $e^{\mp ix} = \cos x \mp i \sin x$.

12. **(a)** Reduce the following integral equation

$$h(x)u(x) = f(x) + \int_a^{b(x)} K(x,t)u(t)dt, \qquad h(x) > 0$$

to the following form:

$$\phi(x) = g(x) + \int_a^{b(x)} k(x,t)\phi(t)dt \tag{E.1}$$

where $k(x,t) = \dfrac{K(x,t)}{\sqrt{h(t)h(x)}}$.

Hint: Divide both sides of (E.1) by the function $\sqrt{h(x)}$ on (a, b).

(b) Show that if $K(x,t)$ is symmetric in (E.1), then the resulting modified kernel $k(x,t)$ in (E.1) is also symmetric.

13. Give the equations that describe the rate of change of the two biological species living together of (1.13) and (1.14) when they are separate (independent). *Hint:* See (2.9)–(2.11).

14. Use the following hints to derive the probability expression $p(x) = e^{-\sigma x}$ for a neutron to travel a distance x without being absorbed. σ is the cross section (of the nuclei) of the material as it appears to the neutron.

Here we shall assume only that the probability function $p(x)$ satisfies

$$p(x_1 + x_2) = p(x_1)p(x_2). \tag{E.1}$$

The following steps are aimed at generating a differential equation in $p(x)$ whose solution is $p(x) = e^{-\sigma x}$.

Let $p(x)$ be the probability that the neutron (particle) moves with no collision, then consider the particle traveling an extra small distance Δx, whereby we have $p(x + \Delta x)$ at $x + \Delta x$. But according to (E.1) we have

$$p(x + \Delta x) = p(x)p(\Delta x). \tag{E.2}$$

Also, with $p(0) = 1$, the (decreasing) $p(\Delta x)$ can very well be approximated by $p(\Delta x) = 1 - \sigma\Delta x$.

$$p(\Delta x) = 1 - \sigma\Delta x, \quad \sigma > 0. \tag{E.3}$$

Use (E.2) with the above approximation of $p(\Delta x)$ in (E.3) to generate a first-order differential equation in $p(x)$, then solve it to find $p(x) = e^{-\sigma x}$. Of course we have the boundary condition $p(0) = 1$ for determining the arbitrary constant in the solution of the first order differential equation.

15. Consider the Fredholm integral equation of the first kind (1.23) in $f(E)$, the neutron spectrum, where the output $g(x)$ is a measured data.

    (a) What is the major difficulty in obtaining an accurate value of $f(E)$.
    Hint: Recall that $g(x)$ is measured data for a fixed number values of x, and most likely a (numerical) differentiation of this data (with its inaccuracy due to the measurement) may be needed to find $f(E)$, as it is often the case for equations of the first kind.

    (b) Assume that the cross section $y = \sigma(E)$ is a monotonically increasing function of the energy. This is to allow writing it's inverse $E = \sigma^{-1}(y)$ as a function of y. Show that the integral equation (1.23) will reduce to finding $F(y)$ in the following equation,

    $$g(x) = \int_a^b e^{-xy} F(y)dy \tag{E.1}$$

    where

    $$F(y) = f(\sigma^{-1}(y))\frac{d}{dy}\sigma^{-1}(y) \tag{E.2}$$

    and

    $$\sigma(E_{\text{min}}) = a, \qquad \sigma(E_{\text{max}}) = b. \tag{E.3}$$

    (c) Give the special case of $\sigma(E)$ that would make the integral (E.1) as the (typical) Laplace transform of $F(y)$.

16. Reduce the initial value problem

    $$\frac{d^2u}{dx^2} + u = 0, \qquad x > 0$$
    $$u(0) = 1, \qquad u'(0) = 0$$

    to a Volterra integral equation. *Hint:* See (1.29)–(1.31) and use (1.32).

17. Reduce the boundary value problem

    $$\frac{d^2u}{dx^2} + \lambda u = 0,$$
    $$u(0) = 0, \qquad u\left(\frac{\pi}{2}\right) = 0$$

    to a Fredholm integral equation. *Hint:* See (1.33)–(1.35) and use (1.36) and (1.37).

1.2 CLASSIFICATION OF INTEGRAL EQUATIONS

As we remarked in the preceding section, it seems that most of the integral equations we have presented fall under two main categories: those with variable limits of integration, such as (1.7), (1.8), (1.10), (1.12), (1.15), (1.20), and (1.32), and those with fixed limits of integration, such as (1.17), (1.19), (1.23), (1.24) and (1.36). These two classes of integral equations are called Volterra[8] and Fredholm[9] integral equations, respectively. As we shall see in Chapter 2, these two classes represent two different sets of problems and require different methods of solution, which we present in Chapters 3 and 5, respectively. In the following we present a more detailed classification of integral equations in order to become familiar with the conditions and the terminology that soon will be used in the formulation of the problems or the construction of their respective solutions.

The most general *linear* integral equation in $u(x)$ can be presented as[10]

$$h(x)u(x) = f(x) + \int_a^{b(x)} K(x,\xi)u(\xi)d\xi \tag{1.39}$$

or in *operational notation*, similar to what we wrote for (1.1) in (1.5),

$$h(x)u(x) - f(x) = (\mathcal{K}u)(x), \qquad a \le x \tag{1.39a}$$

where $\mathcal{K}$ defines the above integration operation on the function u in (1.39). The equation (1.39) is called a *Volterra integral equation* when $b(x) = x$,

$$h(x)u(x) = f(x) + \int_a^{x} K(x,\xi)u(\xi)d\xi. \tag{1.40}$$

When $h = 0$ it is called a *Volterra equation of the first kind*,

$$-f(x) = \int_a^{x} K(x,\xi)u(\xi)d\xi \tag{1.41}$$

and is called a *Volterra equation of the second kind* when $h(x) = 1$,

$$u(x) = f(x) + \int_a^{x} K(x,\xi)u(\xi)d\xi. \tag{1.42}$$

It is clear that Abel's equation (1.20) is a Volterra equation of the first kind, whereas the equation for torsion of a wire (1.15) is a Volterra equation of the second kind. The initial value problem equation (1.32) is also a Volterra equation of the second kind, with

$$f(x) = 1 + \int_0^{x} (x-\xi)g(\xi)d\xi \quad \text{and} \quad K(x,\xi) = \lambda(x-\xi) \tag{1.43}$$

[8]Volterra's important work in this area was done in 1884-1896.

[9]Fredholm's important contribution was made in 1900-1903.

[10]Note that for the rest of the text we will be using other variables of integration, such as t and y, in addition to ξ, in (1.39). Also, we may refer to an integral equation only as "equation."

after comparing (1.32) with (1.42).

The integral equation (1.39) is called a *Fredholm integral equation* when $b(x) = b$, a constant,

$$h(x)u(x) = f(x) + \int_a^b K(x,\xi)u(\xi)d\xi. \tag{1.44}$$

It is also called a *Fredholm equation of the first and second kind* when $h(x) = 0$ and $h(x) = 1$, respectively,

$$-f(x) = \int_a^b K(x,\xi)u(\xi)d\xi, \tag{1.45}$$

$$u(x) = f(x) + \int_a^b K(x,\xi)u(\xi)d\xi. \tag{1.46}$$

Examples of Fredholm equations of the first kind are the hanging chain equation (1.17), the Laplace transform (1.25) in $u(x)$, the Fourier transform (1.26) in $u(x)$, and the inverse Fourier transform (1.27) in $U(\lambda)$. Examples of Fredholm equations of the second kind are the equation of the deflection of a rotating shaft in $y(x)$ (1.19), which corresponds to $f(x) = 0$ in (1.46), and (the boundary value problem) equation (1.36) (with $f(x) = 0$ in (1.46)).

The Volterra and Fredholm integral equations look very similar except for the limits $b(x) = x$ and $b(x) = b$ in (1.40) and (1.44), respectively. However, as we remarked before, they have a different origin and will require different methods of solution. We illustrate in Section 2.4 and Section 2.5 how the Volterra and Fredholm integral equations are representations of initial and boundary value problems, respectively.

In the case of either Volterra equation (1.40) or the Fredholm equation (1.44), the integral equation is termed *homogeneous* when $f(x) \equiv 0$,

$$h(x)u(x) = \int_a^x K(x,\xi)u(\xi)d\xi, \tag{1.40h}$$

$$h(x)u(x) = \int_a^b K(x,\xi)u(\xi)d\xi. \tag{1.44h}$$

An example of a homogeneous Volterra equation is the Bernoulli equation [(E.1) of Example 2 in the last section] in $f(x)$

$$kxf(x) = \int_0^x f(\xi)d\xi \tag{1.47}$$

while the deflection of a rotating shaft (1.19) in $y(x)$,

$$y(x) = \omega^2 \int_0^l \rho(\xi)F(x,\xi)y(\xi)d\xi \tag{1.19}$$

is a homogeneous Fredholm equation.

An integral equation, like other equations, is termed *linear* in $u(x)$ if when $u_1(x)$ and $u_2(x)$ are solutions to its associated homogeneous case in $u(x)$, then their linear combination $c_1u_1(x) + c_2u_2(x)$ is also a solution to that homogeneous integral equation. For example, the integral equation (1.15) for the torsion of a wire,

$$m(t) = h\omega(t) + \int_{-\infty}^{t} \phi(t,\tau)\omega(\tau)d\tau \tag{1.15}$$

is a linear equation in $\omega(t)$, as we show in the next example, while the following integral equation

$$u(x) = \int_a^b K(x,t)u^2(t)dt$$

is nonlinear in $u(t)$. We will illustrate the rather obvious nonlinearity of this integral equation in part b) of the next example.

We may mention here that this book covers only linear integral equations. This is with the exception of a brief general discussion of *nonlinear* integral equations of the, somewhat, theoretical treatment, in the (optional) Chapter 6.

Example 4 a) Linear Integral Equations.

The integral equation (1.15), of the torsion of a wire, is linear in $\omega(t)$, since when $\omega_1(t)$ and $\omega_2(t)$ are solutions to its associated homogeneous case (1.15h),

$$0 = h\omega(t) + \int_{-\infty}^{t} \phi(t,\tau)\omega(\tau)d\tau \tag{1.15h}$$

we have

$$0 = h\omega_1(t) + \int_{-\infty}^{t} \phi(t,\tau)\omega_1(\tau)d\tau \tag{E.1}$$

$$0 = h\omega_2(t) + \int_{-\infty}^{t} \phi(t,\tau)\omega_2(\tau)d\tau. \tag{E.2}$$

Hence if we multiply (E.1) by c_1 and (E.2) by c_2 and add we obtain

$$0 = h[c_1\omega_1(t) + c_2\omega_2(t)] + \int_{-\infty}^{t} \phi(t,\tau)[c_1\omega_1(\tau) + c_2\omega_2(\tau)]d\tau \tag{E.3}$$

which says now that $\omega(t) = c_1\omega_1(t) + c_2\omega_2(t)$ satisfies (1.15h) and hence this linear combination of $\omega_1(t)$ and $\omega_2(t)$ is also a solution of the homogeneous equation (1.15h). We may note here that the most general integral equation (1.39) of this section,

$$h(x)u(x) = f(x) + \int_a^{b(x)} K(x,\xi)u(\xi)d\xi, \tag{1.39}$$

is linear in $u(x)$ and hence almost all the integral equations in this section and this book (Chapters 3 to 5) are linear (see Exercises 3 for a few examples of nonlinear integral equations). Also, in Chapter 6 we have a brief introduction to the basic theory of linear as well as nonlinear integral equations.

b) Nonlinear Integral Equations.

In contrast to the linear integral equation in part a), we will show that the following (homogeneous) integral equation is nonlinear in $u(x)$,

$$u(x) = \int_a^b k(x,t)u^2(t)dt. \tag{E.4}$$

Of course, our familiarity with linear systems like linear algebraic equations and linear differential equations tells us that the presence of the quadratic term $u^2(t)$ inside the above integral results in the equation being nonlinear in $u(t)$. However to further illustrate the way of proving linearity we will show here that if $u_1(t)$ and $u_2(t)$ are solutions to (E.4), then their linear combination $c_1u_1(t) + c_2u_2(t)$ is not a solution to (E.4). We proceed as we did in part a) by assuming $u_1(t)$ and $u_2(t)$ as solutions of (E.4) to have

$$u_1(t) = \int_a^b k(x,t)u_1^2(t)dt, \tag{E.5}$$

$$u_2(t) = \int_a^b k(x,t)u_2^2(t)dt. \tag{E.6}$$

If we multiply (E.5) by c_1 and (E.6) by c_2 and add we have

$$c_1u_1(t) + c_2u_2(t) = \int_a^b k(x,t)[c_1u_1^2(t) + c_2u_2^2(t)]dt \tag{E.7}$$

where we see clearly that the linear combination $c_1u_1(t) + c_2u_2(t)$ [of the two solutions $u_1(t)$ and $u_2(t)$ of (E.4)] is not a solution to (E.4), as required if it is to be a linear integral equation. Hence (E.4) is a nonlinear integral equation in $u(t)$.

The function $K(x,\xi)$ in (1.40) is called the *kernel or nucleus* of the integral equation. An integral equation is termed *singular* if the range of integration is infinite or the kernel $K(x,\xi)$ becomes infinite in the range of integration. The Fourier integral in $u(x)$ of (1.26) is singular, because the range of integration is infinite $(-\infty,\infty)$, while Abel's equation (1.20) is singular because the kernel $1/\sqrt{y-\eta}$ becomes infinite in the range of integration $(0,y)$ at $\eta = y$.

For the unbounded kernel singular integral equations, there are two important classes that we should differentiate between, since their methods of solution are completely different. An integral equation with kernel $K(x,t) = k(x,t)/|x-t|^\alpha$, $0 < \alpha < 1$, where $k(x,t)$ is bounded, is termed a *weakly singular* equation, or that its kernel $K(x,t)$ is with *weak singularity*. The other class of singular integral equations is that of *strong singularity* with kernel $K(x,t) = k(x,t)/(x-t)$, where $k(x,t)$ is bounded. These are called kernels with strong singularity or with *Cauchy singular* kernel.

The generalized Abel integral equation

$$f(x) = \int_0^x \frac{u(\xi)}{(x-\xi)^\alpha} d\xi, \qquad 0 < \alpha < 1 \tag{1.48}$$

is singular since its kernel $\frac{1}{(x-\xi)^\alpha}$ is singular at $\xi = x$, and it is also with weak singularity corresponding to $\alpha = \frac{1}{2}$.

In the case of $K(x,\xi) = K(x-\xi)$ in (1.40), that is, when the kernel depends on the difference $x-\xi$, which is what we will call a *difference kernel*, this Volterra equation of the first kind assumes a Laplace type of convolution product, that we shall discuss in detail in Section 1.4.1. Such type equations yield themselves to the Laplace transform method of solution that we shall illustrate in Sections 1.4.1 and 3.2.1. The following singular Fredholm equation of the first kind with difference kernel,

$$f(x) = \int_{-\infty}^{\infty} K(x-\xi)u(\xi)d\xi \tag{1.49}$$

assumes the Fourier type of convolution product, which will be discussed in Section 1.4.2, thus suggests a Fourier transform method of solution. These two examples may illustrate the different methods used for solving Volterra and Fredholm integral equations. In Section 1.4 we present the Laplace and Fourier integral transforms and illustrate their methods of solving integral equations with difference kernels.

Exercises 1.2

1. Classify each of the following integral equations according to whether it is a Fredholm or a Volterra integral equation. Also determine whether it is homogeneous, singular, and so on.

   (a) $u(x) = x - \sin x + e^x(x-1) + \int_0^x [\sin x - e^x(x-t)]u(t)dt$

   (b) $u(x) = e^x - \lambda \int_0^1 G(x,\xi)u(\xi)d\xi,$

   $$G(x,\xi) = \begin{cases} \xi(1+x), & 0 \le x \le \xi \\ x(1+\xi), & \xi \le x \le 1 \end{cases}$$

   (c) $\dfrac{1}{1+x^2} = \displaystyle\int_{-\infty}^{\infty} \frac{u(\xi)}{x-\xi}d\xi$

   (d) $f(x) = \displaystyle\int_x^1 \frac{u(\xi)}{(\xi^2 - x^2)^{\frac{1}{2}}}d\xi$

   (e) $x = \displaystyle\int_0^x \frac{u(t)}{\sqrt{x-t}}dt$

   (f) $u(x) = \lambda \displaystyle\int_{-\infty}^{\infty} e^{-|x-\xi|}u(\xi)d\xi$

   (g) $u(x) + \lambda \displaystyle\int_0^1 (x^2\xi + x\xi^2)u(\xi)d\xi = g(x)$

2. Classify each integral equation of Problems 9 to 11 and 12 in Exercises 1.1 according to whether it is a Fredholm or a Volterra equation. Also determine whether it is homogeneous, singular, and so on.

3. Show whether or not (the *homogeneous* parts) of the following integral equations are linear.

   Hint: For the proof of linearity, according to the definition and paralleling Example 4 a), we consider only the homogeneous version of the integral equation [i.e., $f(x) \equiv 0$ in (1.42) or (1.46)].

   (a) The integral equation (1.19) in $y(x)$ of the small deflection of a rotating shaft.

   (b) The integral equation in $u(x)$,

$$u(x) = f(x) + \int_0^x K(x,\xi)u^2(\xi)d\xi$$

   (c) The most general integral equation (1.39) in $u(x)$ of this section.

   (d) The integral equation in $s(x)$,

$$s(x) = g(x) + \int_a^b K(x,\xi)s(\xi)\frac{ds}{d\xi}d\xi$$

4. Consider the following integral equation representation of what is termed the "Dirichlet problem," then classify it as an integral equation in the charge density distribution function $\rho(\vec{t})$.

   This problem deals with finding the linear charge density distribution $\rho(\vec{t})$ along a (more general) smooth closed contour C that causes a given, or desired, potential distribution $f(\vec{t})$ in the interior of the closed curve C. It is represented as the following integral equation in $\rho(\vec{t})$,

$$f(\vec{t}) = \rho(\vec{t}) - \frac{1}{\pi}\int_C \frac{\rho(\vec{\tau})\cos\theta_{\vec{t},\vec{\tau}}}{\tau}ds$$

   where $\vec{t}$ and $\vec{\tau}$ represent the position vectors of points in the interior and on the curve C, respectively, $\vec{r} = \vec{\tau} - \vec{t}$ is the vector distance between such points, $r = ||\vec{r}||$, $\vec{n}$ is the unit exterior normal vector to C at $\vec{\tau}$, and ds is an arc length increment of C.

5. Determine the class of singularity for the kernel of each of the following integral equations

   (a) The Abel problem in (1.20).

   (b) The generalized Abel problem (1.48).

   (c) The problems in Exercises 1(c), (d), and (e).

6. Classify the following integral equation in $u(t)$

$$\int_0^1 \frac{u(t)dt}{t-x} = 1 - 2x, \qquad 0 < x < 1 \tag{E.1}$$

7. **(a)** Show that the kernel

$$K(x,t) = k(x,t) \ln |x-t|$$

where $k(x,t)$ is bounded, is with weak singularity.

Hint: Write

$$K(x,t) = k(x,t) \frac{|x-t|^{\beta} \ln |x-t|}{|x-t|^{\beta}}, \qquad 0 < \beta < 1 \tag{E.1}$$

and use L'Hospital rule on $|x-t|^{\beta} \ln |x-t|$ as $t \to x$ to show that it is bounded and tends to zero as $t \to x$.

(b) Determine the type of singularity of the integral equation,

$$f(x) = \frac{1}{\pi} \int_0^1 \phi(t) \ln \left| \frac{x-t}{x+t} \right| dt, \qquad 0 < x < 1 \tag{E.2}$$

Hint: See part (a).

(c) Show that from the integral equation (E.2) of part (b) we can generate an integral equation of the first kind with Cauchy kernel.

Hint: Differentiate the equation with respect to x, and allow the interchange of differentiation and integration.

8. Consider the integral equation of the second kind in $u(x)$

$$f(x) = u(x) - \lambda \int K(x,t)u(t)dt. \tag{E.1}$$

Show that to the linear combination $c_1 u_1(x) + c_2 u_2(x)$ of the solutions to (E.1),

$$f_i(x) = u_i(x) - \lambda \int K(x,t)u_i(t)dt, \qquad i = 1, 2, \tag{E.2}$$

there corresponds the nonhomogeneous term $c_1 f_1(x) + c_2 f_2(x)$ in (E.1).

Hint: Substitute $u_1(t)$, $u_2(t)$ as solutions corresponding to $f_1(x)$ and $f_2(x)$ in (E.1), then multiply the resulting first and second equations (for u_1 and u_2) by c_1 and c_2, respectively, and add.

9. **(a)** Show that if $\{\phi_i(t)\}_{i=1}^{n}$ is a set of solutions to the (linear) homogeneous Fredholm integral equation,

$$\phi(x) = \lambda \int_a^b K(x,t)\phi(t)dt \tag{E.1}$$

corresponding to λ_i, i.e.,

$$\phi_i(x) = \lambda_i \int_a^b K(x,t)\phi_i(t)dt \tag{E.2}$$

then the linear combination

$$\sum_{i=1}^{n} c_i \phi_i(t) \tag{E.3}$$

of such solutions is also a solution to (E.1).

Hint: Multiply both sides of (E.2) by c_i, sum from $i = 1$ to n, and invoke (E.2) for the integral to give $\frac{\phi_i(x)}{\lambda_i}$.

1.3 SOME IMPORTANT IDENTITIES AND BASIC DEFINITIONS

In this section we will derive and illustrate very basic identities that are needed to facilitate the analysis of reducing an important class of initial value problems and boundary value problems to Volterra and Fredholm integral equations, respectively, and vice versa. The latter topics are presented in Sections 2.4 and 2.5, respectively. This includes a basic identity that reduces the repeated integrations, necessary for integrating higher order derivatives, to a single integral. The other identity is the generalized *Leibnitz rule* for differentiating integrals (with variable limits of integration), which is needed for reducing an integral equation to a differential equation. The rest of the section is devoted to few very basic definitions.

In Chapter 2 we present initial and boundary value problems associated with linear differential equations and, usually, homogeneous auxiliary conditions, to show how they can be represented by Volterra and Fredholm integral equations, respectively. In doing so we need to perform a number of integrations. For example, in the second-order differential equation of the form $d^2y/dx^2 = F(x)$, we can integrate twice to obtain

$$\frac{dy}{dx} = \int_a^x F(\xi)d\xi + c_1$$

$$y(x) = \int_a^x \int_a^\xi F(t)dtd\xi + c_1 x + c_2. \tag{1.50}$$

Note how we had to change the variable of integration (ξ to t in the inner integral) to keep x, the independent variable, as the limit of the last integration.

1.3.1 Multiple Integrals Reduced to Single Integrals

The double integral (1.50) can be reduced to a single integral:

$$\int_a^x \int_a^\xi F(t)dtd\xi = \int_a^x (x-t)F(t)dt. \tag{1.51}$$

This can be proved by two methods. The first is integrating by parts, letting $dv = d\xi$ and $u(\xi) = \int_a^\xi F(t)dt$ in (1.51), and knowing that $du/d\xi = F(\xi)$.

$$\begin{aligned} \int_a^x \int_a^\xi F(t)dtd\xi &= \left[\xi \int_a^\xi F(t)dt\right]_a^x - \int_a^x \xi F(\xi)d\xi \\ &= x\int_a^x F(t)dt - 0 - \int_a^x \xi F(\xi)d\xi \\ &= \int_a^x (x-\xi)F(\xi)d\xi = \int_a^x (x-t)F(t)dt \end{aligned} \tag{1.51}$$

where we replaced ξ by t in the last integral since ξ and t are only dummy variables of integration (with the same limits a to x).

The second method involves exchanging the two integrals. We will also illustrate this method since it is often used. The domain of the double integral (1.50) is shown in Figure 1.8, where the integration over t first, then ξ, is indicated by the solid arrows. When the integration is interchanged (i.e., when we integrate with respect to ξ then t), as indicated by the dashed arrows, we obtain

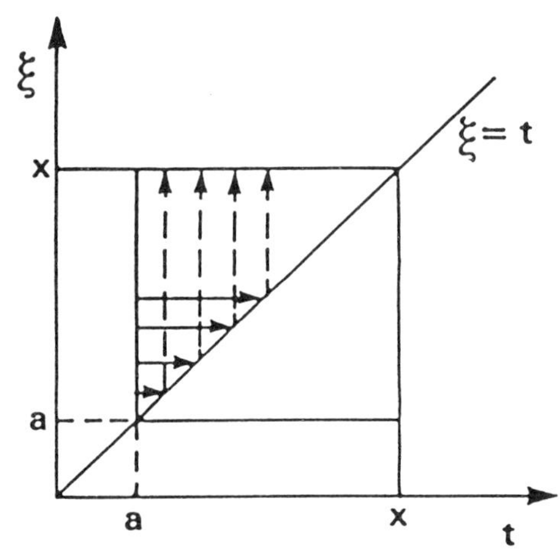

Fig. 1.8 Domains for performing the integration in (1.51) with respect to t first (solid lines) or with respect to ξ first (dashed lines).

$$\begin{aligned} \int_a^x \int_a^\xi F(t)dtd\xi &= \int_a^x F(t)\left[\int_t^x d\xi\right] dt \\ &= \int_a^x (x-t)F(t)dt \end{aligned} \tag{1.51}$$

after evaluating the simple integral $\int_t^x d\xi = x - t$.

Example 5

Reduce the differential equation

$$\frac{d^2y}{dx^2} = \lambda y(x) \tag{E.1}$$

to an integral equation.

We let $d^2y/dx^2 = F(x)$ and integrate once with respect to x to obtain

$$\frac{dy}{dx} = \int_a^x F(\xi)d\xi + c_1 \tag{E.2}$$

and if we integrate again as in (1.50), we have

$$y(x) = \int_a^x \int_a^\xi F(t)dtd\xi + c_1x + c_2 = \int_a^x (x-t)F(t)dt + c_1x + c_2 =$$

$$= \int_a^x (x-\xi)F(\xi)d\xi + c_1x + c_2 \tag{E.3}$$

after using (1.51). But from (E.1),

$$F(x) = \frac{d^2y}{dx^2} = \lambda y(x) \tag{E.4}$$

which we can substitute in (E.3) to obtain the integral equation

$$y(x) = \lambda \int_a^x (x-\xi)y(\xi)d\xi + c_1x + c_2. \tag{E.5}$$

The identity (1.51) serves for second-order differential equations, but for nth-order derivatives we need the following generalization of (1.51) for reducing the resulting n repeated integrations to a single integral,

$$\int_a^x \int_a^{x_1} \cdots \int_a^{x_{n-1}} F(x_n)dx_n \cdots dx_1 = \frac{1}{(n-1)!}\int_a^x (x-x_1)^{n-1}F(x_1)dx_1 \tag{1.52}$$

which can be proved in the same way (see Exercise 6).

1.3.2 Generalized Leibnitz Formula

Once we obtain an integral equation for the initial or boundary value problem, it becomes natural to inquire whether this integral equation indeed satisfies the original differential equation. For differentiating an integral equation (1.39) with variable limit of integration $b(x)$, we need the following *generalized Leibnitz formula*

$$\frac{d}{dx}\int_{\alpha(x)}^{\beta(x)} F(x,y)dy = \int_{\alpha(x)}^{\beta(x)} \frac{\partial F}{\partial x}(x,y)dy$$

$$+F(x,\beta(x))\frac{d\beta}{dx} - F(x,\alpha(x))\frac{d\alpha}{dx}. \tag{1.53}$$

It is valid if both $F(x, y)$ and its partial derivative $\frac{\partial F}{\partial x}$ are continuous functions of x and y, and if both $\frac{d\alpha}{dx}$ and $\frac{d\beta}{dx}$ are continuous. With such conditions, one should be very cautious when it comes to differentiating integral equations with singular kernels. For example, the above conditions are clearly not satisfied in the case of Abel's integral equation,

$$f(x) = \int_0^x \frac{u(\xi)}{(x-\xi)^\alpha}d\xi, \qquad 0 < \alpha < 1 \tag{1.48}$$

since here $F(x, \xi) = u(\xi)/(x - \xi)^\alpha$ is unbounded at $x = \xi$.

The rule in (1.53) is a generalization of *the fundamental theorem of integral calculus,*

$$\frac{d}{dx}\int_a^x F(y)dy = F(x). \tag{1.54}$$

To prove (1.53), we let

$$\phi(\alpha,\beta,x) = \int_{\alpha(x)}^{\beta(x)} F(x,y)dy \tag{1.55}$$

and

$$\frac{\partial f}{\partial y}(x,y) = F(x,y) \tag{1.56}$$

where then

$$\phi(\alpha,\beta,x) = \int_{\alpha(x)}^{\beta(x)} \frac{\partial f}{\partial y}(x,y)dy = f(x,\beta) - f(x,\alpha). \tag{1.57}$$

So we will use the chain rule on $\phi(\alpha, \beta; x)$ as function of three variables $\alpha(x)$, $\beta(x)$ and x,

$$\frac{d\phi}{dx} = \frac{\partial\phi}{\partial x} + \frac{\partial\phi}{\partial\beta}\frac{d\beta}{dx} + \frac{\partial\phi}{\partial\alpha}\frac{d\alpha}{dx} \tag{1.58}$$

and allow partial differentiation with respect to x inside the integral of (1.55), giving us

$$\frac{\partial\phi}{\partial x} = \frac{\partial}{\partial x}\int_{\alpha(x)}^{\beta(x)} F(x,y)dy = \int_{\alpha(x)}^{\beta(x)} \frac{\partial F}{\partial x}(x,y)dy \tag{1.59}$$

since $\partial\phi/\partial x$ here means keeping α and β as constants. Also, if we use $\phi(\alpha,\beta,x) = f(x,\beta) - f(x,\alpha)$ from (1.57), we have

$$\frac{\partial\phi}{\partial\beta} = \frac{\partial}{\partial\beta}[f(x,\beta) - f((x,\alpha)] = \frac{\partial f(x,\beta)}{\partial\beta} - \frac{\partial f(x,\alpha)}{\partial\beta} = F(x,\beta) - 0 \quad (1.60)$$

and

$$\frac{\partial\phi}{\partial\alpha} = \frac{\partial}{\partial\alpha}[f(x,\beta) - f(x,\alpha)] = 0 - \frac{\partial f(x,\alpha)}{\partial\alpha} = -F(x,\alpha). \quad (1.61)$$

So if we combine (1.59), (1.60), and (1.61) in (1.58), we have

$$\frac{d\phi}{dx} = \int_{\alpha(x)}^{\beta(x)} \frac{\partial F}{\partial x}(x,y)dy + F(x,\beta)\frac{d\beta}{dx} - F(x,\alpha)\frac{d\alpha}{dx} \quad (1.62)$$

which is (1.53).

Example 6

Verify that the solution of the Volterra integral equation

$$y(x) = \lambda\int_a^x (x-\xi)y(\xi)d\xi + c_1x + c_2 \quad (E.1)$$

of Example 5 is the solution of the differential equation

$$\frac{d^2y}{dx^2} = \lambda y(x). \quad (E.2)$$

To do this we differentiate $y(x)$ in (E.1) once to obtain

$$\frac{dy}{dx} = \lambda\int_a^x y(\xi)d\xi + c_1 \quad (E.3)$$

after using (1.53) on the integral in (E.1) with $\alpha(x) = a$, $\beta(x) = x$, and $K(x,\xi) = x - \xi$. If we now differentiate (E.3) using (1.53) again, or its special case (1.54), we obtain (E.2):

$$\frac{d^2y}{dx^2} = \lambda y(x). \quad (E.2)$$

Another similarly important case where we will need the generalized Leibnitz rule (1.53) is when we reduce a Fredholm integral equation to its equivalent boundary value problem associated with a differential equation, which we hope is a familiar one to solve. This is attained by differentiating the integral equation, as we did in Example 6, until we reduce it to a differential equation and then seek the boundary conditions needed from the integral equation. For example, the homogeneous Fredholm integral equation

$$u(x) = \lambda\int_0^1 K(x,t)u(t)dt \quad (E.4)$$

with the kernel

$$K(x,t) = \begin{cases} x(1-t), & 0 \le x \le t \\ t(1-x), & t \le x \le 1 \end{cases} \tag{E.5}$$

can be reduced to the boundary value problem in u,

$$\frac{d^2u}{dx^2} + \lambda u = 0 \tag{E.6}$$

$$u(0) = 0 \tag{E.7}$$

$$u(1) = 0 \tag{E.8}$$

which we will leave as an exercise. [See Exercise 5 of this section or Example 6 of Section 2.5, equations (E.1)–(E.11).]

In looking at the final result of the *generalized Leibnitz formula* (1.53) of the last section

$$\frac{d}{dx}\int_{\alpha(x)}^{\beta(x)} F(x,y)dy = \int_{\alpha(x)}^{\beta(x)} \frac{\partial F}{\partial x}(x,y)dy$$
$$+F(x,\beta(x))\frac{d\beta}{dx} - F(x,\alpha(x))\frac{d\alpha}{dx} \tag{1.53}$$

we can basically interpret it in the direction of a rule that resulted in allowing us to enter the differentiation operation inside the integral on the left side of (1.53), as a partial differentiation, as seen in the integration term on the right side of (1.53). This, we may term now, as some type of interchange of the two basic operations of differentiation and integration. In mathematical analysis, and especially its applications, one faces many situations of such interchange of many very basic mathematical operations. A summary and illustration of the main theorems that allow such an interchange are found in Jerri [1992, pp. 99–104, pp. 377–382.]

1.3.3 Convergence of Integrals and Basic Definitions

As is expected, when we deal with improper integrals, we must assure the convergence (and, sometimes a certain type) for the individual integrals. Very familiar situations are when we deal with the *Laplace transform* of $f(x)$ on $(0,\infty)$,

$$F(s) = \mathcal{L}\{f\} = \int_0^\infty e^{-sx} f(x)dx \tag{1.63}$$

and the *Fourier transform* of $f(x)$ on $(-\infty,\infty)$.

$$F(\lambda) = \mathcal{F}\{f\} = \int_{-\infty}^{\infty} e^{-i\lambda x} f(x)dx \tag{1.64}$$

since they are defined as improper integrals. In the case of the Laplace transform, especially, there is a very reasonable and applicable class of functions $f(x)$ which guarantees the existence of its Laplace transform as the improper integrals in (1.63). This class of functions is described as (i) *sectionally continuous* on each bounded interval $0 < x < R$, and (ii) of *exponential order* as $x \to \infty$, i.e., $|f(x)|$ does not grow faster than $Me^{\alpha x}$, where M and α are constants (or that there exist positive numbers M and A such that $|f(x)| < Me^{\alpha x}$ for all $x > A$.)

Definition 1 $f(x)$ is called *sectionally (or piecewise) continuous* on an interval $a < x < b$ if this interval can be subdivided by a finite number of points $a = x_0 < x_1 < x_2 < \cdots < x_n = b$ into n subintervals $x_{i-1} < x < x_i$, $i = 1, 2, 3, \cdots, n$),

(i) the function $f(x)$ is continuous on each of the subintervals: $x_{i-1} < x < x_i$, $i = 1, 2, 3, \cdots, n$ and

(ii) $f(x)$ approaches a finite limit as x approaches the limits of the subinterval, x_{i-1} and x_i, from the interior.

Figure 1.9 illustrates a function $f(x)$ which is sectionally continuous on the interval (a, b); that is, it is continuous on each of the open subintervals (a, x_1), (x_1, x_2), and (x_2, b). Note, for example, that the left- and right-hand limits $f(x_2-)$ and $f(x_2+)$, as x approaches x_2, are not equal, and we say that $f(x)$ has a *jump discontinuity* at x_2 of magnitude $J = f(x_2+) - f(x_2-)$.

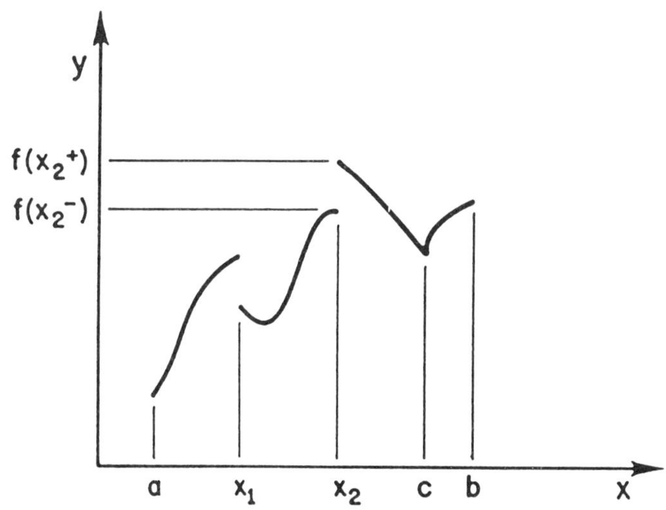

Fig. 1.9 A sectionally continuous function $f(x)$ on (a, b) with two jump discontinuities at x_1 and x_2. From Jerri [1992], courtesy of Marcel Dekker Inc.

For the theory of Fourier series and integrals, we shall need a more restricted class of functions than the above piecewise continuous functions, namely the *piecewise smooth* functions.

Definition 2 A piecewise continuous function in an interval $a < x < b$ is termed piecewise (or sectionally) smooth if, in addition to being piecewise continuous,

(1) its first derivative df/dx is continuous on each of the subintervals $x_{i-1} < x < x_i$, $i = 1, 2, 3, \cdots, n$, and

(2) df/dx approaches a finite limit as x approaches the limits of the subinterval x_{i-1} and x_i from the interior; i.e., there exists $f'(x_{i-1}+), f'(x_i-)$ for each $(x_{i-1}, x_i), i = 1, 2, 3, \cdots, n$.

For example, the function in Figure 1.9 is piecewise continuous on (a, b) but it is not piecewise smooth because of condition (2) above in the subinterval (c, b), where its derivative df/dx does not approach a limit $f'(c+)$ as x approaches the end point c from the right. For completeness, we may mention that the function is sectionally smooth on (a, c), and it is smooth on (a, x_1).

Definition 3 $f(x)$ is termed of *exponential growth*, or of *exponential order*, as $x \to \infty$ if there exists a constant α such that $e^{-\alpha x}|f(x)|$ is bounded for all x greater than a finite number A. In other words, $|f(x)| \leq Me^{\alpha x}$ for $x > A$, with M, α constants, and we say $f(x)$ is $O(e^{\alpha x})$. For example, the function $f(x) = e^{3x}\sin x$ is of exponential order $O(e^{3x})$ since with $M = 1, |f(x)| = |e^{3x}\sin x| \leq e^{3x}$ for all x. However, functions like $f(x) = e^{3x^2}$ are not of exponential order. Polynomial functions like $f(x) = x^n$ are clearly of exponential order with $\alpha > 0$.

In the following Example 7, we will illustrate how easy it is to prove the existence of the Laplace transform $F(s)$ in (1.63) for the class of functions $f(x)$, which are sectionally continuous and of exponential order. This Example will constitute the proof of Theorem 1 in Section 1.4.1.

Example 7 The Existence of Laplace Transform[11]

If $f(x)$ is

(i) sectionally continuous on the interval $0 \leq x \leq A$, and

(ii) of exponential order $e^{\alpha x}$, that is, $|f(x)| \leq Me^{\alpha x}$ for $x > A$,

then the Laplace transform $F(s)$ of $f(x)$ in (1.63) exists for $s > \alpha$.

Proof

$$F(s) = \int_0^\infty e^{-sx} f(x)dx = \int_0^A e^{-sx} f(x)dx + \int_A^\infty e^{-sx} f(x)dx. \qquad (E.1)$$

The first integral on the finite interval $(0, A)$ clearly converges since $e^{-sx}f(x)$ is bounded for the sectionally continuous $f(x)$. For the convergence of the second integral, we will use the result of comparison of improper integrals, along with the exponential growth of $f(x)$, to show that it converges provided that $s > \alpha$.

$$\begin{aligned} \left|\int_A^\infty e^{-sx} f(x)dx\right| &\leq \int_A^\infty |e^{-sx} f(x)|dx \leq \int_A^\infty e^{-sx}|f(x)|dx \\ &\leq M\int_A^\infty e^{-(s-\alpha)x}dx < \infty \text{ for } s > \alpha \end{aligned} \qquad (E.2)$$

[11]Optional

where in the last integral we used $|f(x)| \leq Me^{\alpha x}$. The last improper integral clearly converges for $s > \alpha$, which concludes our proof.

In this fashion, we have shown not only that the Laplace transform exists by proving that $\int_0^\infty e^{-sx} f(x)dx < \infty$, for $s > \alpha$ but also that $e^{-sx} f(x)$ is absolutely integrable, i.e., $\int_0^\infty |e^{-sx} f(x)|dx < \infty$, for $s > \alpha$.

We must also note that the above two conditions (i) and (ii), that $f(x)$ be sectionally continuous on $(0, A)$ and of exponential growth as $x \to \infty$, are sufficient but not necessary. An example of $f(x)$ not sectionally continuous on $(0, \infty)$ is $f(x) = \frac{1}{x^{\frac{1}{2}}}$ which is infinite as $x \to 0$. This means that the first of the above two sufficient conditions (i) is not satisfied. However it can be shown, with the aid of using the gamma function, that the Laplace transform of $1/x^{\frac{1}{2}}$ does exist as $\mathcal{L}\left\{\frac{1}{x^{\frac{1}{2}}}\right\} = \sqrt{\frac{\pi}{s}}$, $s > 0$. [See the definition of the gamma function in (1.74) and the Laplace transform pair for $\nu = -\frac{1}{2}$ in (1.79), and Exercise 1(b) (and 4(b)) of Section 1.4.]

Example 8 An Important Necessary Condition of Laplace Transform

A very important necessary condition for the existence of the Laplace transform $F(s)$, of the above class of functions of Example 7, is that $F(s)$ must vanish as s approaches infinity. This can be easily shown when we write

$$\lim_{s\to\infty} F(s) = \lim_{s\to\infty} \int_0^\infty e^{-st} f(t)dt, \quad s > \alpha. \tag{E.1}$$

Then if we "formally" allow the interchange of the limit process with the integration, an operation that is valid[12] for $f(t)$ in the class of the functions in Example 7, $|f(t)| < Me^{\alpha t}$, we have

$$\lim_{s\to\infty} |F(s)| \leq M \int_0^\infty \lim_{s\to\infty} e^{-(s-\alpha)t} dt = 0, \qquad s > \alpha.$$

since, clearly, $\lim_{s\to\infty} e^{-(s-\alpha)t} = 0$ for $s > \alpha$.

We have this example to illustrate one of the difficulties with solving integral equations of the first kind. Here we can look at finding the inverse of the Laplace transform $f(t) = \mathcal{L}^{-1}\{F(s)\}$ of $F(s)$ as attempting to solve the following integral equation of the first kind in $f(t)$.

$$F(s) = \int_0^\infty e^{-st} f(t)dt, \qquad s > \alpha.$$

So, to search for a reasonable solution $f(t)$ such as that of the (large) class of functions of Example 7, we must be very careful (or fussy) about what is assigned above as $F(s)$. Well, $F(s)$ must at least vanish as s approaches infinity to satisfy the necessary condition shown in the above Example 8. So if we are given $F(s) = \frac{s+1}{s}$,

[12]The condition for allowing the above interchange of the two operations is very close to Lebesgue convergence theorem. See Jerri [1992, p. 99, Theorem 2.10].

we know that there exists no solution $f(t)$ for the above integral equation in the class of functions described in Example 7. In other words, there is no such function $f(t)$ in the domain of the Laplace transform operator that is mapped to the given $F(s) = \frac{s+1}{s}$! So, if we write the general integral equation of the first kind in $u(t)$

$$g(x) = \int_a^b K(x,t)u(t)dt,$$

we must be aware not to have $g(x)$ assigned arbitrarily. This means that the theory of such equations must be checked thoroughly, for such and other difficulties, before going after a solution. In Section 5.4 we present some essentials of this subject and its several, possibly chronic, difficulties for the Fredholm integral equations of the first kind. In Section 3.2 we only present possible methods of solution for Volterra integral equations of the first kind.

Exercises 1.3

1. **(a)** Verify that

$$y(x) = \frac{1}{b}\int_0^x \sin b(x-t)f(t)dt \tag{E.1}$$

is a solution of the following initial value problem

$$\frac{d^2y}{dx^2} + b^2 y = f(x), \tag{E.2}$$

$$y(0) = y'(0) = 0. \tag{E.3}$$

Hint: See Example 6.

(b) Reduce the initial value problem (E.2) and (E.3) to a Volterra integral equation. Note that in (E.1) we have the solution to the problem (E.2) and (E.3), and not the integral equation representation of it that we are seeking here.

Hint: See Example 5 then invoke the initial conditions of (E.3) to determine the arbitrary constants c_1 and c_2.

(c) Verify your answer in part (b) by reducing it to the initial value problem (E.2) and (E.3). *Hint:* The use of the generalized Leibnitz rule (1.53) is very helpful in differentiating the integral of the integral equation as the answer of part (b).

2. **(a)** Verify that the Volterra integral equation

$$y(x) = \cos x - x - 1 - \int_0^x (x-t)y(t)dt \tag{E.1}$$

reduces to the initial value problem

$$\frac{d^2y}{dx^2} + y = -\cos x, \tag{E.2}$$

$$y(0) = 0, \qquad y'(0) = -1. \tag{E.3}$$

Hint: See Example 6.

(b) Reduce the initial value problem in (E.2) and (E.3) to the Volterra integral equation in (E.1). *Hint:* See Example 5 and problem 1(b).

3. **(a)** Use $F(x) = d^2u/dx^2$ to reduce the differential equation

$$\frac{d^2u}{dx^2} = \lambda u(x) + g(x), \quad x > 0 \tag{E.1}$$

of an initial value problem to an integral equation.

Hint: See Example 5.

(b) Verify your answer in part (a) by showing that $u(x)$ in the integral equation satisfies the differential equation (E.1).

4. Reduce the integral equation

$$u(x) = \lambda \int_0^\infty e^{-|x-t|} u(t)dt$$

to a differential equation.

Hint: write

$$u(x) = \lambda \left[\int_0^x e^{-(x-t)} u(t)dt + \int_x^\infty e^{-(t-x)} u(t)dt \right]$$

and then differentiate twice.

5. **(a)** Differentiate the Fredholm integral equation

$$u(x) = \lambda \int_0^1 K(x,t)u(t)dt, \tag{E.1}$$

$$K(x,t) = \begin{cases} x(1-t), & 0 \le x \le t \le 1 \\ t(1-x), & 0 \le t \le x \le 1 \end{cases} \tag{E.2}$$

to reduce it to a differential equation. *Hint:* Write the integral equation with its explicit kernel (E.2) as

$$\begin{aligned} u(x) &= \lambda \int_0^x (1-x)tu(t)dt + \lambda \int_x^1 x(1-t)u(t)dt \\ &= \lambda(1-x) \int_0^x tu(t)dt + \lambda x \int_x^1 (1-t)u(t)dt \end{aligned} \tag{E.3}$$

and differentiate, realizing that each term in the right side of (E.3) is a product of two functions of x, and use (1.53) for differentiating the integrals.

(b) Use the form (E.3) to find the two boundary conditions for $u(x)$ at $x = 0$ and $x = 1$ as $u(0) = 0$ and $u(1) = 0$.

(c) Solve the resulting boundary value problem associated with the differential equation of part (a) and the boundary conditions in part (b).

Hint: You have the two linearly independent solutions $\sin\sqrt{\lambda}x$ and $\cos\sqrt{\lambda}x$, for the differential equation, to use in a linear combination to satisfy the two boundary conditions $u(0) = 0$, $u(1) = 0$. Here you will end up with an "infinite" set of solutions (called the "eigenfunctions" or "characteristic" functions of the boundary value problem). These solutions are associated with the discrete values $\lambda_n, n = 1, 2, \cdots, n$ of the parameter λ in (E.1), (which are called the eigenvalues or "characteristic" values of the boundary value problem). (See also Example 6 of Section 2.5.)

6. Prove the result (1.52) for $n = 3$ of the triple integration.

Hint: consider the triple integral $\int_a^x \int_a^\xi \int_a^t F(y)dydtd\xi$, and let $G(t) = \int_a^t F(y)dy$, use (1.51) for the double integral $\int_a^x \int_a^\xi G(t)dtd\xi$, then use the same integration by parts (with respect to ξ) on the resulting double integral $\int_a^x \int_a^\xi (x-\xi)F(y)dyd\xi = \int_a^x (x-\xi)[\int_a^\xi F(y)dy]d\xi$ with $u(\xi) = \int_a^\xi F(y)dy$ and $dv(\xi) = (x-\xi)d\xi$.

1.4 LAPLACE, FOURIER, AND OTHER TRANSFORMS

In this section we present a brief summary of some important properties of Laplace and Fourier transforms. These transforms are very useful for solving certain initial and boundary value problems associated with differential equations and partial differential equations with constant coefficients. Also, as we shall soon show, the Laplace and Fourier transforms are used for solving Volterra and (singular) Fredholm integral equations with difference kernels, respectively. These transforms may also be applied to integro-differential equations. For detailed treatment of these and other transforms, see Jerri [1992] and Sneddon [1972].

Other singular Fredholm integral equations, as results of finding the inverse of integral transforms such as the Hilbert and Mellin transforms, will be studied and illustrated with the help of the following analysis of Fourier and Laplace transforms. As was mentioned in the preface, the treatment here is elementary as for the first undergraduate course. However, for the interest of the reader who wants to go to some reasonable mathematical rigor, we have in this edition spelled out the basic results clearly and carefully as theorems.

1.4.1 The Laplace Transform

The Laplace transform of the function $f(x)$ defined on $(0, \infty)$ is

$$F(s) \equiv \mathcal{L}\{f\} = \int_0^\infty e^{-sx} f(x)dx. \tag{1.63}$$

We have already defined in Section 1.3 the (usual) class of functions $f(x)$ for which the above improper Laplace integral exists. This is being the class of *sectionally continuous* (Definition 2) and of *exponential order* (Definition 4). We then proved such existence in Example 7, and which we shall repeat only the statement here as Theorem 1 on the existence of Laplace transform for such class of functions.

Theorem 1 The Existence of the Laplace Transform

In the following statement we will use $f(t)$ instead of $f(x)$, since, as it shall become very clear shortly, we reserve x for the real part of the complex number $z = x + iy$ for $F(z)$ (the extension of $F(s)$ to $F(z)$). If $f(t)$ is (as in Example 7)

(i) sectionally continuous on the interval $0 \leq t \leq A$ and

(ii) of exponential order $e^{\alpha t}$, i.e., $|f(t)| \leq Me^{\alpha t}$ for $t > A$,

then, the Laplace transform $F(s)$ of $f(t)$ in (1.63) exists for $s > \alpha$. The proof is done in details in Example 7. It is clear that the equation defining Laplace transform in (1.61) represents a *Fredholm integral equation of the first kind* in $f(x)$ with kernel $K(s,t) = e^{-st}$, which is *singular* since the integral is with an infinite limit. To speak about the *inverse* of the Laplace transform in (1.63), that is $f(t) = \mathcal{L}^{-1}\{F(s)\}$, in our present notion of integral equations, is to embark on the attempt to solve the *singular integral equation of the first kind* (1.63) in $f(t)$. As is the case for solving most singular integral equations of this type, the tools of complex analysis are employed. For our purpose in this book, where we don't require formal preparation in functions of complex variables, we will be satisfied with the following clear statement of the result. We will, however, follow this by a more appropriate formula at the level of this book, but, possibly due to its impracticability as shown in (1.67), it is not much referred to in the discussion of the Laplace transform in almost all textbooks. Such formula (1.67) uses only differentiation, and without resorting to complex variables. The solution $f(t)$ to the singular integral equation (1.63) is given as the *inverse Laplace transform* of $F(s)$, which we shall state the conditions for its existence in Theorem 2,

$$f(t) = \mathcal{L}^{-1}\{F\} = \frac{1}{2\pi i} \lim_{L\to\infty} \int_{\gamma-iL}^{\gamma+iL} e^{\sigma t} F(\sigma)d\sigma, \quad \gamma > \text{Real } \{z_i\}, \tag{1.65}$$

where $\{z_i\}$ are the singularities of $F(z)$. The above integral is a complex line integral of $F(z)$, $z = x + iy$, taken along a vertical line in the complex plane at $x = \gamma$ where $z = \sigma = \gamma + iy$, and where the location $x = \gamma$ is to the right of all singularities $\{z_i\}$ of $F(z)$. For example the Laplace transform of $f(t) = e^{2t}$, $0 < t < \infty$ is the real valued function $F(s) = \frac{1}{s-2}$. For the inversion formula (1.65) we extend $F(s)$ analytically to the complex plane as $F(z) = \frac{1}{z-2} = \frac{1}{x+iy-2} = \frac{1}{x-2+iy}$, and note that it has only one singularity at $z_1 = 2$, thus we take the vertical complex line integral to the right of the real part of z_1 which is $x_1 = 2$, $\gamma > 2$. The derivation of the Laplace transform inversion formula (1.65) involves relating the Laplace transform to the Fourier transform of *causal* functions ($f(t) \equiv 0, t < 0$), where the definition of both transforms is extended to complex variables. In this introductory book we don't assume preparation in complex variables, and in our next very brief discussion, and a statement of an important theorem, we will only use something like the above basic elements of complex numbers.

In Theorem 1 we stated conditions for the existence of the Laplace transform (1.63),

$$F(s) = \int_0^\infty e^{-st} f(t)dt = \mathcal{L}\{f\} \tag{1.63}$$

as an improper integral, which is of utmost importance to the theory of Laplace transform as an *integral transform*. What concerns us here in the topic of *integral equations* is to see the Laplace transform (1.63) as a *singular integral equation* in $f(t)$, $0 < t < \infty$. Thus, before we mention the solution to such an equation as in the Laplace inversion formula (1.65), we should assert the existence of such a solution. This is covered in the following Theorem 2 on the existence of the inverse Laplace transform in (1.65) as the solution to the singular integral equation (1.63) in $f(t)$. Note that we are using here t instead of x as the variable for $f(t)$ in (1.65) because we need to use x as the real part of the complex variable $z = x + iy$, since the theorem needs the extension of the definition of the Laplace transform $F(s)$ to complex numbers $F(z) = F(x + iy)$ i.e. the variable x now stands for the s in $F(s)$ of (1.63).

Theorem 2 The Existence of the Inverse Laplace Transform (as a solution to the singular integral equation (1.63)

If $F(z)$, $z = x + iy$ is the Laplace transform of any function $f(t)$ of exponential order $O(e^{x_0 t})$, where $f(t)$ and $f'(t)$ are sectionally continuous in each interval $0 \le t \le T$, then the (inversion) integral of $F(z)$ in (1.65) along any line $x = \gamma$, where $\gamma > x_0$, exists and represents $f(t)$,

$$\mathcal{L}^{-1}\{F(z)\} = f(t), \quad t > 0. \tag{1.66}$$

At any point t_0, where $f(t)$ is discontinuous, the inversion integral represents the mean value $\frac{1}{2}[f(t_0+) + f(t_0-)]$; when $t = 0$ it has the value $\frac{1}{2}f(0+)$, and when $t < 0$ it has the zero value. We should note here that the conditions of this theorem on $f(t)$ as a solution of the singular integral equation

$$F(s) = \int_0^\infty e^{-st} f(t)dt \tag{1.63}$$

are not so stringent, since in the applications, for example, $f(t)$ may be the displacement of mechanical vibrations or electrical current, and we can easily impose "sectional continuity" and "of exponential order" on the displacement $f(t)$ and its derivative $\frac{df}{dt}$.

This version of the theorem for the existence of the solution $f(t)$ of the singular integral equation (1.63) is what we considered the appropriate one for this book from among other theorems, whose statements involve complex analysis. However for our purpose of solving the integral equation (1.63) in $f(t)$, we would like to have the conditions to be put on the given function $F(s)$, and not its analytic extension $F(z)$. This is exactly the advantage of the other, not well quoted in texts, form of solution to (1.63) as we shall present in (1.67). What remains about the solution of (1.63) is its uniqueness. As it is the case for all integral transforms, their inverse is not unique in the sense that two solutions $f_1(t)$ and $f_2(t)$ of (1.63) could differ at any finite set of points $t_1, t_2, \cdots, t_n$, or even at an infinite set of points $t_1, t_2, \cdots$, and still give the same $F(s)$ (see Exercise 3(b)). For the proof of Theorem 2, and the other theorems, see Churchill [1972].

Another Formula for the Inverse Laplace Transform

A not much seen formula (in textbooks) for the inverse Laplace transform $f(t) = \mathcal{L}^{-1}\{F(s)\}$ is the following,[13]

$$f(t) = \lim_{k\to\infty} \left\{ \left[\frac{(-1)^k}{k!} \frac{d^k F}{ds^k} \left(\frac{k}{s} \right) \right] \cdot \left(\frac{k}{s} \right) \right\}. \tag{1.67}$$

We may remark here that though this formula is appealing on first sight, it is very demanding, the possible reason that it is scarcely mentioned in texts compared to (1.65). It is obvious that this formula puts the burden on the given function $F(s)$, which is more suitable for us as we look for an inverse Laplace transform $f(t)$ as the solution of the integral equation of the first kind

$$F(s) = \int_0^\infty e^{-st} f(t)dt.$$

First it requires $F(s)$ to have very high order derivatives in $\frac{d^k}{ds^k}F(\frac{k}{s})$, and for large arguments $\frac{k}{s}$ as k becomes very large. This may also illustrate another difficulty for finding the solution of integral equations of the first kind. What makes this problem worse, is that we often have the known output $F(s)$ as a finite number of data, along with these points inaccuracy of their measurements. So, the derivative of this data with its inaccuracy will have its own error, and for higher derivatives such errors will be compounded to render the result useless!

This, as we planned it, should illustrate again the "inherited" difficulties of the integral equations of the first kind. So it is not the problem of formulas (1.65) or (1.67), where they "innocently" require the best of (the input) functions $F(s)$: as infinitely differentiable (analytic) functions as seen in (1.67), or what is hidden in (1.65) as the requirement of $F(z)$ being analytic except for an "isolated" finite or infinite number of points in the complex plane.

[13]From Wing [1991, p. 8], which is attributed to Post and Widder [Widder, 1946].

Properties, Pairs of the Laplace Transform

The most important property of the Laplace transform is known for solving differential equations, where it transforms the differential operation $\frac{df}{dx}$ on $f(x)$ to an algebraic operation $sF(s) - f(0)$ on its Laplace transform $F(s)$. This can be shown next by using (1.63), performing one integration by parts and assuming that $\lim_{x\to\infty} e^{-sx}f(x) = 0$ [i.e., $f(x)$ is with exponential growth $e^{\alpha x}$ as $x \to \infty$ and $s > \alpha$]:

$$\int_0^\infty e^{-sx}\frac{df}{dx}dx = f(x)e^{-sx}\Big|_0^\infty + s\int_0^\infty e^{-sx}f(x)dx$$

$$= -f(0) + sF(s) = sF(s) - f(0). \tag{1.68}$$

(We may note that $f(0)$ here is $f(0+)$ since $f(t)$ is defined on $(0, \infty)$, and we have $\lim_{x\to 0+} f(t) \equiv f(0)$.) A more precise statement of this "formal" result is given as the following Theorem 3.

Theorem 3 The Laplace Transform of Derivatives

Let $f(x)$ be a real function which is

(i) continuous for $x \geq 0$ and of exponential order $e^{\alpha x}$, and let

(ii) df/dx be sectionally (piecewise) continuous in every finite closed interval $0 \leq x \leq A$. Then $\mathcal{L}\{df/dx\}$ exists for $s > \alpha$ and (1.68) results,

$$\mathcal{L}\left\{\frac{df}{dx}\right\} = sF(s) - f(0), \qquad s > \alpha. \tag{1.68}$$

By the same method it can be shown that

$$\mathcal{L}\left\{\frac{d^2f}{dx^2}\right\} = s^2F(s) - sf(0) - f'(0) \tag{1.69}$$

which we shall leave as an exercise [see Exercise 6(a)]. These results can be extended to higher derivatives, and as seen in (1.68) and (1.69) we must supply the proper initial conditions on $f(x)$.

The above results, starting with Theorem 3, show the advantage of Laplace transform in reducing differential equations (with constant coefficients) in $f(x)$, $0 < x < \infty$, and its given appropriate initial conditions to an algebraic equation in the Laplace transform $F(s)$. These are of general interest in methods of applied mathematics, but what concerns us here, when dealing with integral equations, should be the result of Laplace transforming an integral of the unknown function. A result in this direction is

$$\mathcal{L}\left\{\int_0^x f(\xi)d\xi\right\} = \frac{F(s)}{s}, \qquad s > 0. \tag{1.70}$$

This pair complements our very important Laplace transform pair of the derivative $\frac{df}{dx}$ as given in Theorem 3,

$$\mathcal{L}\left\{\frac{df}{dx}\right\} = sF(s) - f(0). \tag{1.68}$$

The result in (1.70) can be proved easily by letting $g(x) \equiv \int_0^x f(\xi)d\xi$, with its Laplace transform $G(s)$, and clearly $g(0) = 0$. From the fundamental theorem of calculus we have $\frac{dg}{dx} = f(x)$, and if we use (1.68) above for $\mathcal{L}\left\{\frac{dg}{dx}\right\}$ we have

$$F(s) = \mathcal{L}\{f(x)\} = \mathcal{L}\left\{\frac{dg}{dx}\right\} = sG(s) - g(0) = sG(s),$$

$$G(s) = \frac{F(s)}{s}, \qquad s > 0$$

In terms of integral equations, these two results (1.68) and (1.70) should prove useful when dealing with some *integro-differential* equations, where the sought unknown function $f(x)$ is operated on by integration as well as differentiation.

A more general result concerning the Laplace transform of integral operations is that of the *Laplace convolution theorem*. This is an extremely useful tool to the important class of *Volterra integral equations with difference kernel*. But before stating the convolution theorem as Theorem 4, we should point out the difficulty facing the Laplace transform (or other similar integral transforms) method when we have to deal with *variable coefficients* differential (or integral) equations. An important result in this direction is

$$\mathcal{L}\{x^n f(x)\} = (-1)^n \frac{d^n}{ds^n} F(s), \tag{1.71}$$

which simply says, that it may be a disadvantage to work with the Laplace transform when dealing with variable coefficient terms in the differential equation to be transformed. This is so, since, for n, a nonnegative integer, a polynomial coefficient of order n in the original equation will result in an nth order differential equation in the Laplace transform space. This result (1.71) can be derived easily when we differentiate the Laplace integral in (1.63)

$$\begin{aligned} \frac{d^n}{ds^n}F(s) &= \frac{d^n}{ds^n}\int_0^\infty e^{-sx} f(x)dx \\ &= \int_0^\infty f(x)\frac{d^n}{ds^n}e^{-sx}dx = \int_0^\infty (-x)^n f(x)e^{-sx}dx, \end{aligned} \tag{1.72}$$

giving the desired result

$$\mathcal{L}\{x^n f(x)\} = (-1)^n \frac{d^n}{ds^n} F(s), \tag{1.71}$$

after allowing the interchange of differentiation with integration.

Again, and before introducing the important convolution theorem of Laplace transform in (1.84) as Theorem 4, it is instructive at this point to have a few illustrations.

In solving initial value problems associated with differential or integral equations we may need to Laplace-transform some familiar functions, for example,

$$\mathcal{L}\{e^{ax}\} = \int_0^\infty e^{-sx}e^{ax}dx = \left|\frac{e^{-(s-a)x}}{-(s-a)}\right|_0^\infty = \frac{1}{s-a}, \qquad s > a. \tag{1.73}$$

Also, after a problem is transformed to an algebraic equation in $F(s)$ and then solved for $F(s)$, we need to transform $F(s)$ back to $f(x)$, the solution of the original problem, which is called the *inverse Laplace transform* of $F(s)$ and is denoted by $f(x) = \mathcal{L}^{-1}\{F(s)\}$. As was discussed earlier then presented in equation (1.65), the direct Laplace transform inversion formula involves complex integration, a topic which is not assumed as prerequisite for the level of this text. Thus, on this level of preparation, and as it is done in all elementary differential equations books where the Laplace transform is used, we will depend on the tabulated values of the Laplace transform (Table 1.1) for the few illustrations in this book and refer the interested reader to books of extensive tables[14] for Laplace transform. For example, if the solution of the Laplace-transformed problem is $F(s) = 1/(s-5)$, then from (1.73), the inverse Laplace transform of $1/(s-5)$ is $f(x) = e^{5x}$, which is the solution of the original problem.

In Table 1.1, $\Gamma(\nu)$ is the *gamma function*, which is defined as

$$\Gamma(\nu) = \int_0^\infty x^{\nu-1}e^{-x}dx, \quad \nu \neq 0, -1, -2, \cdots \tag{1.74}$$

and where it can be shown that $\Gamma(\nu+1) = \nu\Gamma(\nu)$, $\Gamma(n+1) = n!$ (for n positive integer), $\Gamma(1) = 0! \equiv 1$, and $\Gamma(\frac{1}{2}) = \sqrt{\pi}$ (see Exercise 3(b)). Also, the *error function* $\text{erf}(x)$ is defined as

$$\text{erf}(x) = \frac{2}{\sqrt{\pi}}\int_0^x e^{-\xi^2}d\xi \tag{1.75}$$

and the *error function complementary* $\text{erfc}(x)$ is defined as

$$\text{erfc}(x) = 1 - \text{erf}(x) = \frac{2}{\sqrt{\pi}}\int_x^\infty e^{-\xi^2}d\xi \tag{1.76}$$

where

$$\int_0^\infty e^{-\xi^2}d\xi = \frac{\sqrt{\pi}}{2}.$$

$J_0(x)$ in Table 1.1 is a *Bessel function* of the first kind of order 0, which is a special case of $J_n(x)$ that is one of the two solutions of the Bessel differential equation,

$$x^2\frac{d^2u}{dx^2} + x\frac{du}{dx} + (x^2 - n^2)u = 0 \tag{1.77}$$

[14]See Roberts and Kaufman [1966], Ditkin and Prudnikov [1965], Erdelyi et al. [1954].

Table 1.1 Laplace Transform Pairs

$f(x)$	$\mathcal{L}\{f(x)\} \equiv F(s) = \int_0^\infty e^{-sx} f(x)dx$
Pairs	
1. $x^\nu, Re\nu > -1$	$\frac{\Gamma(\nu+1)}{s^{\nu+1}}, \quad s > 0$
2. e^{ax}	$\frac{1}{s-a}, \quad s > a$
3. $\sin ax$	$\frac{a}{s^2+a^2}$
4. $\cos ax$	$\frac{s}{s^2+a^2}$
5. $\text{erf}c\left(\frac{a}{2\sqrt{x}}\right)$	$\frac{1}{s}e^{-a\sqrt{s}}, \quad a > 0$
6. $\text{erf}(\sqrt{x})$	$\frac{1}{s\sqrt{s+1}}, \quad s > -1$
7. $J_0(ax)$	$\frac{1}{\sqrt{s^2+a^2}}$
Operations	
8. $cf(x)$	$cF(s)$
9. $f_1(x) + f_2(x)$	$F_1(s) + F_2(s)$
10. $f(ax)$	$\frac{1}{a}F\left(\frac{s}{a}\right)$
11. $\frac{d^n f}{dx^n}$	$s^n F(s) - \sum_{k=0}^{n-1} s^k f^{(n-k-1)}(0)$
12. $\frac{\partial}{\partial y} f(x,y)$	$\frac{\partial}{\partial y} F(s,y)$
13. $x^n f(x)$	$(-1)^n \frac{d^n}{ds^n} F(s)$
14. $\int_0^x f(\xi)d\xi$	$\frac{1}{s}F(s)$
15. $e^{-ax} f(x)$	$F(s+a)$
16. $\int_0^x f_1(x-\xi) f_2(\xi) d\xi \equiv f_1 * f_2$	$F_1(s)F_2(s)$

which is bounded at $x = 0$. The series representation of $J_n(x)$ is

$$J_n(x) = \sum_{k=0}^{\infty} \frac{(-1)^k (\frac{x}{2})^{n+2k}}{k!(n+k)!}. \tag{1.78}$$

The first Laplace transform pair in Table 1.1,

$$\mathcal{L}\{x^\nu\} = \frac{\Gamma(\nu+1)}{s^{\nu+1}}, \qquad \nu > -1 \tag{1.79}$$

is very important and can be proved easily with the aid of (1.74). The pair

$$\mathcal{L}\{e^{-ax} f(x)\} = F(s+a) \tag{1.80}$$

is easily proved since

$$\begin{aligned} \mathcal{L}\{e^{-ax} f(x)\} &= \int_0^\infty e^{-sx} e^{-ax} f(x)dx \\ &= \int_0^\infty e^{-(s+a)x} f(x)dx = F(s+a) \end{aligned}$$

after using the definition of the Laplace transform (1.63). The two Laplace transform pairs

$$\mathcal{L}\{\sin ax\} = \frac{a}{s^2+a^2} \tag{1.81}$$

$$\mathcal{L}\{\cos ax\} = \frac{s}{s^2+a^2} \tag{1.82}$$

can be proved by direct integration if we use the *Euler identities*,

$$\sin ax = \frac{e^{iax} - e^{-iax}}{2i} \quad \text{and} \quad \cos ax = \frac{e^{iax} + e^{-iax}}{2}. \tag{1.83}$$

Example 9 Find the inverse Laplace transform of

$$F(s) = \frac{1}{s^2+9} + \frac{1}{s(s+1)}. \tag{E.1}$$

In Table 1.1 we note that the Laplace transform and its inverse are linear operations,

$$\mathcal{L}\{f_1(x) + f_2(x)\} = F_1(s) + F_2(s). \tag{E.2}$$

Hence

$$\begin{aligned} f(x) = \mathcal{L}^{-1}\{F(s)\} &= \mathcal{L}^{-1}\left\{\frac{1}{s^2+9} + \frac{1}{s(s+1)}\right\} \\ &= \mathcal{L}^{-1}\left\{\frac{1}{s^2+9}\right\} + \mathcal{L}^{-1}\left\{\frac{1}{s(s+1)}\right\}. \end{aligned} \tag{E.3}$$

From (1.81) it is clear that

$$\mathcal{L}^{-1}\left\{\frac{1}{s^2+9}\right\} = \frac{1}{3}\sin 3x \tag{E.4}$$

but it is not so clear how to find

$$\mathcal{L}^{-1}\left\{\frac{1}{s(s+1)}\right\}.$$

However, if we write the partial fraction of $1/s(s+1)$,

$$\frac{1}{s(s+1)} = \frac{1}{s} - \frac{1}{s+1}, \tag{E.5}$$

we can again use the linearity property of the inverse Laplace transform to write

$$\begin{aligned}\mathcal{L}^{-1}\left\{\frac{1}{s(s+1)}\right\} &= \mathcal{L}^{-1}\left\{\frac{1}{s} - \frac{1}{s+1}\right\} = \mathcal{L}^{-1}\left\{\frac{1}{s}\right\} \\ &\quad -\mathcal{L}^{-1}\left\{\frac{1}{s+1}\right\} = 1 - e^{-x}\end{aligned} \tag{E.6}$$

after consulting (1.79) with $\nu = 0$ and (1.80) with $a = 1$. Hence, from (E.4) and (E.6), the final solution to (E.3) is

$$\begin{aligned} f(x) &= \mathcal{L}^{-1}\left\{\frac{1}{s^2+9}\right\} + \mathcal{L}^{-1}\left\{\frac{1}{s(s+1)}\right\} \\ &= \frac{1}{3}\sin 3x + 1 - e^{-x}\end{aligned} \tag{E.7}$$

The (Laplace) Convolution Product

The most important Laplace transform pair used for solving Volterra integral equation with difference kernel is

$$\mathcal{L}\left\{\int_0^x f_1(x-\xi)f_2(\xi)d\xi\right\} = F_1(s)F_2(s). \tag{1.84}$$

The integral

$$\int_0^x f_1(x-\xi)f_2(\xi)d\xi \equiv (f_1 * f_2)(x) \tag{1.85}$$

is called the Laplace *convolution product* of the two functions $f_1(x)$ and $f_2(x)$ and is denoted by $f_1 * f_2$. The result in (1.84) is the *convolution theorem* for the Laplace transform.

Even though the $*$ in (1.85) represents different convolution multiplication for the different integral transforms, we shall nevertheless use the same symbol, as most books do, without a fear of confusion.

The following Example 10 illustrates the use of (1.84) for solving a Volterra integral equation with difference kernel. This Example is followed by a precise statement of the result in (1.84) that constitutes the *convolution theorem* for Laplace transform as Theorem 4.

Example 10 Use the Laplace transform to find the solution of the following Volterra integral equation of the first kind with difference kernel $K(x-t) = e^{x-t}$

$$\sin x = \int_0^x e^{x-t} u(t)dt. \tag{E.1}$$

Before we Laplace-transform both sides of (E.1), we recognize that the right-hand side is in the Laplace convolution product form (1.84) with $f_1(x) = e^x$ and $f_2(x) = u(x)$. Now we Laplace-transform both sides of (E.1) to obtain

$$\begin{aligned} \mathcal{L}\{\sin x\} &= \frac{1}{s^2+1} = \mathcal{L}\left\{\int_0^x e^{x-t}u(t)dt\right\} \\ &= \mathcal{L}\{e^x * u(x)\} = \frac{1}{s-1}U(s) \end{aligned} \tag{E.2}$$

after using (1.81) for

$$\mathcal{L}\{\sin x\} = \frac{1}{s^2+1},$$

the convolution theorem (1.84), (1.73) for $\mathcal{L}\{e^x\} = 1/(s-1)$, and letting $\mathcal{L}\{u(x)\} = U(s)$. Hence, from (E.2), the Laplace transform of (E.1) is

$$\frac{1}{s^2+1} = \frac{U(s)}{s-1}. \tag{E.3}$$

From this we find

$$U(s) = \frac{s-1}{s^2+1} = \frac{s}{s^2+1} - \frac{1}{s^2+1},$$

$$u(x) = \mathcal{L}^{-1}\left\{\frac{s}{s^2+1} - \frac{1}{s^2+1}\right\} = \mathcal{L}^{-1}\left\{\frac{s}{s^2+1}\right\} - \mathcal{L}^{-1}\left\{\frac{1}{s^2+1}\right\} = \cos x - \sin x$$

after consulting (1.82) and (1.81) for the Laplace transform of $\cos x$ and $\sin x$, respectively.

The Convolution Theorem for Laplace Transform

Now, we shall elaborate more on the important convolution theorem for Laplace transform as was briefly introduced in (1.84) and (1.85) since it is used for solving the type of Volterra integral equations whose integral is in the form of the Laplace convolution product as illustrated in the above Example 10. First we introduce again a bit more refined definition of the convolution product $f_1 * f_2$ of (1.85) associated with the Laplace transform, which is essential for the statement of the convolution theorem.

Definition 4 Let $f_1(x)$ and $f_2(x)$ be causal (vanish identically for $x < 0$) and defined on $(0, \infty)$; then their *(Laplace) convolution product* is defined as

$$(f_1 * f_2)(x) \equiv \int_0^x f_1(x-\xi)f_2(\xi)d\xi. \tag{1.85}$$

It is easy to show that this convolution product is *commutative*, that is,

$$\begin{aligned} f_1 * f_2 &= f_2 * f_1, \\ (f_1 * f_2)(x) &= \int_0^x f_1(x-\xi) f_2(\xi) d\xi \\ &= \int_0^x f_1(\eta) f_2(x-\eta) d\eta \equiv (f_2 * f_1)(x) \end{aligned} \tag{1.86}$$

after letting $x - \xi = \eta$, and where we used the fact that $f_2(x)$ is a causal function where $f_2(x-\eta) \equiv 0$ for $\eta > x$.

We are now in a position to state more precisely the convolution theorem for the Laplace transform as the result in (1.84) and (1.85).

Theorem 4 The Convolution Theorem for Laplace Transform

If $F_1(s)$ and $F_2(s)$ are, respectively, the Laplace transform of $f_1(x)$ and $f_2(x)$, which are in the class of functions we considered in Theorem 1, by being sectionally continuous on each interval $0 \le x \le A$, and of exponential order $e^{\alpha x}$ as $x \to \infty$ (i.e. $|f(x)| \le Me^{\alpha x}$ for $x > B$). Then, the Laplace transform of the convolution product $(f_1 * f_2)(x)$ exists as $F_1(s)F_2(s)$ for $s > \alpha$ i.e.,

$$\mathcal{L}\{(f_1 * f_2)(x)\} = F_1(s)F_2(s), \quad s > \alpha. \tag{1.84}$$

Example 11 Laplace Transform Inverse — Use of the Convolution Theorem

Find

$$\mathcal{L}^{-1}\left\{\frac{1}{s(s^2+1)}\right\}.$$

We have here a product of two Laplace transform $F_1(s) = 1/s$ and $F_2(s) = 1/(s^2+1)$, where according to (1.73) and (1.81) their corresponding inverse Laplace transform are $f_1(x) = 1$ and $f_2(x) = \sin x$. Hence, according to (1.84) the result is the convolution product of 1 and $\sin x$, that is,

$$\begin{aligned} \mathcal{L}^{-1}\left\{\frac{1}{s(s^2+1)}\right\} &= \int_0^x 1 \sin(x-\xi) d\xi = \cos(x-\xi)\Big|_0^x = 1 - \cos x \\ \mathcal{L}^{-1}\left\{\frac{1}{s(s^2+1)}\right\} &= 1 - \cos x. \end{aligned} \tag{E.1}$$

1.4.2 Fourier Transforms

The Fourier *exponential* transform of the function $f(x)$ defined on $(-\infty, \infty)$ is

$$F(\lambda) = \mathcal{F}\{f\} = \int_{-\infty}^{\infty} e^{-i\lambda x} f(x) dx. \tag{1.87}$$

It is clear that (1.87) represents a (singular) Fredholm integral equation of the first kind in $f(x)$ with kernel $K(\lambda, x) = e^{-i\lambda x}$, and so we should be interested in solving for $f(x)$ as the inverse Fourier transform $f(x) = \mathcal{F}^{-1}\{F\}$. Fortunately, and in contrast with the Laplace transform, the Fourier exponential transform has a simple and symmetric formula for its inverse,

$$f(x) = \mathcal{F}^{-1}\{F\} = \frac{1}{2\pi}\int_{-\infty}^{\infty} e^{i\lambda x} F(\lambda)d\lambda. \tag{1.88}$$

We should note that there are a number of variations,[15] with minor modifications, for the definition of the Fourier transform and its inverse. This usually depends on the field of the text or research reference, where such notation, usually, is the most convenient for that particular subject. For example in physics books, we see the Fourier transform and its inverse written as

$$F(p) = \frac{1}{\sqrt{2\pi}}\int_{-\infty}^{\infty} e^{-ipx} f(x)dx, \tag{1.89}$$

$$f(x) = \frac{1}{\sqrt{2\pi}}\int_{-\infty}^{\infty} e^{ipx} F(p)dp \tag{1.90}$$

where x is the *spatial* variable, and p is the variable in the *momentum* (Fourier) space.

Indeed we shall need in Section 2.7 to extend the above definition (1.89) and (1.90) to functions of three variables in the spatial (vector) variable $\vec{r} = \vec{i} + \vec{j}y + \vec{k}z$ (or $\vec{r} = \vec{i}x_1 + \vec{j}x_2 + \vec{k}x_3$) for $f(\vec{r})$, and the momentum (or frequency) vector variable $\vec{\lambda} = \vec{i}\lambda_1 + \vec{j}\lambda_2 + \vec{k}\lambda_3$ for $F(\vec{\lambda})$. This will be done for the mathematical modeling of Schrödinger equation of quantum physics in three dimensions as a Fredholm integral equation in the wave function $\Psi(\vec{\lambda})$ of the three-dimensional momentum space. The complete details are left for Section 2.7. Here, we note that as we rely on a source or reference for Schrödinger equation which is in physics texts, it is most convenient for us to quote the results in that particular notation, which is based on (1.89) and (1.90) for physics books. So in this notation, the Fourier transform of the wave function $\psi(\vec{r})$ in three dimensional space is $F(\vec{\lambda})$ in the three dimensional momentum space, which is

$$F(\vec{\lambda}) = \frac{1}{(2\pi)^{\frac{3}{2}}}\int_{-\infty}^{\infty}\int_{-\infty}^{\infty}\int_{-\infty}^{\infty} e^{-i\vec{\lambda}\cdot\vec{r}} f(\vec{r})dx_1dx_2dx_3, \tag{1.91}$$

$$F(\vec{r}) = \frac{1}{(2\pi)^{\frac{3}{2}}}\int_{-\infty}^{\infty}\int_{-\infty}^{\infty}\int_{-\infty}^{\infty} e^{i\vec{\lambda}\cdot\vec{r}} F(\vec{\lambda})d\lambda_1d\lambda_2d\lambda_3 \tag{1.92}$$

where $\vec{r} = \vec{i}x_1 + \vec{j}x_2 + \vec{k}x_3$ and $\vec{\lambda} = \vec{i}\lambda_1 + \vec{j}\lambda_2 + \vec{k}\lambda_3$.

Now, we may even see either of (1.87) or (1.88) as singular Fredholm integral equation in $f(x)$ for (1.87) and in $F(\lambda)$ for (1.88). Since the two integrals of (1.87) and (1.88) are symmetric in $f(x)$ and $F(\lambda)$, we may have exact similar theorems for the existence of the Fourier transform $F(\lambda)$ of (1.87) with condition on $f(x)$ and for the existence of $f(x)$ of (1.88) with condition on $F(\lambda)$. We will state the first theorem on the existence of Fourier transform $F(\lambda)$ in (1.87).

[15] For such varied notations in books and references of different fields, see Jerri [1992, pp. 129, 156].

Theorem 5 Existence of the Fourier Transform of Absolutely Integrable Functions

If $f(x)$ in (1.87) is absolutely integrable, i.e., $\int_{-\infty}^{\infty} |f(x)|dx < \infty$, its Fourier transform $F(\lambda)$ exists and, moreover, it is continuous.

According to this theorem and the symmetry between the Fourier transform and its inverse, one may think of the same type theorem for the existence of $f(x)$ in (1.88) as the solution to the singular Fredholm integral equation (1.87) in $f(x)$. What is needed here, of course, is that $F(\lambda)$ in (1.88) is absolutely integrable, thus Theorem 5 is satisfied for the convergence of the integral of (1.88) for its $f(x)$ to exist. However, it turned out that, in general, this is not the case for $F(\lambda)$ of the absolutely integrable $f(x)$, since the Fourier transform $F(\lambda)$, of the absolutely integrable function $f(x)$, is not necessarily absolutely integrable. To give an example, we consider the function

$$f(x) = \begin{cases} e^{-x}, & x \geq 0 \\ 0, & x < 0 \end{cases}$$

which is absolutely integrable on $(-\infty, \infty)$

$$\int_{-\infty}^{\infty} |f(x)|dx = \int_{0}^{\infty} e^{-x}dx = 1$$

with Fourier transform

$$F(\lambda) = \int_{0}^{\infty} e^{-i\lambda x}e^{-x}dx = \int_{0}^{\infty} e^{-(1+i\lambda)x}dx = -\frac{1}{1+i\lambda}e^{-i\lambda x}e^{-x}\bigg|_{0}^{\infty} = \frac{1}{1+i\lambda}$$

after knowing that $e^{-i\lambda x} = \cos\lambda x - i\sin\lambda x$ is bounded at $x = \infty$. $F(\lambda)$ here is a complex-valued function whose absolute value $|F(\lambda)| = \sqrt{F(\lambda)\overline{F(\lambda)}}$ where $\overline{F(\lambda)}$ is the complex conjugate of $F(\lambda)$, which is obtained from $F(\lambda)$ by replacing i by $-i$,

$$|F(\lambda)| = \sqrt{F(\lambda)\overline{F(\lambda)}} = \sqrt{\frac{1}{(1+i\lambda)(1-i\lambda)}} = \frac{1}{\sqrt{1+\lambda^2}}.$$

This $F(\lambda)$ is not absolutely integrable, since

$$\int_{-\infty}^{\infty} |F(\lambda)|d\lambda = \int_{-\infty}^{\infty} \frac{1}{\sqrt{1+\lambda^2}}d\lambda = \sinh^{-1}\lambda\bigg|_{\lambda=-\infty}^{\infty} = \infty - (-\infty) = \infty$$

So, if we look at the symmetric form of the inverse Fourier transform (1.88), we cannot be sure of the existence (of $f(x)$) of this integral for such an input $F(\lambda)$, which is the output of (1.87) as the Fourier transform of an absolutely integrable function $f(x)$. The reason is that $F(\lambda)$ here is, in general, not necessarily absolutely integrable to guarantee the integral in (1.88) to exist, according to Theorem 5, and define $f(x)$. But, in practice, we would like to Fourier-transform back and forth from $f(x)$ to $F(\lambda)$ in (1.87) and then in a very symmetric way from $F(\lambda)$ to $f(x)$

via (1.88). Indeed, when we speak of signals, $f(x)$ is the representation in the time space while $F(\lambda)$ is its representation in the frequency, or Fourier, space. In quantum mechanics $f(x)$ is in the coordinate space, while $F(\lambda)$ is the representation in the momentum (λ) space.

To be able to utilize the Fourier transform (1.87) and its inverse (1.88) as convergent integrals, we must restrict our class of transformed functions $f(x)$ to more than just absolutely integrable. One of the simplest versions of such restrictions, which satisfies our needs here and which we shall adopt in this book, is that $f(x)$ must be *sectionally smooth* in addition to being *absolutely integrable* on $(-\infty, \infty)$. This is the statement of the *Fourier integral Theorem* 6 that we shall state next after recalling Definition 2 of sectionally (or piecewise) *smooth* functions.

Definition 2 A piecewise continuous function in an interval $a < x < b$ is termed *piecewise (or sectionally) smooth*, if in addition to being piecewise continuous,

i) its first derivative df/dx is continuous on each of the subintervals $x_{i-1} < x < x_i$, $i = 1, 2, 3, \cdots, n$. ii) $\dfrac{df}{dx}$ approaches a finite limit as x approaches the limits of the subinterval x_{i-1} and x_i from the interior; i.e., there exists $f'(x_{i-1}+)$, $f'(x_i-)$ for each (x_{i-1}, x_i), $i = 1, 2, 3, \cdots, n$.

Theorem 6 The Fourier Integral Theorem (the Fourier Transform Inversion Formula)

If $f(x)$ is a piecewise smooth function on every finite interval of the real line $(-\infty, \infty)$ and is absolutely integrable on $(-\infty, \infty)$, then

$$\begin{aligned} f(x) &= \frac{1}{2\pi}\int_{-\infty}^{\infty} d\lambda e^{i\lambda x}\int_{-\infty}^{\infty} f(\xi)e^{-i\lambda\xi}d\xi \\ &= \frac{1}{2\pi}\int_{-\infty}^{\infty} d\lambda \int_{-\infty}^{\infty} f(\xi)e^{i\lambda(x-\xi)}d\xi \end{aligned} \tag{1.93}$$

converges to $\frac{1}{2}[f(x+) + f(x-)]$. It converges to $f(x)$ at x where $f(x)$ is continuous. The detailed proof is found (among other references) in the author's book on integral and discrete transforms [Jerri, 1992].

We should note that the above double integral in (1.93) is equivalent to the integral

$$\lim_{A\to\infty} \frac{1}{\pi}\int_0^A d\lambda \int_{-\infty}^{\infty} f(\xi)\cos\lambda(x-\xi)d\xi = \frac{1}{2}[f(x+) + f(x-)] \tag{1.94}$$

which is what we usually see in most statements of the Fourier integral formula. Such an equivalence is shown easily when we use the Euler identity $e^{i\lambda(x-\xi)} = \cos\lambda(x-\xi) + i\sin\lambda(x-\xi)$ in the inner integral of (1.93), and recognize the zero contribution of the (odd function) $\sin\lambda(x-\xi)$ to the outer integral.

The above analysis including the two Theorems 5 and 6 should make clear, we hope, that it is one thing to have conditions on $f(x)$ to guarantee the convergence of its Fourier integral to represent $F(\lambda)$ in (1.87) and another to have conditions on

$f(x)$ or $F(\lambda)$ for the existence of the solution $f(x)$ of the singular Fredholm integral equation of the first kind (1.87) in $f(x)$. The essence here is that we are given the general class of functions for which the given function $F(\lambda)$ of (1.87) and its sought solution $f(x)$ belong. The question that still remains is how the form of the solution to (1.87) was constructed or derived as another Fourier integral in (1.88) or in the Fourier integral formula (1.93). The rigorous proof for the construction of (1.88) or (1.93) is somewhat long, and can be found with necessary details in the author's book on the subject of transforms (see also the reference at the end of Chapter 2 therein). There are, however, other methods that if presented in a fast or simple way, would need either more of a mathematical background like *generalized functions* or lack the rigor in justifying a number of assumed limiting processes.

The familiar method found in most books on the undergraduate level, does not need sophisticated concepts, but it does slide over some very important justifications of passing to the limits. Such justifications, if done very properly, may even make such proof longer than the above mentioned detailed one.

Fourier Sine and Cosine Transforms

Next, we will turn to the two very special cases of Fourier transforms, namely those for $f(x)$ being *odd or even functions*, which will result in the Fourier sine and cosine transforms, respectively. We present these transforms to illustrate again, that their inverses represent solutions of very familiar singular Fredholm integral equations of the first kind in $f(x)$. We will state the existence of their solutions (the inverse transform) as simple Corollaries 1 and 2 to the Fourier integral Theorem 6. The fact that the Fourier exponential transform of an odd (even) function will reduce to Fourier sine (cosine) transform is left for an exercise.

We define the *Fourier sine integral* for $f(x)$ on $(0, \infty)$ as

$$\mathcal{F}_s\{f\} = F_s(\lambda) = \int_0^\infty f(x) \sin \lambda x dx \tag{1.95}$$

and we see it as a *singular Fredholm integral equation of the first kind* with kernel $K(x, \lambda) = \sin \lambda x$. The solution $f(x)$ to (1.95) is the *inverse* Fourier sine transform

$$f(x) = \mathcal{F}_s^{-1}\{F_s\} = \frac{2}{\pi} \int_0^\infty F_s(\lambda) \sin \lambda x d\lambda \tag{1.96}$$

whose existence and form can be justified as a special case of the Fourier integral Theorem 6 as its following Corollary 1. The proof, is easy to establish from Theorem 6.

Corollary 1 (to Theorem 6) The Inverse Fourier Sine Transform

If $f(x)$ is an absolutely integrable function on $(0, \infty)$, and is piecewise smooth on every finite interval of $(0, \infty)$, then

$$\frac{1}{2}[f(x+) + f(x-)] = \frac{2}{\pi} \int_0^\infty \sin \lambda x d\lambda \int_0^\infty f(\xi) \sin \lambda \xi d\xi. \tag{1.97}$$

Next, we define the *Fourier cosine integral* of $f(x)$ on $(0, \infty)$ as

$$\mathcal{F}_c\{f\} = F_c(\lambda) = \int_0^\infty f(x) \cos \lambda x dx \tag{1.98}$$

which is another *singular Fredholm integral equation of the first kind* in $f(x)$. As was done for the Fourier sine integral (1.95) the existence and the form of the solution $f(x)$ to (1.98) (as the *inverse* Fourier cosine transform),

$$f(x) = \mathcal{F}_c^{-1}\{f\} = \frac{2}{\pi} \int_0^\infty F_c(\lambda) \cos \lambda x d\lambda \tag{1.99}$$

can be established easily with the aid of Theorem 6 as its following second corollary.

Corollary 2 (to Theorem 6) The Inverse Fourier Cosine Transform

If $f(x)$ is absolutely integrable on $(0, \infty)$, and is piecewise smooth on every finite interval of $(0, \infty)$, then

$$\frac{1}{2}[f(x+) + f(x-)] = \frac{2}{\pi} \int_0^\infty \cos \lambda x d\lambda \int_0^\infty f(\xi) \cos \lambda \xi d\xi. \tag{1.100}$$

We remark here that $f(x)$ in (1.88) represents the solution of the Fredholm integral equation (1.87) in $f(x)$. Although (1.88) offers a very direct way of evaluating the inverse Fourier transform, the integration may still be an involved one. Hence, for the few illustrations that we have here, we will depend on the tabulated values of Fourier transform presented in Table 1.2 and refer the interested reader to more extensive tables.[16] Some of these tabulated pairs can be obtained by simple integration. For example:

$$\begin{aligned} \mathcal{F}\{e^{iax} f(x)\} &= \int_{-\infty}^\infty e^{iax} f(x) e^{-i\lambda x} dx \\ &= \int_{-\infty}^\infty e^{-i(\lambda - a)x} f(x) dx = F(\lambda - a) \end{aligned}$$

which is the 8th entry in Table 1.2.

Example 12

The Fourier transform of the function

$$f(x) = \begin{cases} A, & |x| \le a \\ 0, & |x| > a \end{cases} \tag{E.1}$$

is

$$\begin{aligned} F(\lambda) &= \int_{-\infty}^\infty e^{-i\lambda x} f(x) dx = A \int_{-a}^a e^{-i\lambda x} dx \\ &= \frac{Ae^{-i\lambda x}}{-i\lambda}\bigg|_{-a}^a = \frac{A}{-i\lambda}\left(e^{-i\lambda a} - e^{i\lambda a}\right) = \frac{2A \sin \lambda a}{\lambda} \end{aligned} \tag{E.2}$$

[16] See Ditkin and Prudnikov [1965], Jerri [1992].

Table 1.2 Fourier Transform Pairs

	$f(x)$	$F(\lambda)$
	Pairs	
1.	$f(x) = \begin{cases} A, & \lvert x\rvert \le a \\ 0, & \lvert x\rvert > a \end{cases}$	$\dfrac{2A\sin\lambda a}{\lambda}$
2.	$e^{i\alpha x^2}$	$\sqrt{\dfrac{\pi}{\alpha}}e^{i\pi/4}e^{-i\lambda^2/4\alpha}$
3.	$\dfrac{J_n(t)}{t^n}$	$\begin{cases} \dfrac{2(1-\lambda^2)^{n-1/2}}{1\cdot 3\cdot 5\cdots(2n-1)} & \lvert\lambda\rvert < 1 \\ 0 & \lvert\lambda\rvert > 1 \end{cases}$
4.	$e^{-\alpha\lvert x\rvert}$	$\dfrac{2\alpha}{\alpha^2+\lambda^2}$
5.	$\dfrac{e^{-\frac{x^2}{4a}}}{\sqrt{4\pi a}}$	$e^{-\lambda^2 a}$, $a > 0$
6.	$\dfrac{1}{x^2+b^2}$	$\frac{\pi}{b}e^{-b\lvert\lambda\rvert}$ $b > 0$
	Operations	
7.	$f(ax)$	$\dfrac{1}{\lvert a\rvert}F\left(\dfrac{\lambda}{a}\right)$
8.	$e^{iax}f(x)$	$F(\lambda - a)$
9.	$f(x - x_0)$	$e^{-ix_0\lambda}F(\lambda)$
10.	$\dfrac{d^n f}{dx^n}$	$(i\lambda)^n F(\lambda)$
11.	$\dfrac{\partial^n f(x,y)}{\partial x^n}$	$(i\lambda)^n F(\lambda, y)$
12.	$\int_{-\infty}^{\infty} f(\xi)d\xi$	$\dfrac{F(\lambda)}{i\lambda}$
13.	$\int_{-\infty}^{\infty} f_1(\xi)f_2(x-\xi)d\xi$	$F_1(\lambda)F_2(\lambda)$
14.	$F(x)$	$2\pi f(-\lambda)$

which is the first entry in Table 1.2.

We may remark again that the inversion (1.88) of the Fourier transform (1.87) may not be easily accessible from the existing Fourier transform tables, or that the given transform $F(\lambda)$ of (1.87) is a set of data. In such cases we have to resort to the numerical approximation of the integral in (1.88). Fortunately, such computations have been greatly reduced with the establishment of the very fast numerical algorithm called the fast Fourier transform (FFT).[17]

Example 13

Find the Fourier exponential transform of

$$g(x) = \frac{\sin ax}{x}. \tag{E.1}$$

Here we note that from Example 10, we have

$$\mathcal{F}\left\{ f(x) = \left\{ \begin{array}{ll} \frac{1}{2}, & |x| \le a \\ 0, & |x| > a \end{array} \right. \right\} = \frac{\sin a\lambda}{\lambda} = F(\lambda). \tag{E.2}$$

In this problem we have $f(x) = (\sin ax)/x$, but from the symmetry of the Fourier transform (1.87) and its inverse (1.88) we have (as indicated in the last entry of Table 1.2)

$$\mathcal{F}\{F(x)\} = 2\pi f(-\lambda). \tag{E.3}$$

Hence, if we use

$$F(x) = \frac{\sin ax}{x} \quad and \quad f(\lambda) = \left\{ \begin{array}{ll} \frac{1}{2}, & |\lambda| \le a \\ 0, & |\lambda| > a \end{array} \right.$$

in (E.3) we obtain

$$\mathcal{F}\left\{ \frac{\sin ax}{x} \right\} = 2\pi f(-\lambda) = \left\{ \begin{array}{ll} \pi, & |\lambda| \le a \\ 0, & |\lambda| > a. \end{array} \right. \tag{E.4}$$

We note here that if we consider the precise statement of convergence in (1.94), the Fourier transform here converges to $\frac{\pi}{2}$ at $|\lambda| = a$, where there is a jump discontinuity with the result in (E.4) becoming a

$$\mathcal{F}\left\{ \frac{\sin ax}{x} \right\} = \left\{ \begin{array}{ll} \pi, & |\lambda| < a \\ 0, & |\lambda| > a \\ \frac{\pi}{2}, & |\lambda| = a. \end{array} \right. \tag{E.5}$$

[17] See Brigham [1974, 1988], Briggs and Henson [1995] and Jerri [1992].

The tabulated Fourier transforms are used for evaluating many basic improper integrals.

As we mentioned earlier, the most important property of the Fourier transform, for solving (singular) Fredholm integral equations with a *difference kernel*, is the *convolution theorem*, which states that

$$\mathcal{F}\left\{\int_{-\infty}^{\infty} f_1(x-\xi)f_2(\xi)d\xi\right\} = F_1(\lambda)F_2(\lambda) \tag{1.101}$$

where $F_1(\lambda) = \mathcal{F}(f_1)$ and $F_2(\lambda) = \mathcal{F}(f_2)$. It is clear that the Fourier convolution product of $f_1(x)$ and $f_2(x)$

$$f_1 * f_2(x) \equiv \int_{-\infty}^{\infty} f_1(x-\xi)f_2(\xi)d\xi \tag{1.102}$$

has limits of integration which are suitable for the (singular) Fredholm integral equation with difference kernel (1.49)

$$f(x) = \int_{-\infty}^{\infty} K(x-\xi)u(\xi)d\xi \tag{1.49}$$

and that $f_1(x-\xi)$ stands for $K(x-\xi)$, the difference kernel.

We shall state the convolution theorem for the Fourier exponential transform (1.87) with precise conditions for its validity, as Theorem 7.

It is easy to show that the convolution product of (1.102) is commutative, distributive over addition, and associative, as stated in the following lemma.

Lemma For $f_1(x)$, $f_2(x)$, and $f_3(x)$ in the class of functions that are bounded, absolutely integrable on $(-\infty, \infty)$, and sectionally continuous on each bounded interval,

(1) $(f_1 * f_2)(x) = (f_2 * f_1)(x)$ (1.103)

(2) $f_1(x) * [f_2(x) + f_3(x)] = (f_1 * f_2)(x) + (f_1 * f_3)(x)$ (1.104)

(3) $f_1 * (f_2 * f_3) = (f_1 * f_2) * f_3$ (1.105)

We leave the (formal) proof as a simple exercise.

Theorem 7 The Fourier Convolution Theorem

If $f_1(x)$ and $f_2(x)$ are in the class of functions that are bounded, absolutely integrable on $(-\infty, \infty)$, and sectionally continuous over each bounded interval-then the Fourier transform of the convolution product exists, and its Fourier transform is the product $F_1(\lambda)F_2(\lambda)$, as in (1.101):

$$\mathcal{F}(f_1 * f_2)(x) = F_1(\lambda)F_2(\lambda) \tag{1.101}$$

where $F_1(\lambda)$ and $F_2(\lambda)$ are the Fourier transforms of $f_1(x)$ and $f_2(x)$, respectively.

Example 14 Inverse Fourier Transform — Application of the Convolution Theorem

Find the inverse Fourier transform of

$$U(\lambda,t)=F(\lambda)e^{-\lambda^2 t}=\mathcal{F}\{u(x,t)\},u(x,0)=f(x),-\infty<x<\infty. \qquad (E.1)$$

We note that this function is a product of $F(\lambda)=F\{f\}$ and $e^{-\lambda^2 t}=\mathcal{F}\left\{\frac{e^{-x^2/4t}}{\sqrt{4\pi t}}\right\}$, $t>0$ (see entry 5 in Table 1.2). So, according to the result (1.101) of the convolution Theorem 7, its inverse Fourier transform $u(x,t)$ is the convolution product of $f(x)$ and $\frac{e^{-x^2/4t}}{\sqrt{4\pi t}}$,

$$\begin{aligned} u(x,t) &= \mathcal{F}^{-1}\{U(\lambda,t)\}=\mathcal{F}^{-1}\left\{F(\lambda)e^{-\lambda^2 t}\right\}=[f(x)]*\left[\frac{e^{-x^2/4t}}{\sqrt{4\pi t}}\right] \\ &= \int_{-\infty}^{\infty} f(\xi)\cdot\frac{e^{-(x-\xi)^2/4t}}{\sqrt{4\pi t}}d\xi \qquad (E.2)\\ u(x,t) &= \frac{1}{\sqrt{4\pi t}}\int_{-\infty}^{\infty} f(\xi)e^{-(x-\xi)^2/4t}d\xi,\ \ t>0. \end{aligned}$$

Parseval's Equality

Another useful property of the Fourier transform is *Parseval's equality*,

$$\frac{1}{2\pi}\int_{-\infty}^{\infty} F_1(\lambda)\overline{F_2(\lambda)}d\lambda=\int_{-\infty}^{\infty} f_1(x)\overline{f_2(x)}dx \qquad (1.106)$$

where $\overline{F}$ is *the complex conjugate* of F, which as we stated earlier is obtained from $F(\lambda)$ by replacing i by $-i$. The important special case of (1.106) for $f_1=f_2=f$,

$$\frac{1}{2\pi}\int_{-\infty}^{\infty} |F(\lambda)|^2 d\lambda=\int_{-\infty}^{\infty} |f(x)|^2 dx \qquad (1.107)$$

can be derived from the convolution theorem (1.101), which we shall leave for Exercise 10 with a clear supporting hint.

Singular Fredholm Integral Equations—Fourier Convolution Product Type

As we had done with the convolution theorem for the Laplace transform (1.84) for solving Volterra integral equations with difference kernels, we will illustrate here the use of the above Fourier convolution theorem (1.101) to solve singular Fredholm integral equations with difference kernels, and where the integral, as required by (1.101) is defined on $(-\infty,\infty)$. To satisfy the theory we shall consider the given functions in the equation as well as the solution to satisfy the existence of the Fourier transform and its inverse as spelled out in Theorems 5 and 6. We should also add here that we must check first a theorem for the existence of the solution of the given integral equation, which we shall discuss in Chapter 5 (as done in Theorems 1-4 in Section 5.1 and Theorem 6 in Section 5.3 for Fredholm equations of the second kind, and in Theorem 5 of Section 5.2 and Theorem 7 of Section 5.4 for Fredholm equations of the first kind). So, the illustration here may be considered as a formal one in the absence of the above required checks for justifying our steps.

Consider the following singular Fredholm integral equation of the first kind in $u(x)$ with the given function $f(x)$ and the difference kernel $k(x-\xi)$,

$$f(x) = \int_{-\infty}^{\infty} k(x-\xi)u(\xi)d\xi \tag{1.108}$$

where the integral is in the Fourier convolution product form (1.102), So if we let $F(\lambda)$, $K(\lambda)$ and $U(\lambda)$ be the Fourier transforms of $f(x)$, $k(x)$ and $u(x)$ respectively, then Fourier-transform both sides of the integral equation (1.108), using the convolution theorem (1.101) on the integral we have an algebraic equation in $U(\lambda)$,

$$\begin{aligned} F(\lambda) &= K(\lambda)U(\lambda), \\ U(\lambda) &= \frac{F(\lambda)}{K(\lambda)}, \qquad K(\lambda) \neq 0 \end{aligned} \tag{1.109}$$

provided that $F(\lambda)$ does not vanish. What remains is to find the solution $u(x)$ as the inverse Fourier transform of $U(\lambda)$, which we will illustrate in the next example.

Example 15 Fourier Transform for Solving a Singular Fredholm Integral Equation

Consider the singular Fredholm integral equation of the second kind in $u(x)$,

$$u(x) = e^{-|x|} + \mu \int_{-\infty}^{\infty} e^{-|x-\xi|}u(\xi)d\xi. \tag{E.1}$$

If we take the Fourier transform of both sides of this equation, using the convolution theorem on the integral in (E.1) and realizing in this special problem that $f(x) = e^{-|x|}$ and the kernel $k(x) = e^{-|x|}$, where from the fourth entry in Table 1.2, we have

$$\mathcal{F}\left\{e^{-|x|}\right\} = \frac{2}{1+\lambda^2}, \tag{E.2}$$

the Fourier transform of (E.1) becomes

$$U(\lambda) = \frac{2}{1+\lambda^2} + \mu\frac{2}{1+\lambda^2}U(\lambda),$$

and

$$U(\lambda) = \frac{2}{\lambda^2+1-2\mu}, \qquad \lambda^2+1-2\mu \neq 0. \tag{E.3}$$

The denominator vanishes for real values of λ when $\mu \geq \frac{1}{2}$ and some real value of λ, so for $\mu < \frac{1}{2}$ we can take the inverse Fourier transform of $U(\lambda)$ in (E.3) to have

$$\begin{aligned} u(x) = \mathcal{F}^{-1}\{U(\lambda)\} &= \frac{1}{2\pi}\int_{-\infty}^{\infty} \frac{2e^{i\lambda x}}{(1-2\mu)+\lambda^2}d\lambda \\ &= \frac{e^{-\sqrt{1-2\mu}|x|}}{\sqrt{1-2\mu}} \end{aligned} \tag{E.4}$$

after using again the fourth entry in Table 1.2 for $e^{-\alpha|x|}$ with $\alpha = \sqrt{1-2\mu}$.

For another illustration of using the Fourier transform for solving homogeneous (singular) Fredholm integral equations (with difference kernel!), we present the next Example 16.

Example 16 Singular Homogeneous Fredholm Equations[18]

In this example we shall consider two homogeneous singular Fredholm equations to further illustrate the use of the Fourier transform, and to point out some main features that differentiate the singular from the nonsingular integral equations.

a) Consider the first equation in $u(x)$

$$u(x) = \mu\int_{-\infty}^{\infty} \epsilon^{-|x-t|}u(t)dt. \tag{E.1}$$

We note that the integral is in the form of the Fourier convolution product, moreover its kernel is the same as that of the equation of the second kind (E.1) in Example 15. So it may be appealing to try the convolution theorem in the same exact way like what we did for Example 15, and we have

$$\begin{aligned} U(\lambda) &= \mu\frac{2}{1+\lambda^2}U(\lambda), \\ U(\lambda) &= \mathcal{F}\{u\}. \end{aligned} \tag{E.2}$$

This equation (E.2) gives a nontrivial solution $U(\lambda)$ only if the parameter μ is restricted such that $\mu = \frac{1+\lambda^2}{2}$. What remains now is to find the function, (or functions) $u(x)$ that satisfies the integral equation (E.1) with $\mu = \dfrac{1+\lambda^2}{2}$. With the knowledge of the basic properties of Fourier transform, one may arrive at such solutions, sometimes, by a kind of inspection. We observe in (E.1) that our solution $u(t)$ is an input that results in an output $u(x)$ which is within the $\dfrac{2}{1+\lambda^2}$ constant factor. If we also recall that the kernel $e^{-|x-t|}$ is shifted by t, where according to the Fourier pair from Table 1.2, we have

$$\mathcal{F}\{f(x-x_0)\} = e^{-ix_0\lambda}F(\lambda). \tag{E.3}$$

[18] Optional

Then if we nominate $u(t) = e^{-i\lambda t}$ for a solution, the integral on the right of (E.1) will represent no more than the Fourier transform of $e^{-|x-t|}$, which according to (E.3) should give us $e^{-ix\lambda}F(\lambda) = e^{-ix\lambda} \cdot \frac{2}{1+\lambda^2}$. Thus (E.1) results in

$$u(x) = \mu \frac{2}{1+\lambda^2} e^{-ix\lambda} \tag{E.4}$$

as solution $u(x) = e^{-i\lambda x}$ to (E.1) provided that the parameter μ (which we shall call an *eigenvalue* later) is restricted to $\mu = \frac{1+\lambda^2}{2}$. We should note in this example of homogeneous singular Fredholm integral equation (E.1) that for every $\mu > \frac{1}{2}$ (i.e., infinity of μ values) we have solution $e^{-i\lambda x}$ to the singular equation (E.1).

b) Here we consider another homogeneous singular Fredholm integral equation in $u(x)$,

$$u(x) = \mu \int_0^\infty u(t) \sin xt dt \tag{E.5}$$

where we note that the integral on the right is a *Fourier sine integral*. So the equation (E.5) puts a Fourier sine integral $U(x)$ as in (1.95)

$$U(x) = \int_0^\infty \sin xt u(t) dt \tag{E.6}$$

and its inverse $u(t)$ as in (1.96)

$$u(t) = \frac{2}{\pi} \int_0^\infty \sin xt U(x) dx \tag{E.7}$$

within a multiple constant, i.e., (E.5) can be obtained from (E.6) if we let $U(x) = \frac{u(x)}{\mu}$. But if this is used in (E.7) we have

$$u(t) = \frac{2}{\pi\mu} \int_0^\infty \sin xt u(x) dx \tag{E.8}$$

an integral equation in the same form as our original equation (E.5), and they will be compatible if their coefficients are equal, i.e., $\mu = \frac{2}{\pi\mu}$, $\mu^2 = \frac{2}{\pi}$ or $\mu = \pm\sqrt{\frac{2}{\pi}}$. Hence the integral equation (E.6) will have *nontrivial* solutions for these two values of μ. To find the possible solutions corresponding to these values may become a long search, and of course a check of the detailed tables of Fourier transforms[19] should be the best route for a first search. It turned out that for any positive constant value of a, the following functions

$$y_1(x) = \sqrt{\frac{\pi}{2}} e^{-ax} + \frac{x}{a^2 + x^2}, \qquad x > 0 \tag{E.9}$$

[19] See [Ditkin and Prudnikov [1965] and Erdelyi et al. [1954].

and

$$y_2(x) = \sqrt{\frac{\pi}{2}}e^{-ax} - \frac{x}{a^2+x^2}, \qquad x > 0 \tag{E.10}$$

are solutions of the integral equation (E.5) for $\mu_1 = \sqrt{\frac{2}{\pi}}$ and $\mu_2 = -\sqrt{\frac{2}{\pi}}$, respectively. This should illustrate that for one value of the parameter $\mu = \mu_1 = \sqrt{\frac{2}{\pi}}$ for example, the integral equation (E.5) has infinity of solutions $y_1(x)$ in (E.9) for all the positive values of the constant a involved in $y_1(x)$.

The above-illustrated two features of i) infinite values of the parameter μ in the singular equation (E.1) and ii) the infinity of solutions corresponding to one value of the parameter $\mu = \sqrt{\frac{2}{\pi}}$ for the other singular equation (E.5) do represent important characteristics of singular Fredholm integral equations.

Fourier Transform of Derivatives

For completeness we present the following result (1.110), which represents one of the most important properties of the Fourier transforms for solving differential equations (with, usually, constant coefficients). This is the transforming of the derivative $\frac{d^2f}{dx^2}$ (or higher derivatives) in the x-space to the algebraic $-\lambda^2F(\lambda)$ in the Fourier $\lambda-$space. Of course, as we mentioned for the Laplace transform, the combination of such results like (1.110) for algebraizing differential operators, and the convolution theorem (1.101), for algebraizing the convolution integral operator, can be used in the attempt of solving integro-differential equations on $(-\infty,\infty)$. We will next state and prove this result for Fourier-transforming the first derivative $\frac{df}{dx}$ as Theorem 8. The cases of higher derivatives like that of (1.110) for $\frac{d^2f}{dx^2}$ (or for $\frac{d^nf}{dx^n}$) will follow easily as a corollary to this theorem.

As in the case of the Laplace transform, the Fourier transform also algebraizes differential operators with constant coefficients, for example, $\mathcal{F}\{d^2f/dx^2\} = -\lambda^2F(\lambda)$ with the conditions of $\lim_{x\to\pm\infty} f(x) = 0$ and $\lim_{x\to\pm\infty} f'(x) = 0$. This can be shown after two repeated integrations by parts and by assuming that $\lim_{x\to\pm\infty} f(x) = 0$ and $\lim_{x\to\pm\infty} \frac{df}{dx} = 0$.

$$\mathcal{F}\left\{\frac{d^2f}{dx^2}\right\} = \int_{-\infty}^{\infty} e^{-i\lambda x}\frac{d^2f}{dx^2}dx = e^{-i\lambda x}\frac{df}{dx}\bigg|_{-\infty}^{\infty} + i\lambda\int_{-\infty}^{\infty} e^{-i\lambda x}\frac{df}{dx}dx$$

$$= 0 + i\lambda\left[e^{-i\lambda x}f(x)\big|_{-\infty}^{\infty} + i\lambda\int_{-\infty}^{\infty} e^{-i\lambda x}f(x)dx\right].$$

$$\mathcal{F}\left\{\frac{d^2 f}{dx^2}\right\} = -\lambda^2 F(\lambda) \tag{1.110}$$

A more precise statement in this direction is the following Theorem 8.

Theorem 8 The Fourier Transform of $\dfrac{df}{dx}$

Let $f(x)$ be continuous and absolutely integrable on $(-\infty, \infty)$, and let $\dfrac{df}{dx}$ be piecewise smooth and absolutely integrable on $(-\infty, \infty)$; then

$$\mathcal{F}\left\{\frac{df}{dx}\right\} = i\lambda F(\lambda). \tag{1.111}$$

Now we state the corresponding result to (1.111) in the Fourier space,

$$\mathcal{F}\{-ixf(x)\} = \frac{dF}{d\lambda} \tag{1.112}$$

or

$$\mathcal{F}^{-1}\left\{\frac{dF}{d\lambda}\right\} = -ixf(x). \tag{1.113}$$

Just as was done formally in (1.72) for the similar Laplace transform pair (1.71), this result is obtained by differentiating the Fourier integral

$$\begin{aligned} F(\lambda) &= \int_{-\infty}^{\infty} e^{-i\lambda x} f(x)dx, \\ \frac{dF}{d\lambda} &= \frac{d}{d\lambda}\int_{-\infty}^{\infty} e^{-i\lambda x} f(x)dx = \int_{-\infty}^{\infty} \frac{d}{d\lambda} e^{-i\lambda x} f(x)dx \\ &= \int_{-\infty}^{\infty} -ixf(x)e^{-i\lambda x}dx = \mathcal{F}\{-ixf(x)\}. \end{aligned} \tag{1.114}$$

This step of interchanging differentiation, with respect to λ, with integration is justified if the middle integral above resulting from such interchange is uniformly convergent to allow such an operation.

Finite(-Limit) Fourier Sine and Cosine Transforms—Fourier Coefficients

The Laplace and Fourier transforms, we introduced up till now, were used on functions defined on an infinite interval $(0, \infty)$ or $(-\infty, \infty)$. We will now introduce the *finite* sine and cosine transforms of functions defined on the finite interval $(0, \pi)$ and with the real variable λ limited to *integer* values.

The *finite sine transform* of $f(x)$, defined on $(0, \pi)$, is

$$F_s(n) = \int_0^{\pi} \sin nx f(x)dx, \qquad n \text{ integer} \tag{1.115}$$

whereas its inverse is the *Fourier sine series* of $f(x)$, $0 < x < \pi$,

$$f(x) = \frac{2}{\pi}\sum_{n=1}^{\infty} F_s(n)\sin nx. \tag{1.116}$$

The *finite cosine transform* of $f(x)$, defined on $(0, \pi)$ is

$$F_c(n) = \int_0^\pi \cos nx f(x)dx, \quad n \text{ integer} \tag{1.117}$$

and its inverse is the *Fourier cosine series* of $f(x), 0 < x < \pi$,

$$f(x) = \frac{1}{\pi}F_c(0) + \frac{2}{\pi}\sum_{n=1}^{\infty} F_c(n)\cos nx. \tag{1.118}$$

We note here that the finite sine transform in (1.115) is related to the Fourier sine coefficients b_n of the Fourier sine series in (1.116) as $F_s(n) = \frac{\pi}{2}b_n$, while the finite cosine transform (1.117) is related to the Fourier cosine coefficients a_n, of the Fourier cosine series in (1.118), as $F_c(n) = \frac{\pi}{2}a_n, n = 1, 2, \cdots,$.

Since it is our main purpose here to bring examples of integral equations, we can also look at the finite-limit Fourier sine and cosine transforms (1.115), (1.117) as *nonsingular* Fredholm integral equations of the first kind in $f(x)$, as compared to the singular equations of the (infinite-limit) Fourier sine and cosine transforms in (1.95), (1.98) with their infinite limit of integration on $(0, \infty)$. The other point noticed is that the nonsingular Fredholm equation of the first kind of the finite sine transform (1.115), for example, is solvable for $f(x), 0 < x < \pi$, in (1.116) as an infinite series in terms of the *discrete values* of the given function $F_s(n)$, while the singular integral equation of the Fourier sine transform (1.95) is solvable (for $f(x), 0 < x < \infty$) as an infinite integral in terms of the *continuous values* of the given function $F_s(\lambda)$ in (1.96). The same analogy can be drawn for the finite cosine transform (1.118) versus the infinite-limit one in (1.98). The singular property of the integral equation and the continuum of λ values are important characteristics of such (not so easy to treat) equations.

The following finite exponential Fourier transform $F(n)$ can also be defined in a similar way, where its inverse $f(x)$ is expressed as an infinite (Fourier) series, and similar conclusions regarding the discreteness of λ are reached.

$$F(n) = \int_{-\pi}^{\pi} e^{-inx} f(x)dx, \tag{1.119}$$

$$f(x) = \frac{1}{2\pi}\sum_{n=-\infty}^{\infty} F(n)e^{inx}, \qquad -\pi < x < \pi. \tag{1.120}$$

We should mention that the above finite transforms are also used in an *operational* way, similar to the Laplace and Fourier transforms, for algebraizing derivatives defined on the finite domain. For example the Fourier sine and cosine transform algebraize *even* order derivatives,

$$\int_0^\pi \sin nx \frac{d^2 f}{dx^2}dx = n\{f(0) - (-1)^n f(\pi)\} - n^2 F_s(n), \tag{1.121}$$

$$\int_0^\pi \cos nx \frac{d^2 f}{dx^2} dx = (-1)^n f'(\pi) - f'(0) - n^2 F_c(n). \tag{1.122}$$

These can be derived by simple twice integration by parts. More importantly, that we will need their above algebraization properties in Section 4.1 for modeling the potential distribution in a charged square plate as Fredholm integral equation in two dimensions (see Exercise 24 of Section 4.1).

We may note that the finite Fourier sine transform of $\frac{d^2 f}{dx^2}$ in (1.121) requires the values of the function $f(0)$ and $f(\pi)$ at the two ends of the interval $(0, \pi)$, while the finite cosine transform requires the derivatives $f'(0)$ and $f'(\pi)$ at the end points of $(0, \pi)$.

The finite exponential transform also has the same operational property for algebraizing all order derivatives of $f(x)$ on the finite interval of its definition, for example

$$\int_{-\pi}^\pi e^{-inx} \frac{df}{dx} dx = (-1)^n \{f(\pi) - f(-\pi)\} + inF(n) \tag{1.123}$$

which can be accomplished easily by using one integration by parts.

To summarize, in this section we showed how integral transforms especially the Laplace and Fourier transforms help in solving differential equations (with constant coefficients), and Volterra and some singular Fredholm integral equations with special (difference) kernels. This is done where each of the differential equations as well as the integral equations in $f(x)$ become simple algebraic equations in their transforms $F(\lambda)$. All our illustrations and applications emphasized this fact. However, and at the same time, a particular important integral transform like the Fourier transform owes its simple applicability to the theory of integral equations, that solved for its inverse $f(x) = \mathcal{F}^{-1}\{F(\lambda)\}$ (or the return to the x-space from the transform λ-space). This is a problem of *singular* Fredholm integral equation of the first kind in $f(x)$,

$$F(\lambda) = \int_{-\infty}^\infty e^{-i\lambda x} f(x) dx \tag{1.87}$$

and whose solution (luckily) was derived to be

$$f(x) = \frac{1}{2\pi} \int_{-\infty}^\infty e^{i\lambda x} F(\lambda) dx. \tag{1.88}$$

So for any integral transform no matter how effective it is, at the end we have to solve for its integral equation that is associated with finding its inverse. For an integral transform to be useful, there must be a balance between the difficulty in finding its inverse and its special properties in efficiently solving differential, integral or integro-differential equations among others.

1.4.3 Other Transforms

Here[20] we introduce a number of other familiar integral (or finite) transforms, for the main purpose of pointing out to that finding the inverse of any of these transforms, is a very clear example of solving an, often, singular Fredholm integral equation of the first kind in the transformed function $f(x)$. Of course, this is besides their use, in parallel to Laplace and Fourier transforms as shown in (1.127) for the case of Hankel transform, and for facilitating the mathematical modeling and solution of some integral equations. We will not cover the latter here, but we will revisit it briefly in Section 2.6.2 to illustrate its use in the integral representation of the electrified disc problem. The detailed modeling and solution of this problem is done in Example 1 of Appendix A.

The Hankel Transform

The Hankel transform $F_n(\lambda)$ of $f(r)$ is defined as

$$F_n(\lambda) = \mathcal{H}_n\{f\} = \int_0^\infty rJ_n(\lambda r)f(r)dr \tag{1.124}$$

where $J_n(x)$ is the Bessel function of the first kind of order n, which is (the bounded solution at $x = 0$) of the two solutions of the Bessel differential equation,

$$x^2\frac{d^2u}{dx^2} + x\frac{du}{dx} + (x^2 - n^2)u = 0, \tag{1.125}$$

$$J_n(x) = \sum_{k=0}^{\infty} \frac{(-1)^k(\frac{x}{2})^{n+2k}}{k!(n+k)!}. \tag{1.126}$$

In a similar operational way to that of the Laplace and Fourier transforms, it can be shown, using (a rather lengthy!) integration by parts[21] (with appropriate boundary conditions), that the Hankel transform algebraizes the Bessel differential equation with variable coefficients (or its following variable coefficient operator), that is,

$$\mathcal{H}_n\left\{\frac{d^2f}{dr^2} + \frac{1}{r}\frac{df}{dr} - \frac{n^2}{r^2}f\right\} = -\lambda^2F_n(\lambda). \tag{1.127}$$

With such an advantage, it remains to transform back $F_n(\lambda)$ in the transform space to the original function $f(r)$ in the physical r-space. This would mean solving the singular integral equation (1.124) in $f(r)$ with the Bessel function kernel, which, fortunately, has been established as the following inverse Hankel transform,

$$f(r) = \mathcal{H}_n^{-1}\{F_n\} = \int_0^\infty \lambda J_n(\lambda r)F_n(\lambda)d\lambda \tag{1.128}$$

[20]Optional

[21]See Jerri [1992, pp. 16–18, Example 1.6].

where we note a perfect symmetry between the Hankel transform (1.124) and its inverse (1.128). We may mention that the derivation of the inverse Hankel transform is done with the help of the Fourier transform of functions $f(x_1, x_2)$ in two dimensions with circular symmetry, i.e., $f(x,y) = f(\sqrt{x^2+y^2}) = f(r)$. This leads us to the simple extension of the Fourier exponential transform of functions of multiple variables as we did in (1.91) and (1.92) for the Fourier transform in three dimensions. The *double* Fourier transform of $f(x,y)$

$$\mathcal{F}_{(2)}\{f(x,y)\} = F(\lambda_1, \lambda_2) = \int_{-\infty}^{\infty}\int_{-\infty}^{\infty} e^{-i\lambda_1 x - i\lambda_2 y} f(x,y)dxdy \tag{1.129}$$

can be seen as an example of higher (two) dimensional integral equation in $f(x,y)$. Its inverse, as the solution to such a two-dimensional (singular) integral equation of the first kind, is

$$\mathcal{F}_{(2)}^{-1}\{F(\lambda_1, \lambda_2)\} = f(x,y) = \frac{1}{4\pi^2}\int_{-\infty}^{\infty}\int_{-\infty}^{\infty} e^{i\lambda_1 x + i\lambda_2 y} F(\lambda_1, \lambda_2)d\lambda_1 d\lambda_2 \tag{1.130}$$

which can be obtained from the Fourier integral formula (1.93) with a rather simple extension to two dimensions. Of course, the higher dimensional Fourier transforms would be used for solving *partial differential equations* by, usually, algebraizing their derivatives with respect to the spatial variables.

Next we present the Hilbert and Mellin transforms, for the main purpose of showing that finding their inverses is a matter of solving singular Fredholm equations of the first kind.

The Hilbert Transform

The *Hilbert transform* of the function $f(x)$ defined on $(-\infty, \infty)$ is

$$F(\lambda) = \mathcal{H}\{f\} = \frac{1}{\pi}P\int_{-\infty}^{\infty} \frac{f(x)dx}{x-\lambda} \tag{1.131}$$

where P refers to the *Cauchy principal value* of the *improper* integral

$$P\int_{-\infty}^{\infty} \frac{f(x)dx}{x-\lambda} \equiv \lim_{\substack{\epsilon\to 0+ \\ A\to\infty}} \left[\int_{-A}^{\lambda-\epsilon} \frac{f(x)}{x-\lambda}dx + \int_{\lambda+\epsilon}^{A} \frac{f(x)}{x-\lambda}dx\right], \tag{1.132}$$

in case there is a value $\lambda\epsilon(-A, A)$ for which the integrand in (1.131) above becomes infinite. When both limits on the right side of (1.132) exist as $\epsilon \to 0$, the integral is convergent and we drop the P to use $\int$ instead of $P\int$.

The inverse of this Hilbert transform $F(\lambda)$ is

$$f(x) = \mathcal{H}^{-1}\{F(\lambda)\} = \frac{1}{\pi}P\int_{-\infty}^{\infty} \frac{F(\lambda)}{x-\lambda}d\lambda. \tag{1.133}$$

We may mention that this inverse Hilbert transform, as a solution to the singular Fredholm integral equation of the first kind (1.131) in $f(x)$, can be derived with the help of the Fourier integral formula.[22]

Example 17 A Hilbert Transform-Type Integral Equation

A very direct simple example of, at least the use of the available Hilbert transform tables, for solving integral equations in the form (1.131) is

$$\frac{1}{\pi}P\int_{-\infty}^{\infty}\frac{u(x)}{x-\lambda} = \sin 2\lambda \tag{E.1}$$

The general method of solving such singular integral equation would involve "heavy" use of complex variables. So, here, and in parallel to what we do for the inverse Laplace transform in the sophomore course on differential equations, we appeal to the available tables of the Hilbert transforms to find the pair

$$\mathcal{H}\{\cos bx\} = \frac{1}{\pi}P\int_{-\infty}^{\infty}\frac{\cos bx dx}{x-\lambda} = -\sin b\lambda. \tag{E.2}$$

Hence a simple comparison of (E.1) and (E.2) shows that (E.1) is a special case of (E.2) with $b = 2$, whence the solution of (E.1) is $u(x) = -\cos 2x$.

The Mellin Transform

The Mellin transform of the function $f(x)$ defined on $(0, \infty)$ is

$$F(\lambda) = \mathcal{M}\{f\} = \int_0^{\infty} x^{\lambda-1} f(x)dx \tag{1.134}$$

The inverse Mellin transform is

$$f(x) = \frac{1}{2\pi i}\lim_{L\to\infty}\int_{\gamma-iL}^{\gamma+iL} x^{-\lambda}F(\lambda)d\lambda, \quad \gamma > \text{Real }\{z_i\}, \; z = x+iy \tag{1.135}$$

which can be seen as the solution to the singular integral equation (1.134). Here in (1.135), as it was the case for the Laplace inversion formula (1.65), γ is taken to be larger than Real $\{z_i\}$, the real part of any singularity z_i of $F(z)$ inside the integral of (1.135) (where $z = x + iy$).

In Exercise 19 we will present the convolution product and its associated Mellin transform convolution theorem,

$$\mathcal{M}\{f_1 * f_2\} = \mathcal{M}\left\{\int_0^{\infty} f_1(\xi) f_2\left(\frac{x}{\xi}\right)\frac{d\xi}{\xi}\right\} = F_1(\lambda)F_2(\lambda) \tag{1.136}$$

where $F_1(\lambda)$ and $F_2(\lambda)$ are the Mellin transforms of $f_1(x)$ and $f_2(x)$, respectively. Then this theorem is used, in parallel to the Laplace transform, for solving special class of integral equations of the first kind, that are in the form of such convolution product.

[22] See Erdelyi et al. [1954].

Exercises 1.4

The Laplace Transform

1. Find the Laplace transform of the following functions (see Table 1.1).

   (a) $x^{1/2}, 0 < x < \infty$

   (b) $x^{-1/2}, 0 < x < \infty$

   (c) $\int_0^x e^{2(x-t)} t dt$

   Hint: See the (Laplace-type) convolution theorem in (1.84).

   (d) $x - \int_0^x (x-t)^2 u(t) dt$

   (e) $\int_0^x e^{2(x-t)} \frac{du}{dt} dt, u(0) = 0$

   Hint: See (1.84) and (1.68).

   (f) $x \sin x$

2. Find the inverse Laplace transform $f(x) = \mathcal{L}^{-1}\{F(s)\}$ of the following functions $F(s)$.

   (a) $\dfrac{1}{s - (\lambda + 1)}$

   (b) $\dfrac{G(s)}{s - (\lambda + 1)}$

   (c) $\dfrac{1}{(s-3)^2 + 5}$

   (d) $\dfrac{1}{(s-3)^2} + \dfrac{1}{s-2} + \dfrac{1}{s}$

   (e) $\dfrac{F(s)}{\sqrt{s}}$

   Hint: Use the (Laplace-type) convolution theorem (1.84).

   (f) $\sqrt{s} F(s)$

   Hint: Let $\sqrt{s}F(s) = s[F(s)/\sqrt{s}] = sH(s)$ and use (1.68) and the result of part (e) for $h(t) = \mathcal{L}^{-1}\{H(s)\}$, noting that $h(0) = 0$ in part (e).

3. **(a)** Show that $\dfrac{1}{s-a}$, $s > a$ is the Laplace transform of the following two functions.

   (i) $f_1(t) = e^{at}$

   and

   (ii) $f_2(t) = \begin{cases} e^{at}, & 0 < t < 3, \quad or \quad t > 3 \\ 1, & t = 3 \end{cases}$

(b) Use the information in part (a) to conclude the nonunique nature of the solution $f(t)$ of the (singular) integral equation of the first kind,

$$\frac{1}{s-a} = \int_0^\infty e^{-st} f(t)dt.$$

4. **(a)** Find the inverse Laplace transform of $F(s) = \frac{1}{s\sqrt{s+1}}$.

   Hint: Use the Laplace convolution theorem (1.84) with $F_1(s) = \frac{1}{s}$ and $F_2(s) = \frac{1}{\sqrt{s+1}}$, where $f_1(x) = 1$, and $f_2(x) = \frac{e^{-x}}{\sqrt{\pi x}}$ remembering the shifting property for $F_2(s)$.

   (b) Use the following identity,

$$\Gamma(\nu)\Gamma(1-\nu) = \frac{\pi}{\sin \pi\nu}$$

   to show that $\Gamma\left(\frac{1}{2}\right) = \sqrt{\pi}$.

5. **(a)** Find the Laplace transform of Abel's integral equation (a Volterra equation of the first kind with difference kernel).

$$1 = \int_0^x \frac{1}{\sqrt{x-\xi}} u(\xi)d\xi.$$

   Hint: See the convolution theorem in (1.84) and Exercise 1(b).

   (b) Solve for $u(x)$.

6. **(a)** Prove the result in (1.69).

   Hint: Let $g(x) = \frac{df}{dx}$, so

$$\mathcal{L}\left\{\frac{d^2f}{dx^2}\right\} = \mathcal{L}\left\{\frac{dg}{dx}\right\} = s\mathcal{L}\{g\} - g(0) = s\mathcal{L}\left\{\frac{df}{dx}\right\} - f'(0)$$

   (b) Attempt to put conditions on $f(x)$ and $\frac{df}{dx}$ to establish a precise statement of a theorem for the formal result in (1.69).

   (c) Use the convolution theorem in (1.84) to prove the result in (1.70).

   Hint: Consider the integral in (1.84) as a convolution product of $f_1(t) = 1$ and $f_2(t) = f(t)$.

7. Find the inverse Laplace transform of

$$F(s) = \frac{1}{s(s^2+1)}$$

by two methods.

(a) By using partial fractions.

Hint: $\dfrac{1}{s(s^2+1)} = \dfrac{A}{s} + \dfrac{B_s + C}{s^2+1}$; find A, B, and C.

(b) By using the convolution theorem.

Hint: $F_1(s) = \dfrac{1}{s},\ F_2(s) = \dfrac{1}{s^2+1}$.

8. **(a)** Show that the following initial value problem of the harmonic oscillator in $u(t), 0 < t < \infty$,

$$\frac{d^2u}{dt^2} + \omega^2 u = 0, \qquad t > 0 \tag{E.1}$$

$$u(0) = 0 \tag{E.2}$$

$$u'(0) = 1 \tag{E.3}$$

is equivalent to the Volterra integral equation

$$u(t) = t + \omega^2 \int_0^t (x - t)u(x)dx. \tag{E.4}$$

Hint: See Example 6 and Exercise 2(a) of Section 1.3.

(b) Verify that

$$u(t) = \frac{\sin \omega t}{\omega}$$

is a solution to both the initial value problem (E.1)–(E.3) and its equivalent Volterra integral equation (E.4).

Hint: You may substitute directly, or use the Laplace transform to solve the initial value problem (E.1)–(E.3); or the convolution theorem (1.84) to solve the integral equation (E.4).

Fourier Transforms

9. **(a)** Prove the following shifting properties of the Fourier transform,

(i) *Shifting in the physical x-space*

$$\mathcal{F}\{f(x-a)\} = e^{-i\lambda a}F(\lambda). \tag{E.1}$$

(ii) *Shifting in the Fourier λ-space*

$$\mathcal{F}\left\{e^{icx}f(x)\right\} = F(\lambda - c). \tag{E.2}$$

(b) Most of the functions we deal with are assumed to be real-valued functions $f(x)$ on $(-\infty,\infty)$. When $f(x)$ is complex-valued, prove the following result, which we need for the use of the Parseval equality in (1.106),

$$\mathcal{F}\{\overline{f(-x)}\} = \overline{\mathcal{F}(\lambda)},$$

where $\overline{f(x)}$ stands for the complex conjugate of $f(x)$, i.e., where each $i=\sqrt{-1}$ in $f(x)$ is replaced by $-i$.

Hint: Take the complex conjugate of the Fourier transform of $f(-x)$, and note that the complex conjugation operation is distributive over addition and multiplication, i.e., $\overline{f_1+f_2}=\overline{f_1}+\overline{f_2}$ and $\overline{f_1 f_2}=\overline{f_1}\ \overline{f_2}$.

10. Use the (Fourier-type) convolution Theorem 7 as in (1.101), which can be written as

$$\frac{1}{2\pi}\int_{-\infty}^{\infty} e^{i\lambda x}F_1(\lambda)F_2(\lambda)d\lambda = \int_{-\infty}^{\infty} f_1(x-t)f_2(t)dt \qquad (E.1)$$

to derive the very important equality for Fourier analysis, namely *Parseval's equality* in (1.107).

Hint: Let $x=0$ in (E.1). Then consider the special case of $f_2(t)=\overline{f_1(-t)}$ in (E.1) with the use of the result in part (b) of Exercise 9 to have $\mathcal{F}\{\overline{f_1(-t)}\} = \mathcal{F}\{f_2(t)\} = \overline{F_1(\lambda)}$.

11. Consider $f(x)$, $-\infty < x < \infty$.

(a) Show that the exponential Fourier transform $F(\lambda)$ of $f(x)$ reduces to a Fourier sine transform like that of (1.95) when $f(x)$ is an *odd* function $f_0(x)$.

Hint: In the integral of (1.87) write $e^{-i\lambda x} = \cos\lambda x - i\ \sin\lambda x$, then recognize that the integrand $f_0(x)\cos\lambda x$ is an odd function where its integral on $(-\infty,\infty)$ vanishes.

(b) Show that the exponential Fourier transform $F(\lambda)$ of the *even* function $f(x)$ reduces to a Fourier cosine transform like that of (1.98).

Hint: See the hint for part (a), where in the present case $f_e(x)\sin\lambda x$ is an odd function, thus its integral on $(-\infty,\infty)$ vanishes.

12. Consider the function of two variables $u(x,y)$ and the Laplacian of this function

$$\nabla^2 u \equiv \frac{\partial^2 u}{\partial x^2} + \frac{\partial^2 u}{\partial y^2}. \qquad (E.1)$$

Show that the double Fourier transform, as given in (1.129), of this Laplacian of $u(x,y)$ is the following algebraic form in $U(\lambda_1,\lambda_2)$ the double Fourier transform of $u(x,y)$, i.e.,

$$\mathcal{F}_{(2)}\left\{\frac{\partial^2 u}{\partial x^2}+\frac{\partial^2 u}{\partial y^2}\right\} = \int_{-\infty}^{\infty}\int_{-\infty}^{\infty} e^{-(i\lambda_1 x + i\lambda_2 y)}\left[\frac{\partial^2 u}{\partial x^2}+\frac{\partial^2 u}{\partial y^2}\right]dxdy$$

$$= -\left(\lambda_1^2 + \lambda_2^2\right) U(\lambda_1, \lambda_2). \tag{E.2}$$

Hint: Start with the Fourier transform of $\frac{\partial^2 u}{\partial x^2}(x, y)$ as a function of x, $\mathcal{F}\left\{\frac{\partial^2 u}{\partial x^2}\right\} = \mathcal{F}\left\{\frac{\partial u_x}{\partial x}\right\} = i\lambda_1 \tilde{U}(\lambda_1, y)$ where $\tilde{U}(\lambda_1, y) = \mathcal{F}\left\{\frac{\partial u}{\partial x}\right\}$. Another integration gives $\mathcal{F}\left\{\frac{\partial u}{\partial x}\right\} = -\lambda_1^2 \tilde{\tilde{U}}(\lambda_1, y)$, where $\tilde{\tilde{U}}(\lambda_1, y) = \mathcal{F}\{u(x, y)\}$. The second Fourier integral of (E.2) with respect to y gives $\mathcal{F}\{\tilde{\tilde{U}}(\lambda_1, y)\} = U(\lambda_1, \lambda_2)$ with the final result of (E.2) as

$$\mathcal{F}_{(2)}\left\{\frac{\partial^2 u}{\partial x^2}\right\} = -\lambda_1^2 U(\lambda_1, \lambda_2).$$

For the second term in (E.2) do the same starting with the Fourier transform of $\frac{\partial^2 u(x, y)}{\partial y^2}$ as a function of y.

13. **(a)** Show that $f(x)$ in its Fourier sine series representation (1.116) is, indeed, a solution of the (nonsingular) Fredholm integral equation of the first kind (1.115) in $f(x)$.

    Hint: Substitute the series (1.116) of $f(x)$ inside the integral of (1.115), interchange the operations of summation and integration, then use the *orthogonality property* of $\{\sin mx\}_{m=1}^{\infty}$ on the interval $(0, \pi)$, i.e.,

$$\int_0^\pi \sin mx \sin nx\, dx = \begin{cases} 0, & n \neq m \\ \frac{\pi}{2}, & n = m. \end{cases} \tag{E.1}$$

    The interchange of the operation of integration with the infinite summation of (1.116) is allowed when $f(x)$ is *square integrable* on $(0, \pi)$, i.e., $\int_0^\pi f^2(x)dx < \infty$.

    (b) By the same method in part (a), show that $f(x)$ of (1.118) is a solution of the Fredholm integral equation (1.117). You will need to use the following, a bit different from (E.1), integration result that relates to the orthogonality of $\{\cos mx\}_{m=0}^{\infty}$ on $(0, \pi)$.

$$\int_0^\pi \cos mx \cos nx\; dx = \begin{cases} 0, & n \neq m \\ \frac{\pi}{2}, & n = m \neq 0 \\ \pi, & n = m = 0 \end{cases} \tag{E.2}$$

14. Consider $\mathcal{F}_{\mathcal{L}}(\lambda)$ and $\mathcal{F}_F(\lambda)$ as, respectively, the Laplace and Fourier transforms of the *causal* function $f(x)$, i.e., $f(x) \equiv 0$ for $x < 0$.

    Show that these transforms are related as

$$\mathcal{F}_F(\lambda) = \mathcal{F}_{\mathcal{L}}(i\lambda).$$

Hint: Note that for the Fourier transform of $f(x)$, its integral on $(-\infty, 0)$ is zero since the causal function $f(x)$ vanishes identically there.

15. **(a)** Find the Fourier exponential transform of the (singular) Fredholm integral equation with difference kernel assuming that the solution does exist.

$$f(x) = g(x) + \int_{-\infty}^{\infty} k(x-\xi) f(\xi) d\xi$$

Hint: Note that the (improper) integral is in the Fourier convolution product form (1.102), so use the (Fourier) convolution theorem in (1.101).

(b) Solve for $F(\lambda)$ in part (a) and then find $f(x) = \mathcal{F}^{-1}\{F(\lambda)\}$, the solution to part (a).

16. Solve the following (singular) Fredholm integral equation with difference kernel

$$\phi(x) = f(x) + \lambda \int_{-\infty}^{\infty} k(x-\xi)\phi(\xi) d\xi, \qquad (E.1)$$

assuming that the solution does exist.

Hint: Note that the integral is in the Fourier convolution product form (1.102), so use the (Fourier) convolution theorem (1.101).

17. Solve the following (singular) indexsingular!Fredholm equation Fredholm equation in $\phi(x)$,

$$\phi(x) = e^{-|x|} + \int_{-\infty}^{\infty} e^{-a|x-t|}\phi(t)dt, \qquad a > 2 \qquad (E.1)$$

Hint: Note that the (improper) integral is in the *Fourier convolution product* form, where the Fourier convolution Theorem 7 as in (1.101) can be used to algebraize the integration operation. Also remember that

$$\mathcal{F}\left\{e^{-a|x|}\right\} = \frac{2a}{a^2+\lambda^2}. \qquad (E.2)$$

18. Solve the following (singular) integral equation in $\phi(x)$,

$$\phi(x) - \mu \int_{-\infty}^{\infty} \frac{\phi(t)dt}{1+(x-t)^2} = p_1(x), \qquad (E.1)$$

where $p_1(x)$ is the gate function

$$p_a(x) = \begin{cases} 1, & |x| \le a \\ 0, & |x| > a \end{cases} \qquad (E.2)$$

for $a = 1$.

Hint: Note that the (improper) integral is in the *Fourier convolution product* form (1.101), and recall the two Fourier pairs

$$\mathcal{F}\left\{e^{-a|x|}\right\} = \frac{2a}{a^2 + \lambda^2} \tag{E.3}$$

and

$$\mathcal{F}\{p_a(x)\} = \frac{2 \sin a\lambda}{\lambda}.$$

Other Transforms

19. **(a)** Find the Mellin transform of

$$f(x) = e^{-ax}, \qquad a > 0.$$

Hint: Use (1.134) and let $ax = z$, then appeal to the definition of the gamma function as given in (1.74).

(b) The Mellin transform-type convolution product of $f_1(x)$ and $f_2(x)$, $0 < x < \infty$; is defined as

$$f_1 * f_2 \equiv \int_0^\infty f_1(\xi) f_2\left(\frac{x}{\xi}\right) \frac{d\xi}{\xi}.$$

The associated convolution theorem is given in (1.136) as

$$\mathcal{M}\left\{\int_0^\infty f_1(\xi) f_2\left(\frac{x}{\xi}\right) \frac{d\xi}{\xi}\right\} = F_1(\lambda) F_2(\lambda), \tag{E.1}$$

where $F_1(\lambda)$ and $F_2(\lambda)$ are the Mellin transforms of $f_1(x)$ and $f_2(x)$, respectively.

Now let $U(\lambda)$ be the Mellin transform of $u(x)$, then use this convolution theorem in (E.1) (and the Mellin transform pair in part (a)) to find the Mellin transform of the following singular integral equation of the second kind in $u(x)$,

$$u(x) = e^{-ax} + \int_0^\infty e^{-\frac{x}{\xi}} u(\xi) \frac{d\xi}{\xi} \tag{E.2}$$

to show that you will have an algebraic equation in $U(\lambda)$.

1.5 BASIC NUMERICAL INTEGRATION FORMULAS

In the preceding section we introduced the Laplace, Fourier, and other integral transforms and illustrated their suitability for solving only special cases of integral equations. In particular, the Laplace and Fourier transforms are compatible with the following Volterra and Fredholm (singular) integral equations with *difference kernels* $K(x-t)$, respectively:

$$u(x) = f(x) + \int_0^x K(x-t)u(t)dt \tag{1.37}$$

$$u(x) = f(x) + \int_{-\infty}^{\infty} K(x-t)u(t)dt. \tag{1.38}$$

Hence, as expected, the methods we introduce in Chapters 3 and 5 for solving the Volterra integral equations and (mostly nonsingular) Fredholm integral equations will depend on the type of equation, in particular its kernel. But in general these and other special analytical methods may fail; then we must resort to other approximate or numerical methods of solution. The approximate method of solution involves approximating the integral equation by another with a known solution which can be made close to the exact solution of the original problem. We must stress that all the methods being it analytic, approximate or numerical must be preceded by some assurance of the "existence" of the sought (unknown) solution, in general. Also, if possible, we shall have some idea about the stability of the solution for integral equations of the first kind in particular, a topic that we shall touch upon briefly in Section 5.4. Such existence theorems for the variety of integral equations covered here, including the singular ones and those of the first kind, are usually stated most precisely in a general abstract mathematical setting, which requires more of the abstract analysis preparation than assumed for this book. However, we will attempt to give a good intuitive explanation and as precise statements as possible for such an important topic of the existence of the solution to a given integral equation. A brief presentation of this "theoretical" topic is given in the (optional) Chapter 6.

In this section we will review the very basic numerical integration formulas. They will be used for the numerical setting of Volterra integral equations in Section 3.3, and Fredholm integral equations in Section 5.5. These formulas include the *trapezoidal rule*, *Simpson's rule*, and the *midpoint formula*. For the level of this elementary book, we present the higher quadrature rules [see (1.140)] only for the interested reader, thus we decided to have them covered in a new (optional) Chapter 7 along with their necessary tables. There we will give a good number of illustrations for the use of these higher quadrature rules for more accurate numerical integration, and their use in the numerical setting and solving linear Volterra as well as Fredholm integral equations. This is done to support the numerical methods of Sections 3.3 and 5.5, where only the basic rules of this section are used.

1.5.1 Basic (Elementary) Integration Formulas

Numerical methods of solutions for integral equations approximate the integral involved. For example, the integral $\int_a^b f(x)dx$ is approximated by a finite sum

$$\int_a^b f(x)dx \approx S_n(x) = \sum_{i=0}^{n} f(x_i)\Delta_i x \tag{1.139}$$

where usually the sample values $f(x_i)$ are equally spaced with the increment $\Delta_i x = \Delta x = \frac{b-a}{n}$ for n equal increments of the interval (a, b). In general $\Delta_i x$ may be variable, but usually for the very basic formulas of elementary numerical methods of integrations, to be discussed soon, the increment is taken as equal $\Delta x = \frac{b-a}{n}$. However, and depending on the particular formula, the ordinates $f(x_i)$ in the approximate sum above may be given a *weight* that is indicated by D_i (instead of just $\Delta_i x$) to be written for (1.139) as $D_i f(x_i)$ for fixed $\Delta x = \frac{b-a}{n}$, instead of the simplest version $f(x_i)\Delta_i x$,

$$\int_a^b f(x)dx \approx S_n(x) = \sum_{i=0}^{n} D_i f(x_i). \tag{1.140}$$

Such *weights* D_i, (or *quadratures*) are equivalent to approximating the function $f(x)$ on the subinterval $\Delta_i x$ by a simple curve. Such a curve is a simple straight line for the *trapezoidal rule* and a parabola for *Simpson's rule* to be discussed next. Other quadratures, where higher degree polynomials or other functions are used, are also available in the literature and are used in books on numerical methods of solving integral equations, which we shall present and illustrate in our detailed discussion of the numerical solution of Volterra and Fredholm integral equations in Sections 7.2 and 7.3, respectively. Numerical analysis is the subject that deals with such an approximation in the most accurate and efficient way. However, for our purpose in this section we shall be satisfied with the very basic formulas used for the numerical integration above. So, we will first present the most familiar formulas of numerical integration: the *trapezoidal rule, Simpson's rule*, and the *midpoint formula*. In the next section we will illustrate primarily the use of the trapezoidal rule for evaluating integrals. We will also give an exercise to illustrate the use of Simpson's rule. In Sections 3.3 and 5.5 (also in Sections 7.2 and 7.3) we will show how the *linear* Volterra and Fredholm integral equations are reduced, respectively, to a *triangular* and a *square* system of $n+1$ linear algebraic equation in the $n+1$ unknowns of the solution (approximate!) samples $u(x_i)$, $i = 0, 1, 2, \cdots, n$. We should, actually, use another symbol $\tilde{u}(x_i)$ to indicate the solution of the linear system as an approximation to the samples $u(x_i)$ of the solution of the corresponding integral equation. But, for simplicity, and in accordance with the usual notation, we will adhere to $u(x_i)$ without much fear of confusion!

The basic formulas we present here to approximate the integral $\int_a^b f(x)dx$ on the interval (a, b) by a partial sum use n equal increments, $\Delta x = (b-a)/n$. It is clear from Figure 1.10 and (1.139) and (1.140) that the form of the particular summation formula will depend on how the increment of area $\Delta_i A$ under the curve is approximated. Such an area has constant width $\Delta x = (b-a)/n$ and hence, its value will depend on how we approximate the height or the value of the function $f(x)$ on the subinterval (x_{i-1}, x_i), as illustrated in Figure 1.10a. This choice, as we shall see next with the following basic integration formulas of the *trapezoidal rule* and the *Simpson rule*, will translate in terms of the weights D_i of (1.140).

a) The *trapezoidal rule* approximates the function on the subinterval (x_{i-1}, x_i) by a straight line passing through the points $(x_{i-1}, f(x_{i-1}))$ and $(x_i, f(x_i))$, and uses

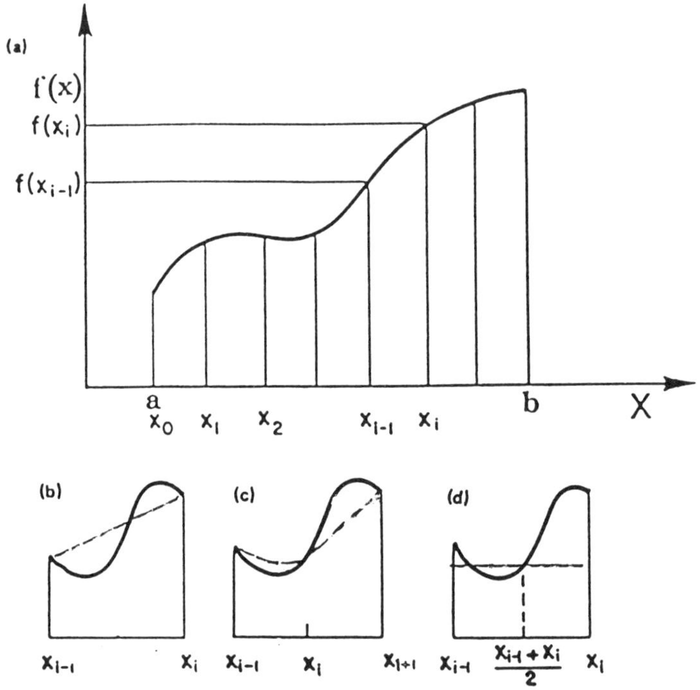

Fig. 1.10 The trapezoidal, Simpson's, and the midpoint rule for numerical approximation of integrals.

the area of a trapezoid with height $(1/2)[f(x_{i-1}) + f(x_i)]$ as shown in Figure 1.10b to approximate the area $\Delta_i A$, which gives

$$\int_a^b f(x)dx \sim \frac{b-a}{n}\left[\frac{1}{2}f(x_0) + f(x_1) + f(x_2) + \cdots \right.$$
$$\left. + f(x_i) + \cdots + f(x_{n-1}) + \frac{1}{2}f(x_n)\right]. \tag{1.141}$$

If we compare this formula with (1.140) we note that the weights $D_0, D_1, D_2, \cdots$, D_{n-1}, D_n given to the ordinates $f(x_0)$, $f(x_1), \cdots, f(x_n)$ are, respectively, $\frac{1}{2}, 1, 1, \cdots, 1, \frac{1}{2}$ multiples of $\Delta x = \frac{b-a}{n}$.

We may mention here that the higher quadrature rules, of numerical integration, that we shall present in Section 7.1 for the very interested reader, are with more elaborate (or fancier!) weights than the above simple ones of halves and ones.

Hence they require their own tables, which we supply in Section 7.1 along with the discussion and illustration of each particular rule.

For completeness, the error $E_T(f)$, involved in the above trapezoidal rule (1.141) for approximating the integral $\int_a^b f(x)dx$,

$$E_T(f) \equiv \int_a^b f(x)dx - \frac{b-a}{n}\left[\frac{1}{2}f(x_0) + f(x_1) + f(x_2) + \cdots + f(x_i) + \cdots + f(x_{n-1}) + \frac{1}{2}f(x_n)\right] \tag{1.142}$$

for a *twice differentiable* function $f(x)$ on $[a, b]$, can be estimated by

$$|E_T(f)| \leq \frac{1}{12} \cdot h^2(b-a)M = \frac{1}{12}\frac{(b-a)^3}{n^2}M \tag{1.143}$$

where $h = \frac{b-a}{n}$, and M is the maximum value of $|f''(x)|$ on $[a, b]$.

b) *Simpson's rule* approximates the function on (x_{i-1}, x_{i+1}) by a parabola that passes through three points—the left point $(x_{i-1}, f(x_{i-1}))$, the middle point $(x_i, f(x_i))$, and the right point $(x_{i+1}, f(x_{i+1}))$ as shown in Figure 1.10c and which results in

$$\int_a^b f(x)dx \sim \frac{b-a}{3n}[f(x_0) + 4f(x_1) + 2f(x_2) + 4f(x_3) + 2f(x_4) + \cdots + 4f(x_{n-1}) + f(x_n)] \tag{1.144}$$

where n is restricted to be an *even* integer. Here we note that the choice for the weights of the ordinates D_i in (1.144) are $1, 4, 2, 4, \cdots, 4, 1$ multiples of the (smaller) constant increments $\Delta x = \frac{b-a}{3n}$ for $n = 4, 6, 8, \cdots$, and for the special case of $n = 2$, the weights are $1, 4$, and 1 multiples of $\frac{b-a}{3n}$ as can be checked easily from Figure 1.10c. The detailed derivation of (1.144) is not as clear as that of (1.142), but can be found in most basic numerical analysis and some calculus texts[23]. The error $E_S(f)$ in the Simpson's rule approximation of the integral $\int_a^b f(x)dx$,

$$E_S(f) \equiv \int_a^b f(x)dx - \frac{b-a}{3n}[f(x_0) + 4f(x_1) + 2f(x_2) + 4f(x_3) + 2f(x_4) + \cdots + 4f(x_{n-1}) + f(x_n)] \tag{1.145}$$

for a *four times differentiable* function $f(x)$ on $[a, b]$, can be estimated by

$$|E_S(f)| \leq \frac{1}{180}h^4(b-a)M = \frac{1}{180}\frac{(b-a)^5}{n^4}M, \tag{1.146}$$

[23]Anton [1995, pp. 473–477]

where $h = \frac{b-a}{n}$ and M is the maximum value of $|\frac{d^4 f}{dx^4}|$ on the interval $[a, b]$. If we compare the errors $E_T(f)$ in (1.143) and $E_S(f)$ in (1.146) for the trapezoidal and Simpson's rules, we see that they depend respectively on $\frac{1}{n^2}$ and $\frac{1}{n^4}$, which says, roughly, that for large number of increments n, the Simpson rule may become superior to the trapezoidal rule. Of course, the function considered and, in particular, its differentiability that determines M in both errors, will play a major role. There are examples, however, where the trapezoidal rule becomes preferable! The illustrations of the above trapezoidal rule and Simpson's rule along with estimates for their error bounds, are the subject of Exercises 1, 2, and 3.

c) The *midpoint formula* approximates the height of the area $\Delta_i A$ by (a fixed) value of the function $f(\frac{x_{i-1}+x_i}{2})$ at the midpoint $\frac{x_{i-1}+x_i}{2}$ that results in

$$\int_a^b f(x)dx \sim \frac{b-a}{n}\left[f\left(\frac{x_0+x_1}{2}\right)+f\left(\frac{x_1+x_2}{2}\right)+\cdots\right.$$
$$\left.+f\left(\frac{x_{i-1}+x_i}{2}\right)+\cdots+f\left(\frac{x_{n-1}+x_n}{2}\right)\right] \tag{1.147}$$

which should be clear from Figure 1.10d.

We may interject here again that in Section 7.1 we will present the other more powerful (or higher order) quadrature rules. There, we will concentrate on a few representatives of the two main groups of quadrature rules, in particular, the *Newton-Cotes* rules, the *repeated* Newton-Cotes rules in part A of Section 7.1, and the *Gauss* quadrature rules in part B of Section 7.1. The Newton-Cotes rules use *equidistant* samples. The simple trapezoidal rule (1.141) and Simpson's rule (1.144) that we presented in this section are no more than two special cases of repeated Newton-Cotes rules of the lowest degrees 1 and 2, respectively. The Gauss quadrature rules, on the other hand, use *nonequidistant* samples, and which, in general, are more efficient as we shall illustrate in part B of Section 7.1 for approximating integrals, and in Section 7.3 for the numerical solution of Fredholm integral equations. Newton-Cotes rules, or a combination of their repeated versions, are more suitable for Volterra integral equations as we shall illustrate in Section 7.2.

The detailed derivation of (1.144) is not as clear as that of (1.141). Still we will have a chance in Section 7.1 to see (1.141) and (1.144) as two special cases of a repeated Newton-Cotes rule of degree 1 and 2, respectively.

In comparing the above three basic numerical integration formulas, we see that the midpoint, the trapezoidal, and Simpson's rule approximate the function on a subinterval, respectively, by polynomials of degree zero (flat top), first degree (straight line), and second degree (parabola). Other quadrature formulas involving higher order polynomials or other special functions are considered, along with their error analysis and illustration in books on numerical methods of integral equations.[24] As we mentioned before, our emphasis in this book is to introduce integral equations

[24]The reader is advised to consult detailed references like Delves and Mohammed [1988], and Baker and Miller [1977] (also Kondo [1991]).

to the student for the first time, with their varied aspects including their numerical approximations, but without necessarily requiring much of abstract analysis. For this purpose, and to clarify our illustrations in Sections 3.3 and 5.5 for the numerical solutions of Volterra and Fredholm equations, respectively, we will stay with the above trapezoidal and Simpson's rule. For particular problems, especially in Section 5.5 on numerical methods of Fredholm integral equations, it is tempting to draw upon higher order quadrature rules but keeping in mind the mathematical level of this book, we have relegated this discussion to Section 7.3 for the very interested reader. We shall also back this treatment with a clear reference to the specific detailed source of such analysis. Sections 3.3 and 5.5 will be devoted to the numerical solution of Volterra and Fredholm integral equations respectively, where only the above basic numerical integration rules are used. In Section 7.1 we will present the higher order quadrature of integration with a number of illustration for numerically evaluating integrals. This is followed by using these rules in Sections 7.2 and 7.3 for the numerical solution of Volterra and Fredholm integral equations, respectively. In Section 7.1 we supply the very necessary tables of the above higher quadrature formulas that are needed for our illustrations. A specific reference is given there for the more detailed tables

A Prelude to the Numerical Approximation Setting of Fredholm Equations

With the above simple introduction of the basic elementary formulas of numerical integration, we may state here that the numerical methods of solution for integral equations involve, for example, that for the general Fredholm integral equation of the second kind,

$$u(x) = f(x) + \int_a^b K(x,t)u(t)dt \tag{1.148}$$

we start by approximating the integral numerically by a partial sum of the form

$$S_n(x) = \sum_{j=0}^{n} K(x,t_j)u(t_j)\Delta_j t. \tag{1.149}$$

Here, as we mentioned above, the points t_j are equally spaced, but can be chosen at one's convenience, and as it may be required by the chosen quadrature formula beyond the two simple ones discussed above. Also Δt may stand for the usual equal increment, but in general, the index j in $\Delta_j t$ may indicate a weight D_j assigned to the ordinates $K(x,t_j)u(t_j)$ (of the integrand) by the particular numerical integration formula used as we discussed for (1.140) and illustrated for the trapezoidal rule (1.141) and Simpson's rule (1.144). In this general sense of different weights D_j for the $n+1$ values $u(t_j)$ in the sum of (1.149), we rewrite it as

$$S_n(x) = \sum_{j=0}^{n} K(x,t_j)D_j u(t_j). \tag{1.150}$$

We shall continue this in Sections 3.3 and 5.5 for the Volterra and Fredholm integral equations respectively, where we use the trapezoidal and Simpson's rule; and in Sections 7.2 and 7.3 where we use the higher quadrature rules.

1.5.2 The Smoothing Effect of Integration

In the future, and in particular with integral equations of the first kind,

$$g(x) = \int K(x,t)u(t)dt \tag{1.151}$$

we note that while the sought solution $u(x)$ may not be of good quality, for a reasonably behaved kernel $K(x,t)$, the result $g(x)$ of the above integration is of better quality. For example, $u(x)$ may have a jump discontinuity in the domain of the integration, while the result $g(x)$ of the integration in (1.151) is continuous. In general, we describe such result as the "smoothing effect" of the integration process, which we shall illustrate next with a simple example. So it goes without saying that while the integration process may cover the bad quality of the integrated function; its inverse operation, namely, the differentiation process will definitely uncover such bad quality. We are also most concerned if the solution of an integral equation is in terms of derivatives of a known function or some integral of it. This happened to be the case with some integral equations of the first kind, for example the famous Abel's problem (1.20)

$$-\sqrt{2g}f(y) = \int_0^y \frac{\phi(\eta)d\eta}{\sqrt{y-\eta}} \tag{1.20}$$

has its solution found, via the use of Laplace transform, in Example 8 of Section 3.2 as $\phi(x)$ in (3.41) in terms of a derivative of an integral of the known function $f(x)$ weighed by $\frac{1}{\sqrt{x-t}}$,

$$\phi(x) = -\frac{\sqrt{2g}}{\pi}\frac{d}{dx}\int_0^x \frac{f(t)}{\sqrt{x-t}}dt. \tag{3.41}$$

Example 16

To give a simple example, let us consider the very special case of the integral equation of the first kind with $K(x,t) = 1, 0 \leq x; 0 \leq t \leq x$,

$$h(x) = \int_0^x u(t)dt. \tag{E.1}$$

Let us assume that we are given $h(x)$ as the following (continuous) roof function on $0 < x < 2a$,

$$h(x) = \begin{cases} x, & 0 < x < a \\ a - x, & a < x < 2a \end{cases} \tag{E.2}$$

as illustrated in Figure 1.11 and we are to find the solution $u(x)$ of (E.1) on $(0, 2a)$. Of course, it is obvious that $u(x)$ can be obtained from simply differentiating the given function $h(x)$,

$$u(x) = \frac{dh}{dx}. \tag{E.4}$$

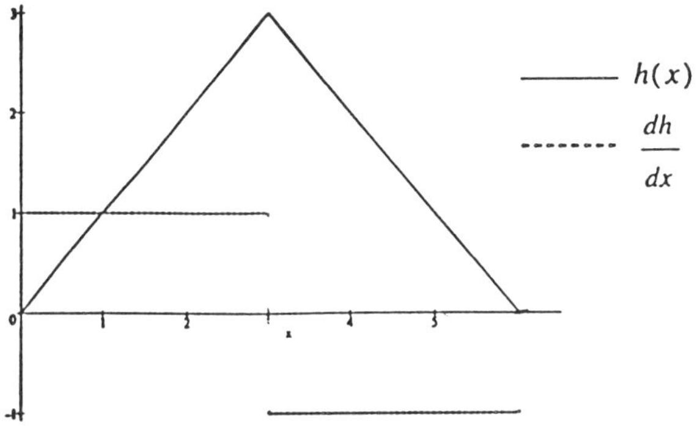

Fig. 1.11 The roof function $h(x)$ of (E.2) and its discontinuous derivative $\frac{dh}{dx}$ of (E.5).

Now a simple look at $\dfrac{dh}{dx}$ in Figure 1.11, and we see the uncovered difficulty due to this differentiation, namely, that the derivative of $h(x)$ does not exist at $x = a$. Instead, and in contrast the continuous $h(x)$, $\frac{dh}{dx}$ is continuous only on $(0, a)$ and $(a, 2a)$, and it has a clear jump discontinuity of size 2 at $x = a$,

$$\frac{dh}{dx} = \begin{cases} 1, & 0 < x < a \\ -1, & a < x < 2a. \end{cases} \tag{E.5}$$

The matter is even more serious when $h(x)$ is given as data, where, of course, it is within the accuracy of the measurement of the data. So, if we are, in principle, after $u(x) = \frac{dh}{dx}$, then we must approximate $\frac{dh}{dx}$ by $\frac{\Delta h}{\Delta x} = \frac{h(x+\Delta x)-h(x)}{\Delta x}$. But this computation for $\frac{\Delta h}{\Delta x}$ will compound the final error in $\frac{\Delta h}{\Delta x}$, when we start with the inaccurate data of $h(x)$.

1.5.3 Interpolation of the Numerical Solutions of Integral Equations

Lagrange Interpolation Formula

As the result of solving an integral equation numerically, with n increments for the integral involved, we obtain $N = n+1$ approximate sample values $\{\tilde{u}_i = \tilde{u}(x_i)\}_{i=1}^{N}$ for the solution $u(x)$. Then it is desirable to interpolate between these *discrete* values to obtain a continuous function $\tilde{u}(x)$ as an approximation to the solution $u(x)$. There are many different formulas for such interpolation. We present here the well known *Lagrange interpolation formula*, which is based on the use of a *polynomial* of degree $n = N - 1$,

$$p_n(x) = a_n x^n + a_{n-1}x^{n-1} + \cdots + a_1 x + a_0, \quad n = N - 1 \tag{1.152}$$

to interpolate between the N discrete values $\{u(x_i)\}_{i=1}^{N}$ that result in the continuous approximation $\tilde{u}(x)$ to the approximated function $u(x)$.

The Lagrange interpolation formula is used for *not necessarily equidistant* samples $\{f(x_i)\}_{i=1}^N$ of $f(x)$, we will use $\tilde{f}(x)$, instead of $f(x)$, for the (approximate) interpolated function to distinguish it from the exact $f(x)$,

$$\tilde{f}(x) = \sum_{j=1}^{N} l_j(x) f(x_j) \tag{1.153}$$

where the function $l_j(x)$ is defined as the following quotient of two products

$$\begin{aligned} l_j(x) &= \frac{(x-x_1)(x-x_2)\cdots(x-x_{j-1})(x-x_{j+1})\cdots(x-x_N)}{(x_j-x_1)(x_j-x_2)\cdots(x_j-x_{j-1})(x_j-x_{j+1})\cdots(x_j-x_N)} \\ &= \frac{\Pi_{i=1}^N (x-x_i) \quad i \neq j}{\Pi_{i=1}^N (x_j-x_i) \quad i \neq j.} \end{aligned} \tag{1.154}$$

(We note here that the factors $(x - x_j)$ and $(x_j - x_j)$ are missing, respectively, in the numerator and denominator of $l_j(x)$, which is indicated in the product notation by $i \neq j$ to say excluding the jth factor in both products.)

An interpolation formula should first give us the sampling points, thus requires from (1.54) that

$$l_j(x_m) = \begin{cases} 1, & m = j \\ 0, & m \neq j \end{cases} \tag{1.55}$$

which is clearly the case. This is so because for $m \neq j$ we have a factor $(x_m - x_m) = 0$ in the numerator but no such factor in the denominator, and all other factors there are nonzero. In the case of $l_j(x_j)$ all factors in both numerator and denominator are the same, and without the factor $(x_j - x_j)$, since it is missing in both places, and the result is $l_j(x_j) = 1$.

Example 17 Interpolating the Numerical Solution of a Volterra Equation

Consider the following Volterra integral equation of the second kind in $u(x)$,

$$u(x) = x - \int_0^x (x-t)u(t)dt \tag{E.1}$$

We note that it is with difference kernel and the integral is in the form of the Laplace convolution product, where we can use the Laplace transform to solve it, similar to what we illustrated in Example 10 of Section 1.4. The exact solution can be easily obtained as $u(x) = \sin x$ using the Laplace transform, and can be quickly verified, after a simple integration by parts for $\int_0^x t \sin t dt$ (and remembering that x in $x - t$ of the integral is considered as constant, since the integration is done with respect to t.)

In Example 9 of Section 3.3, we use only four increments on the interval $[0, 4]$ to try to find the five numerically approximated values of the solution $\tilde{u}(x_j)$, $j = 1, 2, 3, 4$ and 5 at the indicated locations $x = 0, 1, 3, 4$ and 4 as shown in Table 1.3.

Table 1.3 Numerical and Exact Solutions of Volterra Integral Equation (E.1) of Example 17

x	0	1	2	3	4
Numerical value of $u(x)$	0	1	1	0	-1
Exact value of $u(x) = \sin x$	0	0.8415	0.9093	0.1411	-0.7568

The table also includes the corresponding exact values $u(x_j) = \sin x_j, j = 1, 2, 3, 4,$ and 5. Figure 1.12 illustrates the comparison between these exact and approximate five values of the solution to (E.1). Note that we graphed the exact values, then purposely connected them with (an exact) graph as a solid line. This is done because we know that the solution is $u(x) = \sin x$ for all $x > 0$. However, for the approximate numerical values $\tilde{u}(x_j)$ we only graphed, what we are sure of, as the approximated five values. So, it is left for some interpolation formula to use these few values and fill between them to give an idea of a continuous approximate solution $\tilde{u}(x)$. Here we appeal to the Lagrange interpolation formula (1.153) and (1.154) to do this job.

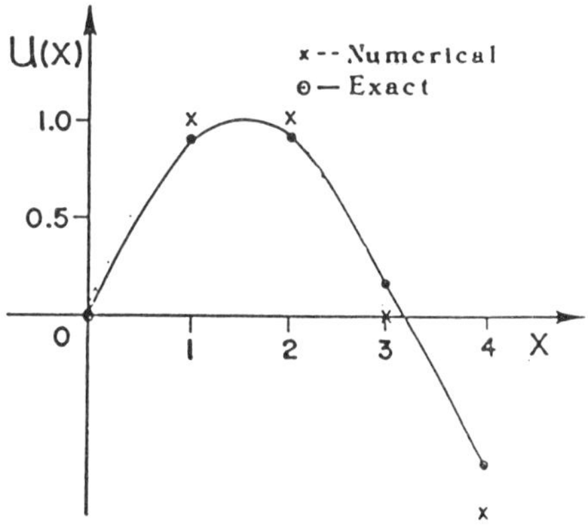

Fig. 1.12 Numerical and exact solutions of Volterra equation (E.1) of Example 17. (Also Example 9 and Table 3.1 in Section 3.3.)

To use the Lagrange interpolation of (1.153) and (1.154), we first prepare $l_j(x)$, $j = 1, 2, \cdots, 5$ of (1.154)

$$
\begin{aligned}
l_1(x) &= \frac{(x-1)(x-2)(x-3)(x-4)}{(0-1)(0-2)(0-3)(0-4)} = \frac{(x-1)(x-2)(x-3)(x-4)}{24} \\
l_2(x) &= \frac{(x)(x-2)(x-3)(x-4)}{(1)(1-2)(1-3)(1-4)} = \frac{(x)(x-2)(x-3)(x-4)}{-6} \\
l_3(x) &= \frac{(x)(x-1)(x-3)(x-4)}{(2)(2-1)(2-3)(2-4)} = \frac{(x)(x-1)(x-3)(x-4)}{4} \\
l_4(x) &= \frac{(x)(x-2)(x-3)(x-4)}{3(3-1)(3-2)(3-4)} = \frac{(x)(x-1)(x-2)(x-3)}{-6} \\
l_5(x) &= \frac{(x)(x-1)(x-2)(x-3)}{4(4-1)(4-2)(4-3)} = \frac{(x)(x-1)(x-2)(x-3)}{24}.
\end{aligned} \tag{E.2}
$$

So, using the above data and these functions $\{l_j(x)\}_{j=1}^5$ of (E.2) in (1.153) we obtain the interpolated approximate solution $\tilde{u}(x)$,

$$
\begin{aligned}
\tilde{u}(x) = (0)\frac{(x-1)(x-2)(x-3)(x-4)}{24} + (1)\frac{x(x-2)(x-3)(x-4)}{-6} + \\
(1)\frac{x(x-1)(x-3)(x-4)}{4} + (0)\frac{x(x-1)(x-2)(x-3)}{-6} \\
+(-1)\frac{x(x-1)(x-2)(x-3)}{24} = \frac{x^4}{24} - \frac{x^3}{4} - \frac{x^2}{24} + \frac{5x}{4}.
\end{aligned} \tag{E.3}
$$

This interpolated approximate solution $\tilde{u}(x)$ is computed and shown in Figure 1.13, where it connects the five numerical sample values $\tilde{u}(x_j)$, $j = 1, 2, 3, 4, 5$, and $\tilde{u}(x)$ is compared with the exact solution $u(x) = \sin x$. Samples of the comparison are $u(0.5) = 0.4794$, $\tilde{u}(0.5) = 0.5859$, $u(2.5) = 0.5985$, $\tilde{u}(2.5) = 0.5859$, $u(3.4) = -0.2555$, $\tilde{u}(3.4) = -0.4896$.

Next we illustrate the use of the Lagrange interpolation formula for two approximate numerical solutions of a Fredholm integral equation of the second kind.

Example 18 Interpolating Two Numerical Solutions of a Fredholm Integral Equation

In Section 5.5 we will cover a simple detailed treatment for the numerical solution of Fredholm integral equations. This is illustrated in Example 20 of Section 5.5.1 for the numerical setting and solution of the following Fredholm integral equation of the second kind

$$
u(x) = \sin x + \int_0^1 (1 - x\cos xt)u(t)dt. \tag{E.1}
$$

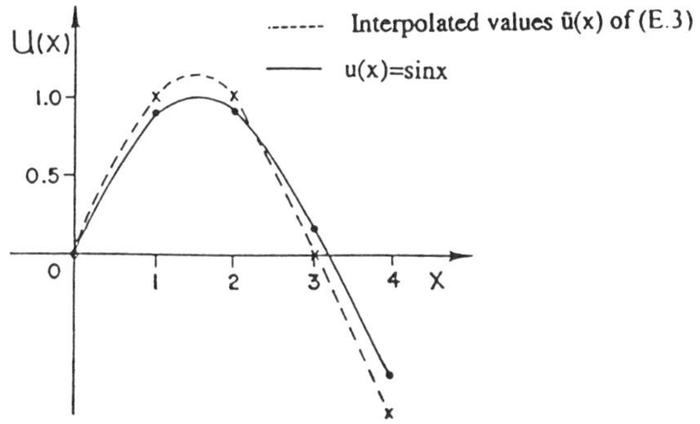

Fig. 1.13 The interpolated approximate solution $\tilde{u}(x)$ of (E.3) and the exact solution $u(x) = \sin x$ of (E.1) in Example 17.

In this Example 20, we consider only three (approximate values $\tilde{u}(x_j), j = 1, 2, 3$ at $x_1 = 0$, $x_2 = 0.5$ and $x_3 = 1.0$. The result is a 3×3 system of algebraic equations in $\tilde{u}(x_j), j = 1, 2, 3$, since we have these three values for the input $\tilde{u}(t_j), j = 1, 2, 3$ inside the approximating sum of the integral, for each $\tilde{u}(x_j), j = 1, 2, 3$ of the output $u(x)$ of (E.1).

The solution of this 3×3 system of algebraic equations is the subject of Exercise 2(a) and 2(b) of Section 5.5, where the trapezoidal rule and the Simpson's rule are used, respectively, for approximating the integral in (E.1). These results are shown in Table 1.4, and are to be compared with the exact solution $u(x) = 1, 0 \le x \le 1$ of (E.1). For such square system of algebraic equations, we leave it (in Section 5.5) for the preparation of the reader to deal with the solution using matrix methods. For the most basic method, we present a review of the Cramer's rule in the next section.

Now we will use the Lagrange interpolation formula (1.153) and (1.154) for both sets of the three samples, then compare with the exact solution $u(x) = 1$.

From Example 20 and Exercise 2 of Section 5.5, we have in Table 1.4 the two sets of approximate sample values $\tilde{u}(x_j)$ of the solution to (E.1),

We first prepare $l_1(x)$, $l_2(x)$ and $l_3(x)$ of (1.154) to be used in the Lagrange interpolation formula (1.153), then we use the two sets of data to find their respective interpolations.

$$\begin{aligned} l_1 &= \frac{(x-0.5)(x-1)}{(-0.5)(-1)} = 2\left(x - \frac{1}{2}\right)(x-1) \\ l_2 &= \frac{x(x-1)}{(0.5)(0.5-1)} = -4x(x-1) \\ l_3 &= \frac{x(x-0.5)}{(1)(1-0.5)} = 2x(x-0.5). \end{aligned} \tag{E.2}$$

If we use the approximate samples values of Exercise 1(a) in Section 5.5 and the above functions $l_j(x)$ in (1.153) we have

Table 1.4 Numerical Solutions of Fredholm Equation (E.1) of Example 18 (as given in the answer to Exercise 2(a), (b) of Section 5.5), using a) the Trapezoidal Rule and b) the Simpson's Rule

j	x_j	$\tilde{u}(x_j)$: Exer. 2(a)	$\tilde{u}(x_j)$: Exer 2(b)
1	0.0	1.013	0.99987
2	0.5	1.009	0.99992
3	1.0	1.021	0.99967

$$\tilde{u}(x) = (1.013)2\left(x - \frac{1}{2}\right)(x-1) - (1.009)4x(x-1) + (1.021)2x\left(x - \frac{1}{2}\right) \tag{E.3}$$

which can be easily computed to find that this $\tilde{u}(x)$ is very close to the exact value of $u(x) = 1$. An example are the few values $\tilde{u}(0.3) = 1.0087$, $\tilde{u}(0.7) = 1.012$ and $\tilde{u}(0.9) = 1.0173$.

When we use the approximate samples values from Exercise 2(b), with the functions $l_j(x)$ of (E.2) in (1.153), we obtain

$$\tilde{u}(x) = (0.99987)2\left(x - \frac{1}{2}\right)(x-1) - (0.99992)4x(x-1) + (0.99967)2x\left(x - \frac{1}{2}\right) \tag{E.4}$$

which interpolates similar to (E.3). An example are the few values $\tilde{u}(0.3) = 0.9999$, $\tilde{u}(0.7) = 0.9999$, $\tilde{u}(0.9) = 0.9997$.

As we mentioned earlier, the numerical methods of solving Volterra and Fredholm integral equations will require be the subjects of Sections 3.3 and 5.5, respectively. There we will illustrate the numerical setting with the aim at a numerical solution for these equations using the simplest numerical integration formulas, namely, the trapezoidal rule (1.141) and Simpson's rule (1.144). In Chapter 7 we will follow such treatment by concentrating on the use of the *higher order quadrature rules*. In particular, the Newton-Cotes rules and the Maclaurin rule will be used for Volterra equations in Section 7.2, while the Gauss quadrature rules are used for Fredholm equations in Section 7.3. The Gauss quadratures will also be consulted and illustrated for their use in finding an approximate solution of one type of *singular* Fredholm integral equations, namely, those whose singularity is due to (only) the limit (or limits) of integration being infinite.

1.5.4 Review of Cramer's Rule

As we mentioned in the last section, the resulting square system of linear algebraic equations, for numerically approximating the Fredholm integral equation, will require some basic knowledge of matrix analysis. We present here a review of the very basic such needed computations, namely, Cramer's rule for solving $N \times N$ system of linear algebraic equations.

Consider the square $N \times N$ system of linear equations in $x_1, x_2, \cdots, x_N$,

$$\begin{aligned} a_{11}x_1 + a_{12}x_2 + \cdots + a_{1N}x_N &= b_1 \\ a_{21}x_1 + a_{22}x_2 + \cdots + a_{2N}x_N &= b_2 \\ \vdots \quad \vdots \quad \vdots \\ a_{N1}x_1 + a_{N2}x_2 + \cdots + a_{NN}x_N &= b_N \end{aligned} \tag{1.156}$$

with its matrix form

$$AX = B \tag{1.157}$$

where $A = [a_{ij}]_{i,j=1,2,\cdots,N}$, $X = [x_i]_{i=1,2,\cdots,N}$, and $B = [b_i]_{i=1,2,\cdots,N}$. Of course, we must remember that the existence of a unique solution to this system in (1.156) [or (1.157)] is strongly dependent on $|A| = \det A$, the determinant of the coefficients matrix A. This can be stated for the nonhomogeneous system ($b \neq 0$), (here 0 is the zero column matrix) that when $|A| \neq 0$, then the system has a *unique solution* (for any values of b), but if $|A| = 0$, then there exist values of b for which there is *no solution*, and other values of b for which there are *infinitely many solutions*. On the other hand, in the case of the *homogeneous* system ($b = 0$), if $|A| \neq 0$, the system has only the *trivial* solution $X = 0$, but when $|A| = 0$, the system has *infinitely many solutions*.

So for $|A| \neq 0$ the unique solution of the nonhomogeneous system ($b \neq 0$) in (1.156) or (1.157) is given by Cramer's rule as

$$x_1 = \frac{1}{|A|} \begin{vmatrix} b_1 & a_{12} & \cdots & a_{1n} \\ b_2 & a_{22} & \cdots & a_{2n} \\ \cdot & \cdot & \cdots & \cdot \\ \cdot & \cdot & \cdots & \cdot \\ \cdot & \cdot & \cdots & \cdot \\ b_n & a_{n2} & \cdots & a_{nn} \end{vmatrix}$$

$$x_2 = \frac{1}{|A|} \begin{vmatrix} a_{11} & b_1 & \cdots & a_{1n} \\ a_{21} & b_2 & \cdots & a_{2n} \\ \cdot & \cdot & \cdots & \cdot \\ \cdot & \cdot & \cdots & \cdot \\ \cdot & \cdot & \cdots & \cdot \\ a_{n1} & b_n & \cdots & a_{nn} \end{vmatrix}$$

$$\vdots$$

$$x_n = \frac{1}{|A|} \begin{vmatrix} a_{11} & \cdots & a_{1n-1} & b_1 \\ a_{21} & \cdots & a_{2n-1} & b_2 \\ \cdot & \cdots & \cdot & \cdot \\ \cdot & \cdots & \cdot & \cdot \\ \cdot & \cdots & \cdot & \cdot \\ a_{n1} & \cdots & a_{nn-1} & b_n \end{vmatrix} \tag{1.158}$$

We also remind that $|A|$, the determinant of the matrix A is computed as

$$|A| = \sum_{j=1}^{N} (-1)^{i+j} a_{ij} |M_{ij}| \tag{1.159}$$

for the expansion about the ith row of $A(i-\text{fixed})$, and

$$|A| = \sum_{i=1}^{N} (-1)^{j+i} a_{ij} |M_{ij}| \tag{1.160}$$

for the expansion about the jth column of $A(j-\text{fixed})$.

$|M_{ij}|$ is the *minor* of the entry a_{ij}, which is the determinant of the (remaining) $(N-1) \times (N-1)$ matrix after deleting the ith row and the jth column of A. Also $(-1)^{i+j}|M_{ij}|$ is called the *cofactor* of a_{ij}.

Exercises 1.5

1. **(a)** Find the exact value of the following definite integral

$$\int_0^1 \frac{dx}{1+x^2}. \tag{E.1}$$

Hint:

$$\int \frac{dx}{1+x^2} = \tan^{-1} x. \tag{E.2}$$

(b) Use the trapezoidal rule with $n = 4$ to approximate the integral in (E.1).

(c) Estimate the error in the trapezoidal rule approximation of the integral in (E.1).

Hint: For the maximum value M of $|\frac{d^2 f}{dx^2}|$ of $f(x) = \frac{1}{1+x^2}$ on $(0,1)$ to be used for the error estimate in (1.47), we must find $\frac{d^3 f}{dx^3} = \frac{24x(1-x^2)}{(1+x^2)^4}$ in the search for a critical point of $f''(x)$. Note in this case $\frac{d^3 f}{dx^3} > 0$ on $(0,1)$, and the minimum value of $f''(x) = \dfrac{-2(1-3x^2)}{(1+x^2)^3}$ occurs as $f''(0) = -2$ at the end point $x = 0$, where $M = |f''(0)| = 2$.

(d) Compare the actual error of this numerical approximation with its estimate (or upper bound) of part (c).

2. **(a)** Find the exact value of the following definite integral

$$\int_0^1 \frac{dx}{1+x}. \qquad (E.1)$$

(b) Use Simpson's rule with $n = 4$ to approximate the integral in (E.1).

(c) Estimate the error in the Simpson's rule approximation of the integral in (E.1).

Hint: Here we need the maximum value of $|\frac{d^4 f}{dx^4}| = |\frac{24}{(1+x)^5}|$ of $f(x) = \frac{1}{1+x}$ on $(0, 1)$, which clearly occurs at $x = 0$, $M = |f^{(4)}(0)| = 24$.

(d) Compare the actual error of this numerical approximation with its estimate (or upper bound) of part (c).

3. Use six increments to approximate the integral

$$\int_3^6 \frac{x dx}{4+x^2}$$

using

(a) the trapezoidal rule.

(b) Simpson's rule.

(c) Find the exact value of the integral, and compare it with the two approximate values in parts (a) and (b).

4. In Exercise 1 of Section 7.2 we solve the following Volterra integral equation

$$u(x) = 1 - 2x + 4x^2 + \int_0^x [3 + 6(x-t) - 4(x-t)^2]u(t)dt \qquad (E.1)$$

numerically.

Use the Lagrange interpolation formula (1.153)–(1.154) to interpolate those numerical values of the solution that are found in the answer to Exercise 1 of Section 7.2. Compare the interpolated result with the exact solution $u(x) = e^x$ of (E.1). (This is the same as Exercise 5 of Section 7.2.)

5. Consider the Fredholm integral equation,

$$u(x) = e^{-x} - \int_0^1 xe^t u(t)dt \qquad (E.1)$$

and the three samples of its numerical solution $u_1 = u(0) = 1$, $u_2 = u(0.5) = 0.3674966$, and $u_3 = u(1) = -0.11089$ as done in Exercise 3a(i) of Section 5.5.

(a) Use the Lagrange interpolation formula (1.153)–(1.154) to find the (approximate) interpolation $\tilde{u}(x)$ of these approximate samples.

(b) Do the same as in part (a) for the eleven samples of the numerical solution of (E.1) as given in the answer to Exercise 3a(ii) of Section 5.5.

2

Modeling of Problems as Integral Equations

In this chapter we formulate, in as much detail as possible, a number of the problems that we presented in Section 1.1 in terms of their natural representation as integral or integro-differential equations. We also show in detail how initial and boundary value problems, associated with linear differential equations, and some specified auxiliary conditions are reduced to *Volterra* and *Fredholm* integral equations, respectively. Finally, we illustrate the formulation of a *mixed* boundary condition for the electrified plate in terms of *dual integral equations*.

The formulation of the first group of problems with integral equations as their natural representation will depend on the realization, demonstrated in Section 1.1, that an integral equation like

$$u(x) = \int_a^b K(x,\xi)u(\xi)d\xi \tag{2.1}$$

relates the present state $u(x)$ to the accumulation (integral) of the changes $K(x,\xi)u(\xi)$ of all its other values $u(\xi)$ for $a < \xi < b$. This is in contrast with the differential equation representation, which gives a local relation. For example,

$$u(x) = k\frac{du}{dx} \tag{2.2}$$

relates the state $u(x)$ to its instantaneous rate of change du/dx, and hence the relation to its values only in the very immediate neighborhood. Indeed, the formulation in terms of derivatives is the essence of the mathematical modeling for the basic natural law of *causality* which made differential equations the important subject it is today. In comparison, we say that the importance of integral equations stems from representing accumulative or *hereditary* situations, where, for example, a state $u(t)$ is affected by the accumulation of changes in all its previous values such as the population-type problems as in (1.8). Of course, the mathematical modeling for problems involving both hereditary and causal principles is quite familiar and is given in terms of integro-differential equations, as in the case of the biological special living together (1.13) and (1.14).

We may remark that what was said here about a state as a function of time and its previous values can be extended to functions of space variables such as the hanging chain, (1.17) and (1.18), and to more than one variable, such as the case of the charge density for potential on a unit disc in (1.38). This accumulation effect for the hanging chain in (1.17) is clearly seen where the shape $f(x)$ at the point $x \epsilon (0, 1)$ is affected by all the forces on the interval of its definition (0,1).

In this chapter we first present the formulation of some problems of population, control, mechanics, and radiation transport, then follow with initial and boundary value problems, dual integral equations, and Schrödinger equations in the three-dimensional momentum space.

2.1 POPULATION DYNAMICS

As mentioned earlier the study of population growth includes the forecasting of any future surge in birthrates, which is of great importance for future planning throughout the world. In this section we formulate the problem of human population growth, the problem of the surge in birthrates, and the problem of two biological species living together.

2.1.1 Human Population

Let the number of people present at time $t = 0$ be n_0. If we look at survival or insurance tables, we find that there is some sort of a survival function $f(t)$ similar to that shown in Figure 2.1, which gives the fraction of people surviving to age t. It is assumed that these people are either male or female. The surviving population $n_s(t)$ at time t is then

$$n_s(t) = n_0 f(t) \tag{2.3}$$

where $n_s(0) = n_0 f(0) = n_0$.

Under normal circumstances there is a continuous addition to the population through new births. If children are born at an average rate $r(t)$, then in a particular time interval $\Delta_i \tau$ about the time τ_i, there are $r(\tau_i)\Delta_i \tau$ children added who, if they survive, will be of age $t - \tau_i$ at time t. But according to Figure 2.1, only a fraction $f(t - \tau_i)$ of these children will survive to age $t - \tau_i$, so the final addition to the population at time t, from the children born in the interval $\Delta_i \tau$ about time τ_i, is

$$f(t - \tau_i) r(\tau_i) \Delta_i \tau.$$

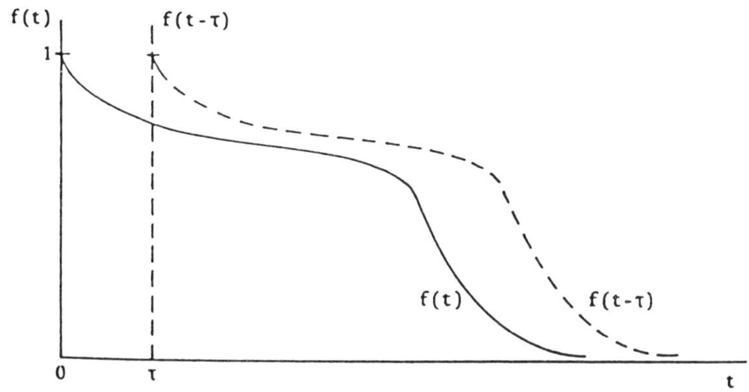

Fig. 2.1 The survival function.

Now if this process is repeated for all the m subintervals of the time interval $(0, t)$, we obtain the partial sum

$$b_m(t) = \sum_{i=1}^{m} f(t - \tau_i) r(\tau_i) \Delta_i \tau \tag{2.4}$$

as the number of people added through new births which, if passed to the limit (as $m \to \infty$), becomes the integral

$$b(t) = \int_0^t f(t - \tau) r(\tau) d\tau. \tag{2.5}$$

If this is added to $n_s(t)$ in (2.3) (the survivors of the initial population), we obtain the total population at time t as

$$n(t) = n_s(t) + b(t) = n_0 f(t) + \int_0^t f(t - \tau) r(\tau) d\tau. \tag{2.6}$$

It is reasonable now to assume that the rate of birthrate $r(t) = \dfrac{dn}{dt}$ is proportional to $n(t)$, the number of the population present at time t,

$$r(t) = kn(t). \tag{2.7}$$

From (2.6) and (2.7) it follows that

$$n(t) = n_0 f(t) + k \int_0^t f(t - \tau) n(\tau) d\tau \tag{2.8}$$

which is a Volterra integral equation of the second kind in $n(t)$ with a difference kernel $kf(t - \tau)$. Such a kernel reminds us of the Laplace method of solution discussed in Section 1.4.1, which we will illustrate in Section 3.1.3.

2.1.2 Biological Species Living Together

Consider the two species with number $n_1(t)$ and $n_2(t)$, respectively, present at time t that we presented in (1.13) and (1.14) of Section 1.1. If the two species are left separate, it is reasonable to assume that their rate of change dn/dt is proportional to $n(t)$, the number found at time t,

$$\frac{dn}{dt} = \alpha n(t). \tag{2.9}$$

Here α is the proportionality constant or the coefficient of increase ($\alpha > 0$) or decrease ($\alpha < 0$), depending on whether there is a growth or decline in the population, respectively. Let us assume that the first species is increasing with coefficient of increase $k_1 (k_1 > 0)$ and the second species is decreasing with coefficient of decrease $-k_2 (k_2 > 0)$. Then if the two species are left separate, the growth of the first is represented by

$$\frac{dn_1}{dt} = k_1 n_1(t), \qquad k_1 > 0 \tag{2.10}$$

and the decline of the second species is represented by

$$\frac{dn_2}{dt} = -k_2 n_2(t), \qquad k_2 > 0. \tag{2.11}$$

Now we are to formulate the state of equilibrium of these two species, when put together, with the assumption that the second species (a predator) will feed on the first (the prey). This will, of course, affect the rate of change of both species; the first (prey) will have a slower rate of growth, while the second (predator) will have a slower rate of decline.

To formulate this situation in terms of a reasonable mathematical model, we start with the model of the separate species (2.10) and (2.11). We then attempt to modify it in two steps to allow for the new situation, where we must introduce factors to decrease k_1 in (2.10) and increase $-k_2$ in (2.11). To start with k_1, the rate of increase of the first species (prey), it is reasonable to assume that its decrease is proportional to $n_2(t)$, the number of the second species (predator) present. Hence k_1 should be modified to k_1'.

$$k_1' = k_1 - \gamma_1 n_2(t) \tag{2.12}$$

where γ_1 is a proportionality constant which depends on the first species. The actual decrease in k_1 is due not only to the presence (feeding) of the second species $n_2(t)$ at the present time t but also to all previous presences (feedings) of $n_2(\tau)$ for the whole time interval $t - T_0 < \tau < t$, where T_0 is the finite heredity duration of both species. If in addition to the present γ_1 factor we have the record of its rate of decrease as $f_1(\tau)$ in previous times $t - T_0 < \tau < t$, then the decrease in k_1 at time t, due to the decrease in the time interval $\Delta\tau$, is $-f_1(t-\tau)n_2(\tau)\Delta\tau$. Here we used $f_1(t-\tau)$, since a species at the present time t had a chance to resist $n_2(\tau)$ at time $t - \tau$ and hence a factor $f_1(t-\tau)$. The total decrease in k_1 in the whole time interval T_0 is

$$\delta k_1 = -\int_{t-T_0}^{t} f_1(t-\tau)n_2(\tau)d\tau. \tag{2.13}$$

If we combine this previous decrease (2.13) with the present decrease $-\gamma_1 n_2(t)$ [as in (2.12)], we obtain

$$k_{1eff} = k_1 - \gamma_1 n_2(t) - \int_{t-T_0}^{t} f_1(t-\tau)n_2(\tau)d\tau \tag{2.14}$$

as the final effectively reduced coefficient of increase of the first species.

By reasoning similar to the above, the coefficient of decrease $-k_2$ of the second species (predator) will increase by $\gamma_2 n_1(t)$ due to present feeding on the first species, and hence $-k_2$ should be replaced by the higher value

$$-k_2' = -k_2 + \gamma_2 n_1(t). \tag{2.15}$$

Again it is not the present good feeding alone which causes the increase in $-k_2$, but also all previous feedings that depended on the presence of the first species $n_1(\tau)$. Hence in the time interval $\Delta\tau$, there is an increase of $f_2(t-\tau)n_1(\tau)\Delta\tau$ and the total increase in the same period T_0 is

$$\delta k_2 = \int_{t-T_0}^{t} f_2(t-\tau)n_1(\tau)d\tau. \tag{2.16}$$

The final (effectively increased) value of $-k_2$ is obtained by adding δk_2 in (2.16) to that of (2.15):

$$-k_{2eff} = -k_2 + \gamma_2 n_1(t) + \int_{t-T_0}^{t} f_2(t-\tau)n_1(\tau)d\tau. \tag{2.17}$$

Now we modify k_1 in (2.10) to its effective reduced value (2.14) and $-k_2$ in (2.11) to its effective increased value (2.17), to obtain the model for the equilibrium state of the two species living together:

$$\frac{dn_1}{dt} = n_1(t)\left[k_1 - \gamma_1 n_2(t) - \int_{t-T_0}^{t} f_1(t-\tau)n_2(\tau)d\tau\right], \quad k_1 > 0 \tag{2.18}$$

$$\frac{dn_2}{dt} = n_2(t)\left[-k_2 + \gamma_2 n_1(t) + \int_{t-T_0}^{t} f_2(t-\tau)n_1(\tau)d\tau\right], \quad k_2 > 0. \tag{2.19}$$

This system of two (nonlinear) integro-differential equations is as we presented them in (1.13) and (1.14).

Exercises 2.1

1. Consider the integral equation (2.8) in $n(t)$; the number of the human population at time t.

   (a) Let $N(s)$ and $F(s)$ be the Laplace transform of $n(t)$ and $f(t)$, respectively. Find the Laplace transform of (2.8) and hence find $N(s)$.

   (b) Assume an exponential-type survival function $f(t) = e^{-ct}$, $c > k > 0$. Find $N(s)$, then solve for $n(t)$ of (2.8) as the inverse Laplace transform of $N(s)$.

2. Consider the problem of radioactive decay where the rate of decrease of the number of atoms is proportional to the number of atoms $n(t)$ present at time t.

   Hint: See (2.9)–(2.11).

   (a) Write the differential equation and its initial condition assuming that n_0 is the initial number of atoms present at time $t = 0$, and remembering that the proportionality constant for the decay being a *negative* number $-k$, $k > 0$.

   (b) Reduce the initial value problem in part (a) to an integral equation.

   (c) Solve the initial value problem in part (a) for $n(t)$.

   (d) Verify that the solution of the initial value problem in part (c) is also a solution of the integral equation in part (b).

3. Consider the human population problem (1.8), and assume that the survival function is described roughly by the exponential function $f(t) = e^{-\frac{t}{T}}$, where T is the average life span of a typical person.

   (a) Write the resulting integral equation and find its Laplace transform and hence $N(s) = \mathcal{L}\{n(t)\}$.

   (b) Use the inverse Laplace transform to find the solution $n(t)$, the population at time t.

   (c) In (2.7) we assumed that the birthrate is proportional to the number present in the population,

   $$r(t) = \frac{dn}{dt} = kn(t) \tag{2.7}$$

   where we can take k as the rate of population variation per capita. Use the result in part (b) to show that

   (i) The population increases in an exponential fashion when $T > 1/k$ (i.e., when the average life span of the typical person is larger than the reciprocal of the per capita rate of change of the population).

   (ii) The population decreases in an exponential fashion when $T < 1/k$.

4. Consider the problem of birthrate and the possibility of a surge in the birthrate, which we considered in (1.9) of Section 1.1.

(a) Let us assume that there are initially b_0 women, and that they will give birth to a female child at a rate $h(t)$ per year. Find their contribution to the female birthrate at time t.

(b) To find the birthrate $b(t)$ at time t, we must add to this birthrate of part (a) the contribution to birthrates of girls born at time $\tau > 0$ when they are at age τ in the range of childbearing age $\alpha < \tau < \beta$. Girls born at time $t-\tau$ will at future time t belong to the birthrate $b(t-\tau)$. If the probability of a girl living to age τ is $\rho(\tau)$ and the probability of the girl at this age giving birth to a female child in a time interval $\Delta\tau$ is $m(\tau)\Delta\tau$, find the contribution to the birthrate $b(t)$ from women in the subinterval $\Delta\tau$ of the range of childbearing age τ [born at $t-\tau$ with birth rate $b(t-\tau)$].

(c) Find the contribution to birthrate $b(t)$ at time t from all women in the childbearing age range $\alpha < \tau < \beta$.

Hint: Pass to the limit for the sum in part (b) to become an integral.

(d) Find the expression for the total birthrate $b(t)$ that includes the contribution of the women present at the initial time $t = 0$.

5. Consider the problem of the surge in birthrate of problem 4, where its birthrate $b(t)$ is governed by the integral equation

$$b(t) = b_0 h(t) + \int_\alpha^\beta b(t-\tau)\rho(\tau)m(\tau)d\tau. \tag{E.1}$$

The integral above represents the contribution to birthrate of girls born at time $t > 0$ when they are at the childbearing age τ, $\alpha < \tau < \beta$.

Use the following detailed steps in parts (a) and (b) to show that the above integral equation can be expressed as the following *Volterra-type* integral equation with *difference kernel*

$$b(t) = \begin{cases} b_0 h(t) + \displaystyle\int_0^t b(t-\tau)\rho(\tau)m(\tau)d\tau, & t \leq \beta \\ \displaystyle\int_0^t b(t-\tau)\rho(\tau)m(\tau)d\tau, & t > \beta. \end{cases} \tag{E.2}$$

(a) For $t \leq \beta$ you can write the integral (E.1) as

$$\begin{aligned} b(t) = b_0 h(t) \quad &+ \int_0^\alpha b(t-\tau)\rho(\tau)m(\tau)d\tau + \int_\alpha^t b(t-\tau)\rho(\tau)m(\tau)d\tau \\ &+ \int_t^\beta b(t-\tau)\rho(\tau)m(\tau)d\tau. \end{aligned} \tag{E.3}$$

The first integral goes to zero since $m(\tau)\Delta\tau$, the probability of a girl bearing a female child at age $t < \alpha$ which is outside the childbearing age, is zero. It is added to help having integration from $\tau = 0$ to $\tau = t$ to fit the Laplace convolution product-type integral as shown in the integral of the first branch in (E.2). Also, the third integral of (E.3) is of zero contribution since $b(t)$ takes care of birthrates for $t > 0$ and it is assumed zero for $t \leq 0$. In other words $b(t-\tau)$ is zero in the third integral, since $t-\tau < 0$ for $t < \tau < \beta$. Hence the first term plus the second integral of (E.3) give the first branch of (E.2).

(b) For $t > \beta$ note that since we are taking the origin of time as the birth of the oldest childbearing woman, then $b_0 h(t) = 0$ for $t > \beta$ in (E.1), since this term now takes care of birthrates to women at age $t > \beta$, which is outside the childbearing range (α, β). With $b_0 h(t) = 0$, we now rewrite (E.1) as

$$\begin{aligned} b(t) &= \int_0^\alpha b(t-\tau)\rho(\tau)m(\tau)d\tau + \int_\alpha^\beta b(t-\tau)\rho(\tau)m(\tau)d\tau \\ &\quad + \int_\beta^t b(t-\tau)\rho(\tau)m(\tau)d\tau \qquad t > \beta, \end{aligned} \tag{E.4}$$

by adding the first and third integrals, which are of zero contribution, since $m(\tau)\Delta\tau = 0$ for $\tau < \alpha$ and $\tau > \beta$. Hence (E.4) gives the second branch of (E.2).

$$b(t) = \int_0^t b(t-\tau)\rho(\tau)m(\tau)d\tau, \qquad t > \beta. \tag{E.5}$$

2.2 CONTROL AND OTHER PROBLEMS

In this section we present the formulation, in terms of integral equations, of two control problems: the problem of how to keep a specified number of machines always available in operating condition and that of controlling the deviation of a steering shaft from its indicated direction.

2.2.1 Mortality of Equipment and Rate of Replacement

The problem of finding the rate dr/dt at which equipment should be replaced, to keep a specified number $f(t)$ in operating condition at any time t, is formulated similar to that of the human population problem of Section 2.1.1. We first assume that we have $s(t)$, the function that determines the number of pieces of new equipment bought at $t = 0$ that survives to time t. If we start with $f(0)$ as the number of new pieces bought at time $t = 0$, then, due to loss or wear, only the fraction $f(0)s(t)$ will survive to time t. To keep a specified number larger than $f(0)s(t)$ at time t we must continuously add equipment at the desired rate from time $t = 0$ to time t. If the desired rate of replacement at which we must add new equipment at time τ is $dr(\tau)/d\tau$, then at time t this equipment will be of age $t-\tau$ with a survival function $s(t-\tau)$ that is dependent on their age $t-\tau$. From $\left(\frac{dr}{d\tau}\right)\Delta\tau$, what we replace in the time interval $\Delta\tau$, only a fraction $s(t-\tau)(dr/dt)\Delta\tau$ will survive to time t. Hence, if these survivals of the continuous replacements are added along the time interval $(0, t)$, we obtain

$$r(t) = \int_0^t s(t-\tau)\frac{dr}{d\tau}d\tau, \qquad t > 0 \tag{2.20}$$

the number of pieces of equipment surviving to time t, which were purchased as replacements during the time $0 < \tau < t$. If we add this to $f(0)s(t)$, the surviving number of pieces of original equipment (new at time $t = 0$), we obtain the (desired) total number of pieces of equipment in operating condition at time t,

$$f(t) = f(0)s(t) + \int_0^t s(t-\tau)\frac{dr}{d\tau}d\tau, \tag{2.21}$$

which is a Volterra integral equation of the first kind in the unknown rate of replacement $\frac{dr}{dt}$.

Exercises 2.2

1. Assume an exponential-type survival function $s(t) = e^{-ct}$ for the integral equation (2.21) in $\frac{dr}{dt}$, the rate at which the equipment must be replaced.

   (a) Use the Laplace transform to solve for the necessary rate of replacement $\frac{dr}{dt}$, in order that we keep a constant number of machines $f(t) = A$ at all times t.

   (b) Verify your answer.

2. Consider the problem (1.16) of the deviation $\phi(t)$ of the steering angle $\theta_s(t)$ of the rotating shaft from a constant direction indicator angle $\theta_i(t) = 1, t > 0$. Only a correction torque proportional to the deviation $\phi(t)$ and another one proportional to the rate of change of the deviation are applied. The rotating shaft starts from rest at a zero angle $\theta_s(0) = 0, \theta_s'(0) = 0$.

   (a) Write the mathematical model for the problem.

   (b) Find the Laplace transform for the equation in part (a).

   (c) Solve for the deviation $\phi(t)$ and hence for $\theta_s(t)$ of the rotating shaft angle. For simplicity let $I = 1, b = 2$, and $a = 1$.

   (d) Use the same method and conditions above to solve for the deviation $\phi(t)$ for the complete problem (1.16) when the accumulation (integral) torque correction term is included. For simplicity let $I = c = 1$ and $a = b = 3$.

3. Electric potential in a disc. The electric potential $u(r,\theta)$ at a point (r,θ) inside a disc of radius a, which is free of charge and where the boundary of the disc $(r = a)$ is kept at a potential $u(a,\theta) = f(\theta)$, is given by the *Poisson integral*

$$u(r,\theta) = \frac{a^2 - r^2}{2\pi} \int_{-\pi}^{\pi} \frac{f(\phi)d\phi}{a^2 + r^2 - 2ar\cos(\theta - \phi)}, \qquad r < a.$$

   (a) State a problem that makes the above an integral equation in $f(\phi)$.
   Hint: Be cautious about $\lim_{r\to a} u(r,\theta)$. For more details, see the derivation of this Poisson formula in (4.69) at the end of Section 4.1 in Chapter 4.

   (b) Show that the potential $f(\theta)$ on the boundary must be distributed in such a way that its average is equal to the value of the potential $u(0,\theta)$ at the center of the disc.
   Hint: $\frac{1}{b-a}\int_a^b f(x)dx$ is the average of $f(x)$ on the interval (a, b).

4. Determining a source of neutrons in an absorbing uniform rod. This exercise relates to our discussion of determining the energy spectrum $f(E)$ of radiation such as neutrons, which we derived in Section 1.1 as the following Fredholm integral equation of the first kind (1.23) in $f(E)$,

$$g(x) = \int_{E_{min}}^{E_{max}} e^{-x\sigma(E)} f(E)dE. \tag{1.23}$$

   Here $g(x)$ is the number of emerging neutrons on the other side of the used slab of uniform thickness x, and $\sigma(E)$ is the absorption cross section of the material of the slab as it appears to the incoming neutrons of energy E.

   (a) Consider now a uniform bar of length b with source of neutrons $f(y)$, $0 < y < b$ to be determined. It is assumed, for a simple model, that the neutrons move only in two directions — right and left — along the rod with constant absorption cross section σ. With the input as $f(y)$, assume that we can measure the output at position x as $h(x)$. (Of course $h(x)$ is a set of measurements for finite number of location points x.) Show that $f(y)$ satisfies the following Fredholm integral equation of the first kind

$$h(x) = \int_0^b e^{-\sigma|x-y|} f(y)dy. \tag{E.1}$$

   Hint: For the neutrons moving — right and left — you must consider the two cases $x > y$ and $x < y$, and for both cases the neutrons must travel distance $|x - y|$.

   (b) Show that an analytic solution to (E.1) can be obtained as

$$f(y) = \frac{1}{2}\left\{\sigma h(y) - \frac{1}{\sigma}h''(y)\right\}. \tag{E.2}$$

Hint: Write the integral in (E.1) as

$$h(x) = \int_a^x e^{-\sigma(x-y)} f(y)dy + \int_x^b e^{-\sigma(y-x)} f(y)dy \qquad (E.3)$$

then differentiate twice using the generalized Leibnitz rule (1.53) to have an expression for $h''(y)$ in terms of $h(y)$ as required in (E.2).

(c) The above analytical solution in (E.2) for the Fredholm equation of the first kind (E.1) is in terms of the measured data $h(y)$ and its second order derivative $h''(y)$. Explain how such an analytic solution in (E.2) of the Fredholm equation of the first kind, as may be anticipated, gives a bad numerical solution for (E.1).

Hint: Recall the inaccuracy in the measured data $h(y)$, and how the numerical differentiation $\left(\frac{\Delta h}{\Delta y} = \frac{h(y+\Delta y) - h(y)}{\Delta y}\right)$ is so sensitive to even small errors in the differentiated function.

2.3 MECHANICS PROBLEMS

In this section we formulate problems dealing with the shape of the hanging chain and Abel's problem.

2.3.1 Hanging Chain

Here we consider the problem of how a variable density $\rho(x)$ must be distributed in the form of a chain or a rope, in order that it may assume a given shape $f(x)$.

First we consider an elastic string under an initial constant tension T_0, and a vertical force F acting at one point. Then we derive the equation for the case of distributed forces along the string, for example, the variable gravitational force due to a variable linear density of the string.

(a) Displacement Due to a Single Vertical Force: Consider the string AB of length l in Figure 2.2 under initial constant tension T_0. (Recall that in Figure 2.2 we take $y(x)$ to be positive in the downward direction of gravity, and that the force of the point mass m is its weight $w = mg$.) Let F be a constant vertical force acting on the string at $x = \xi$ to displace it by a small vertical distance $y(\xi)$ which is very small compared to ξ. If we equate the vertical forces, assuming that the tension is constant (T_0) along the string, we have

$$T_0 \sin\phi + T_0 \sin\theta = F. \qquad (2.22)$$

But for small ϕ and θ we have $\sin\phi \sim \tan\phi = y(\xi)/\xi$ and $\sin\theta \sim \tan\theta = y(\xi)/(l-\xi)$; hence (2.22) becomes

$$T_0 \frac{y(\xi)}{\xi} + T_0 \frac{y(\xi)}{l-\xi} = F,$$

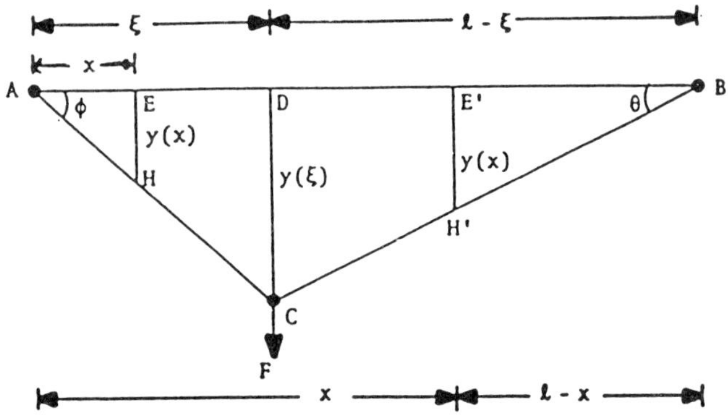

Fig. 2.2 Displacement due to a single vertical force F at ξ. From Jerri [1982, 1986], courtesy of COMAP, Inc.

$$y(\xi) = \frac{F}{T_0 l}\xi(l - \xi). \tag{2.23}$$

To find the displacement $y(x)$ at any point x we consider the similar triangles ACD and AHE for $x \le \xi$ where $y(x)/y(\xi) = x/\xi$, which when combined with $y(\xi)$ from (2.23) gives

$$y(x) = \frac{x}{\xi}y(\xi) = \frac{F}{T_0 l}x(l - \xi), \qquad 0 \le x \le \xi. \tag{2.24}$$

For $\xi \le x \le l$ we use the similar triangles CBD and $H'BE'$, where $y(x)/y(\xi) = (l - x)/(l - \xi)$ and (2.23) to give

$$y(x) = \frac{l - x}{l - \xi}y(\xi) = \frac{F}{T_0 l}\xi(l - x), \qquad \xi \le x \le l. \tag{2.25}$$

So from (2.24) and (2.25) the displacement $y(x)$ for $0 \le x \le l$, due to the single vertical force F at $x = \xi$, is

$$y(x) = FG(x, \xi) = F\begin{cases} \dfrac{x(l - \xi)}{T_0 l}, & 0 \le x \le \xi \\ \dfrac{\xi(l - x)}{T_0 l}, & \xi \le x \le l. \end{cases} \tag{2.26}$$

It is important to note the two branches of the function $G(x, \xi)$ (2.26) where the first branch satisfies the boundary condition $y(0) = 0$, for the first end of the elastic string at $x = 0$ to be fixed; while the second branch satisfies the boundary condition $y(l) = 0$ for a fixed second end at $x = l$. This is a very familiar occurrence when finding the integral representation of boundary value problem as Fredholm integral equations. We will see a function similar to $G(x, y)$ of (2.26) appearing as the kernel in the integral equation to satisfy, the already incorporated, boundary conditions.

Such function is termed *Green's function*, which is discussed in details in Chapter 4, in preparation for the Fredholm integral equations of Chapter 5.

(b) Displacement Due to Distributed Vertical Force: We now consider the vertical force not at one point $x = \xi$ only, but distributed continuously along the string; for example, the gravitational force due to the variable linear density $\rho(\xi)$ of a string. For such a string the gravitational force acting on the element $\Delta\xi$ of the string is $\Delta F(\xi) = g\rho(\xi)\Delta\xi$. According to (2.26), the resulting displacement due to this single force on $\Delta\xi$ is

$$\Delta y(x) = \Delta F(\xi)G(x,\xi) = G(x,\xi)g\rho(\xi)\Delta\xi \tag{2.27}$$

where $G(x,\xi)$ is given by (2.26).

The total displacement due to the gravity force along the whole string is obtained by superimposing all these displacements (2.27) of the elements of the string, or in other words, integrating from $\xi = 0$ to $\xi = l$,

$$y(x) = g\int_0^l G(x,\xi)\rho(\xi)d\xi \tag{2.28}$$

where $G(x,\xi)$ is given by (2.26). This is a Fredholm integral equation of the first kind in $\rho(x)$ that relates how the linear density $\rho(\xi)$ must be distributed along the string so that the string may assume the prescribed shape $y(x)$.

Example 1

We illustrate here how the simple case of constant density $\rho(\xi) = c$ determines the expected (parabolic) shape for the string.

If we use (2.28) for $y(x)$, (2.26) for $G(x,\xi)$, and let $\rho(\xi) = c$, we obtain

$$y(x) = g\int_0^x \frac{c\xi(l-x)}{T_0 l}d\xi + g\int_x^l \frac{cx(l-\xi)}{T_0 l}d\xi. \tag{E.1}$$

We note here how the second and first branches of (2.26) are used for the first and second integrals of (E.1), respectively. Evaluating the two integrals in (E.1) gives

$$\begin{aligned} y(x) &= \frac{cg}{T_0 l}(l-x)\left.\frac{\xi^2}{2}\right|_0^x + \frac{cg}{T_0 l}x\left.\left(l\xi - \frac{\xi^2}{2}\right)\right|_x^l \\ &= \frac{cg}{T_0 l}(l-x)\frac{x^2}{2} + \frac{cgx}{T_0 l}\left[\left(l^2 - \frac{l^2}{2}\right) - \left(lx - \frac{x^2}{2}\right)\right] \\ &= \frac{cg}{T_0 l}\left(\frac{lx^2}{2} - \frac{x^3}{2} + \frac{xl^2}{2} - lx^2 + \frac{x^3}{2}\right) \\ &= \frac{cgx(l-x)}{2T_0}. \end{aligned}$$

2.3.2 Sliding a Bead Along a Wire: Abel's Problem

As we mentioned in Section 1.1, Abel's problem is one of the earliest problems modeled as an integral equation. It deals with finding the path $y(x)$ in the vertical xy-plane (Figure 2.3) along which a particle, under the influence of gravity and starting from rest at y_0, must move in order that it descends a distance y_0 in a prescribed time $t = f(y_0)$.

To simplify the problem, we consider the path of the particle to be known when we know α, the angle that the tangent to the path makes with the x axis. In this case $dy/dx = \tan\alpha$, so $dy/ds = -\sin\alpha$,[1] where $v = ds/dt$.

For a particle starting from rest at $y = y_0$, under gravity, the velocity v at y is governed by

$$v^2 = 2g(y_0 - y),$$

$$v = \frac{ds}{dt} = \sqrt{2g(y_0 - y)} \tag{2.29}$$

where g is the acceleration of gravity. To have the desired expression for dt, we write

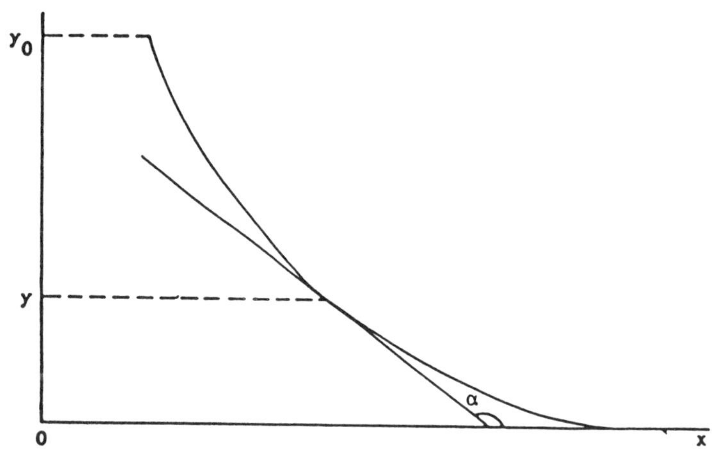

Fig. 2.3 Abel's problem.

$$\frac{dy}{dt} = \frac{dy}{ds}\frac{ds}{dt} = -\sqrt{2g(y_0 - y)}\sin\alpha,$$

$$dt = \frac{-dy}{\sqrt{2g(y_0 - y)}\sin\alpha}. \tag{2.30}$$

[1]Note that $dy/dx < 0$, $\tan\alpha < 0$; $dy/ds < 0$, $\sin\alpha > 0$, $\pi/2 < \alpha < \pi$.

Realizing that α depends on y, we let $1/\sin\alpha = \phi(y)$ in (2.30); then

$$dt = \frac{-\phi(y)dy}{\sqrt{2g(y_0 - y)}} \tag{2.31}$$

and integrate from the initial time of descent $t(y_0) = f(y_0)$ to the final time $t(y = 0) = 0$.

$$t|_{t(y_0)}^{t(0)} = -\int_{y_0}^{0} \frac{\phi(y)dy}{\sqrt{2g(y_0 - y)}},$$

$$0 - t(y_0) = -f(y_0) = -\int_{y_0}^{0} \frac{\phi(y)dy}{\sqrt{2g(y_0 - y)}}. \tag{2.32}$$

Hence (2.32) is the final integral equation in $\phi(y)$ that relates the form of the path $\phi(y)$ to the predetermined time of descent $f(y_0)$ of the particle,

$$-\sqrt{2g}f(y_0) = \int_0^{y_0} \frac{\phi(y)dy}{\sqrt{y_0 - y}}. \tag{2.33}$$

To avoid having the variable y_0 looks like a constant, we replace the two variables y_0 and y by y and η, respectively, to write (2.33) in the form of Abel's integral equation (1.20)

$$-\sqrt{2g}f(y) = \int_0^{y} \frac{\phi(\eta)d\eta}{\sqrt{y - \eta}}. \tag{1.20}$$

We note that taking the final time $t(y = 0) = 0$, we are making a negative initial time $t(y_0) = f(y_0) < 0$.

Example 2

For illustration we will consider the simple case of finding the path in a vertical plane along which the particle must move from rest at $y = y_0$ so that it reaches the ground in (the usual free-falling body) time

$$t = \sqrt{\frac{2y_0}{g}} \tag{E.1}$$

where we expect a vertical path for the fall, i.e., $\alpha = 90°$ in Figure 2.3.

Let us note that this is a very special case with $t = f(y_0) = -\sqrt{2y_0/g}$, which can be solved by using the simple laws of motion since $y_0 = 1/2gt^2$, $t = \sqrt{2y_0/g}$. If we substitute $t = -\sqrt{2y_0/g}$ for $f(y_0)$ in (2.33), noting that the final time is $t = 0$, which necessitates a negative initial time, we have

$$\sqrt{\frac{4gy_0}{g}} = 2\sqrt{y_0} = \int_0^{y_0} \frac{\phi(y)dy}{\sqrt{y_0 - y}}. \tag{E.2}$$

This Abel integral equation may be solved for $\phi(y)$ by using the Laplace transform (see Exercise 4); but for simplicity, we will use the result of Example 1, Section 1.1, which gives $u(x) = 1/2$ as a solution of Abel's equation,

$$x^{1/2} = \int_0^x \frac{u(\xi)d\xi}{\sqrt{x-\xi}}.$$

Hence we may use $\phi(y) = 2/2$ in (E.2) or $\phi(y) = 1 = 1/(\sin\alpha)$. This gives $\alpha = 90°$, which says that the path is vertical.

Exercises 2.3

1. Let the deflection of an elastic string of length l at point x_1 due to a unit force (load) at x_2 be $K(x_1, x_2)$.

   (a) Give the equation that represents the total deflection $D_{1,2}(x)$ at a point x due to a load $L_1(x)$, applied at the middle of the elastic string, and another load $L_2(x)$, applied at $x = \frac{2l}{3}$.

   (b) Give the equation that represents the total deflection $D(x)$ due to a continuous load $L(x) = \rho(x)$, as a result of the string's variable linear density $\rho(x)$.

2. Use (2.27) and (2.28) to find the approximate shape of the string when two thin beads with a constant density of 1 and length $l/20$ and $l/12$ are placed along $(l/5, l/4)$ and $(2l/3, 3l/4)$ of the string, respectively.[2]

   Hint: Use the weight of the bead as the force at the bead's center of gravity.

3. Determine $y(x)$, the shape of the string in (2.28), when the linear density is given by

$$\rho(x) = cx(l-x).$$

4. Use the Laplace transform to solve for the path $\phi(y)$ in Abel's integral equation (2.33), to verify that the path is vertical.

5. Rederive Abel's Tautochrone integral equation (1.21) in $\phi(y) = -\frac{ds}{dy}$ which governs the path along which a particle, starting from (x_0, y_0), must slide (Figure 2.4) to reach the origin in *constant* time $T = f(y_0)$ that is independent of the starting point (x_0, y_0). *Hint:* Use conservation of energy:

$$\frac{1}{2}m\left(\frac{ds}{dt}\right)^2 = mg(y_0 - y)$$

   and note that $\frac{ds}{dy} = \sqrt{1 + (\frac{dx}{dy})^2}$.

[2]The very detailed solution of this problem in five pages is found in "The Student's Solution Manual" to accompany this book [Jerri, 1999]. See the end of the preface for more information.

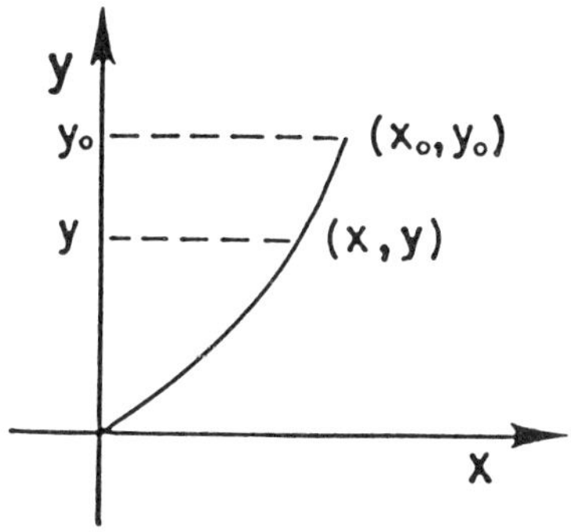

Fig. 2.4 Abel's Tautochrone problem.

6. Consider the Tautochrone problem (1.21) that was presented in Section 1.1. Use conservation of energy,

$$\frac{1}{2}m\left(\frac{ds}{dt}\right)^2 = mg(y_0 - y), \qquad \frac{ds}{dt} < 0,$$

where m is the mass and g the gravity acceleration, to drive the integral equation (1.21)

$$\sqrt{2g}T = \int_0^{y_0} \frac{F'(y)}{\sqrt{y_0 - y}} dy \tag{1.21}$$

where $s = F(y)$, noting that

$$\left(\frac{ds}{dy}\right)^2 = \left(\frac{dF}{dy}\right)^2 = 1 + \left(\frac{dx}{dy}\right)^2.$$

Hint: Note that $t = 0$ to $t = T$ correspond to $y = y_0$ to $y = 0$, respectively, $\frac{ds}{dt} < 0$ that requires a minus sign, and $\frac{ds}{dy} = \sqrt{1 + \left(\frac{dx}{dy}\right)^2}$.

2.4 INITIAL VALUE PROBLEMS REDUCED TO VOLTERRA INTEGRAL EQUATIONS

To illustrate in detail how an initial value problem associated with a linear differential equation and, usually, homogeneous initial conditions reduces to a Volterra integral equation, we consider the following example, which we have already presented in (1.29) and (1.32). This will be followed by the initial value problem associated with the general second-order differential equation.

Example 3

$$\frac{d^2y}{dx^2} = \lambda y(x) + g(x),^3 \tag{1.29}$$

$$y(0) = 1 \tag{1.30}$$

$$y'(0) = 0 \tag{1.31}$$

First we note that we have a nonhomogeneous initial condition in (1.30). However, the whole initial value problem (1.29)–(1.31) can be easily reduced to one with homogeneous initial conditions, by making a simple change of variables $u(x) = y(x) + 1$.

Now we start with the same steps used in Example 5 of Section 1.3. So we let

$$\frac{d^2y}{dx^2} = F(x) \tag{$E.1$}$$

and integrate once with respect to x to obtain

$$\frac{dy}{dx} = \int_0^x F(\xi)d\xi + c_1. \tag{$E.2$}$$

Integrating again gives

$$y(x) = \int_0^x \int_0^\xi F(t)dtd\xi + c_1x + c_2. \tag{$E.3$}$$

To reduce the double integral of (E.3) to a single integral, we use the identity (1.51)

$$\int_a^x \int_a^\xi F(t)dtd\xi = \int_a^x (x-t)F(t)dt = \int_a^x (x-\xi)F(\xi)d\xi \tag{1.51}$$

in (E.3) to give

$$y(x) = \int_0^x (x-\xi)F(\xi)d\xi + c_1x + c_2. \tag{$E.4$}$$

To find the arbitrary constants c_1 and c_2, we apply the initial condition (1.30) on (E.4),

$$y(0) = 1 = 0 + c_2, \qquad c_2 = 1$$

and the initial condition (1.31) on (E.2),

$$y'(0) = 0 = 0 + c_1, \qquad c_1 = 0.$$

Hence (E.4) becomes

$$y(x) = 1 + \int_0^x (x-\xi)F(\xi)d\xi. \tag{$E.5$}$$

[3]For a more general initial value problem, see the first edition of this book, p. 61.

From (E.1) and (1.29) we have

$$\frac{d^2y}{dx^2} = F(x) = \lambda y(x) + g(x). \tag{E.6}$$

If we substitute this value for $F(x)$ in (E.5), we obtain

$$y(x) = \lambda \int_0^x (x-\xi)y(\xi)d\xi + 1 + \int_0^x (x-\xi)g(\xi)d\xi \qquad (1.32), (2.34)$$

which is a Volterra integral equation of the second kind in $y(x)$.

Another Volterra integral equation in $F(x) = d^2y/dx^2$ is obtained when we substitute $y(x)$ from (E.5) in the original differential equation (1.29).

$$F(x) = \lambda \left[1 + \int_0^x (x-\xi)F(\xi)d\xi\right] + g(x),$$

$$F(x) = \lambda \int_0^x (x-\xi)F(\xi)d\xi + \lambda + g(x). \qquad (E.7), (2.35)$$

The two integral equations (1.32) and (E.7), in $y(x)$ and $F(x) = d^2y/dx^2$, respectively, represent the same initial value problem in $y(x)$. Concerning the method of solution, there seems to be a choice between solving (1.32) for the direct value of $y(x)$ or solving (E.7) for $F(x) = d^2y/dx^2$, then integrating this twice to obtain $y(x)$.

The above method and the resulting two forms of the Volterra integral equation in (1.32) and (E.7) are illustrated in the following example.

Example 4 Reduce the initial value problem in $y(x)$

$$\frac{d^2y}{dx^2} + y = \cos x \tag{E.1}$$

$$y(0) = 0 \tag{E.2}$$

$$y'(0) = 0 \tag{E.3}$$

to a Volterra integral equation in: (a) $u(x) = d^2y/dx^2$ and (b) $y(x)$.

First we let $u(x) = d^2y/dx^2$ then integrate once to have

$$\frac{dy}{dx} = \int_0^x u(t)dt + c_1 \tag{E.4}$$

with $c_1 = 0$ after using the initial condition (E.3). We integrate this result again, using the identity (1.51) for the double integration to have

$$y(x) = \int_0^x (x-t)u(t)dt + c_2 \tag{E.5}$$

where we also have $c_2 = 0$ from using the initial condition (E.2),

$$y(x) = \int_0^x (x-t)u(t)dt. \tag{E.6}$$

For the final result (E.6) to be an integral equation, we have two choices:

(a) To make this result as an integral equation in $u(x)$. In this case we use (E.1) to have $y(x) = \cos x - d^2y/dx^2 = \cos x - u(x)$ to substitute for the $y(x)$ term outside the integral for (E.6) to become a Volterra integral equation of the second kind in $u(x) = d^2y/dx^2$

$$\cos x - u(x) = \int_0^x (x-t)u(t)dt,$$

$$u(x) = \cos x - \int_0^x (x-t)u(t)dt. \tag{E.7}$$

(b) To have (E.6) as an integral equation in $y(x)$, we substitute for $u(t)$ inside the integral of (E.6) in terms of $y(t)$ via (E.1), where $u(t) = y''(t) = -y(t) + \cos t$,

$$y(x) = \int_0^x (x-t)[-y(t) + \cos t)]dt,$$

$$\begin{aligned} y(x) &= \int_0^x (x-t)\cos t dt + \int_0^x (t-x)y(t)dt \\ &= x\int_0^x \cos t dt - \int_0^x t\cos t dt + \int_0^x (t-x)y(t)dt, \end{aligned} \tag{E.8}$$

$$y(x) = 1 - \cos x + \int_0^x (t-x)y(t)dt$$

after evaluating the first two integrals.

Exercises 2.4

1. Reduce the initial value problem

$$\frac{d^2y}{dx^2} + y = -\cos x,$$

$$y(0) = 0, \qquad y'(0) = -1$$

(of Exercise 2 in Section 1.3) to the Volterra integral equation in $y(x)$,

$$y(x) = \cos x - x - 1 - \int_0^x (x-t)y(t)dt.$$

2. Reduce the initial value problem in $y(x)$

$$\frac{d^2y}{dx^2} + y = 0 \tag{E.1}$$

$$y(0) = 0 \tag{E.2}$$

$$y'(0) = 1 \tag{E.3}$$

(a) to Volterra integral equation in $u(x) = d^2y/dx^2$.

(b) to Volterra integral equation in $y(x)$.

Hint: (a) let $u(x) = d^2y/dx^2$, integrate it twice, using the identity (1.51) and the initial conditions (E.3) and (E.2), then substitute for $y(x)$ outside in terms of $d^2y/dx^2 = u(x)$ from (E.1). (b) In the integral for $y(x)$ in part (a), substitute for $u(x) = d^2y/dx^2$ in terms of $y(x)$ from (E.1).

3. Reduce the initial value problem

$$\frac{d^2y}{dx^2} - \frac{dy}{dx}\sin x + e^x y = x, \tag{E.1}$$

$$y(0) = 1, \tag{E.2}$$

$$y'(0) = -1 \tag{E.3}$$

(a) to the Volterra integral equation in $u(x) = \dfrac{d^2y}{dx^2}$,

$$u(x) = x - \sin x + e^x(x-1) + \int_0^x [\sin x - e^x(x-t)]u(t)dt \tag{E.4}$$

Hint: See the derivation of (E.7) in Example 3 and (E.7) in Example 4.

(b) to the Volterra integral equation in $y(x)$,

$$y(x) = \frac{x^3}{6} - x + \int_0^x [\sin x + (t-x)(e^t + \cos t)]y(t)dt. \tag{E.5}$$

4. Reduce the initial value problem in $y(x)$,

$$\frac{d^2y}{dx^2} + x\frac{dy}{dx} + y = 0, \quad x > 0 \tag{E.1}$$

$$y(0) = 1 \tag{E.2}$$

$$y'(0) = 0 \tag{E.3}$$

to Volterra integral equation in $u(x) = \dfrac{d^2y}{dx^2}$.

Hint: Let $u(x) = \frac{d^2y}{dx^2}$, integrate it once using the initial condition (E.3) to obtain $\frac{dy}{dx}$, then integrate again using the identity (1.51) and the initial condition (E.2) to obtain $y(x)$. Last substitute these $\frac{dy}{dx}$ and $y(x)$, in (E.1), with their resulting dependence (inside their corresponding integrals) on $u(x) = \frac{d^2y}{dx^2}$.

2.5 BOUNDARY VALUE PROBLEMS REDUCED TO FREDHOLM INTEGRAL EQUATIONS

To illustrate how a boundary value problem associated with a differential equation may be represented by a Fredholm integral equation, we consider the following example of (1.33)–(1.35).[4]

Example 5

$$\frac{d^2y}{dx^2} = \lambda y(x), \qquad a < x < b \tag{1.33}$$

$$y(a) = 0 \tag{1.34}$$

$$y(b) = 0 \tag{1.35}$$

We proceed to integrate (1.33), in the same manner as that followed in Example 4, which gives

$$\frac{dy}{dx} = \lambda \int_a^x y(t)dt + c_1. \tag{E.1}$$

Integrating again and using the integral identity (1.51) gives

$$y(x) = \lambda \int_a^x \int_a^\xi y(t)dtd\xi + c_1 x + c_2 = \lambda \int_a^x (x-t)y(t)dt + c_1 x + c_2. \tag{E.2}$$

For simplicity, we leave the variable of the final integration in (E.2) as t instead of ξ.

To evaluate the arbitrary constants c_1 and c_2, we employ the boundary condition (1.34),

$$y(a) = 0 = 0 + c_1 a + c_2, \qquad c_2 = -c_1 a \tag{E.3}$$

and the boundary condition (1.35),

$$y(b) = 0 = \lambda \int_a^b (b-t)y(t)dt + c_1 b - c_1 a. \tag{E.4}$$

Hence, from (E.3) and (E.4), we have

$$c_1 = \frac{\lambda}{a-b} \int_a^b (b-t)y(t)dt \tag{E.5}$$

[4] For a more general boundary value problem, see the first edition of this book, p. 66.

and

$$c_2 = -c_1 a = -\frac{a\lambda}{a-b}\int_a^b (b-t)y(t)dt. \tag{E.6}$$

So, if we use these values of c_1 and c_2 in (E.2), the final integral representation of the boundary value problem (1.33)–(1.35) is

$$y(x) = \lambda\int_a^x (x-t)y(t)dt + \lambda\frac{x-a}{b-a}\int_a^b (t-b)y(t)dt. \tag{E.7}$$

This integral equation in $y(x)$ can now be rearranged to result in the form of the Fredholm integral equation (1.36) with its kernel $K(x,t)$ defined in (1.37). This is done by writing the second integral as two parts on the intervals $[a,x]$ and $[x,b]$, where the first part will then be combined with the first integral of (E.7),

$$\begin{aligned} y(x) = {} & \lambda\int_a^x (x-t)y(t)dt + \lambda\frac{x-a}{b-a}\int_a^x (t-b)y(t)dt \\ & +\lambda\frac{x-a}{b-a}\int_x^b (t-b)y(t)dt = \lambda\int_a^x \frac{(x-b)(t-a)}{b-a}y(t)dt \\ & +\lambda\int_x^b \frac{(x-a)(t-b)}{b-a}y(t)dt. \end{aligned} \tag{E.8}$$

If we now define the kernel $K(x,t)$ as

$$K(x,t) = \begin{cases} \dfrac{(x-b)(t-a)}{(b-a)}, & a \le t \le x \\ \dfrac{(x-a)(t-b)}{b-a}, & x \le t \le b \end{cases} \tag{E.9}$$

then the last two integrals in (E.8) can be combined as

$$\lambda\int_a^x K(x,t)y(t)dt + \lambda\int_x^b K(x,t)y(t)dt = \lambda\int_a^b K(x,t)y(t)dt. \tag{E.10}$$

Hence (E.8), and in turn (E.7), reduce to the homogeneous Fredholm integral equation

$$y(x) = \lambda\int_a^b K(x,t)y(t)dt \qquad (1.36), (2.36)$$

where $K(x,t)$ is given by (E.9), which is (1.37)

$$K(x,t) = \begin{cases} \dfrac{(x-b)(t-a)}{(b-a)}, & a \le t \le x \\ \dfrac{(x-a)(t-b)}{(b-a)}, & x \le t \le b. \end{cases} \qquad (1.37), (2.37)$$

We want to stress again the equivalence of the homogeneous boundary value problem (1.33)–(1.35) with the homogeneous Fredholm integral equation (1.36) and its kernel in (1.37), since we may sometimes resort to solving the boundary value problem to obtain the solutions for its equivalent homogeneous Fredholm equation. As we mentioned at the end of Example 6 and in Exercise 5 of Section 1.3, this will require us to differentiate the integral equation in order to find its corresponding differential equation (with its boundary conditions), which, hopefully, is easier to solve. This development is illustrated in detail in the following Example 6. The need for such development will become evident when we study the methods of solving nonhomogeneous Fredholm integral equations in Chapter 5, where the solutions of the homogeneous equation are essential for the development. A list of homogeneous boundary value problems with their equivalent homogeneous Fredholm integral equations and, of course, their respective kernels (Green's functions) is given in Appendix B.

Example 6 Reduce the following homogeneous integral equation to a boundary value problem:

$$y(x) = \lambda \int_0^1 K(x,t)y(t)dt, \tag{E.1}$$

$$K(x,t) = \begin{cases} x(1-t), & 0 \le x \le t \\ t(1-x), & t \le x \le 1. \end{cases} \tag{E.2}$$

As we remarked on the function $G(x,\xi)$ of (2.26) for the hanging chain, we can again see very clearly the appearance of this kernel $K(x,t)$ in (1.37) with its first and second branches that satisfy the two boundary conditions at the two boundary points $x = a$ and $x = b$, respectively. This $K(x,t)$ is the Green's function of the boundary value problem (1.33)–(1.35) that, effectively, reduced it to the Fredholm integral equation in (1.37). The *Green's function* is the subject of Chapter 4.

Here we have two ways of doing the problem. The first is to recognize the kernel (E.2) as a special case of (E.9) or (2.37) in Example 5, with $a = 0$ and $b = 1$, and hence the integral equation is a special case of (2.37) which is equivalent to the boundary value problem (1.33)–(1.35), with $a = 0, b = 1$,

$$\frac{d^2y}{dx^2} = -\lambda y(x), \qquad 0 < x < 1 \tag{E.3}$$

$$y(0) = 0, \qquad y(1) = 0. \tag{E.4}$$

There is an *infinity* of solutions for this boundary value problem, as we shall show soon, which are $y_n(x) = \sin n\pi x$, $n = 0, 1, 2, \cdots$, and in turn they are the solutions of the homogeneous Fredholm equation (E.1) and (E.2).

The second method is that in the absence of an integral equation like (2.37) with which to compare (E.1), we may have to keep differentiating the integral equation, hoping that it will finally reduce to a familiar differential equation. In this case, we write (E.1) as

$$y(x) = \lambda \int_0^x t(1-x)y(t)dt + \lambda \int_x^1 x(1-t)y(t)dt \tag{E.5}$$

and note how we used the second branch of $K(x,t)$ in the first integral of (E.5) on the interval $(0,x)$ and the first branch in the second integral of (E.5) on the interval $(x,1)$. Next we must realize that both integrals in (E.5) have variable limits and that their integrands are functions of x; hence, in general, we should use the Leibnitz rule (1.53) in differentiating them. However, in the special case at hand, we can factor the x dependence out of the integrals

$$y(x) = \lambda(1-x)\int_0^x ty(t)dt + \lambda x\int_x^1 (1-t)y(t)dt. \tag{E.6}$$

Now each term is a product of two functions of x and we can use the fundamental theorem of calculus on the integrals. If we differentiate (E.6) once, we obtain

$$\begin{aligned} y'(x) &= -\lambda\int_0^x ty(t)dt + \lambda(1-x)xy(x) + \lambda\int_x^1 (1-t)y(t)dt - \lambda x(1-x)y(x) \\ &= \lambda\int_x^1 (1-t)y(t)dt - \lambda\int_0^x ty(t)dt \end{aligned} \tag{E.7}$$

and if we differentiate again (using the fundamental theorem of calculus), we obtain

$$\begin{aligned} y''(x) &= -\lambda(1-x)y(x) - \lambda xy(x) \\ &= -\lambda y(x) + \lambda xy(x) - \lambda xy(x) = -\lambda y(x) \end{aligned} \tag{E.8}$$

or

$$y'' + \lambda y(x) = 0 \tag{E.9}$$

which is the desired differential equation.

To obtain the boundary conditions, we let $x = 0$ and $x = 1$ in (E.5) to obtain, respectively,

$$y(0) = 0 \tag{E.10}$$

$$y(1) = 0, \tag{E.11}$$

or we can see this easily by substituting $x = 0$ and $x = 1$ in (E.1) to obtain (E.10) and (E.11) with the help of the definition of the kernel (Green's function) $K(x,t)$ in (E.2), which vanishes at $x = 0$ due to its first branch and at $x = 1$ from its second branch.

It is now instructive to solve the resulting simple boundary value problem (E.9)–(E.11), where the general solution of the differential equation (E.9) is

$$y(x) = c_1\cos\sqrt{\lambda}x + c_2\sin\sqrt{\lambda}x. \tag{E.12}$$

For this solution to satisfy the boundary conditions (E.10) and (E.11) we must have

$$\begin{aligned} y(0) &= c_1 + 0 = 0, \quad c_1 = 0 \\ y(1) &= c_2\sin\sqrt{\lambda} = 0, \quad \sqrt{\lambda} = n\pi, \qquad n = 0, \mp 1, \mp 2, \mp 3, \cdots. \end{aligned}$$

Hence the (nontrivial) solutions to the boundary value problems (E.9)–(E.11) are

$$y_n(x) = c_2\sin n\pi x, \quad \lambda_n = n^2\pi^2, \quad n = 1, 2, \cdots \tag{E.13}$$

which are also the nontrivial solutions (eigenfunctions) of the Fredholm integral equation (E.1) and (E.2). The subject of eigenfunctions and eigenvalues is very important for the development of the solution to Fredholm integral equations of the second kind, especially with regard to the existence of such solutions as we shall discuss and illustrate in Section 5.1.2. The preliminaries of the eigenfunctions as solutions to Sturm-Liouville problem,[5] and their use as the orthogonal functions for the Fourier series expansion is done in Section 4.1.2. Such expansion facilitates the representation of the Green's function, which we shall use in Section 4.2 to reduce a boundary value problem to a Fredholm integral equation, which is (the more direct) equivalent way to what we are doing in this section. For other boundary problems of interest with their eigenfunctions and eigenvalues, see Appendix B.

For now, the set of the nontrivial solutions of the homogeneous problem (E.9)–(E.11) $\{\sin n\pi x\}$ are called the *characteristic functions* or *eigenfunctions* and $\lambda_n = n^2\pi^2$ are the *characteristic values* or *eigenvalues* of the kernel $K(x,t)$ of the (equivalent) homogeneous Fredholm integral equation (E.1) and (E.2).

As mentioned before, we shall see in Chapter 5 that the solution of a nonhomogeneous Fredholm integral equation will, as expected, depend on the solutions of its associated homogeneous equation. The solutions of the homogeneous equation are the classical solutions of the homogeneous case of the boundary value problem (1.33)–(1.35). Example 6 is a special case, and for quick reference we tabulate in Appendix B a number of familiar homogeneous boundary value problems, their Green's functions, and their corresponding homogeneous Fredholm integral equations with the Green's function as their kernel. The verification of these results is the subject of a number of examples and exercises in this chapter and in Chapters 4 and 5.

In this section, our illustrations covered only homogeneous differential equations. The nonhomogeneous differential equations case should follow easily, as was done in part (b) of Example 4 for the initial value problem and its resulting nonhomogeneous Volterra integral equation. (Also, see Example 8 of Section 4.2 and many of the exercises in Sections 4.1 and 4.2.)

Exercises 2.5

1. Reduce the boundary value problem

$$\frac{d^2y}{dx^2} = \lambda y(x), \qquad 0 < x < b$$

$$y(0) = 0,$$

$$y(b) = 0$$

to a Fredholm integral equation.

[5] A very important general boundary value problem, whose solutions, with some regularity conditions, are orthogonal functions.

Hint: See Example 5.

2. Consider the kernel $K(x,t)$ in (E.9) of Example 5, as a function of x, for a fixed value of t.

   (a) Show that $K(x,t)$ is continuous at $x = t$.

   (b) Show that $\partial K/\partial x$ is discontinuous at $x = t$, indeed it has a jump discontinuity of 1, i.e., $\partial K/\partial x\,|_{x>t} - \partial K/\partial x|_{x<t} = 1$ as x increases through t.

   (c) Show that on the two subintervals $a \leq x < t$ and $t < x \leq b$, the kernel $K(x,t)$ satisfies the differential equation $\partial^2 K/\partial x^2 = 0$, also $K(x,t)$ satisfies the boundary conditions by vanishing at the end points $x = a$ and $x = b$.

3. **(a)** Reduce the homogeneous Fredholm integral equation

$$u(x) = \lambda \int_0^1 K(x,t)u(t)dt, \tag{E.1}$$

$$K(x,t) = \begin{cases} \dfrac{\sinh x \sinh(t-1)}{\sinh 1}, & 0 \leq x \leq t \\ \dfrac{\sinh t \sinh(x-1)}{\sinh 1}, & t \leq x \leq 1 \end{cases} \tag{E.2}$$

   to a boundary value problem.

   Hint: See Example 6.

   (b) Solve the resulting boundary value problem to find the characteristic functions (eigenfunctions) and the characteristic values (eigenvalues) of the kernel in (E.2) of part (a).

   (c) Verify that the solutions in part (b) satisfy the integral equation in part (a).

4. **(a)** Verify that the *nonlinear* integral equation in $u(x)$,

$$u(x) = \int_0^1 K(x,\xi)\phi(\xi, u(\xi))d\xi \tag{E.1}$$

   with the kernel

$$K(x,\xi) = \begin{cases} \dfrac{1}{6}\xi^2(1-x)^2(3x-\xi-2x\xi), & 0 \leq \xi \leq x \\ \\ \dfrac{1}{6}x^2(1-\xi)^2(3\xi-x-2x\xi), & x \leq \xi \leq 1 \end{cases} \tag{E.2}$$

   is equivalent to the boundary value problem associated with the displacement $u(x)$ of a vibrating beam.

$$\frac{d^4u}{dx^4} = \phi(x, u(x)), \qquad 0 < x < 1 \tag{E.3}$$

$$u(0) = 0, \qquad u'(0) = 0 \qquad (E.4), (E.5)$$

$$u(1) = 0, \qquad u'(1) = 0 \qquad (E.6), (E.7)$$

Hint: See Exercise 3(b) in Section 1.3.

(b) Attempt to reduce the *nonlinear* boundary value problem (E.3)–(E.7) to the nonlinear Fredholm integral equation (E.1) and (E.2).

Hint: Follow the steps of Example 5 where you do four integrations utilizing the (general) identity (1.52), and using the four boundary conditions in (E.4)–(E.7) for determining the four arbitrary constants of the four integrations.

2.6 MIXED BOUNDARY CONDITIONS: DUAL INTEGRAL EQUATIONS

In this section we illustrate the formulation of a mixed boundary condition, where the function is given on one part of the boundary and its integral equation derivative is given on the other part, which will result in a dual integral equation representation. We consider the problem of an electrified plate in the half plane, where we will use the Fourier transform of Section 1.4.2 to reduce its mixed boundary condition to a condition of solving dual integral equations.

2.6.1 Electrified Infinite Plane

We consider here the potential distribution $u(x, y)$ in the half plane ($x > 0$), due to the presence of a plate of width $2a$ (see Fig. 2.5) placed along the y axis with center at the origin and extending in the z direction. The plate is kept at a potential $u(0, y) = g(y)$, $-a < y < a$ and where the rest of the yz plane is assumed to be insulated (i.e., $\partial u(0, y)/\partial x = 0$, $|y| > a$).

The potential distribution in free space here is independent of z, thus as $u(x, y)$ it is governed by the following *Laplace equation* in two dimensions

$$\frac{\partial^2 u}{\partial x^2} + \frac{\partial^2 u}{\partial y^2} = 0, \qquad x > 0, \qquad -\infty < y < \infty \tag{2.38}$$

and the mixed boundary condition is represented as

$$u(0, y) = g(y), \qquad -a < y < a \tag{2.39}$$

$$\frac{\partial u(0, y)}{\partial x} = 0, \qquad |y| > a. \tag{2.40}$$

Since the domain of the problem in y is $-\infty < y < \infty$, we will use the Fourier exponential transform (1.87), which we discussed in Section 1.4.2. We let

$$U(x, \lambda) = \mathcal{F}\{u(x, y)\} = \int_{-\infty}^{\infty} u(x, y) e^{-i\lambda y} dy \tag{2.41}$$

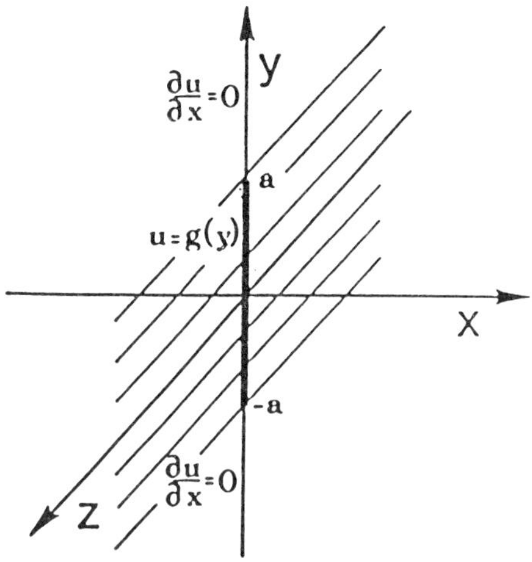

Fig. 2.5 The electrified plate.

and Fourier-transform the Laplace equation (2.38), using the Fourier transform pair in (1.110) on the second partial derivative with respect to y in (2.38), to obtain

$$\frac{d^2U}{dx^2} - \lambda^2 U(x, \lambda) = 0, \tag{2.42}$$

which is a second-order differential equation whose general solution can easily be obtained as

$$U(x, \lambda) = A(\lambda)e^{-|\lambda|x} + B(\lambda)e^{|\lambda|x}. \tag{2.43}$$

Here the inverse Fourier transform (1.88) of $e^{|\lambda|x}$ (as a function of y) does not exist to guarantee a finite solution $u(x, y)$. Hence we must let $B(\lambda) = 0$ in (2.43),

$$U(x, \lambda) = A(\lambda)e^{-|\lambda|x}. \tag{2.44}$$

To find the arbitrary function $A(\lambda)$ in (2.44), we must have a condition on $U(x, \lambda)$ at $x = 0$, which should come naturally from the Fourier transform of the condition on $u(x, y)$ at $x = 0$ through (2.41). But our mixed condition (2.39)–(2.40) at $x = 0$ is not suitable for the required substitution in (2.41), since it is given as a function for $|y| < a$ and as a derivative of the function for $|y| > a$. Instead, we leave $A(\lambda)$ in (2.44) for the moment and we proceed to find $u(x, y)$, the solution of our problem, as the inverse Fourier transform (1.88) of (2.44),

$$\begin{aligned} u(x, y) &= \frac{1}{2\pi}\int_{-\infty}^{\infty} U(x, \lambda)e^{i\lambda y}\, d\lambda \\ &= \frac{1}{2\pi}\int_{-\infty}^{\infty} A(\lambda)e^{-|\lambda|x}e^{i\lambda y}\, d\lambda. \end{aligned} \tag{2.45}$$

Now, we can apply the mixed boundary conditions (2.39) and (2.40) on $u(x, y)$ above to obtain

$$u(0, y) = g(y) = \frac{1}{2\pi} \int_{-\infty}^{\infty} A(\lambda) e^{i\lambda y} d\lambda,$$

$$2\pi g(y) = \int_{-\infty}^{\infty} A(\lambda) e^{i\lambda y} d\lambda, \qquad |y| < a \tag{2.46}$$

and

$$\frac{\partial u}{\partial x}(0, y) = 0 = \frac{1}{2\pi} \int_{-\infty}^{\infty} -|\lambda| A(\lambda) e^{i\lambda y} d\lambda,$$

$$0 = \int_{-\infty}^{\infty} |\lambda| A(\lambda) e^{i\lambda y} d\lambda, \qquad |y| > a. \tag{2.47}$$

(2.46) and (2.47) are dual integral equations in $A(\lambda)$, where they must be solved for $A(\lambda)$ of (2.45) to obtain the solution $u(x, y)$ of the electrified plate problem (2.38)–(2.40).

We remark here that the boundary value problem (2.38)–(2.40) also describes the steady-state temperature distribution $u(x, y)$ in the xy plane, due to a given temperature on the segment $|y| < a$ of the y axis and where the rest of the y axis is completely insulated. Another physical problem that is represented by the boundary value problem above is that of the steady irrotational flow of a perfect fluid through the opening $|y| < a$ of an infinite wall along the y axis.

2.6.2 Electrified Disc

As we have seen in the problem above, the mixed boundary condition (2.39) and (2.40) for the potential distribution in two dimensions $u(x, y)$ was reduced to dual integral equations (2.46) and (2.47). Here we consider the same type of mixed boundary condition for the potential distribution in three-dimensions. This is due to a given potential on a disk of unit radius in the xy plane, with center at the origin and with the rest of the xy plane being completely insulated (Fig. 2.6). This problem also fits the steady-state temperature distribution in three-dimensional space due to a given temperature on the unit disk and where the rest of the xy plane is completely insulated. To describe such boundary conditions, it is important that we use cylindrical coordinates (r, θ, z) for the potential distribution $u(r, \theta, z)$, and hence we write the Laplace equation in cylindrical coordinates,

$$\frac{\partial^2 u}{\partial r^2} + \frac{1}{r}\frac{\partial u}{\partial r} + \frac{1}{r^2}\frac{\partial^2 u}{\partial \theta^2} + \frac{\partial^2 u}{\partial z^2} = 0. \tag{2.48}$$

The special case we consider here is when the potential on the disk is cylindrically symmetric and hence the potential should be independent of the angle $u(r, \theta, z) = u(r, z)$; so the Laplace equation (2.48) in $u(r, z)$ becomes

$$\frac{\partial^2 u}{\partial r^2} + \frac{1}{r}\frac{\partial u}{\partial r} + \frac{\partial^2 u}{\partial z^2} = 0, \quad 0 < r < x, \quad 0 < z < \infty. \tag{2.49}$$

In the very special case of constant potential u_0 on the unit disk with the rest of the xy plane being insulated (see. Fig. 2.6), this mixed boundary condition becomes

$$u(r,0) = u_0, \qquad 0 \le r < 1 \tag{2.50}$$

$$\frac{\partial u}{\partial z}(r,0) = 0, \qquad 1 < r < \infty. \tag{2.51}$$

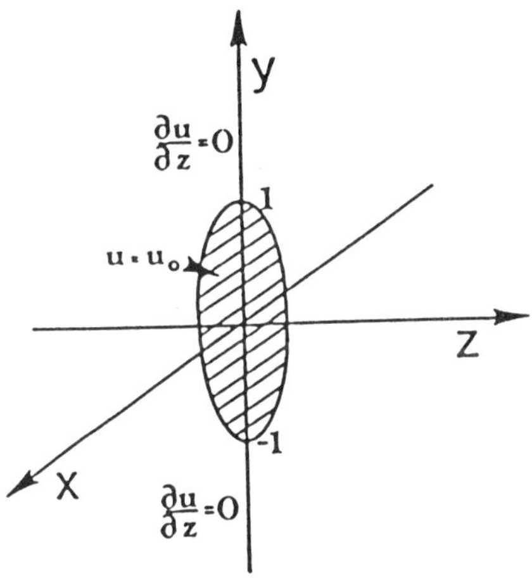

Fig. 2.6 The electrified disc.

We mention again that the boundary value problem (2.49)–(2.51) represents two other physical problems: that of steady-state temperature distribution due to a given constant temperature on the unit disk with the rest of the xy plane completely insulated, and the steady irrotational flow of a perfect fluid through a circular aperture in a rigid wall. In comparison with (2.38), where we used the Fourier transform to algebraize its partial derivative with respect to y, and reduce it to the ordinary differential equation (2.42), equation (2.49) needs the *Hankel transform* to algebraize the differentiation with respect to r. The subject of the Hankel transform was presented very briefly in Section 1.4.3, and its use will be illustrated in Appendix A, where the problem of the electrified disk is discussed and the corresponding dual integral equations are derived as (E.8) and (E.9) in Example 1 (of Appendix A), where a final solution $u(r, z)$ is found in (E.11) of the example.

Exercises 2.6

1. Use the following two integrals involving the Bessel function $J_0(\lambda r)$:

$$\int_0^\infty \frac{\sin\lambda}{\lambda} J_0(\lambda r)d\lambda = \frac{\pi}{2}, \qquad 0 \le r < 1$$

$$\int_0^\infty \sin \lambda J_0(\lambda r) d\lambda = 0, \qquad 1 < r < \infty$$

to solve for $A(\lambda)$ in the following dual-integral equations:

$$u_0 = \int_0^\infty \lambda J_0(\lambda r) A(\lambda) d\lambda, \qquad 0 \le r < 1$$

$$0 = \int_0^\infty -\lambda^2 J_0(\lambda r) A(\lambda) d\lambda, \qquad 1 < r < \infty.$$

Hint: Compare the respective integrals.

2. Fluid flow through an aperture in a rigid plane wall.

   (a) Develop the problem of the steady flow $v(x, y)$ of a perfect fluid in an aperture of width 2a with center at 0 in a rigid plane wall. See the first problem (2.38)–(2.40) of this section.

   (b) Develop the same problem in $v(r, z)$ for a circular operature of radius a.

3. Write the dual integral equations

$$\int_0^\infty \lambda B(\lambda) \sin \lambda x d\lambda = 0, \qquad x > 1 \tag{E.1}$$

$$\int_0^\infty B(\lambda) \sin \lambda x d\lambda = f(x), \quad 0 < x < 1 \tag{E.2}$$

   in $B(\lambda)$ as one (singular) integral equation of the first kind.

2.7 INTEGRAL EQUATIONS IN HIGHER DIMENSIONS

In Section 1.4.2 we indicated that the Fourier transform can be extended to three-dimensions, where we gave such Fourier transforms pair in (1.91) and (1.92) (with the notation used in physics texts). We mentioned there that we shall need such extension to model the Schrödinger (partial differential) equation in the three-dimensional physical space as a Fredholm integral equation in the three-dimensional *momentum space*. Of course, we are also after the Fourier transform's most important operational property, namely, that it algebraizes (linear) derivatives with constant coefficients, which generalizes to the three-dimensional case as we shall see in the derivation of (2.60). Next we present the detailed analysis of using the three-dimensional Fourier transform to give the integral representation of *Schrödinger equation* as a Fredholm integral equation in the three-dimensional momentum space.

2.7.1 Schrödinger Equation as an Integral Equation in the Three-Dimensional Momentum Space

In quantum mechanics the *wave function* $\psi(\vec{r})$, $\vec{r} = \vec{i}x + \vec{j}y + \vec{k}z = (x, y, z)$ is governed by the Schrödinger equation,

$$-\frac{\hbar^2}{2m}\nabla^2\psi(\vec{r}) + U(\vec{r})\psi(\vec{r}) = E\psi(\vec{r}), \tag{2.52}$$

as a partial differential equation with a spatial differential operator, namely, the *Laplacian* ∇^2, where $\nabla^2\psi(\vec{r}) \equiv \frac{\partial^2\psi}{\partial x^2} + \frac{\partial^2\psi}{\partial y^2} + \frac{\partial^2\psi}{\partial z^2}$. Here $m, \hbar, U$, and E are the mass, the Planck constant, the potential energy, and the total energy, respectively. With $a^2 = -\frac{2m}{\hbar^2}E$ and $v(\vec{r}) = -\frac{2m}{\hbar^2}U(\vec{r})$, Schrödinger equation is written as

$$-\nabla^2\psi(\vec{r}) + a^2\psi(\vec{r}) = v(\vec{r})\psi(\vec{r}) \tag{2.53}$$

which is for one body system with interaction potential $v(\vec{r})$. If we consider a *many body system* with interaction potential $v(\vec{r}, \vec{\acute{r}})$ over the whole physical space, Schrödinger equation generalizes to[6]

$$-\nabla^2\psi(\vec{r}) + a^2\psi(\vec{r}) = \int_{-\infty}^{\infty}\int_{-\infty}^{\infty}\int_{-\infty}^{\infty} v(\vec{r}, \vec{\acute{r}})\psi(\vec{\acute{r}})d^3\vec{\acute{r}},$$

$$d^3\vec{\acute{r}} = d\acute{x}d\acute{y}d\acute{z} \tag{2.54}$$

which can be seen as an *integro-(partial) differential equation* in three-dimensions. As we shall see soon, and as mentioned in Section 1.4.2, the Fourier transform in three-dimensions is used to reduce the differentiation operation to an algebraic one, and the above problem (2.54) becomes an integral equation in the Fourier (frequency, or momentum) space. This is usually done because in the momentum space we can easily describe the quantum states, and besides, we have the possibility of a better computational and analytical hold on a variety of such problems when represented as integral equations. The Fourier transform in one dimension was discussed in Section 1.4.2. For the present three-dimensional problem, we need the three-dimensional Fourier transform, which represents a simple extension that we presented in (1.91) and (1.92). Of course, in this section our interest is to merely present clear statements of many examples as integral equations, leaving pursuing the details for the interested reader as we continue covering such details in the rest of the book.

With $\vec{\lambda} = \frac{\vec{p}}{\hbar} = (\lambda_1, \lambda_2, \lambda_3)$ as the wave number (multiplied by 2π), or the momentum in three-dimensions, let $\Psi(\vec{\lambda})$ be the (triple) Fourier transform of the wave function $\psi(\vec{r})$ in the physical space. With such Fourier transformation, the Schrödinger equation of $\psi(\vec{r})$ in the three-dimensional $\vec{r}$-physical space will reduce to

[6]For more details, see Arfken [1970, p. 722].

the following (homogeneous) Fredholm integral equation of $\Psi(\vec{\lambda})$ in the momentum (or wave number) $\vec{\lambda}$-space,

$$(\lambda^2 + a^2)\Psi(\vec{\lambda}) = \int_{-\infty}^{\infty}\int_{-\infty}^{\infty}\int_{-\infty}^{\infty} V(\vec{\lambda}, \vec{\acute{\lambda}})\Psi(\vec{\acute{\lambda}})d^3\vec{\acute{\lambda}}, \tag{2.55}$$

$$\lambda^2 = |\vec{\lambda}|^2 = \lambda_1^2 + \lambda_2^2 + \lambda_3^2$$

where the integration is done over the whole three-dimensional momentum space, $d^3\vec{\acute{\lambda}} = d\acute{\lambda}_1 d\acute{\lambda}_2 d\acute{\lambda}_3$, $a^2 = \frac{-2m}{\hbar^2}E$, and $\lambda^2 = |\vec{\lambda}|^2 = \lambda_1^2 + \lambda_2^2 + \lambda_3^2$. $V(\vec{\lambda}, \vec{\acute{\lambda}})$ is the three-dimensional Fourier transform of the interaction energy $v(\vec{r}, \vec{\acute{r}})$.

Now that we have presented a detailed treatment of the Fourier transform in one dimension in Section 1.4.2, which can be extended easily to higher dimensions, we are in a position to establish the above Fredholm integral equation representation (1.52) of the Schrödinger equation, which is the subject of the following example.

Schrödinger Equation in Momentum Space

In this example we will return to the Schrödinger (partial differential) equation (1.49) in the wave function $\psi(\vec{r})$ of the physical space

$$-\frac{\hbar^2}{2m}\nabla^2\psi(\vec{r}) + U(\vec{r})\psi(\vec{r}) = E\psi(\vec{r}), \tag{2.52}$$

and use the three-dimensional Fourier transform[7] to reduce this partial differential equation to its integral representation as the (homogeneous) Fredholm integral equation (2.55) in the wave function $\Psi(\vec{\lambda})$ of the momentum space

$$(\lambda^2 + a^2)\Psi(\vec{\lambda}) = \int_{-\infty}^{\infty}\int_{-\infty}^{\infty}\int_{-\infty}^{\infty} V(\vec{\lambda}, \vec{\acute{\lambda}})\Psi(\vec{\acute{\lambda}})d^3\vec{\acute{\lambda}}. \tag{2.55}$$

Since the Schrödinger equation is found mostly in physics books, we will adopt the usual notation used there for the Fourier transform and its inverse,

$$\mathcal{F}\{f\} = F(\lambda) = \frac{1}{\sqrt{2\pi}}\int_{-\infty}^{\infty} e^{-i\lambda x} f(x)dx, \tag{2.56}$$

$$\mathcal{F}^{-1}\{F\} = f(x) = \frac{1}{\sqrt{2\pi}}\int_{-\infty}^{\infty} e^{i\lambda x} F(x)dx \tag{2.57}$$

where we notice more symmetry in distributing the multiplicative constant $\frac{1}{\sqrt{2\pi}}$ for the Fourier transform and its inverse. Also to use this with our above notation for the three-dimensional Fourier transform, we use $\vec{\lambda} = \vec{i}\lambda_1 + \vec{j}\lambda_2 + \vec{k}\lambda_3$ for the wave

[7] More on the higher dimensional Fourier transform can be found in Jerri [1992] and Sneddon [1972].

number vector instead of the $\vec{k}$ so as not to confuse the latter with the unit vector $\vec{k}$ in $\vec{r} = \vec{i}x_1 + \vec{j}x_2 + \vec{k}x_3$,

$$\mathcal{F}_{(3)}\{f\} = F(\vec{\lambda}) = \frac{1}{(2\pi)^{\frac{3}{2}}} \int_{-\infty}^{\infty}\int_{-\infty}^{\infty}\int_{-\infty}^{\infty} e^{-i\vec{\lambda}\cdot\vec{r}} \cdots$$

$$\cdots f(\vec{r})dx_1 dx_2 dx_3 \tag{2.58}$$

$$\mathcal{F}_{(3)}^{-1}\{F\} = f(\vec{r}) = \frac{1}{(2\pi)^{\frac{3}{2}}} \int_{-\infty}^{\infty}\int_{-\infty}^{\infty}\int_{-\infty}^{\infty} e^{+i\vec{\lambda}\cdot\vec{r}} \cdots$$

$$\cdots F(\vec{\lambda})d\lambda_1 d\lambda_2 d\lambda_3. \tag{2.59}$$

If we now let $\Psi(\vec{\lambda}) = \mathcal{F}_{(3)}\{\psi(\vec{r})\}$ as the wave function in the momentum $\vec{\lambda}$-space, and apply the three-dimensional Fourier transform to the Schrödinger equation (2.54), we have

$$\frac{1}{(2\pi)^{\frac{3}{2}}} \int_{-\infty}^{\infty}\int_{-\infty}^{\infty}\int_{-\infty}^{\infty} \left[-\frac{\partial^2\psi(\vec{r})}{\partial x^2} - \frac{\partial^2\psi(\vec{r})}{\partial y^2} - \frac{\partial^2\psi(\vec{r})}{\partial z^2} + a^2\psi(\vec{r})\right] e^{-i\vec{\lambda}\cdot\vec{r}}$$

$$dx_1 dx_2 dx_3$$

$$= \frac{1}{(2\pi)^{\frac{3}{2}}} \int_{-\infty}^{\infty} \underbrace{\cdots}_{6} \int_{-\infty}^{\infty} v(\vec{r}, \acute{\vec{r}})\psi(\acute{\vec{r}})e^{-i\vec{\lambda}\cdot\vec{r}} dx_1' dx_2' dx_3' dx_1 dx_2 dx_3.$$

For the left-hand side we employ a simple extension of the Fourier transform important property in (1.110) to three-dimensions to algebraize the three partial differentiation terms to $-(\lambda_1^2 + \lambda_1^2 + \lambda_3^2)\Psi(\vec{\lambda}) = -\lambda^2\Psi(\vec{\lambda})$, and where the fourth term $a^2\psi(\vec{r})$ will simply be transformed to $a^2\Psi(\vec{\lambda})$,

$$(\lambda^2 + a^2)\Psi(\vec{\lambda}) = \frac{1}{(2\pi)^{\frac{3}{2}}} \int_{-\infty}^{\infty} \underbrace{\cdots}_{6} \int_{-\infty}^{\infty} v(\vec{r}, \acute{\vec{r}})\psi(\acute{\vec{r}})e^{-i\vec{\lambda}\cdot\vec{r}} \cdots$$

$$\cdots dx_1' dx_2' dx_3' dx_1 dx_2 dx_3. \tag{2.60}$$

To have an integral equation in $\Psi(\vec{\lambda})$ we must write $\psi(\acute{\vec{r}})$ inside the above integral as an inverse Fourier transform of $\Psi(\acute{\vec{\lambda}})$,

$$(\lambda^2 + a^2)\Psi(\vec{\lambda}) = \frac{1}{(2\pi)^3} \int_{-\infty}^{\infty}\int_{-\infty}^{\infty}\int_{-\infty}^{\infty} \underbrace{\cdots}_{6} \int_{-\infty}^{\infty}\int_{-\infty}^{\infty} \underbrace{\cdots}_{3} \int_{-\infty}^{\infty} v(\vec{r}, \acute{\vec{r}})$$

$$e^{-i\vec{\lambda}\cdot\vec{r} + i\acute{\vec{\lambda}}\cdot\acute{\vec{r}}} \cdot \Psi(\acute{\vec{\lambda}})d\acute{\lambda}_1 d\acute{\lambda}_2 d\acute{\lambda}_3 dx_1' dx_2' dx_3' dx_1 dx_2 dx_3. \tag{2.61}$$

Now we can see the six-dimensional integration over $\vec{r}$ and $\acute{\vec{r}}$ as the three-dimensional Fourier transform $V(\vec{\lambda}, \acute{\vec{\lambda}})$ of the interaction energy $v(\vec{r}, \acute{\vec{r}})$ (as a function of the two spatial vectors $\vec{r}$ and $\acute{\vec{r}}$), and where $V(\vec{\lambda}, \acute{\vec{\lambda}})$ stands for the interaction energy in the momentum space,

$$V(\vec{\lambda}, \vec{\acute{\lambda}}) = \frac{1}{(2\pi)^3} \underbrace{\int_{-\infty}^{\infty} \cdots \int_{-\infty}^{\infty}}_{6} v(\vec{r}, \vec{\acute{r}}) e^{-i\vec{\lambda}\cdot\vec{r} + i\vec{\acute{\lambda}}\cdot\vec{\acute{r}}} dx_1' dx_2' dx_3' dx_1 dx_2 dx_3. \tag{2.62}$$

With this definition of $V(\vec{\lambda}, \vec{\acute{\lambda}})$, the equation (2.61) becomes the desired (homogeneous) Fredholm integral equation in the wave function $\Psi(\vec{\lambda})$ of the momentum space

$$(\lambda^2 + a^2)\Psi(\vec{\lambda}) = \int_{-\infty}^{\infty} \int_{-\infty}^{\infty} \int_{-\infty}^{\infty} V(\vec{\lambda}, \vec{\acute{\lambda}}) \Psi(\vec{\acute{\lambda}}) d\acute{\lambda}_1 d\acute{\lambda}_2 d\acute{\lambda}_3, \tag{2.63}$$

which is what we were set to derive as (2.55).

3

Volterra Integral Equations

In Section 1.2 we presented the Volterra integral equation (1.39) in $u(x)$,

$$h(x)u(x) = f(x) + \int_a^x K(x,\xi)u(\xi)d\xi \tag{1.39}$$

which we termed *the second kind* when $h(x) = 1$,

$$u(x) = f(x) + \int_a^x K(x,\xi)u(\xi)d\xi \tag{1.42}$$

and *the first kind* when $h(x) = 0$,

$$-f(x) = \int_a^x K(x,\xi)u(\xi)d\xi. \tag{1.41}$$

It is interesting to note that Volterra started working on solving such integral equations in 1884, which is before the name integral equations was given to them by du Bois-Reymond in 1888. Volterra's serious study began in 1896.[1]

In Chapter 2 we formulated a number of problems that resulted in Volterra integral equations. This included the human population problem (2.8), the mortality of equipment problem (2.21), and Abel's problem (2.33).

In Example 3 of Section 2.4 we showed how an initial value problem,

$$\frac{d^2y}{dx^2} = \lambda y(x) + g(x) \tag{1.29}$$

[1] Anderson and DeHoog [1980]. See Bôcher [1914] for Volterra's collected work in 1884–1896.

$$y(0) = 1 \tag{1.30}$$

$$y'(0) = 0 \tag{1.31}$$

is reduced to Volterra integral equation of the second kind

$$y(x) = \lambda \int_0^x (x-\xi)y(\xi)d\xi + 1 + \int_0^x (x-\xi)g(\xi)d\xi \tag{1.32}$$

with the nonhomogeneous term

$$f(x) = 1 + \int_0^x (x-\xi)g(\xi)d\xi.$$

In this chapter we present the basic methods and techniques for solving Volterra integral equations, with primary emphasis on illustrating such methods rather than proving them. We may recall that a number of the aforementioned Volterra integral equations had difference kernels and hence were suitable for the Laplace transform method of solution, which we illustrated in a number of examples and exercises in Chapters 1 and 2. The Laplace transform method and other methods are discussed and illustrated in the following section. The basic theorems are stated precisely, and few of them are proved.

3.1 VOLTERRA EQUATIONS OF THE SECOND KIND

3.1.1 Resolvent Kernel Method: Neumann Series

The solution of the Volterra integral equation of the second kind,

$$u(x) = f(x) + \lambda \int_a^x K(x,\xi)u(\xi)d\xi \tag{3.1}$$

may often appear as an integral,

$$u(x) = f(x) + \lambda \int_a^x \Gamma(x,\xi;\lambda)f(\xi)d\xi \tag{3.2}$$

in terms of the given function $f(x)$, where $\Gamma(x,\xi;\lambda)$ is called the *resolvent kernel* of the integral equation (3.1).

When $K(x,\xi)$ and $f(x)$ in (3.1) are both continuous, it is easy to construct the resolvent kernel $\Gamma(x,\xi;\lambda)$ for (3.1) in terms of the following *Neumann series*:[2]

$$\Gamma(x,\xi;\lambda) = \sum_{n=0}^{\infty} \lambda^n K_{n+1}(x,\xi) \tag{3.3}$$

[2]This (successive substitution) method of C. Neumann came about 30 years later after the work of Abel and Liouville in 1823–1839.

where $K_{n+1}(x,\xi)$, the *iterated kernel*, is evaluated as follows:

$$K_{n+1}(x,\xi) = \int_{\xi}^{x} K(x,y)K_n(y,\xi)dy \tag{3.4}$$

and $K_1(x,y) \equiv K(x,y)$.

This is easily shown by assuming the following infinite series form for the solution $u(x)$:

$$u(x) = u_0(x) + \lambda u_1(x) + \lambda^2 u_2(x) + \cdots \tag{3.5}$$

and substituting it in (3.1) to obtain

$$\begin{aligned} &u_0(x) + \lambda u_1(x) + \lambda^2 u_2(x) + \cdots \\ &= f(x) + \lambda \int_a^x K(x,\xi)u_0(\xi)d\xi + \lambda^2 \int_a^x K(x,\xi)u_1(\xi)d\xi + \cdots \end{aligned} \tag{3.6}$$

with the assumption of good enough convergence of the infinite series in (3.5) that allows the exchange of its summation with the integration operation of (3.1) that led to (3.6).

Now we equate the coefficients of each λ of the same power on both with the assumption of good enough convergence of the infinite series in (3.5) that allows the exchange of its summation with the integration operation of (3.1) that led to (3.6), sides of (3.6); for example, the coefficients of λ^0, λ^1, λ^2 in (3.6) are equated to give (3.7), (3.8), and (3.9), respectively

$$u_0(x) = f(x) \tag{3.7}$$

$$u_1(x) = \int_a^x K(x,\xi)u_0(\xi)d\xi \tag{3.8}$$

$$u_2(x) = \int_a^x K(x,\xi)u_1(\xi)d\xi \tag{3.9}$$

$$\vdots$$

$$u_n(x) = \int_a^x K(x,\xi)u_{n-1}(\xi)d\xi. \tag{3.10}$$

So if we substitute $u_0(x) = f(x)$ from (3.7) in (3.8),

$$u_1(x) = \int_a^x K(x,\xi)f(\xi)d\xi \tag{3.11}$$

then use this resulting value of $u_1(x)$ in (3.9),

$$u_2(x) = \int_a^x K(x,\xi)\int_a^{\xi} K(\xi,t)f(t)dtd\xi \tag{3.12}$$

and interchange the integrals in (3.12), keeping in mind the change in the limits of integration as illustrated in Figure 1.8, we obtain

$$\begin{aligned} u_2(x) &= \int_a^x f(t) \left[\int_t^x K(x,\xi)K(\xi,t)d\xi \right] dt \\ u_2(x) &= \int_a^x f(t)K_2(x,t)dt \end{aligned} \tag{3.13a}$$

$$= \int_a^x K_2(x,\xi)f(\xi)d\xi \tag{3.13b}$$

after defining the iterated kernel $K_2(x,\xi)$ according to (3.4).

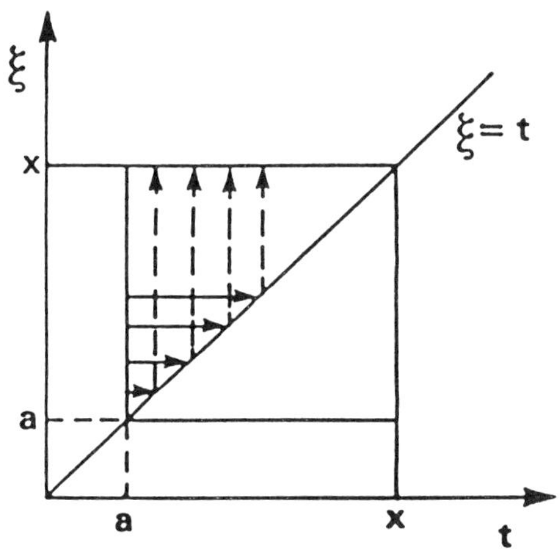

Fig. 1.8 Domains for performing the integration in (1.51) with respect to t first (solid lines) or with respect to ξ first (dashed lines).

Just as the $K(x,t)$ in (3.11) is taken as $K_1(x,t)$ to give $u_1(x)$, the inside integral in (3.13a) defines $K_2(x,t)$, the *iterated kernel*,

$$\begin{aligned} K_2(x,t) &= \int_t^x K(x,\xi)K(\xi,t)d\xi \\ &= \int_t^x K(x,\xi)K_1(\xi,t)d\xi \end{aligned} \tag{3.14}$$

to give $u_2(x)$ in (3.13b), since $K_1(\xi,t) \equiv K(\xi,t)$. In general, following the same steps in (3.11)–(3.14), we can derive the general term for the iterated kernel,

$$K_{n+1}(x,t) = \int_t^x K(x,\xi)K_n(\xi,t)d\xi, \tag{3.15}$$

that gives the general term $u_{n+1}(x)$ of (3.5),

$$u_{n+1}(x) = \int_a^x K_{n+1}(x,\xi)f(\xi)d\xi. \tag{3.16}$$

The final solution is then (formally) obtained from $u(x)$ in (3.5) with $u_0(x) = f(x)$ as in (3.7), $u_1(x)$ in (3.11), $u_2(x)$ in (3.13b) and so on for $u_n(x)$ as in (3.16),

$$\begin{aligned} u(x) &= f(x) + \lambda \int_a^x K_1(x,\xi) f(\xi) d\xi + \lambda^2 \int_a^x K_2(x,\xi) f(\xi) d\xi \\ &\quad + \cdots + \lambda^n \int_a^x K_n(x,\xi) f(\xi) d\xi + \cdots \\ &= f(x) + \lambda \int_a^x [K_1(x,\xi) + \lambda K_2(x,\xi) + \cdots \\ &\quad + \lambda^n K_n(x,\xi) + \cdots] f(\xi) d\xi \\ &= f(x) + \lambda \int_a^x \left[\sum_{n=0}^{\infty} \lambda^n K_{n+1}(x,\xi) \right] f(\xi) d\xi \\ &= f(x) + \lambda \int_a^x \Gamma(x,\xi;\lambda) f(\xi) d\xi \end{aligned} \tag{3.17}$$

which is what was sought in (3.2) and (3.3) to constitute the solution to the Volterra integral equation (3.1) via the present method of the iterated kernels (3.15) for constructing the resolvent kernel $\Gamma(x,\xi;\lambda)$ in (3.3).

Of course, as we mentioned above, these are only the formal steps of the method, which lack the mathematical justification for the convergence of the resolvent kernel infinite series (3.3). This includes proving the general case of the iterated kernel $K_{n+1}(x,t)$ in (3.15). The first part concerning the form of $K_{n+1}(x,t)$ in (3.15) can be accomplished by mathematical induction, which we leave for an exercise. Before offering the major part of a proof of the main result, for $u(x)$ of (3.1) and (3.2) to be the unique solution of (3.1), we shall give a precise statement for it as the following Theorem 1, then illustrate the above method with Example 1. This is followed by a very similar iterated kernels method that we will pursue to prove Theorem 1 in detail. Last we show that $u(x)$ in (3.2) and (3.3) or (3.17) does indeed satisfy the Volterra equation (3.1) to qualify as its solution.

Theorem 1 "The Volterra integral equation of the second kind (3.1) in $u(x)$,

$$u(x) = f(x) + \lambda \int_a^x K(x,\xi) u(\xi) d\xi \tag{3.1}$$

with $f(x)$ integrable on $[a,b]$ and the kernel $K(x,\xi)$ integrable in the triangle, $a \le x \le b$, $a \le \xi \le x$, has *a unique* bounded solution given by $u(x)$ in (3.2) with the resolvent kernel $\Gamma(x,\xi;\lambda)$ represented by the infinite series (3.3), which is convergent for all values of λ."

Example 1 Find the resolvent kernel to solve the Volterra integral equation of the second kind

$$u(x) = f(x) + \lambda \int_0^x e^{x-t} u(t) dt. \tag{E.1}$$

Here we have

$$K_1(x,t) \equiv K(x,t) = e^{x-t}. \tag{E.2}$$

So if we use $K(x,\xi) = e^{x-\xi}$ and $K_1(\xi,t) = K(\xi,t) = e^{\xi-t}$ in (3.4), we obtain

$$\begin{aligned} K_2(x,t) &= \int_t^x K(x,\xi)K_1(\xi,t)d\xi = \int_t^x e^{x-\xi}e^{\xi-t}d\xi = \int_t^x e^{x-t}d\xi \\ &= e^{x-t}\int_t^x d\xi = (x-t)e^{x-t}. \end{aligned} \tag{E.3}$$

Similarly, from (3.4) with $n = 2$, we have

$$\begin{aligned} K_3(x,t) &= \int_t^x K(x,\xi)K_2(\xi,t)d\xi \\ &= \int_t^x e^{x-\xi}(\xi-t)e^{\xi-t}d\xi = \int_t^x (\xi-t)e^{x-t}d\xi \\ &= e^{x-t}\int_t^x (\xi-t)d\xi = e^{x-t}\left(\frac{\xi^2}{2} - t\xi\right)\Big|_t^x \end{aligned} \tag{E.4}$$

$$= e^{x-t}\left[\frac{x^2-t^2}{2} - t(x-t)\right] = \frac{(x-t)^2}{2}e^{x-t}.$$

These calculations can be continued to find that

$$K_{n+1}(x,t) = \frac{(x-t)^n}{n!}e^{x-t}. \tag{E.5}$$

Hence from (3.3) and (E.5) the resolvent kernel for (E.1) is

$$\begin{aligned} \Gamma(x,t;\lambda) &= K_1(x,t) + \lambda K_2(x,t) + \lambda^2 K_3(x,t) + \cdots + \lambda^n K_{n+1}(x,t) + \cdots \\ &= e^{x-t} + \lambda(x-t)e^{x-t} + \lambda^2\frac{(x-t)^2}{2}e^{x-t} + \cdots + \lambda^n\frac{(x-t)^n}{n!}e^{x-t} \\ &\quad + \cdots \\ &= e^{x-t}\left[1 + \lambda(x-t) + \lambda^2\frac{(x-t)^2}{2!} + \cdots + \lambda^n\frac{(x-t)^n}{n!} + \cdots\right] \\ &= e^{x-t}e^{\lambda(x-t)} = e^{(1+\lambda)(x-t)} \end{aligned} \tag{E.6}$$

after realizing that the series in brackets is the Maclaurin series of $e^{\lambda(x-t)}$.

So from (3.2) and the resolvent kernel (E.6), the solution to (E.1) is

$$u(x) = f(x) + \lambda\int_0^x e^{(1+\lambda)(x-t)}f(t)dt. \tag{E.7}$$

It is not often that the series representation of $\Gamma(x,t;\lambda)$ will converge to an expression in closed form [such as $e^{(1+\lambda)(x-t)}$ of (E.6)]. We presented such a special case to illustrate the basic method. In practice, we may have to evaluate numerically a finite number of terms of the Neumann series (3.3), which gives only an approximation of the resolvent kernel $\Gamma(x,t;\lambda)$.

An Iterative Approach

Another very similar method that depends on the above iterated kernels (3.4) and generates the same resolvent kernel (3.3) is discussed next. The advantage seems to be its transparency for the need of proving its convergence and that it is an *iterative* process from the first step. If we look at the left side $u(x)$ of (3.1) as the output of the integral equation while $u(\xi)$ inside the integral as the input, the method would take the whole right-hand side of (3.1) with its two terms, which is the output, and use it as an input inside the integral, i.e., it starts as an iterative process to give

$$\begin{aligned}
u(x) &= f(x) + \lambda \int_a^x K(x,\xi)\left[f(\xi) + \lambda\int_a^\xi K(\xi,t)u(t)dt\right]d\xi \\
&= f(x) + \lambda\int_a^x K_1(x,\xi)f(\xi)d\xi + \lambda^2\int_a^x\int_a^\xi K(x,\xi)\cdot \\
&\quad K(\xi,t)u(t)dtd\xi \\
&= f(x) + \lambda\int_a^x K_1(x,\xi)f(\xi)d\xi + \lambda^2\int_a^x K_2(x,t)u(t)dt
\end{aligned} \tag{3.18}$$

after using the definition of the iterated kernel $K_2(x,t)$ as it was generated in (3.4). The difference between this method and the above former one, associated with the infinite series (3.5) as a starting point, is that in (3.18) we still see clearly the unknown function $u(t)$ involved in the integral of the third term in (3.18). If we input this $u(x)$, of the right-hand side of (3.18), inside the integral of (3.1) again, we easily obtain

$$\begin{aligned}
u(x) = f(x) \quad &+\lambda\int_a^x K_1(x,t)f(t)dt + \lambda^2\int_a^x K_2(x,t)f(t)dt \\
&+\lambda^3\int_a^x K_3(x,t)u(t)dt,
\end{aligned} \tag{3.19}$$

remembering again the definition of the iterated kernels $K_2(x,t)$ and $K_3(x,t)$ of (3.4). If this process is repeated n times we have

$$u(x) = f(x) + \lambda\int_a^x K_1(x,t)f(t)dt + \lambda^2\int_a^x K_2(x,t)f(t)dt$$

$$+\cdots+\lambda^n\int_a^x K_n(x,t)f(t)dt + \lambda^{n+1}\int_a^x K_{n+1}(x,t)u(t)dt \tag{3.20}$$

where the iterated kernel $K_{n+1}(x,t)$ is defined in the same way as in (3.4), and where we can still see the unknown function $u(t)$ involved inside the above last integral of (3.20), which clearly hinders this expression from being a solution to (3.1). The way to show (3.20) becoming a solution is to show that such a series converges as $n\to\infty$ (see Exercise 10 for the detailed steps of the proof). This, of course, means that the $n+1$st term, as the integral involving $u(t)$, will vanish as an (obvious) necessary condition of the convergence of the series. Hence we have the solution

$$\begin{aligned} u(x) &= f(x) + \lambda \int_a^x \left[\sum_{n=0}^{\infty} \lambda^n K_{n+1}(x,t) \right] f(t)dt \\ &= f(x) + \lambda \int_a^x \Gamma(x,t;\lambda) f(t)dt, \end{aligned} \tag{3.21}$$

which is the same as (3.17) (or (3.2)) of the first method (that started with the Neumann series (3.5)).

We will prove here that the resulting function $u(x)$, as represented by the convergent series (3.21) with its resolvent kernel, does indeed satisfy the Volterra equation (3.1) to qualify as its solution. We do this by substituting the expression for $u(x)$ of (3.21) in the integral on the right hand side of (3.1),

$$u(x) = f(x) + \lambda \int_a^x K(x,\xi) u(\xi) d\xi \tag{3.1}$$

to have

$$f(x) + \lambda \int_a^x K(x,\xi) \left[f(\xi) + \lambda \int_a^\xi \Gamma(\xi,t;\lambda) f(t)dt \right] d\xi$$

$$= f(x) + \lambda \int_a^x K(x,\xi) f(\xi) d\xi + \lambda^2 \int_a^x \int_a^\xi K(x,\xi)\Gamma(\xi,t;\lambda) f(t) dt d\xi. \tag{3.22}$$

Now, this result must be shown to reduce to the same expression of $u(x)$ in (3.21). If we interchange the two integrals in the above double integral term [as was done in (3.12) to (3.13a)], we have

$$\begin{aligned} &\lambda^2 \int_a^x \int_a^\xi K(x,\xi)\Gamma(\xi,t;\lambda) f(t) dt d\xi \\ &= \lambda^2 \int_a^x f(t) \left[\int_t^x K(x,\xi)\Gamma(\xi,t;\lambda) d\xi \right] dt \\ &= \lambda^2 \int_a^x f(t) \left[\int_t^x K(x,\xi) \sum_{n=0}^{\infty} \lambda^n K_{n+1}(\xi,t) d\xi \right] dt \\ &= \lambda \int_a^x f(t) \left[\sum_{n=0}^{\infty} \lambda^{n+1} \int_t^x K(x,\xi) K_{n+1}(\xi,t) d\xi \right] dt \\ &= \lambda \int_a^x f(t) \left[\sum_{n=0}^{\infty} \lambda^{n+1} K_{n+2}(x,t) \right] dt \\ &= \lambda \int_a^x \left[\sum_{m=1}^{\infty} \lambda^m K_{m+1}(x,\xi) \right] f(\xi) d\xi \end{aligned} \tag{3.23}$$

after using the property of the iterated integral as in (3.12)–(3.13a) to arrive at $K_{n+2}(x,t)$ above, then substituting for the dummy variable of integration by $t = \xi$ and that of the summation by $m = n + 1$. If we substitute this result (3.23) for the double integral term in (3.22) we have

$$\begin{aligned} & f(x) + \lambda \int_a^x K(x,\xi) f(\xi) d\xi + \lambda \int_a^x \left[\sum_{m=1}^{\infty} \lambda^m K_{m+1}(x,\xi) \right] f(\xi) d\xi \\ &= f(x) + \lambda \int_a^x \left[K(x,\xi) + \sum_{m=1}^{\infty} \lambda^m K_{m+1}(x,\xi) \right] f(\xi) d\xi \\ &= f(x) + \lambda \int_a^x \left[\sum_{m=0}^{\infty} \lambda^m K_{m+1}(x,\xi) \right] f(\xi) d\xi \\ &= f(x) + \lambda \int_a^x \Gamma(x,\xi;\lambda) f(\xi) d\xi \end{aligned} \tag{3.21}$$

which is the same expression of (3.21) that we started with as a (nominated) solution inside the integral on the right-hand side of the Volterra equation of the second kind (3.1).

3.1.2 Method of Successive Approximations (Iterations)

Another very well known method of solving the Volterra integral equation of the second kind (3.1) is to start with substituting a zeroth approximation $u_0(x)$ in the integral [of (3.1) with $\lambda = 1$] to obtain a first approximation $u_1(x)$,

$$u_1(x) = f(x) + \int_0^x K(x,t) u_0(t) dt. \tag{3.24}$$

Then this $u_1(x)$ is substituted again for $u(x)$ in the integral of (3.1) to obtain a second approximation $u_2(x)$,

$$u_2(x) = f(x) + \int_0^x K(x,t) u_1(t) dt.$$

This process can be continued to obtain the nth approximation,

$$u_n(x) = f(x) + \int_0^x K(x,t) u_{n-1}(t) dt. \tag{3.25}$$

Then we have to determine whether $u_n(x)$ approaches the solution $u(x)$ as n increases.

This *successive approximation* (or *iterative*) method applies just as well to Fredholm integral equations of the second kind. It is even tried for the nonlinear Volterra and Fredholm equations of the second kind.

It turns out that if $f(x)$ is continuous for $0 \le x \le a$ and if $K(x,t)$ is also continuous for $0 \le x \le a$ and $0 \le t \le x$, then it can be proved that the sequence

$u_n(x)$ in (3.25) will converge to the solution $u(x)$ of (3.1). We state this result as the following Theorem 2, which we shall follow immediately with a detailed illustrative Example 2.

Theorem 2 "If for the Volterra integral equation of the second kind (3.1) we have $f(x)$ continuous for $0 \le x \le a$ and $K(x,t)$ is also continuous in the triangle $0 \le x \le a, 0 \le t \le x$, then the successive approximations sequence $u_n(x)$ in (3.25) converges to the solution $u(x)$ of 3.1."

Example 2 Successive Approximations. Use the method of successive approximations to solve the Volterra integral equation of the second kind,

$$u(x) = x - \int_0^x (x-t)u(t)dt. \tag{E.1}$$

We first note that the above Theorem 2 applies to this problem, since $f(x) = x$ is continuous for any x, and the kernel $K(x,t) = -(x-t)$ is also continuous in x and t for all their values.

We may remark here that there is always an advantage in making a reasonable zeroth approximation, a matter that becomes clearer after solving a number of problems. In this case we may start with $u_0(t) = 0$ in the integral of (E.1) to obtain $u_1(x)$ according to (3.25),

$$u_1(x) = x - 0, \tag{E.2}$$

so if we substitute this $u(t) = u_1(t) = t$ inside the integral of (E.1), we obtain

$$\begin{aligned} u_2(x) &= x - \int_0^x (x-t)u_1(t)dt \\ &= x - \int_0^x (x-t)t\,dt = x - \left(\frac{xt^2}{2} - \frac{t^3}{3}\right)\Bigg|_0^x \\ &= x - \frac{x^3}{3!}. \end{aligned} \tag{E.3}$$

Now

$$\begin{aligned} u_3(x) &= x \int_0^x (x-t)u_2(t)dt = x - \int_0^x (x-t)\left(t - \frac{t^3}{6}\right)dt \\ &= x - x\left(\frac{t^2}{2} - \frac{t^4}{24}\right)\Bigg|_0^x + \left(\frac{t^3}{3} - \frac{t^5}{30}\right)\Bigg|_0^x \\ &= x - x\left(\frac{x^2}{2} - \frac{x^4}{24}\right) + \left(\frac{x^3}{3} - \frac{x^5}{30}\right) = x - \frac{x^3}{2} + \frac{x^5}{24} + \frac{x^3}{3} - \frac{x^5}{30} \\ &= x - \frac{x^3}{6} + \frac{x^5}{120} = x - \frac{x^3}{3!} + \frac{x^5}{5!}. \end{aligned} \tag{E.4}$$

From (E.2), (E.3), and (E.4) it looks clear now that if we continue this process, we obtain the $n+1$st approximation $u_{n+1}(x)$ as

$$u_{n+1}(x) = x - \frac{x^3}{3!} + \frac{x^5}{5!} - \cdots + (-1)^n \frac{x^{2n+1}}{(2n+1)!}, n = 0, 1, 2, \cdots, n, \qquad (E.5)$$

which is obviously the nth partial sum of the Maclaurin series of $\sin x$,

$$\sin x = \sum_{n=0}^{\infty} (-1)^n \frac{x^{2n+1}}{(2n+1)!}. \qquad (E.6)$$

Hence the solution to (E.1) is

$$u(x) = \lim_{n \to \infty} u_{n+1}(x) = \sin x. \qquad (E.7)$$

As in Example 1, we remark again that it is not very often that a general sequence $u_n(x)$ will converge to such a simple function as $\sin x$, and we may have to resort to approximate numerical methods for evaluating the resulting partial sum (E.5) or in general its integral representation (3.25). We leave it as an exercise to verify that $\sin x$ is a solution to (E.1) by performing the direct integration (Exercise 8).

3.1.3 Laplace Transform Method: Difference Kernel

When the kernel $K(x, \xi)$ depends only on the difference $x - \xi$, it is termed the *difference kernel*, $K(x, \xi) = K(x - \xi)$. The following Volterra equation of the second kind with difference kernel $K(x - \xi)$,

$$u(x) = f(x) + \lambda \int_0^x K(x - \xi) u(\xi) d\xi \qquad (3.26)$$

now has an integral in the form of the Laplace *convolution product* (1.85)

$$\int_0^x K(x - \xi) u(\xi) d\xi = K * u.$$

Hence, as we presented and illustrated in Section 1.4.1, the Volterra equation with difference kernel (3.26) lends itself to the Laplace method of solution. So if we Laplace-transform (3.26), letting $U(s)$, $F(s)$, and $\mathcal{K}(s)$ be the Laplace transform of $u(x)$, $f(x)$, and $K(x)$, respectively, and realize from the convolution theorem (1.84) that $\mathcal{L}\{K * u\} = \mathcal{K}U$, we obtain

$$U(s) = F(s) + \lambda \mathcal{K}(s) U(s) \qquad (3.27)$$

$$U(s) = \frac{F(s)}{1 - \lambda \mathcal{K}(s)}, \qquad \lambda \mathcal{K}(s) \neq 1. \qquad (3.28)$$

The solution $u(x)$ of (3.26) is then the inverse Laplace transform of $U(s)$ in (3.28),

$$u(x) = \mathcal{L}^{-1} \left\{ \frac{F(s)}{1 - \lambda \mathcal{K}(s)} \right\}, \qquad \lambda \mathcal{K}(s) \neq 1 \qquad (3.29)$$

which can be evaluated with the aid of Laplace transform pairs (Table 1.1), as we illustrate in the following example.

Example 3 Use the Laplace transform to solve the problem of Example 1,

$$u(x) = f(x) + \lambda \int_0^x e^{x-t} u(t)dt. \qquad (E.1)$$

The kernel $K(x, t) = e^{x-t}$ is a difference kernel; if we Laplace-transform (E.1) using $\mathcal{K}(s) = \mathcal{L}\{K(x)\} = \mathcal{L}\{e^x\} = 1/(s-1)$ in (3.28), we obtain

$$\begin{aligned} U(s) &= F(s) + \lambda \cdot \frac{1}{s-1} U(s), \\ U(s) &= \frac{F(s)}{1 - [\lambda/(s-1)]} = F(s)\frac{s-1}{s-1-\lambda} \\ &= F(s)\frac{s-1-\lambda+\lambda}{s-1-\lambda} = F(s) + \lambda\frac{F(s)}{s-1-\lambda} \\ &= F(s) + \lambda\frac{F(s)}{s-(\lambda+1)}. \end{aligned} \qquad (E.2)$$

So the solution $u(x)$ of (E.1) is the inverse Laplace transform of (E.2),

$$\begin{aligned} u(x) &= \mathcal{L}^{-1}\left\{F(s) + \lambda\frac{F(s)}{s-(\lambda+1)}\right\} \\ &= \mathcal{L}^{-1}\{F(s)\} + \lambda\mathcal{L}^{-1}\left\{\frac{1}{s-(\lambda+1)}F(s)\right\} \\ &= f(x) + \lambda\mathcal{L}^{-1}\left\{\frac{1}{s-(\lambda+1)}F(s)\right\}. \end{aligned} \qquad (E.3)$$

If we use the convolution theorem (1.84) on the last term in (E.3) with $\mathcal{L}\{e^{(\lambda+1)x}\} = 1/[s-(\lambda+1)]$ from (1.80) (or the second entry in Table 1.1) we obtain

$$\begin{aligned} u(x) &= f(x) + \lambda e^{(\lambda+1)x} * f(x) \\ &= f(x) + \lambda \int_0^x e^{(\lambda+1)(x-t)} f(t)dt \end{aligned} \qquad (E.4)$$

which is the same result (E.7) of Example 1, where the solution was obtained by using the resolvent kernel-Neumann series method.

Example 4 Use the Laplace transform to solve the problem of Example 2,

$$u(x) = x - \int_0^x (x-t)u(t)dt. \qquad (E.1)$$

The kernel $K(x, t) = x - t$ is a difference kernel; we use Laplace transform on (E.1) to obtain

$$U(s) = \frac{1}{s^2} - \frac{1}{s^2}U(s) \qquad (E.2)$$

[using the convolution theorem (1.84) and $\mathcal{L}\{x\} = 1/s^2$ from (1.79) (or the first entry in Table 1.1) with $\nu = 1$]. From (E.2) we have

$$U(s) = \frac{1/s^2}{1 + 1/s^2} = \frac{1}{s^2 + 1} \qquad (E.3)$$

and if we use the Laplace transform pair $\mathcal{L}\{\sin ax\} = \frac{a}{s^2 + a^2}$ of (1.81), we obtain

$$u(x) = \mathcal{L}^{-1}\left\{\frac{1}{s^2 + 1}\right\} = \sin x \qquad (E.4)$$

which is the solution (E.6) of Example 2.

Integro-differential equations

When the function $u(x)$ in an equation is involved in a differentiation as well as an integration operations, the equation is called an *integro-differential equation.* For example,

$$\frac{d^2u}{dx^2} = f(x) + \int_0^x K(x,t)u(t)dt. \qquad (3.30)$$

We note here that when $K(x,t)$ is a difference kernel, the right-hand side of (3.30) is amenable to the Laplace transformation, while the second derivative on the left-hand side, according to (1.69), needs the initial conditions $u(0)$ and $u'(0)$ for its Laplace transformation.

Example 5 Integro-differential Equation. Solve the following initial value problem associated with the integro-differential equation

$$\frac{d^2u}{dx^2} = e^{2x} - \int_0^x e^{2(x-t)}\frac{du}{dt}dt \qquad (E.1)$$

$$u(0) = 0 \qquad (E.2)$$

$$u'(0) = 0. \qquad (E.3)$$

We apply the Laplace transform on (E.1) using (1.69)

$$\mathcal{L}\left\{\frac{d^2u}{dx^2}\right\} = s^2U(s) - su(0) - u'(0) \qquad (E.4)$$

on the left side, the convolution theorem (1.84) for the integral, (1.73) for the term e^{2x}, and (1.68)

$$\mathcal{L}\left\{\frac{du}{dx}\right\} = sU(s) - u(0) \qquad (E.5)$$

for the derivative inside the integral, to obtain

$$s^2U(s) - su(0) - u'(0) = \frac{1}{s-2} - \frac{1}{s-2}[sU(s) - u(0)]. \qquad (E.6)$$

Now we use the initial conditions (E.2) and (E.3) in (E.6) to have

$$\begin{aligned} s^2U(s) &= \frac{1}{s-2} - \frac{sU(s)}{s-2} \\ U(s) &= \frac{1}{s(s-1)^2} = \frac{1}{(s-1)^2} - \frac{1}{s-1} + \frac{1}{s} \end{aligned} \tag{E.7}$$

after using partial fractions for the last line. Hence the solution to (E.1)–(E.3) is the inverse Laplace transform of (E.7),

$$\begin{aligned} u(x) &= \mathcal{L}^{-1}\left\{\frac{1}{(s-1)^2} - \frac{1}{s-1} + \frac{1}{s}\right\} \\ &= xe^x - e^x + 1 \end{aligned}$$

where for the first term we used (1.80) [or (1.71)].

Exercises 3.1

1. Find the resolvent kernel $\Gamma(x, t; \lambda)$ to solve the Volterra equation of the second kind
$$u(x) = f(x) + \lambda \int_0^x u(t)dt.$$
Hint: The kernel here is $K(x, t) \equiv 1$.

2. Find the resolvent kernel to solve the Volterra integral equation of the second kind
$$u(x) = g(x) + \lambda^2 \int_0^x (x-t)u(t)dt.$$

3. Use the Laplace transform method to solve Exercises 1 and 2. *Hint:* For Exercise 1 note that $\int_0^x u(t)dt$ is (a very special case) of the Laplace convolution product type with $f_1(t) = 1$, $f_2(t) = u(t)$; you can also use the pair $\mathcal{L}\{\int_0^x f(x)\} = \frac{1}{s}F(s)$ in the fourteenth entry of Table 1.1. For Exercise 2, note that $\sinh x = -i \sin ix$, i.e., $\mathcal{L}\{\sinh \lambda x\} = \dfrac{\lambda}{s^2 - \lambda^2}$.

4. Use the method of successive approximations [let $u_0(x) \equiv 0$] to solve Exercise 1 for

   (a) $f(x) = 1$, $\lambda = 1$

   (b) $f(x) = x$, $\lambda = 1$

5. Use the method of successive approximations to solve Exercise 2 for $g(x) = 1$ and $\lambda^2 = 1$. Let $u_0(x) \equiv 1$.

6. Use Laplace transform to solve the *integro-differential equation* in $u(x)$ with the given initial conditions

$$\frac{d^2u}{dx^2} - 2\frac{du}{dx} + u(x) = \cos x - 2\int_0^x \cos(x-t)\frac{d^2u}{dt^2}dt - 2\int_0^x \sin(x-t)\frac{du}{dt}dt,$$

$$u(0) = 0, \qquad u'(0) = 0.$$

7. Verify your answers for Exercises 4, 5, and 6.

8. Verify that $u(x) = \sin x$ is a solution of the Volterra integral equation (E.1) of Example 2.

9. Solve the following Volterra integral equation in $u(x)$ by reducing it to a first-order differential equation in $F(x) = \int_0^x \xi u(\xi)d\xi$,

$$u(x) = x + \int_0^x x\xi u(\xi)d\xi. \tag{E.1}$$

Hint: Note that $u(x) = x + xF(x)$,

$$\frac{dF}{dx} = \frac{d}{dx}\int_0^x \xi u(\xi)d\xi = xu(x) = x[x + xF(x)]. \tag{E.2}$$

Solve (E.2) for $F(x)$, and for the constant of integration substitute in the original integral equation (E.1).

10. **(a)** Prove that the last term in (3.20) vanishes as $n \to \infty$ by showing that the infinite series (3.21) converges absolutely and uniformly. Do the proof by justifying the following steps (i) to (iii) with detailed hints:

(i) Show that

$$|K_{n+1}(x,t)| < M^{n+1}\frac{|x-t|^n}{n!} \tag{E.1}$$

where M is the upper bound of $|K(x,t)|$.

Hint: See

$$\begin{aligned} |K_2(x,t)| &= \left|\int_t^x K(x,\xi)K(\xi,t)dt\right| \\ &\le \int_t^x |K(x,\xi)|\,|K(\xi,t)|\,dt \le M^2|x-t| \end{aligned}$$

and

$$\begin{aligned} |K_3(x,t)| &= \int_t^x K(x,\xi)K_2(\xi,t)dt| \le \int_t^x |K(x,\xi)|\,|K_2(\xi,t)|\,d\xi \\ &\le M\int_t^x M^2|\xi-t|d\xi \le M^3\frac{|x-t|^2}{2}. \end{aligned}$$

(ii) With the result (E.1), show that the $n+1$st term in (3.20) is bounded as follows

$$\left|\lambda^{n+1}\int_a^x K_{n+1}(x,t)u(t)dt\right| < m\frac{[M|\lambda||x-a|]^{n+1}}{(n+1)!} \tag{E.2}$$

where m is the upper bound of $|u(t)|$.

(iii) With the result (E.2), show that the sequence in (3.20) converges, and as such its $n+1$st term tends to zero as $n\to\infty$.

Hint: The $n+1$st term is dominated by an nth term of the following series

$$e^{M|\lambda(x-a)|} = \sum_{n=0}^{\infty}\frac{[M|\lambda||x-a|]^n}{n!}$$

which is absolutely and uniformly convergent for all $|\lambda(x-a)|$, as it is clear from a simple ratio test, remembering the presence of the $n!$ in the denominator of the above sequence.

(b) What is the essential property in this method that helped the most in easing the proof of the convergence.

11. Write a Volterra integral equation of the second kind in the function of the two variables $u(x,y)$ in analogy to that of (3.1) in $u(x)$.

    Hint: See (1.38) for a parallel, and note that it is a Fredholm integral equation in two dimensions.

3.2 VOLTERRA INTEGRAL EQUATIONS OF THE FIRST KIND

We should mention at the outset that integral equations of the first kind present their own difficulties as we shall allude to toward the end of this section. In Section 5.4 we will discuss with some details the topic of Fredholm integral equations of the first kind.

For the special case of a Volterra equation of the first kind,

$$f(x) = \lambda\int_0^x K(x,t)u(t)dt \tag{3.31}$$

with kernel $K(x,t)$ such that $K(x,x)\neq 0$ (and is differentiable with respect to x), we will show next that it can be reduced to Volterra equation of the second kind (whose solution, in general, is much more tractable!) If we differentiate both sides of (3.31) with respect to x using the Leibnitz rule (1.53) on the integral, we have

$$\frac{df}{dx} = \lambda\int_0^x \frac{\partial K(x,t)}{\partial x}u(t)dt + \lambda K(x,x)u(x). \tag{3.32}$$

This can easily be rewritten as

$$u(x) = \frac{1}{\lambda K(x,x)} \frac{df}{dx} - \int_0^x \frac{1}{K(x,x)} \frac{\partial K(x,t)}{\partial x} u(t)dt, \quad K(x,x) \neq 0 \tag{3.33}$$

as a Volterra integral equation of the second kind with the nonhomogeneous term

$$g(x) = \frac{1}{\lambda K(x,x)} \frac{df}{dx}$$

and the (new) kernel

$$H(x,t) = \frac{-1}{K(x,x)} \frac{\partial K(x,t)}{\partial x}.$$

Thus

$$u(x) = g(x) + \int_0^x H(x,t)u(t)dt. \tag{3.34}$$

So when $K(x,x) \neq 0$ in (3.31), we can reduce it to a Volterra equation of the second kind and solve it using one of the methods that we discussed in Section 3.1.

Example 6 Solve the following Volterra equation of the first kind after reducing it to a Volterra equation of the second kind:

$$\sin x = \int_0^x e^{x-t} u(t)dt. \tag{E.1}$$

The kernel $K(x,t) = e^{x-t}$ does not vanish when $x = t$ [i.e., $K(x,x) = e^0 = 1$] and hence according to (3.33) with $f(x) = \sin x$, $\lambda = 1$, we have $H(x,t) = -\frac{\partial}{\partial x}(e^{x-t}) = -e^{x-t}$,

$$u(x) = \cos x - \int_0^x e^{x-t} u(t)dt. \tag{E.2}$$

This is a Volterra equation of the second kind which happens to be a special case of the problem (E.1) in Example 1 with $f(x) = \cos x$ and $\lambda = -1$. According to the result (E.7) of Example 1 or (E.4) of Example 3,

$$u(x) = f(x) + \lambda \int_0^x e^{(1+\lambda)(x-t)} f(t)dt \tag{E.3}$$

we have

$$\begin{aligned} u(x) &= \cos x - \int_0^x \cos t dt \\ &= \cos x - [\sin t]_0^x = \cos x - \sin x. \end{aligned} \tag{E.4}$$

When $K(x,x) \equiv 0$, (3.32) is still a Volterra equation of the first kind. However, if $\frac{\partial K}{\partial x}(x,x) \neq 0$ in (3.32), differentiating again will result in a Volterra equation of the second kind. If these attempts fail, the methods of solution become more involved, which we shall allude to briefly toward the end of this section. An exception to this is the special case when (3.32) has a *difference kernel* and hence the method of Laplace transform can be employed, as we shall discuss and illustrate next.

3.2.1 Volterra Integral Equation of the First Kind with a Difference Kernel—Laplace Transform Method

When the Volterra integral equation of the first kind (3.31) has a difference kernel $K(x,t) = K(x-t)$,

$$f(x) = \lambda \int_0^x K(x-t)u(t)dt \tag{3.35}$$

we apply the Laplace transform on (3.35) as we did for (3.26), to obtain

$$F(s) = \lambda K(s)U(s),$$

$$U(s) = \frac{1}{\lambda}\frac{F(s)}{K(s)} \tag{3.36}$$

and the solution to (3.35) is

$$u(x) = \frac{1}{\lambda}\mathcal{L}^{-1}\left\{\frac{F(s)}{K(s)}\right\}. \tag{3.37}$$

Example 7 Volterra Equation of First Kind with a Difference Kernel. Solve the integral equation

$$\sin x = \lambda \int_0^x e^{x-t}u(t)dt \tag{E.1}$$

by using Laplace transform.

This Volterra equation of the first kind (E.1) is with difference kernel $K(x,t) = e^{x-t}$; if we Laplace-transform it, recalling the (Laplace) convolution theorem in (1.84) for the integral of (E.1), and using the Laplace transform pairs in (1.73) and (1.81),

$$\mathcal{L}\{K(x)\} = \mathcal{L}\{e^x\} = \frac{1}{s-1}, \qquad \mathcal{L}\{\sin x\} = \frac{1}{s^2+1}$$

for (E.1) we have

$$\frac{1}{s^2+1} = \lambda\frac{1}{s-1}U(s)$$

$$U(s) = \frac{1}{\lambda}\frac{1/(s^2+1)}{1/(s-1)} = \frac{1}{\lambda}\frac{s-1}{s^2+1} = \frac{1}{\lambda}\left(\frac{s}{s^2+1} - \frac{1}{s^2+1}\right). \tag{E.2}$$

To obtain the solution $u(x)$ of (E.1), we find the inverse Laplace transform of (E.2),

$$u(x) = \frac{1}{\lambda}\mathcal{L}^{-1}\left\{\frac{s}{s^2+1} - \frac{1}{s^2+1}\right\} \tag{E.3}$$

and with the use of the two Laplace transform pairs in (1.82) and (1.81) we obtain

$$u(x) = \frac{1}{\lambda}(\cos x - \sin x) \tag{E.4}$$

which is the result of Example 6 when $\lambda = 1$.

A Main Difficulty of the Integral Equations of the First Kind

The above Example 7 seems to go very well for a solution via the Laplace transformation. However if we change the problem a little, we will uncover the first main difficulty with solving integral equations of the first kind. This is described in that we are not sure that there is a solution (input) $u(x)$ for (3.31) that corresponds to any given function (output) $f(x)$. In other words, when we write (3.31) with operator notation $f = \mathcal{K}\{u\}$, there *may not* be a function $u(x)$ in the domain of the (integral) operator $\mathcal{K}$ that is mapped to the given function $f(x)$ of (3.31). We will show here that in the above example, the function $f(x) = \sin x$ must have been selected very carefully (according to guidelines from advanced theory) to guarantee the existence of the (obtained) solution (E.4) of (E.1). It is after this guarantee that the Laplace transform method of solution got a smooth sailing for the above Volterra integral equation of the first kind in (E.1) with (Laplace) convolution product integral type.

So let us assume that, instead of $f(x) = \sin x$ in (E.1) (with $\lambda = 1$) of the above example, we were given $f(x) = 1$. Then if we don't know about the above difficulty, we can proceed, formally, with the same above Laplace transform method, where we only have the left side $\mathcal{L}\{\sin x\} = \frac{1}{s^2+1}$ of the first equation (E.2) (with $\lambda = 1$) in the above Example 7 changed to $\mathcal{L}\{1\} = \frac{1}{s}$ to result in

$$\frac{1}{s} = \frac{1}{s-1}U(s)$$

$$U(s) = \frac{\frac{1}{s}}{\frac{1}{s-1}} = \frac{s-1}{s} = 1 - \frac{1}{s}.$$

However, for such result of $U(s) = \frac{s-1}{s} = 1 - \frac{1}{s}$, there exists no Laplace transform inverse for the first term 1 of $U(s) = 1 - \frac{1}{s}$. This is an obvious accepted conclusion, since the important necessary condition for the Laplace transform $F(s)$ (of a large class of functions as described in Theorem 1.1 of Section 1.4) is that it must vanish as s approaches infinity, and the above $F(s) = 1$ does not. This important necessary condition was shown in Example 8 of Section 1.3.

Abel's Generalized Integral Equation

The Abel's integral equation (1.20)

$$-\sqrt{2g}f(x) = \int_0^x \frac{\phi(t)dt}{\sqrt{x-t}} \tag{3.38}$$

is a Volterra equation of the first kind with difference kernel $K(x,t) = 1/\sqrt{x-t}$; hence we may use the Laplace transform to solve it.

The generalized Abel integral equation

$$f(x) = \int_0^x \frac{u(t)}{(x-t)^\alpha}dt, \qquad 0 < \alpha < 1 \tag{3.39}$$

is also of the convolution type and, for an appropriate $f(x)$!, can be solved by using Laplace transform (see our comment following Example 7 concerning conditions on

$f(x)$ in (3.31) to guarantee a solution to Volterra equations of the first kind (3.31).) In the next example we use the Laplace transform to solve (3.38) and leave the solution of (3.39), which is

$$u(x) = \frac{\sin \alpha\pi}{\pi} \frac{d}{dx} \int_0^x (x-t)^{\alpha-1} f(t)dt, \qquad 0 < \alpha < 1 \tag{3.40}$$

as an exercise (see Exercise 3).

Example 8 Abel's Integral Equation. Use the Laplace transform to solve Abel's equation (3.38).

We let $F(s)$ and $\Phi(s)$ be the Laplace transform of $f(x)$ and $\phi(x)$, respectively, and use $\mathcal{L}\{K(x)\} = \mathcal{L}\{1/\sqrt{x}\} = \sqrt{\pi/s}$ from (1.79) with $\nu = -\frac{1}{2}$, noting that $\Gamma(1/2) = \sqrt{\pi}$, to obtain

$$-\sqrt{2g}F(s) = \sqrt{\frac{\pi}{s}}\Phi(s) \tag{E.1}$$

$$\Phi(s) = -\sqrt{\frac{2g}{\pi}}\sqrt{s}F(s). \tag{E.2}$$

Hence the solution to Abel's equation (3.38) is

$$\phi(x) = -\sqrt{\frac{2g}{\pi}}\mathcal{L}^{-1}\{\sqrt{s}F(s)\}. \tag{E.3}$$

In trying to use the convolution theorem for $\mathcal{L}^{-1}\{\sqrt{s}F(s)\}$ in (E.3) we have difficulty (impossibility!) in finding $\mathcal{L}^{-1}\{\sqrt{s}\}$, which actually does not exist, since as we explained in Example 8 of Section 1.3, an important necessary condition for a "legitimate" Laplace transform $F(s)$ is that it must vanish as $s \to \infty$, i.e., $\lim_{s\to\infty} F(s) = 0$. This is also seen from the condition $\nu > -1$ on the Laplace transform pair (1.79), which prevents us from letting $\nu = -3/2$ to use the pair for $s^{1/2}$. On the other hand, (1.79) with $\nu = -\frac{1}{2}$ gives the inverse Laplace transform of $s^{-1/2}$ as $1/\sqrt{\pi x}$; we may multiply and divide the right-hand side of (E.2) by $\sqrt{s}$ to obtain

$$\Phi(s) = -\sqrt{\frac{2g}{\pi}} s \frac{1}{\sqrt{s}} F(s) \tag{E.2}$$

$$= -\sqrt{\frac{2g}{\pi}} sH(s), \quad H(s) \equiv \frac{F(s)}{\sqrt{s}} \tag{E.3}$$

and we can easily use the convolution theorem to write

$$\mathcal{L}^{-1}\{H(s)\} = \mathcal{L}^{-1}\left\{\frac{1}{\sqrt{s}}F(s)\right\} = \frac{1}{\sqrt{\pi}} \int_0^x \frac{f(t)}{\sqrt{x-t}}dt = h(x). \tag{E.4}$$

We are still after $\mathcal{L}^{-1}\{\Phi(s)\}$, so according to (1.68),

$$\begin{aligned}\mathcal{L}^{-1}\{sH(s)-h(0)\} &= \frac{dh}{dx} \\ &= \mathcal{L}^{-1}\{sH(s)\}-\mathcal{L}^{-1}\{h(0)\}=\frac{dh}{dx}\end{aligned} \tag{E.5}$$

$$=\mathcal{L}^{-1}\{sH(s)\}=\frac{dh}{dx} \tag{E.6}$$

[since from (E.4) (or on physical grounds) we have $h(0)=0$].

Finally, we use $h(x)$ from (E.4) in (E.6) to obtain $\phi(x)$ [the inverse Laplace transform of $\Phi(s)=-\sqrt{2g/\pi}[sH(s)]$ in (E.2)],

$$\begin{aligned}\phi(x)=-\sqrt{\frac{2g}{\pi}}\frac{dh}{dx} &= -\sqrt{\frac{2g}{\pi}}\frac{d}{dx}\left[\frac{1}{\sqrt{\pi}}\int_0^x \frac{f(t)}{\sqrt{x-t}}dt\right] \\ &= \frac{-\sqrt{2g}}{\pi}\frac{d}{dx}\int_0^x \frac{f(t)}{\sqrt{x-t}}dt,\end{aligned} \tag{3.41}$$

which is the solution to (3.38).

Another Difficulty for Integral Equations of the First Kind

Again this method and the solution in (3.41) for Abel's problem (3.38), as Volterra equation of the first kind, seems to flow very well. However, we may still point out to the "hidden" difficulties in trying to solve integral equations of the first kind. For example, does the integral in (3.41) exist for an arbitrary $f(x)$. Abel's first problem of the Tautochrone (1.21) is for the case of $f(x)$ being constant T, which is very safe, since the integral of (3.41) does exist for $f(x)=1$,

$$\int_0^x (x-t)^{-\frac{1}{2}}dt=-2(x-t)^{\frac{1}{2}}\Big|_{t=0}^{x}=-2\left[0-x^{\frac{1}{2}}\right]=2x^{\frac{1}{2}}.$$

This result of the integration is a nice continuous function for $x\geq 0$, but what is needed for the solution (3.41) is to differentiate this result of the integration. To prepare for this second type of difficulty for equations of the first kind, we remind how integration is a "smoothing process," while differentiation, as the inverse operation of integration, would uncover the "rough spots" (discontinuities, that were smoothened by the integration process). This smoothing of the integration operation was discussed then illustrated in Example 16 of Section 1.5.2 for the (continuous) roof function $h(x)$, and how the differentiation operation uncovers the jump discontinuity in $\frac{dh}{dx}$ as illustrated in Figure 1.11. In our example with $f(x)=T$

$$\phi(x)=-\frac{\sqrt{2g}}{\pi}T\frac{d}{dx}x^{\frac{1}{2}}=-\frac{\sqrt{2g}T}{\pi}\cdot\frac{1}{2}x^{-\frac{1}{2}}$$

which is unbounded at $x=0$. So, solutions of the Volterra integral equations of the first kind that involve related differentiation of the given function $f(x)$ would inherit what the differentiation operation may uncover. This is even more serious if $f(x)$ is given as *data*, which of course has the inaccuracy of the measurement. So, for the above result in (3.41) we have to do numerical integration with its own added error of approximating the integral. On the top of that, this inaccurate numerical result has to be numerically differentiated, which compounds the error for the desired practical solution $\phi(x)$ of (3.38), instead of writing the formal solution as in (3.41). This difficulty of the equations of the first kind is considered to be very serious, because even if we know that a solution exists, we may only get useless inaccurate data for it. Such situations are described by "a small change in the input data $f(x)$ may produce a very large change in the sought output (solution) $u(x)$" of the integral equation of the first kind,

$$f(x) = \int K(x,t)u(t)dt,$$

whereby the solution is termed "unstable." Such difficulties and warning were, of course, not available to Abel around the year 1826. This is taken historically to be a good luck that Abel was not aware of such deterrents, and went ahead to get the solutions, while "not knowing enough to be daunted".[3]

We shall return to this subject of the troubles with integral equations of the first kind when we discuss the Fredholm integral equations of the first kind in Section 5.4.

Exercises 3.2

1. **(a)** Reduce the Volterra integral equation of the first kind

$$x = \int_0^x \cos(x-t)u(t)dt$$

to a Volterra integral equation of the second kind.

(b) Use the Laplace transform to solve the resulting integral equation in part (a).

(c) Use the Laplace transform to solve the problem in part (a) directly, i.e, without having to reduce it to Volterra equation of the second kind.

(d) Use the result in part (b) to verify it as a solution to the integral equation in part (a).

(e) Do parts (a)–(c) above for the Volterra integral equations of the first kind in Exercises 3 and 4 of Section 1.1.

2. Solve the Volterra integral equation of the first kind after reducing it to an equation of the second kind,

$$x = \int_0^x 3^{x-t}u(t)dt.$$

[3]See Lonseth [1977] and Anderson and DeHoog [1980].

Hint: Write $3^t = e^{t \ln 3}$, $\mathcal{L}\{3^t\} = 1/(s - \ln 3)$.

3. Use the Laplace transform to solve the generalized Abel integral equation (3.39),

$$f(x) = \int_0^x \frac{u(t)dt}{(x-t)^\alpha}, \qquad 0 < \alpha < 1$$

Hint: To simplify the answer, use the relation $\Gamma(x)\Gamma(1-x) = \pi/(\sin \pi x)$.

4. Solve each of the following Volterra integral equations of the first kind after reducing them to Volterra equations of the second kind.

(a) $\frac{x^2}{2} = \int_0^x (1 - x^2 + t^2)u(t)dt$

(b) $e^{x^2/2} - 1 = \int_0^x \sin(x-t)u(t)dt$

Hint: For part (b) avoid the Laplace transform (because $\mathcal{L}\{e^{\frac{x^2}{2}}\}$ does not exist, and note that $K(x, x) = \sin 0 = 0$.

5. Consider the following examples of Volterra and Fredholm integral equations with the same kernel,

$$f(x) = \int_0^x \sin(x+\xi)u(\xi)d\xi \tag{E.1}$$

$$f(x) = \int_0^1 \sin(x+\xi)u(\xi)d\xi. \tag{E.2}$$

Show how these two equations illustrate that the existence for a solution to the Fredholm integral equation of the first kind (E.2) has much harder restriction on the class of the given function $f(x)$, for (E.2) to have a solution, than that of the Volterra equation of the first kind (E.1).

Hint: Expand $\sin(x + \xi)$ in its two terms ($\sin(x + \xi) = \sin x \cos \xi$ +$\cos x$ sin ξ), then examine the integrals over ξ in both equations to see that (E.2) requires the very restrictive condition that its $f(x)$ on the left must be of the form $A \sin x + B \cos x$, A, B constants; while (E.1) allows $f(x)$ to be in the more relaxed form $f(x) = g(x) \sin x + h(x) \cos x$. The reason for the difference is the fixed limits of integration in (E.2) versus a variable limit x in (E.1).

6. Consider the Volterra integral equation of the first kind (3.31) in $u(x)$, which after differentiation (with $K(x, x) \neq 0$) had resulted in the Volterra integral equation of the second kind (3.34) in $u(x)$. Assuming that $\frac{\partial K(x,y)}{\partial y}$ exists and is continuous and that $K(x, x) \neq 0$ for all $x \epsilon [a, b]$, use an integration method by letting $\phi(x) \equiv \int_0^x u(t)dt$ to also reduce the equation of the first kind to another one of the second kind in $\phi(x)$,

$$\phi(x) - \int_0^x \frac{\frac{\partial}{\partial t}K(x,t)}{K(x,x)}\phi(t)dt = \frac{f(x)}{\lambda K(x,x)}, \qquad x\epsilon[a,b].$$

Hint: Use integration by parts on (3.31) letting $U(t) = K(x,t)$ and $dV(t) = u(t)dt$, whereby $V(t) = \int u(t)dt = \phi(t)$.

3.3 NUMERICAL SOLUTION OF VOLTERRA INTEGRAL EQUATIONS

In the preceding two sections we presented exact and approximate methods for solving Volterra integral equations. We must recognize that the illustrations we presented there were of a very special form to suit the method and are simple enough to result in a familiar form of solution without very lengthy computations. For more general types of problems, we often resort to approximate methods where the integral equation is replaced or approximated by another one which is closely related and can be handled by the usual methods, and hopefully, with solutions close to those of the original equations. When such methods are not feasible, we have to resort to numerical methods which are also approximate methods and where the integral in the equation is approximated by a sum of N terms. As a result, the integral equation may be reduced to a set of N equations in the N unknowns $u(x_i), i = 0, 1, 2, \cdots, N-1$, the samples of the approximate solution. To illustrate this method clearly we present simple examples, some of which were solved by the exact methods so that we have a chance to compare them with the numerically evaluated (approximate) results. For the same reason we will, at this stage, use one of the simplest and most familiar methods of numerical integration, the trapezoidal rule (1.141), which we have already presented in Section 1.5 along with Simpson's rule (1.144) and the midpoint formula (1.147). For the level of this introductory text, we leave the *higher order quadrature rules* for the interest of the reader. They are discussed and illustrated with a good number of examples and exercises in Chapter 7. There, we also include the tables of the quadrature rules that are necessary for their use in the examples and the exercises.

3.3.1 Numerical Approximation Setting of Volterra Equations

Here we consider the Volterra integral equation of the second kind,

$$u(x) = f(x) + \int_a^x K(x,t)u(t)dt \tag{3.42}$$

with its noted *variable upper limit* of integration x as compared to the *fixed* upper limit b of the Fredholm equation (1.148),

$$u(x) = f(x) + \int_a^b K(x,t)u(t)dt. \tag{1.148}$$

Indeed, this is the major difference in classifying these two main (different) classes of integral equations. So, as expected, this will affect their theories and methods of solutions, and in particular the present numerical approximation by a linear system of equations. As expected, and shall soon become very clear, the numerical setting and the approximation of the integral for the Volterra equation (3.42) will result in the coefficient matrix of the linear system of equations being a (lower) *triangular* one, which is exactly due to the variable upper limit x of the integration in (3.42). This should be clear, since in (3.42) the kernel $K(x,t) \equiv 0$ for $t > x$ as the integrand can be considered identically zero above its upper limit of integration x in (3.42). So for the discrete case we have $K(x_i, t_j) = K_{ij} = 0$ for $j > i$. Of course, a system of linear equations with such a natural triangular coefficient matrix is easy to solve, as we shall illustrate in Example 9. This is especially when compared with the *square* system of equations for the Fredholm integral equation, as we shall discuss and illustrate in Section 5.5.1 [see (5.118) and Example 20].

We will subdivide the interval of integration (a, x) into n $(= N - 1)$ equal subintervals of width $\Delta t = (x_n - a)/n, n \geq 1$, where x_n is the end point we choose for x; we shall set $t_0 = a$ and $t_j = a + j\Delta t = t_0 + j\Delta t$. Since we will be using either t or x as the independent variable for the solution u, we will call $x_0 = t_0 = a$, $x = x_n = t_n$ and $x_i = x_0 + i\Delta t = a + i\Delta t = t_i$, or in short, $x_i = t_i$. We will refer to the value of the function $f(x)$ at x_i as $f(x_i) = f_i$, the value of kernel $K(x,t)$ at (x_i, t_j) as $K(x_i, t_j) \equiv K_{ij}$, and the (approximate) value of the solution $u(x)$ at x_i or t_i as $u(t_i) \equiv u(x_i) \equiv u_i$. $K(x_i, t_j)$ clearly vanishes for $t_j > x_i$ as the integration ends at $t_j \leq x_i$. Note that the particular value $u(x_0) = f(a)$ according to (3.42). So if we use the trapezoidal rule with n subintervals to approximate the integral in the Volterra integral equation (3.42), we have

$$\int_a^x K(x,t)u(t)dt \sim \Delta t \left[\frac{1}{2} K(x, t_0)u(t_0) + K(x, t_1)u(t_1) + \cdots \right.$$

$$\left. + K(x, t_{n-1})u(t_{n-1}) + \frac{1}{2} K(x, t_n)u(t_n) \right],$$

$$\Delta t = \frac{t_j - a}{j} = \frac{x - a}{n}, \quad t_j \leq x, \quad j \geq 1, \quad x = x_n = t_n \tag{3.43}$$

and the integral equation (3.42) is then approximated by the sum

$$u(x) = f(x) + \Delta t \left[\frac{1}{2} K(x, t_0)u(t_0) + K(x, t_1)u(t_1) + \cdots \right.$$

$$\left. + \cdots + K(x, t_{n-1})u(t_{n-1}) + \frac{1}{2} K(x, t_n)u(t_n) \right],$$

$$t_j \leq x, \quad j \geq 1, \quad x = x_n = t_n. \tag{3.44}$$

The integration in (3.43) is over t, $a \leq t \leq x$; thus for $t > x$ we take $K(x,t) \equiv 0$, $K(x_i, t_j) = 0$ for $t_j > x_i$.

Of course, we realize here that the solution desired in (3.43) is an approximate solution of (3.42) since there is an error involved in replacing the integral in (3.42) by the $N = n + 1$ terms of the trapezoidal rule (1.141). If we consider (the same) $n + 1$

sample values of $u(x)$, $u(x_i) = u_i$, $i = 0, 1, 2, \cdots, n$, equation (3.44) will become a set of $N = n + 1$ equations in $u(x_i)$ (or u_i) [note that $u(x_0) = f(x_0)$ since the integral in (3.42) vanishes for $x = x_0 = a$],

$$\begin{aligned} u(x_0) &= f(x_0) \\ u(x_i) &= f(x_i) \qquad +\Delta t\left[\frac{1}{2}K(x_i,t_0)u_0 + K(x_i,t_1)u_1 + \cdots \right. \\ &\qquad \left. + \cdots + K(x_i,t_{j-1})u_{j-1} + \frac{1}{2}K(x_i,t_j)u_j\right], \\ &\qquad i = 1, 2, \cdots, n, \qquad t_j \le x_i \end{aligned} \tag{3.45}$$

where we note again that $K(x_i, t_j) = 0$ for $t_j > x_i$ since the integration in (3.43) stops at $t_j = x_i$. The system of equations in (3.45) can be written in a more compact form as

$$\begin{aligned} u_0 &= f_0 \\ u_i &= f_i + \Delta t\left[\frac{1}{2}K_{i0}u_0 + K_{i1}u_1 + \cdots + K_{i,j-1}u_{j-1} + \frac{1}{2}K_{ij}u_j\right], \\ i &= 1, 2, \cdots, n, \qquad K_{ij} = K(x_i, t_j), \qquad j \le i \end{aligned} \tag{3.46}$$

which are $N = n + 1$ equations in u_i, the approximation to the solution $u(x)$ of (3.42) at $x_i = a + i\Delta t$ for $i = 0, 1, 2, \cdots, n$.

If we transfer all the terms involving the solution u_i to the left side of (3.46), leaving only the nonhomogeneous part f_i on the right side, then write all the $n + 1$ equations for u_i, $i = 0, 1, 2, \cdots, n$, we have the following *triangular* system of equations:

$$\begin{aligned} &u_0 && = f_0 \\ &-\frac{\Delta t}{2}K_{10}u_0 + \left(1 - \frac{\Delta t}{2}K_{11}\right)u_1 && = f_1 \\ &-\frac{\Delta t}{2}K_{20}u_0 - \Delta t K_{21}u_1 + \left(1 - \frac{\Delta t}{2}K_{22}\right)u_2 && = f_2 \\ &-\frac{\Delta t}{2}K_{30}u_0 - \Delta t K_{31}u_1 - \Delta t K_{32}u_2 + \left(1 - \frac{\Delta t}{2}K_{33}\right)u_3 && = f_3 \\ &\vdots \quad \vdots \quad \vdots \\ &-\frac{\Delta t}{2}K_{n0}u_0 - \Delta t K_{n1}u_1 - \Delta t K_{n2}u_2 - \cdots + \left(1 - \frac{\Delta t}{2}K_{nn}\right)u_n && = f_n \end{aligned} \tag{3.47}$$

as a system of $n + 1$ equations in the $n + 1$ desired unknowns $u_0, u_1, \cdots, u_n$. We note that the form of this set of equations is a very special and desired one since the solutions can be obtained by repeated substitution, starting with $u_0 = f_0$ from the first equation of (3.47), which can be substituted in the second equation to obtain u_1,

$$-\frac{\Delta t}{2}K_{10}u_0 + \left(1 - \frac{\Delta t}{2}K_{11}\right)u_1 = f_1 = -\frac{\Delta t}{2}K_{10}f_0 + \left(1 - \frac{\Delta t}{2}K_{11}\right)u_1,$$

$$u_1 = \frac{f_1 + (\Delta t/2)K_{10}f_0}{1 - (\Delta t/2)K_{11}}. \tag{3.48}$$

Then this value u_1 is substituted in the third equation of (3.47) to obtain u_2, and this substitution process can be continued until we obtain u_n. With this particular triangular system for the Volterra equation, we will be encouraged in the following Example 9 to find the solution. As we remarked earlier, this is in contrast to the square system of equations of a Fredholm equation of Section 5.5 as illustrated in Example 20 of that section, where the solution of the system is not as easy and straightforward as the above one.

Example 9 Numerical Solution of Volterra Equation. Use the numerical method described above to find the approximate values of the solution for the following Volterra equation at $x = 0, 1, 2, 3,$ and 4; then compare these values with the exact solution $u(x) = \sin x$,

$$\begin{aligned} u(x) &= x - \int_0^x (x - t)u(t)dt \\ &= x + \int_0^x (t - x)u(t)dt. \end{aligned} \tag{E.1}$$

Here we have $f(x) = x$, $K(x,t) = t - x$ for $t \le x = 0, 1, 2, 3, 4$ and is zero for $t > x = 0, 1, 2, 3, 4$, and $a = 0$ with $u(0) = 0$. We also have $n = 4$ and hence $\Delta t = (4 - 0)/4 = 1$. So from (3.47) the five equations in u_0, u_1, u_2, u_3, and u_4 are

$$u_0 = f_0 = 0 \tag{E.2}$$

$$-\frac{1}{2}K_{10}u_0 + \left(1 - \frac{1}{2}K_{11}\right)u_1 = f_1 = 1 \tag{E.3}$$

$$-\frac{1}{2}K_{20}u_0 - K_{21}u_1 + \left(1 - \frac{1}{2}K_{22}\right)u_2 = f_2 = 2 \tag{E.4}$$

$$-\frac{1}{2}K_{30}u_0 - K_{31}u_1 - K_{32}u_2 + \left(1 - \frac{1}{2}K_{33}\right)u_3 = f_3 = 3 \tag{E.5}$$

$$-\frac{1}{2}K_{40}u_0 - K_{41}u_1 - K_{42}u_2 - K_{43}u_3 + \left(1 - \frac{1}{2}K_{44}\right)u_4 = f_4 = 4 \tag{E.6}$$

Hence if we substitute in (E.3)–(E.6) the values for $K_{10} = K(1,0) = 0 - 1 = -1$, $K_{11} = 0$, $K_{20} = -2$, $K_{21} = -1$, $K_{22} = 0$, $K_{30} = -3$, $K_{31} = -2$, $K_{32} = -1$, $K_{33} = 0$, $K_{40} = -4$, $K_{41} = -3$, $K_{42} = -2$, $K_{43} = -1$, and $K_{44} = 0$, we obtain

$$
\begin{array}{ll}
u_0 = 0 & \\
\frac{1}{2}u_0 + u_1 = 1, & u_1 = 1 - \frac{1}{2}u_0 = 1 - 0 = 1 \\
u_0 + u_1 + u_2 = 2, & u_2 = 2 - u_0 - u_1 = 2 - 0 - 1 = 2 - 1 = 1 \\
\frac{3}{2}u_0 + 2u_1 + u_2 + u_3 = 3, & u_3 = 3 - \frac{3}{2}u_0 - 2u_1 - u_2 = 3 - 2 - 1 = 0 \\
2u_0 + 3u_1 + 2u_2 + u_3 + u_4 = 4, & u_4 = 4 - 2u_0 - 3u_1 - 2u_2 - u_3 \\
 & \quad = 4 - 0 - 3 - 2 - 0 = -1
\end{array}
$$

after substituting u_0 from (E.2) for obtaining u_1 in (E.3), and so on. So the numerical approximation to the sample values of the solution are $u_0 = 0$, $u_1 = 1$, $u_2 = 1$, $u_3 = 0$, and $u_4 = -1$, which are compared to the exact values $u(x) = \sin x$ as $u_0 = \sin 0 = 0$, $u_1 = \sin 1 = 0.8415$, $u_2 = \sin 2 = 0.9093$, $u_3 = \sin 3 = 0.1411$, and $u_4 = \sin 4 = -0.7568$ as illustrated in Table 3.1 and Figure 3.1.

Table 3.1 Numerical and Exact Solutions of Volterra Integral Equation

x	0	1	2	3	4
Numerical value of $u(x)$	0	1	1	0	−1
Exact value of $u(x) = \sin x$	0	0.8415	0.9093	0.1411	−0.7568

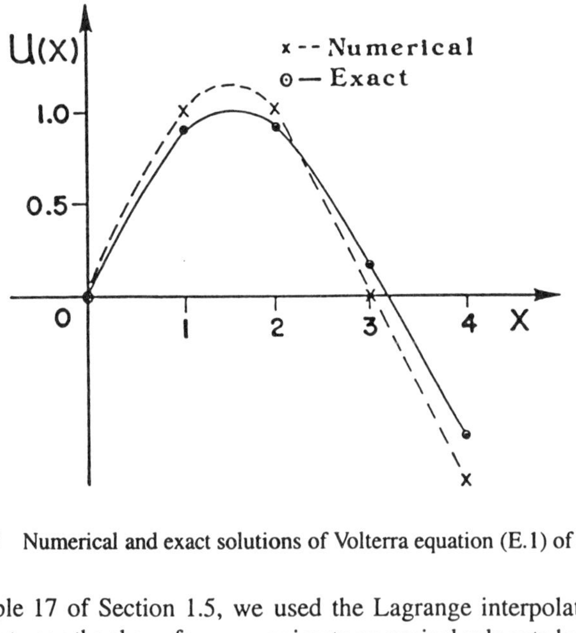

Fig. 3.1 Numerical and exact solutions of Volterra equation (E.1) of Example 9.

In Example 17 of Section 1.5, we used the Lagrange interpolation formula to interpolate between the above four approximate numerical values to have a continuous curve that connects them, which was illustrated in Figure 1.13, and which resembles the dotted line in Figure 3.1.

We may return to (3.42) and (3.47) and emphasize again how the numerical method reduced, or more precisely approximated, the Volterra equation of the second kind (3.42) to a (*lower triangular*) system of $N = n+1$ equations in $N = n+1$ unknowns as in (3.47), where N is the number of approximated sample values u_i of the solution desired. Now we recognize that the set of equations (3.47) can be written in a matrix notation form

$$\mathcal{K}U = F \tag{3.49}$$

where $\mathcal{K} = (K_{ij})$ is the $(n+1) \times (n+1)$ matrix of the coefficients of the system of equations (3.47), $U = (u_i)$ is the column matrix of the sample solutions, and $F = (f_i)$ is the column matrix of the sample values of the nonhomogeneous part $f(x)$ in (3.47). The symbolic matrix form (3.49) can be written explicitly as

$$\begin{bmatrix} 1 & 0 & 0 & & \cdots & 0 \\ -\frac{\Delta t}{2}K_{10} & 1-\frac{\Delta t}{2}K_{11} & 0 & 0 & \cdots & 0 \\ -\frac{\Delta t}{2}K_{20} & -\Delta t K_{21} & 1-\frac{\Delta t}{2}K_{22} & & \cdots & 0 \\ -\frac{\Delta t}{2}K_{30} & -\Delta t K_{31} & -\Delta t K_{32} & 1-\frac{\Delta t}{2}K_{33} & & 0 \\ \vdots & \vdots & \vdots & \vdots & & \vdots \\ -\frac{\Delta t}{2}K_{n0} & -\Delta t K_{n1} & -\Delta t K_{n2} & & \cdots & 1-\frac{\Delta t}{2}K_{nn} \end{bmatrix} \bullet$$

$$\begin{bmatrix} u_0 \\ u_1 \\ u_2 \\ u_3 \\ \vdots \\ u_n \end{bmatrix} = \begin{bmatrix} f_0 \\ f_1 \\ f_2 \\ f_3 \\ \vdots \\ f_n \end{bmatrix} \tag{3.50}$$

which can be verified as (3.47) by performing the simple matrix multiplication. This would mean that the essence of the numerical method is to reduce the Volterra integral equation to a matrix equation. Familiarity with the powerful tools of matrix theory would give us a more efficient way of solving the integral equation numerically.

We should note here again that the numerical method for solving Fredholm integral equations, which we shall discuss in Section 5.5, will follow in the same way; however, it results in a *square* rather than the present *triangular* system of simultaneous equations. Even more important is how the theory regarding the existence of solutions for the system of equations will shed light on the existence of solutions for the Fredholm integral equation. Since we intend to keep this text on the undergraduate level by assuming mainly a basic calculus preparation, we will keep our exercises on this level and leave it for each reader to obtain the result in an efficient way depending on his or her preparation in matrix calculus.

In Chapter 7 (towards the end of Section 7.2), and with the help of its higher quadrature rules, we will also have the chance to make a brief comment and illustrate the numerical solution for a particular class of *singular* Volterra integral equations. They are those equations whose singularity is due to the infinite limit of integration. An example is the integral equation of the torsion of a wire (1.15) in the torsion function $\omega(t)$,

$$m(t) = h\omega + \int_{-\infty}^{t} \phi(t,\tau)\omega(\tau)dt. \tag{1.15}$$

Exercises 3.3

1. Consider the Volterra equation in Example 9

$$u(x) = x - \int_0^x (x-t)u(t)dt$$

(a) Use the trapezoidal rule[4] to solve for $u(x)$ in the interval $(0, 4)$ with enough sample values to compare with the exact solution $u(x) = \sin x$.

Hint: Use $n = 8$.

(b) Tabulate or graph the numerical and exact solutions for comparison and note if there is improvement over those in Example 9.

2. Consider the Volterra equation of Example 3 with $\lambda = 2$ and $f(x) = x$,

$$u(x) = x + 2\int_0^x e^{x-t}u(t)dt \tag{E.1}$$

(a) Find the exact solution. *Hint:* See Example 1 for $\lambda = 2$ and $f(x) = x$, and in particular (E.7) for the solution.

(b) Solve (E.1) numerically for $0 \le x \le 5$. *Hint:* Find five or nine sample values with $n = 4$ or 8, respectively.

(c) Compare the approximate numerical solution in part (b) with the exact one in part (a). Graph both results.

[4]As was done for the development (3.43)–(3.47), the trapezoidal rule is used for the exercises of this section.

3. Consider the Volterra equation of the first kind of Example 6,

$$\sin x = \int_0^x e^{x-t} u(t) dt. \tag{E.1}$$

(a) Find the solution numerically for $0 \leq x \leq 2\pi$. *Hint:* Reduce it to a Volterra integral equation of the second kind as in (E.2) of Example 6,

$$u(x) = \cos x - \int_0^x e^{x-t} u(t) dt \tag{E.2}$$

then use the method of this section as illustrated in (3.46) or (3.47). Find nine sample points with $n = 8$.

(b) Compare the numerical values of part (a) with the exact values of Example 6.

(c) Attempt to find the numerical solution of the Volterra equation of the first kind (E.1) for 3 sample values ($n = 2$) directly [i.e., without reducing it to that of the second kind (E.2)]. *Hint:* Follow the steps from (3.43) to (3.47) for

$$0 = f(x) + \int_0^x K(x,t) u(t) dt \tag{E.3}$$

instead of (3.42).

4

The Green's Function

4.1 CONSTRUCTION OF THE GREEN'S FUNCTION

In Example 5 of Section 2.5 we illustrated how a boundary value problem

$$\frac{d^2y}{dx^2} = \lambda y(x), \qquad a < x < b \tag{1.33}$$

$$y(a) = 0 \tag{1.34}$$

$$y(b) = 0 \tag{1.35}$$

reduced to a Fredholm integral equation

$$y(x) = \lambda \int_a^b K(x,t)y(t)dt \qquad (1.36)(2.36)$$

with the kernel $K(x,t)$ as

$$K(x,t) = \begin{cases} \dfrac{(x-b)(t-a)}{(b-a)}, & a \le t < x \le b \\ \dfrac{(x-a)(t-b)}{(b-a)}, & a \le x < t \le b \end{cases} \qquad (1.37)(2.37)$$

where we referred to it as Green's function. Also, in Example 6 of Section 2.5, we showed how a similar Fredholm integral equation, with kernel as a special case of the above one in (2.37), reduces to a boundary value problem such as the above one in (1.33)–(1.35). In this chapter we will consider a more general boundary value problem, which can be reduced more readily to a Fredholm integral equation with the help of the Green's function associated with such a problem. Next we shall familiarize ourselves with the Green's function and the very basic (elementary) methods of constructing it.[1]

The Green's function method is one of the most important methods for solving boundary value problems associated with *nonhomogeneous* ordinary or partial differential equations. In this chapter we use the Green's function (in Section 4.2) to show again how a boundary value problem is reduced to a Fredholm integral equation with the Green's function as its kernel. First we present methods for constructing the Green's function for boundary value problems associated with nonhomogeneous differential equations. Of central importance to this development is the study of a very important special type of boundary value problem, *the Sturm-Liouville problem*. We will give a brief presentation of this problem and show how its solutions are used in an infinite series for another way of constructing the Green's function. In Section 4.2 we will use the Green's functions to reduce boundary value problems to a Fredholm integral equation with the Green's function as its kernel. A brief discussion with an illustration, of reducing boundary value problems associated with *partial differential* equations to two-dimensional Fredholm integral equations is given at the end of the this section (in Section 4.1.4). The illustration involves the potential distribution in a charged unit disc with grounded rim (see (4.62)–(4.69). Another very related illustration is that of the potential distribution in a charged square with grounded edges, which is the subject of Exercise 24 of this section. In this exercise we have very detailed instructions for using *the finite Fourier sine transform* (see (1.115), (1.116), and (1.121)) to reduce the partial differential equation with boundary conditions (in two variables) to a nonhomogeneous ordinary differential equation with its boundary conditions. Then the Green's function of this section is used to solve the latter problem.

4.1.1 Nonhomogeneous Differential Equations

Consider the boundary value problem associated with a *nonhomogeneous* ordinary differential equation of second-order,

$$A_0(x)\frac{d^2y}{dx^2} + A_1(x)\frac{dy}{dx} + A_2(x)y \equiv Ly = f(x),^2 \quad a < x < b \tag{4.1}$$

$$\alpha_1 y(a) + \alpha_2 y'(a) = 0 \tag{4.2}$$

$$\beta_1 y(b) + \beta_2 y'(b) = 0 \tag{4.3}$$

where L stands for the differential operator (with $A_j(x)$ as real-valued functions with continuous derivatives up to the order $2-j$, $j = 0, 1, 2$ on $[a, b]$, and $A_0(x) \neq 0$ on $[a, b]$), and $\alpha_1, \alpha_2, \beta_1$ and β_2 are constants.

[1] For a complete treatment of the Green's function, the interested reader may consult Stakgold [1979].

[2] In some books $-f(x)$ instead of $f(x)$ is written for the nonhomogeneous term of (4.1), which will bring a $(+)$ sign instead of the $(-)$ in the final solution (4.5) of (4.1)–(4.3).

To solve this boundary value problem we usually attempt to find a particular solution $y_p(x)$ for (4.1) and the general solution $y_h(x)$ for its corresponding *homogeneous* equation,

$$A_0(x)\frac{d^2y}{dx^2} + A_1(x)\frac{dy}{dx} + A_2(x)y = 0. \tag{4.4}$$

The general solution $y_g(x)$ of (4.1),

$$y_g = y_p + y_h$$

is the superposition of the two solutions of the linear equations (4.1) and (4.4). In general, it is difficult to find the particular solution for any arbitrary nonhomogeneous term $f(x)$ of (4.1).

The Green's function method represents a general method of solving the boundary value problem (4.1)–(4.3) where the solution is given as

$$y(x) = -\int_a^b G(x,t)f(t)dt \tag{4.5}$$

an integral in terms of the given nonhomogeneous term $f(x)$ and the Green's function $G(x,t)$. Note that some texts use $-G(x,t)$ instead of $G(x,t)$ in (4.5), it is to make up for using $-f(x)$ instead of $f(x)$ for the nonhomogeneous term of (4.1). The basic reason is convenience, which will become clear in the examples.

Before we illustrate the construction of the Green's function, there is an important particular case of the differential operator L in (4.1) with consequences that will shed more light on the Green's function of (4.5), such as its symmetric property $G(x,t) = \overline{G(t,x)}$, and which will aid a great deal in the method of constructing it. Here $\bar{G}$ is the *complex conjugate* of G, which we shall take as G since we will often work with $G(x,t)$ as a real-valued function. The particular form of the differential operator L in (4.1) is that of the *self-adjoint* form, which means that $(vLu - uLv)dx$ must be an exact differential $dg \equiv (vLu - uLv)dx$ for any two functions $u(x)$ and $v(x)$ operated on by L. When the differential operator L of (4.1) is a self-adjoint one, we will show that its associated Green's function $G(x,t)$ in (4.5), for the boundary value problem (4.1)–(4.3), is *symmetric* [i.e. $G(x,t) = G(t,x)$] (see (4.25)).

We should point out now that while second-order differential operators can be written in a self adjoint form; in general, this is not the case for differential operators of order $n > 2$.

A very important example in applied mathematics of a self-adjoint differential operator is the following second-order one[3]

$$Lu \equiv \frac{d}{dx}\left[r(x)\frac{du}{dx}\right] - q(x)u(x), \quad r(x) > 0 \tag{4.6}$$

[3]Here we should use L^* instead of the same L of (4.1), but since we are going to work mainly with the above L^*, we shall designate it, for simplicity, as L.

which is used with the following well-known *Sturm-Liouville* problem (4.7)–(4.9),

$$Lu + \lambda\rho(x)u(x) \equiv \frac{d}{dx}\left[r(x)\frac{du}{dx}\right] + [-q(x) + \lambda\rho(x)]u(x) = 0, \quad \rho(x) > 0 \tag{4.7}$$

$$\alpha_1 u(a) + \alpha_2 u'(a) = 0 \tag{4.8}$$

$$\beta_1 u(b) + \beta_2 u'(b) = 0. \tag{4.9}$$

with some (usual) regularity conditions on the coefficients in (4.6) [or (4.1)] as spelled out at the beginning of Section 4.1.3 [and following (4.1)]. We may mention here that λ in (4.7) is called "the eigenparameter", and when solutions $u_n(x)$ are found for the problem (4.7)–(4.9), they are termed "the eigenfunctions" that correspond to the eigenvalues $\lambda = \lambda_n$ in (4.7). In this sense the boundary value problem (4.7)–(4.9) represents an important example of "an eigenvalue problem" that we shall return to in Section 4.1.3. To be on the careful side we should use L_s instead of L for the above self-adjoint differential operator to differentiate it from the operator L used in (4.1). However, since we are going to work mainly with the above one in (4.6) we shall, for simplicity, use L.

In the next example we illustrate that the differential operator L in (4.6) is self-adjoint, and follow it by a method of constructing the Green's function.

Example 1 Self-Adjoint Operator

To show that the Sturm-Liouville differential operator L in (4.6) is in the self-adjoint form, we must show that $(vLu - uLv)dx$ is an exact differential [i.e., $(vLu - uLv)dx = dg$].

If we use L from (4.6) above, we have

$$\begin{aligned} vLu - uLv &= v\frac{d}{dx}[r(x)u'] + v[-q(x)]u - u\frac{d}{dx}[r(x)v'] - u[-q(x)]v \\ &= v\frac{d}{dx}[r(x)u'] - u\frac{d}{dx}[r(x)v'] \\ &= vru'' + vr'u' - urv'' - ur'v' \\ &= r\{vu'' - uv''\} + r'\{vu' - uv'\}. \end{aligned} \tag{E.1}$$

But

$$\begin{aligned} \frac{d}{dx}[r(x)\{vu' - uv'\}] &= r(x)\{v'u' + vu'' - u'v' - uv''\} + r'(x)\{vu' - uv'\} \\ &= r\{vu'' - uv''\} + r'\{vu' - uv'\} \end{aligned} \tag{E.2}$$

Hence from (E.1) and (E.2) we conclude that

$$vLu - uLv = \frac{d}{dx}[r(x)\{vu' - uv'\}],$$

$$(vLu - uLv)dx = d[r(x)\{vu' - uv'\}] = dg \tag{E.3}$$

which means that $(vLu - uLv)dx$ is an exact differential with $g(x) = r(x)\{vu' - uv'\}$, and hence can be integrated to give

$$\int_a^b (vLu - uLv)dx = [r(x)(vu' - uv')]\bigg|_a^b . \tag{E.4}$$

From now on we will assume the self-adjoint form of the second-order differential operator L of (4.6) instead of that in (4.1).

In this example we merely showed that the form in (4.6) is the self-adjoint form. Indeed it can be shown that any second-order differential operator such as L in (4.1) can be made self-adjoint by multiplying it by

$$p(x) = \frac{r(x)}{A_0(x)} = \frac{1}{A_0(x)} \exp\left(\int \frac{A_1(x)}{A_0(x)} dx\right), \qquad A_0(x) \neq 0 \tag{E.5}$$

which we shall leave for an exercise with a detailed hint [see Exercise 25(a)]. Moreover, in general, differential operators of order $n > 2$ are not necessarily self-adjoint. To illustrate this point we state that the simple third-order differential operator L of $Lu = \frac{d^3u}{dx^3} + u$ is not self-adjoint, while the fourth-order operator L in $Lu = \frac{d^4u}{dx^4} + u$ is self adjoint. We will show the first case here, and leave the case of the fourth-order differential operator for an exercise (see Exercise 25(b)).

To show that L in $Lu = u''' + u$ is not self-adjoint, we write $vLu - uLv$, then add and subtract terms, which result in only the major parts of $(vLu - uLv)dx$ as a sum of exact differentials,

$$\begin{aligned}
vLu - uLv &= vu''' - uv''' \\
&= vu''' + v'u'' - v'u'' - v''u' + v''u' + v'''u - v'''u - uv''' \\
&= (vu'')' - (v'u')' + (v''u')' - 2v'''u \\
&= \frac{d}{dx}[vu'' - v'u' + v''u'] - 2v'''u.
\end{aligned}$$

Hence $(vLu - uLv)dx$ is not an exact differential because of the last term .

4.1.2 Construction of the Green's Function — Variation of Parameters Method

In this section we use the method of *variation of parameters* to construct the Green's function $G(x,t)$ for the integral representation (4.5) of the solution of the nonhomogeneous boundary value problem (4.1)–(4.3) with L as in (4.6). The result will show clearly that the Green's function associated with the self-adjoint differential operator L of (4.6) is symmetric. With the aid of this and other basic properties of the Green's function we are often able to construct the Green's function without having to go through the full details of the analytic method. We will illustrate both methods with simple examples.

The method of constructing the Green's function [and hence solving the nonhomogeneous problem 4.1 (with L as in (4.6)), (4.2) and (4.3)] depends primarily on the solutions of the *associated homogeneous* problem (4.7)–(4.9) in the sense that they both have the same differential operator L.

Let $v_1(x)$ and $v_2(x)$ be two linearly independent solutions of the associated homogeneous equation (4.7). The variation of parameters method assumes the form

$$u(x) = w_1(x)v_1(x) + w_2(x)v_2(x) \tag{4.10}$$

for the solution $u(x)$ of the nonhomogeneous problem (4.1), where the unknown variable coefficients (parameters) $\omega_1(x)$ and $\omega_2(x)$ are to be determined via this method.

For now we will assume that neither of the solutions $v_1(x)$ or $v_2(x)$ of (4.7) satisfies both boundary conditions (4.2) at $x = a$ and (4.3) at $x = b$. A simple intuitive reason for this assumption can be found by looking at the shape $y(x)$ of the hanging chain in Figure 2.2, where the solution of the nonhomogeneous problem with its nonhomogeneous external force $F(x)$ consists of two different straight lines, the one for $0 \leq x < \xi$ satisfying (only) the boundary condition $y(0) = 0$ at $x = 0$, and the one for $\xi < x \leq l$ satisfying the boundary condition $y(l) = 0$ at the other end $x = l$.

The analytical reason behind this assumption—of not allowing either of $v_1(x)$ or $v_2(x)$ to satisfy both boundary conditions—is that if one of them does say $v_2(x)$, then by following the same method of construction of the solution as that we are about to use, we can show that we end up with an extra condition,

$$\int_a^b v_2(x)f(x)dx = 0 \tag{4.11}$$

that contributes to the nonuniqueness of the final solution $u(x)$. To stay with our aim, of constructing the Green's function, we would rather not deal with (4.11) for the present, and we leave it as an exercise [see exercise 20(a) with it's very detailed leading steps].

To prepare for making $u(x)$ of (4.10) a solution to the nonhomogeneous equation $Lu = f$ with the self-adjoint differential operator L as in (4.6), we first find the derivative of $u(x)$ in (4.10),

$$u'(x) = w_1(x)v_1'(x) + w_2(x)v_2'(x) + w_1'(x)v_1(x) + w_2'(x)v_2(x). \tag{4.12}$$

A very important step in the method of variation of parameters is to reduce the expected second-order differentiation of the operator L, on the unknown functions (parameters) $w_1(x)$ and $w_2(x)$ to a first-order differentiation. This is accomplished by assigning to zero the last two terms involving $w_1'(x)$ and $w_2'(x)$ in (4.12),

$$w_1'(x)v_1(x) + w_2'(x)v_2(x) = 0 \tag{4.13}$$

leaving $u'(x)$ of (4.12) free of the first derivatives $w_1'(x)$ and $w_2'(x)$,

$$u'(x) = w_1(x)v_1'(x) + w_2(x)v_2'(x). \tag{4.14}$$

When this $u'(x)$ is substituted in the nonhomogeneous second-order differential equation $Lu = f$ of (4.1) with L as in (4.6), the result is a linear equation (4.15) in $w_1'(x)$ and $w_2'(x)$, which we can combine with the first equation (4.13) to solve for $w_1'(x)$ and $w_2'(x)$,

$$\begin{aligned}
Lu &= \frac{d}{dx}[r(x)\{w_1(x)v_1'(x) + w_2(x)v_2'(x)\}] \\
&\quad -q\{w_1(x)v_1(x) + w_2(x)v_2(x)\} = f(x) \\
&= w_1(x)\frac{d}{dx}[r(x)v_1'(x)] + w_1'(x)r(x)v_1'(x) \\
&\quad +w_2(x)\frac{d}{dx}[r(x)v_2'(x)] + w_2'(x)r(x)v_2'(x) \\
&\quad +w_1(x)[-q(x)]v_1(x) + w_2(x)[-q(x)]v_2(x) = f(x) \\
&= w_1(x)\left\{\frac{d}{dx}\left[r(x)\frac{dv_1}{dx}\right] - q(x)v_1(x)\right\} \\
&\quad +w_2(x)\left\{\frac{d}{dx}\left[r(x)\frac{dv_2}{dx}\right] - q(x)v_2(x)\right\} \\
&\quad +w_1'(x)r(x)v_1'(x) + w_2'(x)r(x)v_2'(x) = f(x) \\
&= 0 + 0 + r(x)[w_1'(x)v_1'(x) + w_2'(x)v_2'(x)] = f(x), \\
&\quad w_1'(x)v_1'(x) + w_2'(x)v_2'(x) = \frac{f(x)}{r(x)}
\end{aligned} \tag{4.15}$$

after using the fact that $v_1(x)$ and $v_2(x)$ are solutions of the homogeneous equation (4.7) to make the foregoing coefficients of both $w_1(x)$ and $w_2(x)$ vanish.

In (4.13) and (4.15) we have the main result of the variation of parameters method as two simultaneous equations in the *first* derivatives $w_1'(x)$ and $w_2'(x)$ of the unknown variable parameters $w_1(x)$ and $w_2(x)$,

$$w_1'(x)v_1(x) + w_2'(x)v_2(x) = 0 \tag{4.13}$$

$$w_1'(x)v_1'(x) + w_2'(x)v_2'(x) = \frac{f(x)}{r(x)}. \tag{4.15}$$

The solutions to these equations are

$$w_1'(x) = \frac{-f(x)v_2(x)}{r(x)[v_1(x)v_2'(x) - v_2(x)v_1'(x)]} \tag{4.16}$$

$$w_2'(x) = \frac{f(x)v_1(x)}{r(x)[v_1(x)v_2'(x) - v_2(x)v_1'(x)]} \tag{4.17}$$

Before we integrate to find $w_1(x)$ and $w_2(x)$ we will take advantage of the fact that the differential operator L of (4.7) is self-adjoint to show that the denominator in (4.16) and (4.17) is a constant.

In the preceding section (Example 1) we showed that L is a self-adjoint operator, which means that for any two twice-differentiable functions u and v, we have

$$(vLu - uLv)dx = d[r(x)\{v(x)u'(x) - u(x)v'(x)\}]$$

as an exact differential.

If we use $v_1(x)$ and $v_2(x)$ here, they are also solutions of $Lu = 0$, which will make the left side vanish,

$$0 = v_1 Lv_2 - v_2 Lv_1 = \frac{d}{dx}[r(x)\{v_1(x)v_2'(x) - v_2(x)v_1'(x)\}]$$

where upon integration we have the desired result,

$$r(x)\{v_1(x)v_2'(x) - v_2(x)v_1'(x)\} = B = \text{const} \tag{4.18}$$

for the denominator in (4.16) and (4.17),

$$w_1'(x) = -\frac{1}{B}f(x)v_2(x) \tag{4.16$'$}$$

$$w_2'(x) = \frac{1}{B}f(x)v_1(x). \tag{4.17$'$}$$

If we integrate these two equations, we obtain the variable coefficinets $w_1(x)$ and $w_2(x)$ of the solution in (4.10),

$$w_1(x) = -\frac{1}{B}\int_{c_1}^{x} f(\xi)v_2(\xi)d\xi \tag{4.19}$$

$$w_2(x) = \frac{1}{B}\int_{c_2}^{x} f(\xi)v_1(\xi)d\xi \tag{4.20}$$

where c_1 and c_2 are arbitrary constants to be determined from the implication of the boundary conditions (4.8), (4.9) on $w_1(x)$, and $w_2(x)$, respectively, as seen in (4.22), (4.23). The result is that these arbitrary constants are chosen as $c_1 = a$ and $c_2 = b$.

To find such appropriate conditions will depend on the earlier basic assumption that neither of the solutions $v_1(x)$ and $v_2(x)$ satisfies both boundary conditions (4.8) at $x = a$ and (4.9) at $x = b$. These boundary conditions are, of course, to be satisfied by the final desired solution $u(x)$ in (4.10) of the problem 4.1 (with L as in (4.6)), (4.2) and (4.3). For the boundary condition (4.8) on $u(x)$ at $x = a$, we have

$$\begin{aligned} \alpha_1 u(a) + \alpha_2 u'(a) &= \alpha_1[w_1(a)v_1(a) + w_2(a)v_2(a)] \\ &\quad + \alpha_2[w_1(a)v_1'(a) + w_2(a)v_2'(a)] = 0 \\ &= w_1(a)[\alpha_1 v_1(a) + \alpha_2 v_1'(a)] + w_2(a)[\alpha_1 v_2(a) \\ &\quad + \alpha_2 v_2'(a)] = 0 \end{aligned} \tag{4.21}$$

where we used (4.14) for $u'(x)$.

If we assume in (4.21) that $v_2(x)$ satisfies the boundary condition (4.8) at $x = a$ [i.e., $\alpha_1 v_2(a) + \alpha_2 v_2'(a) = 0$] while $v_1(x)$ does not [i.e., $\alpha_1 v_1(a) + \alpha_2 v_1'(a) \neq 0$], then (4.21) gives

$$w_1(a)[\alpha_1 v_1(a) + \alpha_2 v_1'(a)] = 0$$

which forces the boundary condition on $w_1(x)$ to be $w_1(a) = 0$, which is what we need for determining the arbitrary constant c_1 involved in the solution $w_1(x)$ of (4.19). If we apply this condition on (4.19), we have

$$w_1(a) = 0 = -\frac{1}{B}\int_{c_1}^{a} f(\xi)v_2(\xi)d\xi = 0$$

which is satisfied if we choose the arbitrary constant $c_1 = a$,

$$w_1(x) = -\frac{1}{B}\int_{a}^{x} f(\xi)v_2(\xi)d\xi \tag{4.22}$$

If, on the other hand, we assume that only $v_1(x)$ satisfies the boundary condition (4.9) at $x = b$, steps similar to those of (4.21) yield

$$\begin{aligned}
\beta_1 u(b) + \beta_2 u'(b) &= \beta_1[w_1(b)v_1(b) + w_2(b)v_2(b)] + \beta_2[w_1(b)v_1'(b)\\
&\quad + w_2(b)v_2'(b)]\\
&= w_1(b)[\beta_1 v_1(b) + \beta_2 v_1'(b)] + w_2(b)[\beta_1 v_2(b) + \beta_2 v_2'(b)]\\
&= 0 + w_2(b)[\beta_1 v_2(b) + \beta_2 v_2'(b)] = 0,\\
&\quad w_2(b) = 0
\end{aligned}$$

since $\beta_1 v_2(b) + \beta_2 v_2'(b) \neq 0$.

If we apply this boundary condition on $w_2(x)$ in (4.20), we have

$$w_2(b) = \frac{1}{B}\int_{c_2}^{b} f(\xi)v_1(\xi)d\xi = 0$$

which is satisfied if we choose the arbitrary constant $c_2 = b$,

$$w_2(x) = \frac{1}{B}\int_{b}^{x} f(\xi)v_1(\xi)d\xi = -\frac{1}{B}\int_{x}^{b} f(\xi)v_1(\xi)d\xi. \tag{4.23}$$

With the variable coefficients $w_1(x)$ in (4.22) and $w_2(x)$ in (4.23), the final solution (4.10), of the nonhomogeneous differential equation 4.1 (with L as in (4.6)) with its associated homogeneous boundary conditions (4.2) and (4.3), becomes

$$
\begin{aligned}
u(x) &= w_1(x)v_1(x) + w_2(x)v_2(x) \\
&= -\frac{1}{B}v_1(x)\int_a^x f(\xi)v_2(\xi)d\xi - \frac{1}{B}v_2(x)\int_x^b f(\xi)v_1(\xi)d\xi \\
&= -\int_a^x \frac{1}{B}v_1(x)v_2(\xi)f(\xi)d\xi - \int_x^b \frac{1}{B}v_2(x)v_1(\xi)f(\xi)d\xi, \\
u(x) &= -\int_a^b G(x,\xi)f(\xi)d\xi
\end{aligned}
\tag{4.24}
$$

where $G(x, \xi)$ is defined as the Green's function with its *two branches*,

$$
G(x,\xi) = \begin{cases} \dfrac{1}{B}v_1(x)v_2(\xi), & \xi \le x \le b \\[2ex] \dfrac{1}{B}v_2(x)v_1(\xi), & a \le x \le \xi. \end{cases}
\tag{4.25}
$$

Basic Properties of the Green's Function

From this expression for $G(x, \xi)$ of (4.25) with the constant B in (4.18), we will show the following basic properties of the Green's function:

(a) It is clear that the Green's function in (4.25) is symmetric, that is,

$$G(x,\xi) = G(\xi, x)$$

This, of course, is dependent on B of (4.18) being a constant, which is a direct consequence of the differential operator L being self-adjoint.

(b) The Green's function satisfies the boundary conditions (4.8) and (4.9) since $v_1(x)$ of its first branch in (4.25) satisfies the condition (4.9) at $x = b$, and $v_2(x)$ in the second branch satisfies the boundary condition (4.8) at $x = a$.

(c) $G(x, \xi)$ is clearly continuous on the interval $a \le x \le b$; however, its derivative $\partial G(x, \xi)/\partial x$ has a jump discontinuity at $x = \xi$ which is

$$
\left.\frac{\partial G}{\partial x}(x,\xi)\right|_{\substack{x=\xi_+ \\ (x>\xi)}} - \left.\frac{\partial G}{\partial x}(x,\xi)\right|_{\substack{x=\xi_- \\ (x<\xi)}} = -\frac{1}{r(\xi)}
\tag{4.26}
$$

where $r(x)$ is the coefficient of $u''(x)$ in (4.7).

This can be proved by the use of the first branch ($x > \xi$) and the second branch ($x < \xi$) of the Green's function (4.25) for the above right-hand and left-hand derivatives of $G(x, \xi)$,

$$
\left.\frac{\partial G}{\partial x}(x,\xi)\right|_{\substack{x=\xi_+ \\ (x>\xi)}} - \left.\frac{\partial G}{\partial x}(x,\xi)\right|_{\substack{x=\xi_- \\ (x<\xi)}}
$$

$$= \frac{1}{B} v_1'(x) v_2(\xi) \Big|_{x=\xi_+} - \frac{1}{B} v_2'(x) v_1(\xi) \Big|_{x=\xi_-}$$

$$= \frac{1}{B} v_1'(\xi) v_2(\xi) - \frac{1}{B} v_2'(\xi) v_1(\xi)$$

$$= \frac{1}{B} [v_1'(\xi) v_2(\xi) - v_2'(\xi) v_1(\xi)]$$

$$= \frac{1}{B} \left[-\frac{B}{r(\xi)} \right] = -\frac{1}{r(\xi)}$$

after using (4.18) for the B constant value of the factor in brackets above. Property (c) warns against expecting a second derivative for $G(x, \xi)$ at $x = \xi$; however, the second derivative does exit away from this point, as $v_1(x)$ and $v_2(x)$ of (4.25) are indeed the solutions of $Lu = 0$. Hence we can conclude that

(d) $G(x, \xi)$ as a function of x satisfies the homogeneous equation, except at $x = \xi$,

$$LG(x, \xi) = 0, \qquad x \neq \xi \tag{4.27}$$

At $x = \xi$, we have $LG(x, y) = \delta(x - y)$, where $\delta(x - y)$ is defined as the *Dirac delta function* satisfying:

i) $\delta(x - \xi) = 0, x \neq y$

ii) $\int_{R_\epsilon} \delta(x - \xi) d\xi = 1, R_\epsilon : |x - \xi| < \epsilon$

iii) $\int_R \delta(x - \xi) F(\xi) d\xi = F(x)$

for arbitrary continuous function $F(x)$ in the region R: $a < x < b$. The first two properties i) and ii) of the Diract delta function $\delta(x - \xi)$ may allow the simple "popular" (not exact!) interpretation that $\delta(x-y)$ is some "distribution," unlike the usual function, which is zero every where, but spikes at $x = \xi$ in a very narrow neighborhood of width 2ϵ around x. Property iii) says that this very narrow spike, effectively selects the single value $F(x)$ of the integrated function $F(\xi)$ besides it inside the integral, where $F(x)$ is seen as the output of the integral $\int_R \delta(x - \xi) F(\xi) d\xi$.

We will start our illustrations by using the direct method of arriving at $G(x, \xi)$ in (4.25), then follow it by an example of using the four important properties **(a)–(d)** of the Green's function for a faster way of constructing it. But, before we do that we would like to discuss how this treatment of the Green's function, which is primarily aimed at solving boundary value problems, can be modified to treat *initial value problems*. This is followed by the discussion and illustration of boundary value problems associated with differential operators of order $n > 2$, and those with mixed boundary conditions that are not covered in the boundary conditions (4.8) and (4.9) of the present Sturm-Liouville problem, associated with the second-order (self-adjoint) differential operator L in (4.6).

Initial Value Problems

Our present treatment of finding the Green's function of the homogeneous boundary value problem (4.7)–(4.9), is usually aimed at solving the same boundary value problem with nonhomogeneous differential equation (4.1)–(4.3), with the particular solution as given in (4.5). In parallel to this treatment we may inquire about the *initial value problem* with the conditions

$$u(a) = 0 \tag{4.28}$$

$$u'(a) = 0 \tag{4.29}$$

and the second-order nonhomogeneous differential equation (4.6).

$$Lu \equiv \frac{d}{dx}\left[r(x)\frac{du}{dx}\right] - q(x)u(x) = f(x) \tag{4.30}$$

with the same differential operator L as in (4.6). We will show, with few modifications of the above method that the function $R(x,\xi)$, similar to the Green's function, is also used in an integral similar to that of (4.24) to give the solution of this initial value problem as

$$u(x) = \int_a^x R(x,\xi)f(\xi)d\xi, \tag{4.31}$$

$$\begin{aligned} R(x,\xi) &= -\frac{1}{B}[v_1(x)v_2(\xi) - v_2(x)v_1(\xi)] \\ &= \frac{1}{B}[v_2(x)v_1(\xi) - v_1(x)v_2(\xi)] \end{aligned} \tag{4.32}$$

where B is a constant as given in (4.18).

For the solution $u(x)$ in (4.10), the first initial condition (4.28) gives

$$u(a) = w_1(a)v_1(a) + w_2(a)v_2(a) = 0, \tag{4.33}$$

and the second initial condition (4.29) on $u'(x)$, after employing (4.14) gives

$$u'(a) = w_1(a)v_1'(a) + w_2(a)v_2'(a) = 0, \tag{4.34}$$

With $v_1v_2' - v_1'v_2 \neq 0$, this system of equations (4.33) and (4.34) in $w_1(a)$ and $w_2(a)$ gives the trivial solution $w_1(a) = 0$, $w_2(a) = 0$. The first result $w_1(a) = 0$ gives $w_1(x)$ as we had already in (4.22),

$$w_1(x) = -\frac{1}{B}\int_a^x f(\xi)v_2(\xi)d\xi \tag{4.35}$$

The second result $w_2(a) = 0$, which is what matters here for the initial value problem is satisfied if we choose the arbitrary constant $c_2 = a$ in (4.20),

$$
\begin{aligned}
w_2(a) &= \frac{1}{B}\int_{c_2}^{a} f(\xi)v_1(\xi)d\xi = 0, \\
w_2(x) &= \frac{1}{B}\int_{a}^{x} f(\xi)v_1(\xi)d\xi.
\end{aligned} \tag{4.36}
$$

So, if we use $w_1(x)$ from (4.35) and $w_2(x)$ from (4.36) in (4.10), we obtain the solution in (4.31) and (4.32)

$$
\begin{aligned}
u(x) &= w_1(x)v_1(x) + w_2(x)v_2(x) \\
&= -\frac{1}{B}\int_a^x [v_1(x)v_2(\xi) - v_2(x)v_1(\xi)]f(\xi)d\xi \\
&= \int_a^x R(x,\xi)f(\xi)d\xi
\end{aligned} \tag{4.37}
$$

where we have $R(x, \xi)$, the Green's function (-like) for the initial value problem (4.30), (4.28), (4.29), as we stated it in (4.32).

Example 2 Green's function (-like) for an initial value problem.

Consider the following initial value problem in $u(x)$,

$$
Lu \equiv \frac{d^2u}{dx^2} + \lambda^2 u = f(x), \qquad x > 0 \tag{E.1}
$$

$$
u(0) = 0 \tag{E.2}
$$

$$
u'(0) = 0 \tag{E.3}
$$

The two linearly independent solutions to the homogeneous equation

$$
\frac{d^2u}{dx^2} + \lambda^2 u = 0, \tag{E.4}
$$

are $v_1(x) = \sin \lambda x$ and $v_2(x) = \cos \lambda x$. Here $r(x) = 1$, and from (4.18) we have

$$
\begin{aligned}
B &= r(x)[v_1(x)v_2'(x) - v_2(x)v_1'(x)] \\
&= 1[-\lambda \sin \lambda x \sin \lambda x - \lambda \cos \lambda x \cos \lambda x] \\
&= -\lambda.
\end{aligned} \tag{E.5}
$$

From (4.32), the Green's function (-like) $R(x, \xi)$ for this initial value problem (E.1)–(E.3) is

$$
\begin{aligned}
R(x,\xi) &= -\frac{1}{B}[v_1(x)v_2(\xi) - v_2(x)v_1(\xi)] \\
&= \frac{1}{\lambda}[\sin \lambda x \cos \lambda \xi - \cos \lambda x \sin \lambda \xi] \\
&= \frac{\sin \lambda(x-\xi)}{\lambda}
\end{aligned} \tag{E.6}
$$

So from (4.37) the final solution to the initial value problem (E.1)–(E.3) is

$$u(x) = \int_0^x \frac{\sin \lambda(x-\xi)}{\lambda} f(\xi) d\xi. \tag{E.7}$$

To verify that this $u(x)$ is the solution to (E.1)–(E.3) we first need to prepare $u'(x)$ for (E.3) and $u''(x)$ for (E.1). For the integrand $g(x, \xi) = \frac{\sin \lambda(x-\xi)}{\lambda} f(\xi)$ in (E.7), it is best that we appeal to the generalized Leibnitz rule of (1.53),

$$\begin{aligned} u'(x) &= \int_0^x \frac{\partial}{\partial x}\left[\frac{\sin \lambda(x-\xi)}{\lambda} f(\xi)\right] d\xi \\ &\quad + \frac{\sin \lambda(x-x)}{\lambda} f(x) \frac{dx}{dx} - \frac{\sin \lambda(x-0)}{\lambda} f(0) \frac{d0}{dx} \\ &= \int_0^x \frac{\lambda \cos \lambda(x-\xi)}{\lambda} f(\xi) d\xi + 0 - 0 \\ &= \int_0^x \cos \lambda(x-\xi) f(\xi) d\xi. \end{aligned} \tag{E.8}$$

$$\begin{aligned} u''(x) &= \int_0^x \frac{\partial}{\partial x}\left[\cos \lambda(x-\xi) f(\xi)\right] d\xi \\ &\quad + \cos \lambda(x-x) f(x) \frac{dx}{dx} - \cos \lambda(x-0) f(0) \frac{d0}{dx} \\ &= -\lambda \int_0^x \sin \lambda(x-\xi) f(\xi) d\xi + f(x) - 0. \end{aligned} \tag{E.9}$$

So if we substitute this $u''(x)$ in (E.1) we have

$$\begin{aligned} u'' + \lambda^2 u &= -\lambda \int_0^x \sin \lambda(x-\xi) f(\xi) d\xi + f(x) \\ &\quad + \lambda^2 \int_0^x \frac{\sin \lambda(x-\xi)}{\lambda} f(\xi) d\xi \\ &= -\lambda \int_0^x \sin \lambda(x-\xi) f(\xi) d\xi + f(x) \\ &\quad + \lambda \int_0^x \sin \lambda(x-\xi) f(\xi) d\xi \\ &= f(x) \end{aligned}$$

where (E.1) is satisfied. To show the first initial condition (E.2), we use $u'(x)$ of (E.8) for $x = 0$ to have

$$u'(0) = \int_0^0 \cos \lambda(0-\xi) f(\xi) d\xi = 0$$

(E.1) is clearly satisfied as we substitute $x = 0$ in (E.7),

$$u(0) = \int_0^0 \frac{\sin \lambda(0-\xi)}{\lambda} f(\xi) d\xi = 0.$$

Higher Order Differential Equations

Before we start illustrating the method of constructing the Green's function for boundary value problems, we will make two remarks concerning very important points. The first is the question of the existence of a unique Green's function for the boundary value problem, a result that we shall state in the next Theorem 1 without proof. The second question regards the possible generalization of this treatment to boundary value problems associated with nth order differential equations.[4] Indeed the existence theorem for the Green's function applies to homogeneous boundary value problems associated with nth order differential operator, where our above problem of second-order becomes a special case. Let us, then, consider the nth order differential operator L_n,

$$L_n y \equiv A_0(x)\frac{d^n y}{dx^n} + A_1(x)\frac{d^{n-1}y}{dx^{n-1}} + \cdots + A_{n-1}(x)\frac{dy}{dx} + A_n y = 0 \tag{4.38}$$

instead of the second-order differential operator L in (4.1), and, instead of the two homogeneous boundary conditions in (4.2) and (4.3) we need the following n (linear, independent) homogeneous boundary conditions, as applied to the solution $y(x)$ and its first $n-1$ derivatives at $x = a$ and $x = b$, i.e.,

$$\begin{aligned} B_k y \; &= \alpha_k y(a) + \alpha_k' y'(a) + \cdots + \alpha_k^{n-1} y^{(n-1)}(a) \\ &+\beta_k y(b) + \beta_k' y'(b) + \cdots + \beta_k^{(n-1)} y^{(n-1)}(b) = 0, \\ &k = 1, 2, \cdots, n \end{aligned} \tag{4.39}$$

where B_k, $k = 1, 2, \cdots, n$ stands for the operators of these n boundary conditions. In (4.38) the coefficients $A_j(x)$, $j = 0, 1, 2, \cdots, n$ are real-valued functions with continuous derivatives up to the order $n - j$ on $[a, b]$, and $A_0(x) \neq 0$ on $[a, b]$.

We will soon list the four basic properties of the Green's function $G(x, \xi)$ associated with the homogeneous boundary value problem (4.38) and (4.39). With such Green's function we are able to obtain a Fredholm integral equation representation (4.41) in $u(x)$ for the following nonhomogeneous problem associated with the nth order differential operator L_n of (4.38),

$$L_n u + \lambda \rho(x) u = f(x), \qquad a \leq x \leq b \tag{4.40}$$

and (the homogeneous) boundary conditions (4.39),

$$u(x) = -\int_a^b G(x,\xi) f(\xi) d\xi + \lambda \int_a^b G(x,\xi)\rho(\xi) u(\xi) d\xi. \tag{4.41}$$

We may mention that a good sign for guarranteeing the existence of a unique Green's function for (4.41), is that the homogeneous problem (4.38) and (4.39) should have no solution but the trivial one. So we will assume that the general homogeneous

[4] Optional

boundary value problem (4.38) and (4.39) has only the trivial solution in order to guarantee the existence of its unique Green's function. As to the basic properties of this Green's function, we must bear in mind that, although the second-order differential operator can always be put in a self-adjoint form, by using a form of an integrating factor [as shown in (E.5) of Example 1], it is, in general, not the case for differential operators of order n larger than two. This is, possibly, the reason for not seeing the symmetry property of the Green's function at the top of the next list. The rest of the properties follow in parallel to those of the second order differential operator. The ordering of the following properties is influenced by bringing to focus the jump discontinuity property of the Green's function in (4.41).

(i) $G(x,\xi)$ is continuous, and so are all its derivatives with respect to x up to the order $(n-2)$ on the interval $a \le x \le b$. This leaves the expected jump discontinuity for its $(n-1)$th derivative as follows:

(ii) The $(n-1)$th derivative of $G(x,\xi)$ with respect to x at the point $x=\xi$ has a jump discontinuity of magnitude $1/A_0(x)$, i.e.,

$$\left.\frac{\partial^{n-1}G(x,\xi)}{\partial x^{n-1}}\right|_{\substack{x=\xi+\\(x>\xi)}} - \left.\frac{\partial^{n-1}G(x,\xi)}{\partial x^{n-1}}\right|_{\substack{x=\xi-\\(x<\xi)}} = -\frac{1}{A_0(x)}. \tag{4.42}$$

with $A_0(x)$ as in (4.38).

(iii) The Green's function satisfies the n homogeneous boundary conditions

$$B_kG = 0, \qquad k = 1, 2, \cdots, n \tag{4.39a}$$

with (the boundary operators) B_k as in (4.39).

(iv) In each of the two subintervals $a \le x < \xi$ and $\xi < x \le b$ the Green's function, as a function of x, satisfies the nth order homogeneous differential equation (4.38).

$$L_nG(x,\xi) = 0, \qquad x \neq \xi. \tag{4.38a}$$

For the lack of space, we will present in Example 5 only the final result of a boundary value problem associated with a *third-order* differential equation.

Example 3 The Hanging Chain or the Shape of Elastic Thread

Consider the problem of the hanging chain under the influence of the external force $F(x)$. This is a static problem where we can show that its displacement $y(x)$ satisfies the differential equation

$$\frac{d^2y}{dx^2} = -F(x) \tag{E.1}$$

with boundary conditions

$$y(0) = 0 \tag{E.2}$$

$$y(l) = 0. \tag{E.3}$$

The justification for the differential equation (E.1) can be taken from the special *static* (time-independent) case of the well-known *wave equation* of the vibrating string in its small vertical displacement $u(x,t)$,

$$\frac{\partial^2 u}{\partial x^2} - \frac{1}{c^2}\frac{\partial^2 u}{\partial t^2} = -F(x) \tag{E.4}$$

where for the time-independent static case $u(x,t) = y(x)$ it becomes

$$\frac{d^2 y}{dx^2} = -F(x) \tag{E.1}$$

c in (E.4) is the velocity of the wave. The differential operator $L = d^2/dx^2$ is self-adjoint as a very special case of L in (4.6) with $r(x) = 1$, $q(x) = 0$, and the two linearly independent solutions of the associated homogeneous differential equation

$$Ly = \frac{d^2 y}{dx^2} = 0 \tag{E.5}$$

are 1 and x. Note how we avoided for the moment calling these solutions $v_1(x)$ and $v_2(x)$. The reason is that, in addition to $v_1(x)$ and $v_2(x)$ being two linearly independent solutions of (E.5), they are also committed [in (4.25)] to satisfying the boundary condition $v_2(0) = 0$ at $x = a = 0$ and $v_1(l) = 0$ at $x = b = l$. We note here that we can have $v_2(x) = x$, which satisfies the first boundary condition; however, $v_1(x) = 1$, as is cannot satisfy the second boundary condition $v_1(l) = 0$. So for $v_1(x)$ we may consider a linear combination $v_1(x) = c_1 x + c_2$ of the two linearly independent solutions x and 1 and choose the arbitrary constants to satisfy the boundary condition $v_1(l) = 0$. An obvious choice is to let $c_2 = l$ and $c_1 = -1$ for $v_1(x) = l - x$, which is still a solution of (E.5), but now it also satisfies the boundary condition $v_1(l) = l - l = 0$.

With $v_2(x) = x$ and $v_1(x) = l - x$ we will find the constant B of (4.18) for the Green's function in (4.25),

$$\begin{aligned} B &= r(x)[v_1(x)v_2'(x) - v_2(x)v_1'(x)] \\ &= 1[(l-x)1 - x(-1)] = 1(l - x + x) = l \end{aligned} \tag{E.6}$$

so that the Green's function of (4.25),

$$G(x,\xi) = \begin{cases} \dfrac{1}{B} v_1(x) v_2(\xi), & 0 \le \xi \le x \le l \\ \dfrac{1}{B} v_2(x) v_1(\xi), & 0 \le x \le \xi \le l \end{cases} \tag{4.25}$$

becomes

$$G(x,\xi) = \begin{cases} \dfrac{1}{l}\xi(l-x), & 0 \le \xi \le x \le l \\ \\ \dfrac{1}{l}x(l-\xi), & 0 \le x \le \xi \le l. \end{cases} \tag{E.7}$$

This is the same form of $G(x,\xi)$ in (2.26) and Figure 2.2 for the shape of an elastic thread (under constant horizontal tension $T_0 = 1$), which is due to a single vertical force of unit magnitude $F = 1$ that is placed at $x = \xi$, $0 \le x \le l$. The derivation of $G(x,\xi)$ in (2.26) was based on simple balance of vertical and horizontal forces, and the geometrical shape as seen in Figure 2.2. We also saw in (2.37) of Example 5 in Chapter 2, a similar Green's function to that of $G(x,\xi)$ in the above equation (E.7). There we showed that the boundary value problem (1.33)–(1.35) reduces to the Fredholm integral equation (2.36) with its kernel as $K(x,t)$ in (2.37). Now we recognize this kernel $K(x,t)$ as the Green's function of the boundary value problem (1.33)–(1.35).

Finally, the solution to the boundary value problem (E.1)–(E.3) is obtained from (4.24) with $f(x) = -F(x)$ and $G(x,\xi)$ as in (E.7),

$$u(x) = \int_0^l G(x,\xi)F(\xi)d\xi. \tag{E.8}$$

As we indicated in the above Example 3, the expression for the Green's function (4.25) is not explicit enough; it is still left for us to make sure that $v_2(x)$ satisfies the boundary conditions (4.8) at $x = a$ only and $v_1(x)$ satisfies the boundary condition (4.9) at $x = b$, only. For the special boundary conditions

$$u(a) = 0 \tag{4.43}$$

$$u(b) = 0 \tag{4.44}$$

we give the more explicit result for the Green's function and leave its derivation for an exercise [see exercise 11(b)]

$$G(x,\xi) - \begin{cases} \dfrac{1}{BD}[v_2(x)v_1(a) - v_1(x)v_2(a)][v_2(\xi)v_1(b) - v_1(\xi)v_2(b)], a \le x \le \xi \\ \\ \dfrac{1}{BD}[v_2(\xi)v_1(a) - v_1(\xi)v_2(a)][v_2(x)v_1(b) - v_1(x)v_2(b)], \xi \le x \le b \end{cases} \tag{4.45}$$

where $D = v_2(a)v_1(b) - v_1(a)v_2(b) \neq 0$, and B as in (4.18).

Again, the condition $D \neq 0$ also guards against $v_1(x)$ and or $v_2(x)$ satisfying both of the boundary conditions (E.9) and (E.10). With this more explicit formula of Green's function in (4.45) we can now solve the problem in Example 3 more directly with $v_1(x) = 1$, $v_2(x) = x$, where $B = 1$ since

$$B = r(x)[v_1(x)v_2'(x) - v_1'(x)v_2(x)] = 1(1 \cdot 1 - 0 \cdot x) = 1$$

and

$$D = v_2(0)v_1(l) - v_1(0)v_2(l) = 0 \cdot 1 - 1 \cdot l = -l.$$

If we substitute these values of $v_1(x) = 1$, $v_2(x) = x$, $B = 1$ and $D = -l$ in (4.45) we have

$$G(x,\xi) = \begin{cases} -\frac{1}{l}[x \cdot 1 - 1(0)][\xi \cdot 1 - 1 \cdot l], & 0 \le x \le \xi \\ \\ -\frac{1}{l}[\xi \cdot 1 - 1(0)][x \cdot 1 - 1 \cdot l], & \xi \le x \le l \end{cases}$$

$$G(x,\xi) = \begin{cases} -\frac{1}{l}x(\xi - l), & 0 \le x \le \xi \\ \\ -\frac{1}{l}\xi(x - l), & \xi \le x \le l \end{cases}$$

$$= \begin{cases} \dfrac{x(l-\xi)}{l}, & 0 \le x \le \xi \\ \dfrac{\xi(l-x)}{l}, & \xi \le x \le l \end{cases}$$

which is the same answer (E.7) as in Example 3.

As we mentioned earlier, we will illustrate next the use of the four basic properties of the Green's function, instead of the explicit formulas (4.25) and (4.45). This is a very useful and familiar method which is often resorted to in the absence of explicit formulas (an example is the case of more complicated higher dimensional geometries). This method will be followed by another more general method of using infinite (orthogonal) series expansion to represent the Green's function.

Example 4 Construction of the Green's Function—Using its Basic Properties
Construct the Green's function for the boundary value problem

$$\frac{d^2y}{dx^2} + b^2 y = f(x), \qquad b \ne 0, \quad 0 < x < 1 \tag{E.1}$$

$$y(0) = 0 \tag{E.2}$$

$$y(1) = 0. \tag{E.3}$$

In this example we consider the case of (E.1) with $b \ne 0$; in Example 3 we illustrated the construction of the Green's function for the important special case of $b = 0$ in (E.1). First we note that the differential operator $L \equiv (d^2/dx^2) + b^2$ is self-adjoint since it is in the form $\frac{d}{dx}\left(\frac{dy}{dx}\right) + b^2 y$ of (4.7) with $r(x) = 1$, $q = -b^2$. Hence the Green's function for (E.1) is symmetric.

In an attempt to construct the Green's function for the boundary value problem above, we first investigate its corresponding homogeneous boundary value problem,

$$\frac{d^2y}{dx^2} + b^2 y = 0, \qquad 0 < x < 1 \tag{E.4}$$

$$y(0) = 0 \tag{E.2}$$

$$y(1) = 0. \tag{E.3}$$

We shall first use property (b) (following (4.25)) of the Green's function satisfying the (homogeneous) boundary conditions (4.8) and (4.9). Clearly, $\sin bx$ and $\cos bx$ are the two linearly independent solutions of (E.4) with $b \neq 0$. We know that $\sin bx$ and $\sin b(1-x)$ are also two linearly independent solutions of (E.4) with the added advantage that $\sin b(1-x)$, instead of $\cos bx$, satisfies the boundary condition (E.3). Hence we may use either $v_1(x) = \sin bx$, $v_2(x) = \cos bx$ or $v_1(x) = \sin bx$, $v_2(x) = \sin b(1-x)$ in a linear combination to construct the Green's function. Here, for convenience, we adopt the latter choice, and use it in (4.10) to write

$$G(x,\xi) = w_1(\xi)\sin bx + w_2(\xi)\sin b(1-x). \tag{E.5}$$

We leave it for the reader to show that to satisfy (E.3), the first choice will give the same result for the Green's function [see Exercise 2(b)].

To apply the boundary conditions (E.2) and (E.3) on (E.5) we must consider two cases:

(a) The case of $0 \le x \le \xi$, where we let $x = 0$ in (E.5) to satisfy (E.2):

$$\begin{aligned} G(0,\xi) &= w_1(\xi)\sin 0 + w_2(\xi)\sin b = w_2(\xi)\sin b = 0, \qquad w_2(\xi) = 0, \\ G(x,\xi) &= w_1(\xi)\sin bx, \qquad 0 \le x \le \xi. \end{aligned} \tag{E.6}$$

(b) The case of $\xi \le x \le 1$, where we let $x = 1$ in (E.5) to satisfy (E.3):

$$\begin{aligned} G(1,\xi) &= w_1(\xi)\sin b + w_2(\xi)\sin 0 = w_1(\xi)\sin b = 0, \qquad w_1(\xi) = 0, \\ G(x,\xi) &= w_2(\xi)\sin b(1-x), \qquad \xi \le x \le 1. \end{aligned} \tag{E.7}$$

The results (E.6) and (E.7) exemplify the two branches of the Green's function [i.e., (E.6) to satisfy the boundary condition at $x = 0$ and (E.7) to satisfy the boundary condition at $x = 1$]. Hence from (E.6) and (E.7) we have

$$G(x,\xi) = \begin{cases} w_1(\xi)\sin bx, & 0 \le x \le \xi \\ w_2(\xi)\sin b(1-x), & \xi \le x \le 1. \end{cases} \tag{E.8}$$

Now we use the symmetry property (a): $G(x,\xi) = G(\xi,x)$ of the Green's function; a clear choice for the arbitrary functions $w_1(\xi)$ and $w_2(\xi)$, to make $G(x,\xi)$ in (E.8) symmetric in x and ξ, is $w_1(\xi) = C\sin b(1-\xi)$ and $w_2(\xi) = C\sin b\xi$; (E.8) becomes

$$G(x,\xi) = \begin{cases} C \sin b(1-\xi)\sin bx, & 0 \le x \le \xi \\ C \sin b\xi \sin b(1-x), & \xi \le x \le 1. \end{cases} \tag{E.9}$$

To evaluate the arbitrary constant C in (E.9) we use property (c) for the jump condition (4.26) of the derivative $\partial G/\partial x$,

$$\left.\frac{\partial G(x,\xi)}{\partial x}\right|_{x=\xi_+} - \left.\frac{\partial G(x,\xi)}{\partial x}\right|_{x=\xi_-} = \frac{-1}{r(\xi)}$$

$$= (C \sin b\xi)(-b \cos b(1-x)\,|_{x=\xi_+} - (C \sin b(1-\xi))(b \cos bx)\big|_{x=\xi_-}$$

$$= -Cb \sin b\xi \cos b(1-\xi) - Cb \sin b(1-\xi)\cos b\xi = -1,$$

$$Cb \sin b(\xi + 1 - \xi) = Cb \sin b = 1, \qquad C = \frac{1}{b \sin b}. \tag{E.10}$$

Note how we used the second and first branches of $G(x,\xi)$ in (E.9) for $x = \xi_+$ and $x = \xi_-$, respectively, since $x = \xi_+ > \xi$ is in the domain $\xi \le x \le 1$ and $x = \xi_- < \xi$ is in the domain $0 \le x \le \xi$. From (E.9) and (E.10) the final form for the Green's function is

$$G(x,\xi) = \begin{cases} \dfrac{\sin b(1-\xi)\sin bx}{b \sin b}, & 0 \le x \le \xi \\ \dfrac{\sin b\xi \sin b(1-x)}{b \sin b}, & \xi \le x \le 1. \end{cases} \tag{E.11}$$

We will leave it as a simple exercise to show that $G(x,\xi)$ of (E.11) satisfies (the initial part of) property (c) of being continuous, and satisfies condition (d) in (4.27) by satisfying the homogeneous boundary value problem (E.4) for $x \ne \xi$, (E.2), and (E.3) (see exercise 2).

Higher Order Differential Equations—An Illustration

To summarize, our above detailed treatment and illustration was concentrated, mainly, on the construction of the Green's functions associated with second-order differential equations on (a,b) and general boundary conditions at $x = a$ and b as given in (4.1)–(4.3). The basic method of variations of parameters was used to generate the Green's function in (4.25) and was illustrated in Example 3 for the simple boundary conditions $y(0) = 0$, $y(l) = 0$. From such analysis we developed four basic properties (a)–(d) of the Green's function, which were also used as a more efficient way of constructing the Green's function as illustrated in Example 4 for another second-order differential operator and the same boundary conditions. In the following Example 5 we will present the final result for a simple *third-order* differential equation with a different set of boundary conditions, and where Theorem 1 is used to first test for the existence of the unique Green's function.

Example 5 Green's Function—A Boundary Value Problem Associated with Third Order Differential Equation

Consider the following boundary value problems, associated with a third-order differential equation, $(Ly \equiv \frac{d^3y}{dx^3})$,

$$\frac{d^3y}{dx^3} + \lambda y = 2x, \qquad 0 < x < 1 \tag{E.1}$$

or

$$\frac{d^3y}{dx^3} = -\lambda y + 2x, \qquad 0 < x < 1$$

$$y(0) = 0 \tag{E.2}$$

$$y(1) = 0 \tag{E.3}$$

$$y'(0) = y'(1). \tag{E.4}$$

The method of Section 2.5, with multiple integration (using the identity (1.52)) and the application of the boundary conditions, is used to reduce this problem to the Fredholm integral equation of the second kind

$$y(x) = \lambda \int_0^1 G(x,t)y(t)dt + \frac{1}{12}x(x-1)(x^2+x-1) \tag{E.5}$$

with the Green's function as

$$G(x,t) = \begin{cases} -\frac{1}{2}x(x-t)(1-t), & 0 \le x \le t \\ \frac{1}{2}t(t-x)(1-x), & t \le x \le 1. \end{cases} \tag{E.6}$$

With the statement made regarding the existence of a unique Green's function, we have the chance now to test it for the problem associated with the third-order differential operator $L = d^3/dx^3$ in the equation

$$Ly \equiv \frac{d^3y}{dx^3} = 0 \tag{E.7}$$

and its homogeneous boundary conditions in (E.2)–(E.4). All we have to do is to show that the problem (E.7), (E.2)–(E.4) has only the trivial solution $y(x) \equiv 0$. The solution to (E.7), after three integrations is

$$y(x) = c_1\frac{x^2}{2} + c_2x + c_3. \tag{E.8}$$

If we use the boundary condition (E.2) we have $y(0) = c_3 = 0$, $y(x) = (c_1/2)x^2 + c_2x$ and from (E.3) we have

$$\begin{aligned} y(1) &= \frac{1}{2}c_1 + c_2 = 0, \qquad c_1 = -2c_2, \\ y(x) &= \frac{1}{2}c_1x^2 - \frac{1}{2}c_1x. \end{aligned}$$

If we use (E.4) on $y'(x) = c_1 x - \frac{1}{2}c_1$ we obtain

$$y'(0) = 0 = y'(1) = \frac{1}{2}c_1, \qquad c_1 = c_2 = 0.$$

So the three conditions result in $c_1 = c_2 = c_3 = 0$. Hence the solution to the homogeneous boundary value problem (E.7), (E.2)–(E.4) is $y(x) \equiv 0$, a trivial solution. Therefore our problem will have a unique Green's function that we shall leave its construction for an exercise.

4.1.3 Orthogonal Series Representation of Green's Function

Next we will develop the method of series representation of the Green's function. Such series expansion of $G(x, \xi)$ is in terms of the solutions $\{u_n(x)\}_{n=1}^{\infty}$ of the associated homogeneous problem (4.7)–(4.9). We will first discuss the basic properties of these functions and their series expansion, which are very necessary for developing this method of constructing the Green's function.

An extremely important result of the Sturm-Liouville problem (4.7)–(4.9) and its self-adjoint operator (as an eigenvalue problem) is that under the conditions, on the (regular) differential operator L of (4.7), that $r(x)$, $r'(x)$, $q(x)$ and $\rho(x)$ are continuous on the closed interval $a \leq x \leq b$, and that $r(x) > 0$, $\rho(x) > 0$ on $[a, b]$, the solutions $\{u_n(x)\}$ (or eigenfunctions) of the Sturm-Liouville problem are *orthogonal*. By orthogonality of $\{u_n(x)\}$ on the interval (a, b) we mean that for any two different solutions $u_n(x)$ and $u_m(x)$ (of (4.7)–(4.9)) the following integral vanishes:

$$\int_a^b \rho(x) u_n(x) \overline{u_m(x)} dx = 0, \qquad n \neq m \tag{4.46}$$

where $\overline{u_m(x)}$ is the complex conjugate of $u_m(x)$.[5] Here $\rho(x)$ is the (weight) function appearing in (4.7) and a and b are the limits of the interval on which the problem (4.7)–(4.9) is defined. Next we illustrate how the orthogonality property of the solutions $\{u_n(x)\}$ can be employed in expanding given functions in an infinite series of these orthogonal functions—hence the name *orthogonal* or *Fourier series expansion*—which we will use in determining the solutions of certain Fredholm equations in Chapter 5 (in particular, Fredholm integral equations with *symmetric kernel* in Section 5.2.)

The importance of the orthogonality of functions is not limited to series expansion but, as we will see in Chapter 5, will be used as a condition for some of the theorems proved concerning the Fredholm equation. For example, we may need to investigate whether the nonhomogeneous part $f(x)$ in the following Fredholm equation

$$u(x) = \lambda \int_a^b K(x, t) u(t) dt + f(x)$$

[5] In most cases we will have real-valued functions $u_m(x)$, where $\overline{u_m(x)} = u_m(x)$.

is orthogonal to the solutions $u_n(x)$ of the associated homogeneous equation

$$u_n(x) = \lambda_n \int_a^b K(x,t)u_n(t)dt.$$

Also, sometimes we may speak of whether $K(x,t)$ is orthogonal to a function or even to itself.

Sturm-Liouville Problem and the Orthogonal (Fourier) Series Expansion

The orthogonality (4.46) of the solutions of the Sturm-Liouville problem can be easily proved, which we leave as an exercise. We concentrate instead on its very important role in applied mathematics for determining the coefficients c_n of the *orthogonal* or *Fourier series* expansion

$$f(x) = \sum_{n=1}^{\infty} c_n u_n(x) \tag{4.47}$$

of (usually sectionally continuous) or square integrable functions $f(x)$, defined on the interval (a,b), in terms of the orthogonal set of functions $\{u_n(x)\}_{n=1}^{\infty}$.

In the next example we show how the orthogonality (4.46) is used in determining the form of the coefficients of the orthogonal series expansion (4.47) as

$$c_n = \frac{\int_a^b \rho(x)f(x)u_n(x)dx}{\int_a^b \rho(x)u_n^2(x)dx}. \tag{4.48}$$

We mention here for future reference that the integral in the denominator of (4.48) is called the *norm square* of the function $u_n(x)$ and is denoted by

$$||u_n||^2 = \int_a^b \rho(x)u_n^2(x)dx. \tag{4.49}$$

When the orthogonal functions $u_n(x)$ are divided by their norm $||u_n||$,

$$\phi_n(x) = \frac{u_n(x)}{||u_n||}$$

the resulting functions $\phi_n(x)$ are called *orthonormal functions*; it is easy to show that their norm is 1,

$$\begin{aligned}\int_a^b \rho(x)\phi_n^2(x)dx &= \int_a^b \rho(x)\frac{u_n^2(x)}{||u_n||^2}dx \\ &= \frac{1}{||u||_n^2}\int_a^b \rho(x)u_n^2(x)dx = \frac{||u_n||^2}{||u_n||^2} = 1\end{aligned}$$

after employing (4.49).

Also, as we mentioned earlier, the solutions $u_n(x)$ to (4.7)–(4.9) are called the *characteristic functions or eigenfunctions* of the differential operator, for example, the operator L_1 in (4.7),

$$L_1 u_n(x) \equiv \frac{d}{dx}\left[r(x)\frac{du_n}{dx}\right] + q(x)u_n(x) = -\lambda_n \rho(x) u_n(x) \tag{4.50}$$

has the characteristic functions or eigenfunctions $u_n(x)$ with λ_n as their corresponding *characteristic values or eigenvalues*. We will often refer to the orthogonal expansion (4.47) and (4.48) as the *eigenfunctions* expansion.

Example 6 Orthogonal (Fourier) Series Expansion

To show that the coefficient c_n of the orthogonal expansion (4.47) of $f(x)$ is given by (4.48), we first write

$$f(x) = \sum_{n=1}^{\infty} c_n u_n(x) \tag{E.1}$$

then we multiply both sides of (E.1) by $\rho(x)u_m(x)$ and integrate from a to b to obtain

$$\begin{aligned} \int_a^b \rho(x)u_m(x)f(x)dx &= \int_a^b \rho(x)u_m(x)\sum_{n=1}^{\infty} c_n u_n(x)dx \\ &= \sum_{n=1}^{\infty} c_n \int_a^b \rho(x)u_m(x)u_n(x)dx \end{aligned} \tag{E.2}$$

after allowing the interchange of integration with the infinite summation on the right-hand side which is permissible when $f(x)$ is square integrable on $(a,b)_\rho$, where $\rho(x)$ is the weight function as in (4.48) [i.e., $\int_a^b \rho(x)f^2(x)dx$ is finite].

Now the integral inside the infinite series is that which describes the orthogonality property of $\{u_n(x)\}_{n=1}^{\infty}$ and vanishes according to (4.46) when $n \neq m$. This leaves the series with only one term, $c_m \int_a^b \rho(x)u_m^2(x)dx$, and hence (E.2) becomes

$$\int_a^b \rho(x)u_m(x)f(x)dx = c_m \int_a^b \rho(x)u_m^2(x)dx,$$

$$c_m = \frac{\int_a^b \rho(x)u_m(x)f(x)dx}{\int_a^b \rho(x)u_m^2(x)dx} \tag{E.3}$$

which is the Fourier coefficient (4.48) of the orthogonal or Fourier series expansion (4.47) of $f(x)$ on $(a,b)_\rho$ in terms of the orthogonal functions $\{u_m(x)\}_{m=1}^{\infty}$. It is important to note the simpler form for c_n when $\{u_n(x)\}$ are chosen orthonormal as $\{\phi_n(x)\}$, whence

$$c_n = \int_a^b \rho(x)\phi_n(x)f(x)dx \tag{E.4}$$

Convergence in the Mean of the Orthogonal Series Expansion

Of great importance to the Fourier series representation of square integrable functions $f(x)$, is the convergence in the mean of the general orthogonal series of (4.47) to $f(x)$ on the interval $(a,b)\rho$, where $\rho(x)$ is the weight function used in the integrals of the Fourier coefficients in (4.48).

Let $S_N(x)$ be the Nth partial sum of the general orthogonal expansion in (4.47),

$$S_N(x) = \sum_{n=1}^{N} c_n u_n(x). \tag{4.51}$$

The series in (4.47) is said to *converge to f(x) in the mean* on the interval $(a,b)\rho$ (with respect to the weight function $\rho(x)$) if

$$\lim_{N\to\infty} \int_a^b \rho(x)|f(x) - \sum_{n=1}^{N} c_n u_n(x)|^2 dx = 0. \tag{4.52}$$

If the orthogonal series (4.47) converges in the mean for every piecewise continuous function $f(x)$ (or square integrable function) on $(a,b)\rho$, then the orthogonal set $\{u_n(x)\}_{n=1}^{\infty}$ is called a *complete* orthogonal set on $(a,b)\rho$. It turns out that the completeness of the orthogonal set $\{u_n(x)\}_{n=1}^{\infty}$ in the series (4.47) is equivalent to allowing integrating such series term by term, a fact that we used in (E.2) of the above example.

Eigenfunction Expansion of the Green's Function

Besides the direct method of using variation of parameters that resulted in (4.25) or (4.45), we will present here another method (for constructing the Green's function for the nonhomogeneous problem)

$$Lu + \lambda u = f(x) \tag{4.53}$$

where L is a self-adjoint operator. For the present method we expand $u(x)$ and $f(x)$ of (4.53) in a Fourier series of the orthonormal[6] eigenfunctions $\{u_k(x)\}$:

$$u(x) = \sum_{k=1}^{\infty} a_k u_k(x), \qquad a_k = \int_a^b u(x)u_k(x)dx \tag{4.54}$$

$$f(x) = \sum_{k=1}^{\infty} b_k u_k(x), \qquad b_k = \int_a^b f(x)u_k(x)dx \tag{4.55}$$

$$Lu_k = -\lambda_k u_k \tag{4.56}$$

of the operator L [see (4.50) with $\rho(x) = 1$].

[6] $\int_a^b u_k^2(x)dx = 1$. Also $\rho(x)$ of (4.49) can be introduced with simple modification.

If we substitute the expansions (4.54) and (4.55) and use (4.56) in (4.53), we, formally, obtain

$$\sum_{k=1}^{\infty} a_k(\lambda - \lambda_k)u_k(x) = \sum_{k=1}^{\infty} b_k u_k(x). \tag{4.57}$$

But since the eigenfunctions $u_k(x)$ are linearly independent, we may equate the coefficients in (4.57) to obtain

$$a_k(\lambda - \lambda_k) = b_k, \qquad a_k = \frac{b_k}{\lambda - \lambda_k}. \tag{4.58}$$

Hence from (4.54) and (4.58) the solution $u(x)$ to the nonhomogeneous equation (4.53) is

$$\begin{aligned} u(x) &= \sum_{k=1}^{\infty} a_k u_k(x) = \sum_{k=1}^{\infty} \frac{b_k}{\lambda - \lambda_k} u_k(x) \\ &= \sum_{k=1}^{\infty} \frac{u_k(x)}{\lambda - \lambda_k} \int_a^b f(t)u_k(t)dt = \int_a^b f(t) \left[\sum_{k=1}^{\infty} \frac{u_k(x)u_k(t)}{\lambda - \lambda_k} \right] dt \end{aligned} \tag{4.59}$$

after using

$$b_k = \int_a^b f(t)u_k(t)dt \tag{4.60}$$

from (4.55) and exchanging summation with integration.

The solution (4.59) can be written in the form (4.5)

$$u(x) = -\int_a^b G(x,t)f(t)dt \tag{4.61a}$$

where $G(x,t)$ is the Green's function

$$G(x,t) = \sum_{k=1}^{\infty} \frac{u_k(x)u_k(t)}{\lambda_k - \lambda}. \tag{4.61b}$$

Example 7 Nonhomogeneous Boundary Value Problem

Solve the following boundary value problem by using the Green's function.

$$\frac{d^2y}{dx^2} + \lambda y = f(x), \qquad 0 < x < 1 \tag{E.1}$$

$$y(0) = 0 \tag{E.2}$$

$$y(1) = 0 \tag{E.3}$$

The orthonormal eigenfunctions of the corresponding homogeneous problem,

$$\frac{d^2y}{dx^2} + \lambda y = 0, \qquad 0 < x < 1 \tag{E.4}$$

are $u_k(x) = \sqrt{2}\sin k\pi x$ and the eigenvalues are $k^2\pi^2$. Hence from (4.61b) the Green's function for (E.1) is

$$G(x,t) = -2\sum_{k=1}^{\infty} \frac{\sin k\pi x \sin k\pi t}{\lambda - (k\pi)^2} \tag{E.5}$$

and from (4.61a) the solution to the boundary value problem (E.1)–(E.3) is

$$u(x) = 2\sum_{k=1}^{\infty} \frac{\sin k\pi x}{\lambda - (k\pi)^2} \int_0^1 f(t) \sin k\pi t dt. \tag{E.6}$$

4.1.4 Green's Function in Two Dimensions

In this section our discussion was centered, mainly, around the Green's function for boundary value problems associated with ordinary differential equations. In the next section, the Green's function will be used to reduce such boundary value problems to Fredholm integral equations in one variable. In higher dimensional problems, we will encounter boundary value problems associated with *partial differential equations*. One of the methods of constructing the Green's function for such boundary value problems is illustrated next for the potential distribution in a unit disc. The resulting Fredholm integral equation, as expected, will involve the unknown in two variables inside a double integral. Integral equations in three dimensions are also illustrated in Section 2.7 for the Schrödinger equation in the (three-dimensional) momentum space, where the three dimensional Fourier transform is used.

There are a variety of methods for constructing the Green's function for boundary value problems in two and three dimensions. However, for the level of this introductory text, we will limit our very brief presentation here to the following illustration. This involves the potential distribution in a charged unit disc with grounded rim, as we shall discuss next. The same type problem of the potential distribution in a square is left for an exercise (Exercise 24) with very detailed leading hints.

Potential Distribution in a Charged Unit Disc (-Poisson Equation)

The potential function $u(r,\theta)$ due to a charge distribution $f(r,\theta)$ on a unit disc, is governed by the following *Poisson equation* (4.62). We use here *polar* coordinates (r,θ) so that the boundary of the disc as a unit circle is represented simply by one of the coordinates being a constant, namely, $r = 1$. In polar coordinates the *Poisson equation* in $u(r,\theta)$ is

$$\nabla^2 u(r,\theta) = -f(r,\theta)$$

$$\equiv \frac{1}{r^2}\left[\frac{\partial}{\partial r}\left(r^2\frac{\partial u}{\partial r}\right) + \frac{1}{\sin\theta}\frac{\partial}{\partial\theta}\left(\sin\theta\frac{\partial u}{\partial\theta}\right)\right] = -f(r,\theta) \tag{4.62}$$

where $f(r,\theta)$ is the charge density. Also, with the help of complex analysis methods, the *Green's function* of this problem takes the form:

$$G(r,\theta;\rho,\phi) = \frac{1}{4\pi}\ln\frac{1 - 2r\rho\cos(\theta-\phi) + r^2\rho^2}{\rho^2 - 2\rho r\cos(\theta-\phi) + \rho^2}. \tag{4.63}$$

So with a *grounded* rim of the unit disc, we have the boundary condition at $r = 1$,

$$u(1,\theta) = 0, \qquad -\pi < \theta < \pi. \tag{4.64}$$

The solution to the boundary value problem of the Poisson equation (4.62) and the boundary condition (4.64) via the Green's function in (4.63), is

$$u(r,\theta) = \int_0^1 d\rho \int_0^{2\pi} G(r,\theta;\rho,\phi) f(\rho,\phi) d\phi. \tag{4.65}$$

The same Green's function in (4.63) is also used to solve another very related and familiar problem, namely the *Dirichlet problem* on a unit disc. This is a boundary value problem governed by the *Laplace equation*.

$$\nabla^2 u(r,\theta) = 0 \tag{4.66}$$

inside the disc (with no charge, $f(r,\theta) \equiv 0$), and where the potential on the rim is given as

$$u(1,\theta) = g(\theta), \qquad -\pi < \theta < \pi. \tag{4.67}$$

The solution to this Dirichlet boundary value problem (4.66) and (4.67) is of the form

$$u(r,\theta) = -\int_0^{2\pi} \left.\frac{\partial G}{\partial \rho}\right|_{\rho=1} g(\phi) d\phi \tag{4.68}$$

$$= \frac{1}{2\pi}\int_0^{2\pi} \frac{1-r^2}{1-2r\cos(\theta-\phi)+r^2} g(\phi) d\phi, \tag{4.69}$$

which is what we presented in (1.24).

Of course equation (4.65) can be made as a Fredholm integral equation in *two* dimensions, when a charge distribution $f(r,\theta)$ is to be found on the disc to affect the *given* desired potential distribution $u(r,\theta)$ there. In the same way we can say that equation (4.69) is a Fredholm integral equation in *one* dimension in the unknown potential function $f(\theta)$ on the rim of the unit disc, that would produce the given desired potential distribution $u(r,\theta)$ in the interior of the unit disc. The integral in (4.69) is the well known *Poisson integral*.

Another example of a Fredholm integral equation in two dimensions can be made of the answer of Exercise 24 (parts e, f in (E.10)), when we are to ask about the required charge distribution $f(x,y)$ on a square that would result in the given desired potential distribution $u(x,y)$ inside the charged square.

Exercises 4.1

1. Solve the following boundary value problems associated with second-order differential equations on the indicated domain and the particular boundary conditions. Note that for infinite domain we must consider the boundedness of the solution as a condition. λ is assumed real in all the following problems.

(a) $y'' + \lambda^2 y = 0, \qquad 0 < x < l$
$y(0) = y(l) = 0$

(b) $y'' + \lambda^2 y = 0, \qquad 0 < x < l$
$y'(0) = y'(l) = 0$

(c) $y'' + \lambda^2 y = 0, \qquad 0 < x < l$
$y'(0) = 0, \qquad y(l) = 0$

(d) $y'' + \lambda^2 y = 0, \qquad 0 < x < \infty$
$y(0) = 0, \qquad |y(x)| < \infty$

(e) $y'' + \lambda^2 y = 0, \qquad -\infty < x < \infty$
$|y(x)| < \infty$

(f) $y'' - \lambda^2 y = 0, \qquad 0 < x < \infty$
$y(0) = 1, \qquad |y(\infty)| < \infty$

(g) $y'' - \lambda^2 y = 0, \qquad 0 < x < l$
$y(0) = 1, \qquad y(l) = 0$

2. **(a)** Verify all the properties of the Green's function

$$G(x, \xi) = \begin{cases} \dfrac{\sin b(1-\xi)\sin bx}{b\sin b}, & 0 \le x \le \xi \\ \dfrac{\sin b\xi \sin b(1-x)}{b\sin b}, & \xi \le x \le 1 \end{cases}$$

of (E.11) in Example 4.

(b) Consider the choice of the two linearly independent solutions $v_1(x) = \sin bx$ and $v_2(x) = \cos bx$ of (E.4) of Example 4. Follow the same steps in Example 4 to show that you obtain the same result for the Green's function in (E.11).

Construct the Green's function associated with the following boundary value problems.

3. $\dfrac{d^2y}{dx^2} = f(x), \qquad 0 < x < \dfrac{\pi}{2}$
$y(0) = 0, \qquad y\left(\frac{\pi}{2}\right) = 0$
Hint: See (E.6)–(E.9) of Example 8 in Section 4.2.

4. $y'' - b^2 y = f(x), \qquad 0 < x < 1$
$y(0) = 0, \qquad y(1) = 0$

5. Use the Green's function to solve the boundary value problem

$$\frac{d^2y}{dx^2} - y = x, \qquad 0 < x < 1$$

$$y(0) = 0, \qquad y(1) = 0$$

Hint: Use (4.5) and the Green's function of Exercise 4 with $b = 1$. For a series solution form see Example 7 with $\lambda = -1$ and $f(x) = x$.

Use the Green's function to solve the following boundary value problems:

6. $$\frac{d^2y}{dx^2} - y = 2\sin \text{ch}1, \qquad 0 < x < 1 \tag{E.1}$$

$$y(0) = 0, \qquad y(1) = 0 \tag{E.2}$$

Hint: See Exercise 6.

7. Reduce the following boundary value problem, associated with *nonlinear* differential equation, to an integral equation.

$$\frac{d^2y}{dx^2} = f(x, y(x)), \qquad 0 < x < 1$$

$$y(0) = 0, \qquad y(1) = 0$$

Hint: Use the Green's function of Exercise 3 with $\pi/2$ replaced by 1.

8. Use the Green's function—like approach in (4.31) and (4.32), (4.28)–(4.30) to find the solution of the following *initial value problem*,

$$\frac{d^2y}{dx^2} + \lambda^2 y(x) = \cos x, \qquad x > 0 \tag{E.1}$$

$$y(0) = 0 \tag{E.2}$$

$$y'(0) = 0 \tag{E.3}$$

9. Let $v_1(x)$ and $v_2(x)$ be the two linearly independent solutions of the homogeneous equation $Lu = 0$ of (4.7).

(a) Verify that the following $u(x)$ is a solution to the associated (same L) nonhomogeneous equation (4.1),

$$u(x) = C_1 v_1(x) + C_2 v_2(x) + \int_0^x R(x, \xi) f(\xi) d\xi, \tag{E.1}$$

$$R(x, \xi) = -\frac{1}{B}[v_1(x)v_2(\xi) - v_2(x)v_1(\xi)] \tag{E.2}$$

and B is as in (4.18).

(b) Use the result in part (a) to derive the explicit form (4.45) of the Green's function associated with this operator L and the particular boundary conditions $u(a) = 0$ and $u(b) = 0$ of (4.43) and (4.44). *Hint:* Apply the two boundary conditions to $u(x)$ in (E.1) and (E.2) to have two simultaneous equations in C_1 and C_2 to be determined.

Reduce the following boundary value problems to Fredholm integral equations. *Hint:* For all these problems take the term λy to the right side of the differential equation, and consider the resulting $-\lambda y(x) + f(x)$ as if it is a nonhomogeneous term associated with the (new) homogeneous equation $\frac{d^2y}{dx^2} = 0$. Also see problems 14 and 15 and Example 5 (for the mixed boundary conditions).

10. $\frac{d^2y}{dx^2} + \lambda y = 2x + 1, \qquad 0 < x < 1$ (E.1)

$y(0) = y'(1)$ (E.2)

$y'(0) = y(1)$ (E.3)

11. $\frac{d^2y}{dx^2} + \lambda y = e^x, \qquad 0 < x < 1$ (E.1)

$y(0) = y'(0)$ (E.2)

$y(1) = y'(1)$ (E.3)

12. $\frac{d^2y}{dx^2} + \frac{\pi^2}{4} y = \lambda y + \cos\frac{\pi x}{2}, \qquad -1 < x < 1$ (E.1)

$y(-1) = y(1)$ (E.2)

$y'(-1) = y'(1).$ (E.3)

Use the Green's function to solve the following boundary value problems.

13. $\frac{d^2y}{dx^2} - y = -2e^x$ (E.1)

$y(0) = y'(0)$ (E.2)

$y(1) = -y'(1).$ (E.3)

Hint: See Example 5 and the above problem 11 for the mixed boundary conditions.

Find the Green's function associated with the following boundary value problems. *Hint:* See Example 5 for the mixed boundary conditions.

14. $\frac{d^2y}{dx^2} = f(x), \qquad 0 < x < 1$ (E.1)

$y(0) = y'(1)$ (E.2)

$y'(0) = y(1).$ (E.3)

15. $\frac{d^2y}{dx^2} = f(x), \qquad 0 < x < 1$ (E.1)

$$y(0) = y'(0) \tag{E.2}$$
$$y(1) = y'(1) \tag{E.3}$$

16. $$\frac{d^2y}{dx^2} + \frac{\pi^2}{4}y = f(x), \qquad -1 < x < 1 \tag{E.1}$$
$$y(-1) = y(1) \tag{E.2}$$
$$y'(-1) = y'(1) \tag{E.3}$$

17. **(a)** For Exericses 3 to 6, verify Theorem 1 concerning the existence of a unique Green's function.

(b) Determine whether or not a unique Green's function exists for the following boundary value problems, and if it does, construct it.

(i) $$\frac{d^2y}{dx^2} = f(x), \qquad 0 < x < 1 \tag{E.1}$$
$$y(0) = y'(1) \tag{E.2}$$
$$y'(0) = y(1) \tag{E.3}$$

(ii) $$\frac{d^2y}{dx^2} = f(x), \qquad 0 < x < 1 \tag{E.4}$$
$$y(0) = y(1) \tag{E.5}$$
$$y'(0) = y'(1) \tag{E.6}$$

(iii) $$\frac{d^2y}{dx^2} + y = f(x), \qquad 0 < x < \pi \tag{E.7}$$
$$y(0) = y(\pi) \tag{E.8}$$

18. In our discussion regarding the construction of the Green's function (4.25), we assumed that neither of the solutions $v_1(x)$ and $v_2(x)$ of the homogeneous problem satisfies both boundary conditions at $x = a$ and $x = b$.

(a) Assume now that $v_2(x)$ does satisfy both conditions, while $v_1(x)$ satisfies neither; follow the same steps as those used in reaching (4.25), with a solution of the form

$$u(x) = \frac{1}{B}v_1(x)\int_{c_1}^{x} f(\xi)v_2(\xi)d\xi - \frac{1}{B}v_2(x)\int_{c_2}^{x} f(\xi)v_1(\xi)d\xi \quad (E.1)$$

as in (4.10), (4.19), and (4.20) to show that this would require that $v_2(x)$ and $f(x)$ must satisfy another (consistency) condition (4.11),

$$\int_a^b v_2(x)f(x)dx = 0 \qquad (4.11)(E.2)$$

Note that the final solution $u(x)$ does, of course, satisfy both boundary conditions. So apply these conditions on $u(x)$ in (E.1) to arrive at two vanishing integrals that add up to that of (E.2).

(b) With the results in part (a), show that the solution is not unique; that is, show that the solution becomes

$$u(x) = Cv_2(x) + \int_a^b G(x,\xi)f(\xi)d\xi \qquad (E.3)$$

where C is still arbitrary and $G(x,\xi)$ is the same as in (4.25). *Hint:* The c_2 in (E.1), which we could not determine because of the assumed boundary conditions for $v_2(x)$ only, is equivalent to the usual arbitrary constant of integration, so write the second integral of (E.1) on (b,x) with the rest of it on (c_2,b) as constant to give you the term $Cv_2(x)$, remembering that $c_1 = a$ in (E.1) according to (4.22).

Construct the Green's function associated with the following boundary value problems.

19.

$$y'' + y = f(x), \qquad 0 < x < 1$$

$$y(0) = y(1), \qquad y'(0) = y'(1)$$

Hint: Consider the form $A\cos(x - \xi + c)$ for $G(x,\xi)$ where A and c are to be determined.

20. Use the Green's function method to solve the boundary value problem

$$\frac{d^2y}{dx^2} + \pi^2 y = \cos \pi x, \qquad 0 < x < 1$$

$$y(0) = y(1), \qquad y'(0) = y'(1)$$

Hint: See Exercise 19.

21. Consider the following boundary value problems, which is obviously a Sturm-Liouville problem [see (4.7)–(4.9)]:

$$\frac{d^2u}{dx^2} + \lambda^2 u = 0, \qquad 0 < x < 1 \qquad (E.1)$$

$$u(0) = 0 \qquad (E.2)$$

$$u(1) = 0 \qquad (E.3)$$

(a) Without solving for the explicit solutions, prove that any two eigenfunctions $u_n(x)$ and $u_m(x)$ of (E.1)–(E.3) corresponding to two different eigenvalues λ_n and λ_m are orthogonal on the interval [0, 1]; that is,

$$\int_0^1 u_n(x)u_m(x)dx = 0, \qquad \lambda_n \neq \lambda_m$$

Hint: Write the differential equation (4.7) for $u_n(x)$ and for $u_m(x)$ and attempt to arrive at

$$(\lambda_n^2 - \lambda_m^2) \int_0^1 u_n(x) u_m(x) dx = 0$$

(b) Solve the boundary value problem.

(c) Verify by direct integration that the solutions in part (b) are othogonal. *Hint:* To show that $\int_0^1 \sin n\pi x \sin m\pi x dx = 0$ for $n \neq m$, use the trigonometric identity $\sin ax \sin bx = (1/2)[\cos(a-b)x - \cos(a+b)x]$ to simplify the integration.

(d) Write the Fourier series expansion in terms of these functions [solutions of part (b)] for the function $f(x) = x$, $0 < x < 1$. *Hint:* In Example 6 use (E.1) for the Fourier series with $u_n(x) = \sin n\pi x$ and use (E.3) for evaluating the present Fourier sine series coefficients c_n for (E.1).

22. *Orthogonal kernels.* The orthogonal property is not limited to solutions of equations only but extends to other functions; for example, two kernels $K(x,t)$ and $L(x,t)$ are termed orthogonal on $\{(x,t) : a \leq x \leq b, a \leq t \leq b\}$ if the following two integrals vanish:

$$\int_a^b K(x,\tau) L(\tau, t) d\tau = 0$$

$$\int_a^b L(x,\tau) K(\tau, t) d\tau = 0$$

(a) Prove that the kernel $K(x,t) = x^3 t^3$ and the kernel $L(x,t) = x^2 t^2$ are orthogonal on $\{(x,t) : 0 \leq x \leq 1, 0 \leq t \leq 1\}$.

(b) Prove that the kernel $K(x,t) = \sin(x - 2t)$ is orthogonal to itself on the square $\{(x,t) : 0 \leq x \leq 2\pi, 0 \leq t \leq 2\pi\}$; that is, show that $\int_0^{2\pi} \sin (x - 2\tau) \sin(\tau - 2t) d\tau = 0$.

23. The Legendre polynomial $P_n(x)$ of degree n is defined by the Rodriques formula as

$$P_n(x) = \frac{1}{2^n n!} \frac{d^n}{dx^n} (x^2 - 1)^n. \qquad (E.1)$$

(a) Use (E.1) to show that the first four Legendre polynomials are

$$P_0(x) = 1, \qquad P_1(x) = x,$$

$$P_2(x) = \frac{1}{2}(3x^2 - 1), \qquad P_3(x) = \frac{1}{2}(5x^3 - 3x)$$

(b) Verify that the Legendre polynomials in part (a) are orthogonal on (-1,1) with respect to a weighting function $\rho(x) = 1$. *Hint:* See (4.46).

(c) Write the first three terms of the Fourier series expansion in terms of the Legendre polynomials in part (a) to approximate the function $f(x) = e^{2x}$ on (-1,1). *Hint:* See (4.47) and (4.48) with $\rho(x) = 1$, $u_n(x) = P_n(x)$, $n = 0, 1, 2$.

(d) Tabulate or graph the approximate series expansion to compare with its exact value $f(x) = e^{2x}$.

24. In reference to our discussion of the potential distribution in a charged unit disc of (4.62)–(4.65), consider now the potential distribution in a square of side length π with charge density $f(x, y)$, and where the edges are grounded. The boundary value problem for the potential distribution $u(x, y)$ in a charged square with side length π as in Figure 4.1 due to a charge density $f(x, y)$, and with all sides being grounded is

$$\frac{\partial^2 u}{\partial x^2} + \frac{\partial^2 u}{\partial y^2} = -f(x, y), \qquad 0 < x < \pi, \qquad 0 < y < \pi \tag{E.1}$$

$$u(0, y) = 0, \qquad 0 < y < \pi \tag{E.2}$$

$$u(\pi, y) = 0, \qquad 0 < y < \pi \tag{E.3}$$

$$u(x, 0) = u(x, \pi) = 0 \tag{E.4}$$

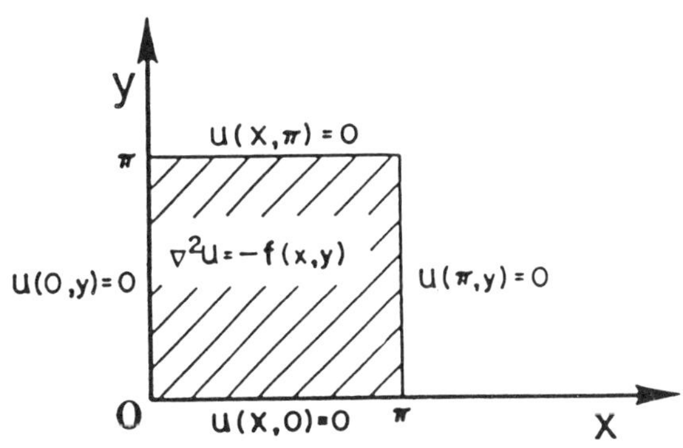

Fig. 4.1 Electric potential in a square plate.

(a) Let $U(n, y)$ and $F(n, y)$ be the finite Fourier sine transforms, as defined in (1.115),

$$U(n, y) = \int_0^\pi u(x, y) \sin nx dx \tag{E.4}$$

$$F(n, y) = \int_0^\pi f(x, y) \sin nx dx \tag{E.5}$$

We are aiming at algebraizing $\frac{\partial^2 u}{\partial x^2}$ the second-order partial derivative with respect to x, to render (E.1) as a second-order (nonhomogeneous) ordinary differential equation in the Fourier sine transform $U(n, y)$. Then we can appeal to the Green's function of the problem in Example 4 to construct $U(n, y)$. Finally we use the *inverse* Fourier sine transform (Fourier sine series) of (1.116) to obtain our original function $u(x, y)$ for the potential on the square.

To accomplish the first part, and in parallel to the use of the Laplace transform for algebraizing differential equations (with initial conditions though!), use the following (operational) property of the finite Fourier sine transform (1.121)

$$\int_0^\pi \sin nx \frac{d^2 f}{dx^2} dx = n\{f(0) - (-1)^n f(\pi)\} - n^2 F_s(n) \qquad (E.6)$$

to transform the partial differential equation of (E.1) in $u(x, y)$ to the following ordinary differential equation in $U(n, y)$,[7]

$$\frac{d^2 U(n, y)}{dy^2} - n^2 U(n, y) = -F(n, y) \qquad (E.7)$$

(b) Show that the boundary conditions (E.2), (E.3) are easily transformed to

$$U(n, 0) = 0 \qquad (E.8)$$

$$U(n, \pi) = 0 \qquad (E.9)$$

(c) Construct the Green's function for the boundary value problem (E.7)–(E.9) in $U(n, y)$. *Hint:* Note that in (E.1)–(E.3) of Example 3 we have the same problem as the above (E.7)–(E.9) except for $b^2 = -n^2$ in (E.7), and the boundary points are 0 and 1 instead of 0 and π in the present boundary conditions of (E.8) and (E.9).

(d) With the help of the Green's function in part (c), find the solution $U(n, y)$ to the boundary value problem (E.7) and (E.8). *Hint:* See (4.1)–(4.3) and (4.5).

(e) Find the solution for the potential $u(x, y)$ in the original boundary value problem (E.1)–(E.3). *Hint:* Use the inverse (finite) Fourier sine transform (Fourier sine series) of (1.116) on $U(n, y)$ of part (d) to find $u(x, y)$.

(f) Attempt to find an expression for the Green's function in two dimensions of the original problem (E.1)–(E.3) as $G(x, y; \xi, \eta)$. *Hint:* In the formal answer of part (d), substitute for $F(n, y)$ (inside the integral representing

[7] For detailed treatment of integral and finite transforms and how to find the proper (compatible) transform for a given boundary value problem, see Jerri [1992].

$U(n, y)$) in terms of its sine integral as in (E.5), then exchange the two integrations with the Fourier series summation to write

$$u(x,y) = \int_0^\pi \int_0^\pi G(x,y;\xi,\eta) f(\xi,\eta) d\xi, \tag{E.10}$$

hence the resulting infinite sine series expression for $G(x, y; \xi, \eta)$.

(g) Derive (E.6) the important property of the finite sine transform. *Hint:* Use two integrations by parts, and recall the definition of $F_s(n)$ in (1.115) as the finite Fourier transform of $f(x)$, $0 < x < \pi$.

25. **(a)** Show that the second-order differential operator L in (4.1) can always be written in the self-adjoint form of L in (4.6). *Hint:* Divide Lu of (4.1) by $A_0(x) \neq 0$, then multiply $\frac{1}{A_0(x)} Lu$ by $r(x) = \exp(\int \frac{A_1(x)}{A_2(x)} dx)$.

(b) Show that the fourth-order differential operator L in $Lu = \frac{d^4u}{dx^4} + u$ is self-adjoint. *Hint:* Write $vLu - uLv$ as we did in Example 1, then add and subtract appropriate terms: $(v'u''' - v'u''' - v''u'' + v''u'' + v'''u'$ -$v'''u')$ to end up with the sum of four exact differentials that can then be written as one exact differential $\frac{d}{dx}[vu''' - v'u'' + v''u'$ -$v'''u] = vLu - uLv$.

4.2 FREDHOLM INTEGRAL EQUATIONS AND THE GREEN'S FUNCTION

In Example 5 of Section 2.5 we used repeated integration to show how a boundary value problem

$$\frac{d^2y}{dx^2} = \lambda y(x), \qquad a < x < b \tag{1.33}$$

$$y(a) = 0 \tag{1.34}$$

$$y(b) = 0 \tag{1.35}$$

reduces to a Fredholm integral equation (2.36) and (2.37). In this section we use the Green's function method to show that the general boundary value problem with parameter λ,

$$A_0(x)\frac{d^2y}{dx^2} + A_1(x)\frac{dy}{dx} + A_2(x)y + \lambda\rho(x)y = h(x), \tag{4.70}$$

$$y(a) = 0 \tag{4.71}$$

$$y(b) = 0 \tag{4.72}$$

reduces to a Fredholm integral equation with the Green's function as its kernel.

To do this we write (4.70) in the form of (4.1) and purposely shift the term $\lambda\rho(x)y$ to the right side, in anticipation of involving it inside the final integral to have an integral equation

$$A_0(x)\frac{d^2y}{dx^2} + A_1(x)\frac{dy}{dx} + A_2(x)y \equiv Ly = h(x) - \lambda\rho(x)y = f(x). \tag{4.73}$$

We note that having a symmetric Green's function $G(x, \xi)$ for (an equivalent problem to that of) the problem (4.73), (4.71) and (4.72) can be easily justified since the differential operator

$$L \equiv A_0(x)\frac{d^2}{dx^2} + A_1(x)\frac{d}{dx} + A_2(x)$$

in (4.73) can be reduced to a self-adjoint form as in (4.6), via multiplying it by a factor $p(x) = \dfrac{r(x)}{A_0(x)}$ of (E.5) in Example 1 (see Exercise 25(a) in Section 4.1).

If we assume that such an operation has already been done to make the above L in (4.73) as a self-adjoint operator, (i.e., L now is in the self adjoint form as given in (4.30)), then according to (4.5), the solution to such equation (4.73) with the boundary conditions (4.71) and (4.72) is

$$\begin{aligned} y(x) &= -\int_a^b G(x,\xi)f(\xi)d\xi \\ &= -\int_a^b G(x,\xi)[h(\xi) - \lambda\rho(\xi)y(\xi)]d\xi \\ &= -\int_a^b G(x,\xi)h(\xi)d\xi + \lambda\int_a^b G(x,\xi)\rho(\xi)y(\xi)d\xi. \end{aligned} \tag{4.74}$$

If we let

$$k(x) = -\int_a^b G(x,\xi)h(\xi)d\xi \tag{4.75}$$

then (4.74) becomes

$$y(x) = k(x) + \lambda\int_a^b G(x,\xi)\rho(\xi)y(\xi)d\xi \tag{4.76}$$

which is a Fredholm integral equation of the second kind with kernel $K(x, \xi) = G(x, \xi)\rho(\xi)$ and a nonhomogeneous term $k(x)$.

This Fredholm equation can be written in a more *symmetric form* if we multiply both sides of (4.76) by $\sqrt{\rho(x)}$,

$$\sqrt{\rho(x)}y(x) = \sqrt{\rho(x)}k(x) + \lambda\int_a^b \sqrt{\rho(x)\rho(\xi)}G(x,\xi)\sqrt{\rho(\xi)}y(\xi)d\xi$$

and let $u(x) = \sqrt{\rho(x)}y(x)$ and $g(x) = \sqrt{\rho(x)}k(x)$,

$$u(x) = g(x) + \lambda\int_a^b \sqrt{\rho(x)\rho(\xi)}G(x,\xi)u(\xi)d\xi. \tag{4.77}$$

As we mentioned in Section 4.1, the Green's function $G(x, \xi)$ is symmetric for such (already made) self-adjoint problem (4.70)–(4.72) as seen in (4.25). It follows that the kernel $K(x, \xi) = \sqrt{\rho(x)\rho(\xi)}G(x, \xi)$ of the resulting Fredholm equation (4.77) is obviously symmetric.

Example 8 Green's Function and Fredholm Equation

Use the Green's function to reduce the boundary value problem

$$\frac{d^2y}{dx^2} + \lambda y = x, \qquad 0 < x < \frac{\pi}{2} \tag{E.1}$$

$$y(0) = 0 \tag{E.2}$$

$$y\left(\frac{\pi}{2}\right) = 0 \tag{E.3}$$

to a Fredholm integral equation.

If we compare this problem with (4.70)–(4.72) we have $h(x) = x$, $\rho(x) = 1$, and $L \equiv d^2/dx^2$, which is self-adjoint and hence the Green's function is symmetric. From (4.76) the integral equation representation is

$$y(x) = -\int_0^{\pi/2} G(x,\xi)\xi d\xi + \lambda \int_0^{\pi/2} G(x,\xi)y(\xi)d\xi \tag{E.4}$$

where we used

$$k(x) = -\int_0^{\pi/2} G(x,\xi)\xi d\xi. \tag{E.5}$$

It remains to construct the Green's function from the corresponding homogeneous boundary value problem

$$\frac{d^2y}{dx^2} = 0, \qquad 0 < x < \frac{\pi}{2} \tag{E.6}$$

$$y(0) = 0 \tag{E.7}$$

$$y\left(\frac{\pi}{2}\right) = 0. \tag{E.8}$$

The solution that satisfies (E.6) and (E.7) is $y(x) = x$, and that which satisfies (E.6) and (E.8) is $y(x) = (\pi/2) - x$; therefore, the symmetric Green's function may be written as two branches,

$$G(x,\xi) = \begin{cases} Cx\left(\frac{\pi}{2} - \xi\right), & 0 \le x \le \xi \\ C\xi\left(\frac{\pi}{2} - x\right), & \xi \le x \le \frac{\pi}{2} \end{cases} \tag{E.9}$$

where the first and second branches satisfy (E.7) and (E.8) in the variable x, respectively.

From the jump discontinuity property (4.26) of $\partial G(x,\xi)/\partial x$ we have

$$\begin{aligned} \frac{\partial G}{\partial x}(x,\xi)\bigg|_{x=\xi_+} - \frac{\partial G}{\partial x}(x,\xi)\bigg|_{x=\xi_-} &= -\frac{1}{r(\xi)} = -1 \\ &= -C\xi - C\left(\frac{\pi}{2} - \xi\right) = \frac{-C\pi}{2} = -1, \quad C = \frac{2}{\pi}. \end{aligned} \tag{E.10}$$

Note how we used the second branch and the first branch of $G(x,\xi)$ in (E.9) for $x=\xi_+$ and $x=\xi_-$, respectively, since $x=\xi_+>\xi$ is in the domain $\xi\le x\le\pi/2$ and $x=\xi_-<\xi$ is in the domain $0\le x\le\xi$. From (E.9) and (E.10) we have

$$G(x,\xi)=\begin{cases}\frac{2}{\pi}x\left(\frac{\pi}{2}-\xi\right), & 0\le x\le\xi\\[2ex] \frac{2}{\pi}\xi\left(\frac{\pi}{2}-x\right), & \xi\le x\le\frac{\pi}{2}\end{cases}\tag{E.11}$$

and the Fredholm integral equation of the second kind (4.76) is

$$y(x)=k(x)+\lambda\int_0^{\pi/2}G(x,\xi)y(\xi)d\xi\tag{E.12}$$

where the kernel $K(x,\xi)=G(x,\xi)$ as given in (E.11) is symmetric and $k(x)$ is

$$\begin{aligned}k(x)&=-\int_0^{\pi/2}G(x,\xi)h(\xi)d\xi=-\int_0^{\pi/2}G(x,\xi)\xi d\xi\\ &=\int_0^x\frac{2}{\pi}\xi\left(x-\frac{\pi}{2}\right)\xi d\xi+\int_x^{\pi/2}\frac{2}{\pi}x\left(\xi-\frac{\pi}{2}\right)\xi d\xi\\ &=\frac{2}{\pi}\left[\frac{\xi^3}{3}\left(x-\frac{\pi}{2}\right)\right]_0^x+\frac{2}{\pi}\left[x\left(\frac{\xi^3}{3}-\frac{\pi\xi^2}{4}\right)\right]_x^{\pi/2}\\ k(x)&=\frac{x^3}{6}-\frac{\pi^2x}{24}.\end{aligned}\tag{E.13}$$

The final Fredholm equation that is equivalent to the boundary value problem (E.1)–(E.3) is

$$y(x)=\frac{x^3}{6}-\frac{\pi^2x}{24}+\lambda\int_0^{\pi/2}G(x,\xi)y(\xi)d\xi\tag{E.14}$$

where the Green's function $G(x,\xi)$ is given in (E.11).

Fredholm Integral Equations in Two Dimensions

For an illustration of reducing boundary value problems associated with partial differential equations, we may consult (4.62)–(4.65) for the potential distribution in a charged unit disc with its rim being grounded. As we had done above, all what we needed there is to construct the Green's function for this two-dimensional boundary value problem. Another very similar illustration is that of the potential distribution on a charged square with its edges being grounded. This was the subject of Exercise 24 of the last section, which is supplied with very detailed leading hints.

For the above two boundary value problems, we have already presented the integral representation of the potential distribution due to a given charge distribution with the help of the Green's function in two dimensions. These results are shown in (4.65) for the charged disc, and in the answer to Exercise 24(e),(f) of the last section for the charged square. So, all what we need to have these problem as Fredholm integral equations in two dimensions, is to ask about the inverse problem, namely to find the charge density $f(x, y)$ (inside the integral) that would affect the given desired potential distribution $u(x, y)$ on the charged square, for example.

Exercises 4.2

1. Reduce the boundary value problem

$$\frac{d^2y}{dx^2} - y = e^x, \qquad 0 < x < 1$$

$$y(0) = 0, \qquad y(1) = 0$$

to an integral equation by first finding the Green's function. *Hint:* See Example 8.

2. Reduce the boundary value problem

$$\frac{d^2y}{dx^2} - \lambda y = x^2, \qquad 0 < x < \frac{\pi}{2}$$

$$y(0) = 0, \qquad y\left(\frac{\pi}{2}\right) = 0$$

to an integral equation. *Hint:* See Example 8.

3. Reduce the boundary value problem

$$\frac{d^3y}{dx^3} + \lambda y = 2x, \qquad 0 < x < 1$$

$$y(0) = 0, \quad y(1) = 0, \quad y'(0) = y'(1)$$

to an integral equation by:

(a) Using the Green's function of Example 5.

(b) First constructing the Green's function starting with the form $a_1(\xi) + xa_2(\xi) + x^2a_3(\xi)$ for $G(x, \xi)$.

For the following boundary value problems, use the Green's function to reduce them to Fredholm integral equations. *Hint:* You may consult the answers of Exercises 4.1 for the appropriate Green's function.

4. $\dfrac{d^2y}{dx^2} + \lambda y = 2x + 1, \qquad 0 < x < 1$

$y(0) = y'(1)$

$y'(0) = y(1)$

Hint: See problem 17(b)i of Exercises 4.1 for the mixed boundary conditions.

5. $\dfrac{d^2y}{dx^2} + \lambda y = e^x$

   $y(0) = y'(0)$

   $y(1) = y'(1)$

   Hint: See problem 15 of Exercises 4.1.

6. $\dfrac{d^2y}{dx^2} + \dfrac{\pi^2}{4}y = \lambda y + \cos\dfrac{\pi x}{2}$

   $y(-1) = y(1)$

   $y'(-1) = y'(1)$ *Hint:* See problem 16 of Exercises 4.1. (for its boundary conditions only.)

7. Write the Fredholm integral equation in the charge density function $f(\rho, \phi)$ that would produce a given potential distribution on a unit disc $u(r, \theta)$, $0 \leq r < 1$, $0 \leq \theta \leq 2\pi$, where the rim of the disc is being grounded. *Hint:* Consult (4.65) and its derivation in (4.62)–(4.65) at the end of the last section.

5

Fredholm Integral Equations

In Section 1.2 we presented the Fredholm integral equation,[1]

$$h(x)u(x) = f(x) + \int_a^b K(x,\xi)u(\xi)d\xi \tag{5.1}$$

which we termed the *second kind* when $h(x) = 1$,

$$u(x) = f(x) + \int_a^b K(x,\xi)u(\xi)d\xi \tag{5.2}$$

and the *first kind* when $h(x) = 0$,

$$-f(x) = \int_a^b K(x,\xi)u(\xi)d\xi. \tag{5.3}$$

When $f(x) \equiv 0$ in (5.2) it becomes the *homogeneous* Fredholm equation,

$$u(x) = \int_a^b K(x,\xi)u(\xi)d\xi. \tag{5.4}$$

We note that the limits a and b of the integrals may be finite or infinite, where the infinite limit makes it a singular equation.

[1] In 1900–1903, Fredholm developed the theory of these integral equations as a limit to the linear system of equations. In 1904 and later, Hilbert established the theory in a rigorous fashion [see Bôcher, 1914].

In Chapters 1 and 2 we formulated a number of problems which resulted in Fredholm integral equations. This included the shape of a hanging chain (2.28) in Section 2.3 and the small deflection of a rotating shaft (1.19) in Section 1.1. In Example 5 of Section 2.5 we illustrated how a boundary value problem,

$$\frac{d^2y}{dx^2} = \lambda y(x), \qquad a < x < b \tag{1.33}$$

$$y(a) = 0 \tag{1.34}$$

$$y(b) = 0 \tag{1.35}$$

reduces to a Fredholm integral equation of the second kind

$$y(x) = \lambda \int_a^b K(x,t)y(t)dt, \tag{1.36}$$

$$K(x,t) = \begin{cases} \dfrac{(x-b)(t-a)}{(b-a)}, & a \le t \le x \\ \dfrac{(x-a)(t-b)}{(b-a)}, & x \le t \le b \end{cases} \tag{1.37}$$

In this chapter we discuss and illustrate a number of exact, approximate, and numerical methods of solving Fredholm integral equations. An important condition, termed the *regularity condition*, for the development of the solutions of Fredholm integral equations, is that the kernel $K(x,t)$ be square integrable in both x and t in the square $\{(x,t) : a \le x \le b, a \le t \le b\}$; that is,

$$\int_a^b \int_a^b |K(x,t)|^2 dx dt = B^2 \tag{5.5}$$

is finite. We must also stress here the relations in the theory of solving the nonhomogeneous Fredholm integral equation of the second kind (5.2) and its corresponding homogeneous equation (5.4) with the (added) important parameter λ,

$$u(x) = \lambda \int_a^b K(x,t)u(t)dt \tag{5.4h}$$

Indeed, the theory very closely parallels that of the theory of systems of nonhomogeneous and homogeneous linear equations. Such a relation is the essence of *Fredholm alternative*, which we state at the end of Section 5.1 and which deals with conditions for the existence of solutions of Fredholm equations. In the next section we present a method for solving the Fredholm integral equation (5.2) when the kernel is of a special form,

$$K(x,t) = \sum_{k=1}^{n} a_k(x)b_k(t) \tag{5.6}$$

a finite sum of products of $a_k(x)$, a function of x only, and $b_k(t)$, a function of t only. Such kernels defined by (5.6) are called *degenerate kernels* or *separable kernels*. In

Section 5.2 we introduce methods of solving another special but very important case of the Fredholm equation with *symmetric kernel* [i.e., $K(x,t) = K(t,x)$]. Section 5.3 is devoted to Fredholm equations of the second kind; Section 5.4 is a new section for this edition to cover in more detail the Fredholm integral equations of the first kind; and Section 5.5 covers basic elements of the numerical (approximate) method of solution. The higher quadratures numerical methods are covered in the added (optional) Chapter 7 in this edition. In each of these sections we illustrate, when appropriate, some approximate methods of solution and methods for determining the eigenvalues of the homogeneous Fredholm equation.

Here we will again be concentrating on the various, mostly, successive approximation (iterative) methods, for constructing a solution. Accurate statements for such results will be stated without the complete proof.

We will start this chapter on Fredholm integral equations with the very special degenerate kernel, as it is easy to illustrate without the need for new tools except for the familiar theory of system of linear equations. This is very important as it represents the fundamental and historical relation of such theory and the theory for Fredholm integral equations, as was done by Fredholm in 1900–1903.

5.1 FREDHOLM INTEGRAL EQUATIONS WITH DEGENERATE KERNEL

5.1.1 Nonhomogeneous Fredholm Equations with Degenerate Kernel

Consider the nonhomogeneous Fredholm equation of the second kind with degenerate kernel $K(x,t) = \sum_{k=1}^{n} a_k(x)b_k(t)$,

$$u(x) = f(x) + \lambda \int_a^b K(x,t)u(t)dt \tag{5.7a}$$

$$= f(x) + \lambda \int_a^b \sum_{k=1}^{n} a_k(x)b_k(t)u(t)dt \tag{5.7b}$$

$$= f(x) + \lambda \sum_{k=1}^{n} a_k(x) \int_a^b b_k(t)u(t)dt \tag{5.7c}$$

after using $K(x,t)$ of (5.6) and exchanging summation with integration. In the following we show how the solution of this Fredholm integral equation with degenerate kernel reduces to solving *a system of linear equations*. If we define c_k as the integral in (5.7c),

$$c_k = \int_a^b b_k(t)u(t)dt \tag{5.8}$$

then (5.7c) becomes

$$u(x) = f(x) + \lambda \sum_{k=1}^{n} c_k a_k(x). \tag{5.9}$$

If we multiply both sides of (5.9) by $b_m(x)$ and integrate from a to b, we produce c_m on the left side,

$$\int_a^b b_m(x)u(x)dx = \int_a^b b_m(x)f(x)dx + \lambda \sum_{k=1}^{n} c_k \int_a^b b_m(x)a_k(x)dx. \tag{5.10}$$

If we define the (new) integrals in (5.10) as

$$f_m = \int_a^b b_m(x)f(x)dx \tag{5.11}$$

and

$$a_{mk} = \int_a^b b_m(x)a_k(x)dx \tag{5.12}$$

then (5.10) becomes

$$c_m = f_m + \lambda \sum_{k=1}^{n} a_{mk}c_k \qquad m = 1, 2, \cdots, n \tag{5.13}$$

which is a set of n linear equations in $c_1, c_2, c_3, \cdots, c_n$. Here f_m and a_{mk} are considered known since we are given $b_m(x)$, $f(x)$, and $a_k(x)$.

So the solution to the Fredholm equation of the second kind (5.2) with degenerate kernel (5.6) reduces to solving for c_m from the system of the n linear equations (5.13) (in the n unknowns c_m, $m = 1, 2, \cdots, n$), since c_m will then be used in the series (5.9) to obtain $u(x)$, the solution of (5.2).

If we use matrix notation, the system of n linear equations (5.13) can be written in the form

$$C = \begin{bmatrix} c_1 \\ c_2 \\ \vdots \\ c_n \end{bmatrix} = \begin{bmatrix} f_1 \\ f_2 \\ \vdots \\ f_n \end{bmatrix} + \lambda \begin{bmatrix} a_{11} & a_{12} & \cdots & a_{1n} \\ a_{21} & a_{22} & & a_{2n} \\ \vdots & \vdots & & \vdots \\ a_{n1} & a_{n2} & \cdots & a_{nn} \end{bmatrix} \begin{bmatrix} c_1 \\ c_2 \\ \vdots \\ c_n \end{bmatrix} \equiv F + \lambda AC \tag{5.14}$$

or as

$$(I - \lambda A)C = F. \tag{5.15}$$

From the theory of linear systems of equations we know that (5.15) has a *unique* solution if the determinant $|I - \lambda A| \neq 0$ and has either infinite or no solution when $|I - \lambda A| = 0$.

Example 1 Solve the Fredholm integral equation

$$u(x) = x + \lambda \int_0^1 (xt^2 + x^2 t)u(t)dt. \tag{E.1}$$

This Fredholm integral equation has a degenerate kernel of the form (5.6),

$$K(x,t) = xt^2 + x^2 t = \sum_{k=1}^{2} a_k(x) b_k(t) \tag{E.2}$$

where $a_1(x) = x$, $a_2(x) = x^2$, $b_1(t) = t^2$, and $b_2(t) = t$. To solve for c_m in (5.13), and hence $u(x)$ of (E.1), we must prepare f_1, f_2 from (5.11) and a_{11}, a_{12}, a_{21}, a_{22} from (5.12). From (E.1) we have $f(x) = x$; hence according to (5.11),

$$f_1 = \int_0^1 b_1(t) f(t) dt = \int_0^1 t^3 dt = \frac{1}{4}$$

$$f_2 = \int_0^1 b_2(t) f(t) dt = \int_0^1 t^2 dt = \frac{1}{3}$$

and the column matrix F of (5.14) becomes

$$F = \begin{bmatrix} f_1 \\ f_2 \end{bmatrix} = \begin{bmatrix} \frac{1}{4} \\ \frac{1}{3} \end{bmatrix}.$$

To prepare the matrix A in (5.14) we use (5.12) to evaluate the elements a_{mk}, with $a_k(x)$ and $b_k(t)$ as in (E.2) for $k, m = 1, 2$,

$$\begin{aligned} a_{11} &= \int_0^1 b_1(t) a_1(t) dt = \int_0^1 t^2 t dt = \int_0^1 t^3 dt = \frac{1}{4} \\ a_{12} &= \int_0^1 b_1(t) a_2(t) dt = \int_0^1 t^2 t^2 dt = \int_0^1 t^4 dt = \frac{1}{5} \\ a_{21} &= \int_0^1 b_2(t) a_1(t) dt = \int_0^1 t t dt = \int_0^1 t^2 dt = \frac{1}{3} \\ a_{22} &= \int_0^1 b_2(t) a_2(t) dt = \int_0^2 t t^2 dt = \int_0^1 t^3 dt = \frac{1}{4}. \end{aligned}$$

Hence $C = F + \lambda AC$ of (5.14) becomes

$$\begin{bmatrix} c_1 \\ c_2 \end{bmatrix} = \begin{bmatrix} \frac{1}{4} \\ \frac{1}{3} \end{bmatrix} + \lambda \begin{bmatrix} \frac{1}{4} & \frac{1}{5} \\ \frac{1}{3} & \frac{1}{4} \end{bmatrix} \begin{bmatrix} c_1 \\ c_2 \end{bmatrix}$$

and if we transfer the matrix product to the left side, we obtain $C - \lambda AC = (I - \lambda A)C = F$,

$$\begin{bmatrix} 1-\frac{\lambda}{4} & -\frac{\lambda}{5} \\ -\frac{\lambda}{3} & 1-\frac{\lambda}{4} \end{bmatrix} \begin{bmatrix} c_1 \\ c_2 \end{bmatrix} = \begin{bmatrix} \frac{1}{4} \\ \frac{1}{3} \end{bmatrix}. \tag{E.3}$$

In general, before solving (E.3) we must evaluate the determinant of the matrix $I - \lambda A$,

$$\begin{aligned} |I - \lambda A| = \begin{vmatrix} 1-\frac{\lambda}{4} & -\frac{\lambda}{5} \\ -\frac{\lambda}{3} & 1-\frac{\lambda}{4} \end{vmatrix} &= \left(1-\frac{\lambda}{4}\right)^2 - \frac{\lambda^2}{15} \\ &= \frac{240 - 120\lambda - \lambda^2}{240}. \end{aligned} \tag{E.4}$$

If $240 - 120\lambda - \lambda^2 \neq 0$, the problem (E.3) has a unique solution for c_1 and c_2 which we can evaluate by finding the inverse, $(I - \lambda A)^{-1}$. As we have only two equations in (E.3),

$$\left(1-\frac{\lambda}{4}\right)c_1 - \frac{\lambda}{5}c_2 = \frac{1}{4} \tag{E.5}$$

$$-\frac{\lambda}{3}c_1 + \left(1-\frac{\lambda}{4}\right)c_2 = \frac{1}{3} \tag{E.6}$$

we can solve them immediately by eliminating one of the unknowns to find

$$c_1 = \frac{60 + \lambda}{240 - 120\lambda - \lambda^2} \tag{E.7}$$

$$c_2 = \frac{80}{240 - 120\lambda - \lambda^2}. \tag{E.8}$$

Once we know c_1 and c_2 we can obtain $u(x)$, the solution to the integral equation (E.1), from (5.9) with $n = 2$, $f(x) = x$, $a_1(x) = x$, and $a_2(x) = x^2$,

$$u(x) = f(x) + \lambda \sum_{k=1}^{n} c_k a_k(x) \tag{5.9}$$

$$\begin{aligned} u(x) &= x + \lambda \sum_{k=1}^{2} c_k a_k(x) \\ &= x + \lambda[c_1 a_1(x) + c_2 a_2(x)] \\ &= x + \lambda \left[\frac{(60+\lambda)x}{240 - 120\lambda - \lambda^2} + \frac{80x^2}{240 - 120\lambda - \lambda^2}\right] \\ &= \frac{(240 - 60\lambda)x + 80\lambda x^2}{240 - 120\lambda - \lambda^2}, \quad 240 - 120\lambda - \lambda^2 \neq 0 \end{aligned} \tag{E.9}$$

As this Example 1 illustrates, this is a clear and simple method for constructing the solution of the *nonhomogeneous* Fredholm integral equation (5.7a). Moreover the condition for the existence of such a solution seems to also be very transparent as $|I - \lambda A| \neq 0$, as we showed in Example 1. This amounts to restricting the parameter λ in (5.7a) not to be a zero of the equation

$$|I - \lambda A| = \frac{240 - 120\lambda - \lambda^2}{240} = 0,$$

i.e., the condition $240 - 120\lambda - \lambda^2 \neq 0$ is sufficient to guarantee the unique solution $u(x)$ as constructed in (E.9) of Example 1. However, this still may raise an immediate question concerning equation (5.7a) when its parameter λ happens to be one of the possible two zeros of the quadratic equation $\lambda^2 + 120\lambda - 240 = 0$, which violates the sufficient condition that guaranteed us the unique solution of (E.9) in our Example 1. Such an important question concerning the existence of solution (or solutions) to nonhomogeneous Fredholm integral equation is the subject of the *Fredholm alternative* as given in Theorems 1 and 2, and illustrated clearly in Examples 3 and 4. The thrust of these theorems depends on our knowledge of solving the associated homogeneous equation of (5.4h),

$$u(x) = \lambda \int_a^b K(x,t)u(t)dt \tag{5.4h}$$

which is the subject of the next section. In that regard the present method will be instrumental in constructing the solutions of the homogeneous equation as illustrated in Example 2. Moreover if the analysis allows a unique solution, or infinity of solutions, for the nonhomogeneous problem in (5.7a), the present method will be used for their construction as illustrated in Example 4.

Indeed, the analysis in the Fredholm alternative applies to more general kernels than the present simple degenerate one of (5.6), but for a simple presentation we will be satisfied with the theory for symmetric kernels, i.e., $K(x,t) = K(t,x)$, which is discussed in Section 5.2. To summarize the purpose of our next illustrations, in regards to the existence of the solution to Fredholm equations of the second kind, the Fredholm alternative will be illustrated for degenerate but nonsymmetric kernel in Example 3, degenerate and symmetric kernel in Example 4 (and symmetric but nondegenerate kernel in Examples 7, 9, and 10 of Section 5.2.)

5.1.2 Fredholm Alternative

Homogeneous Fredholm Equations

We consider here the homogeneous case [i.e., when $f(x) = 0$ in (5.7a)] of the Fredholm integral equation with degenerate kernel

$$\begin{aligned} u(x) &= \lambda \int_a^b K(x,t)u(t)dt \\ &= \lambda \sum_{k=1}^n a_k(x) \int_a^b b_k(t)u(t)dt. \end{aligned} \tag{5.16}$$

We will follow the same steps as those we used for the nonhomogeneous equation (5.7a), to reduce (5.16) to

$$u(x) = \lambda \sum_{k=1}^{n} c_k a_k(x) \tag{5.17}$$

and then to a system of n *homogeneous* equations in c_m,

$$c_m = \lambda \sum_{k=1}^{n} a_{mk} c_k, \qquad m = 1, 2, \cdots, n \tag{5.18}$$

or in matrix notation,

$$(I - \lambda A)C = 0 \tag{5.19}$$

instead of the nonhomogeneous system of linear equations in (5.13) and (5.14). Here A and C are defined as in (5.14).

From the theory of systems of linear equations we can conclude that if $|I - \lambda A| \neq 0$, then the only solution to the homogeneous equation (5.19) is the trivial solution $C \equiv 0$. By using (5.17), the solution to the homogeneous Fredholm equation (5.16) is the trivial solution $u(x) \equiv 0$. On the other hand, when $|I - \lambda A| = 0$, then (5.19) has nontrivial solution. This leads us to discuss next the subject of *eigenvalues and eigenfunctions* of the homogeneous problem.

Eigenvalues and eigenfunctions

For the homogeneous Fredholm integral equation with the kernel $K(x, t)$,

$$u(x) = \lambda \int_a^b K(x, t) u(t) dt \tag{5.20}$$

the parameter $\lambda \neq 0$ for which (5.20) does have a nontrivial solution (i.e., $u(x) \neq 0$) is called the *eigenvalue or characteristic value* of the homogeneous equation (5.20) or, in short, the eigenvalue of the kernel $K(x, t)$ in (5.20). The nontrivial solutions $u_j(x) \neq 0$ corresponding to the eigenvalues λ_j are called the *eigenfunctions or characteristic functions* of (5.20), or in short, the eigenfunctions of the kernel $K(x, t)$.

In this sense, then, the eigenvalues of (5.20) are the solutions of $|I - \lambda A| = 0$, since if λ is not the solution of this equation, then $|I - \lambda A| \neq 0$, and hence (5.18) and in turn (5.20) have the trivial solution. There may exist more than one eigenfunction $\psi_j(x)$ corresponding to a specific eigenvalue λ_j. The number p of such (linearly independent) eigenfunctions $\psi_{j+1}(x), \psi_{j+2}(x), \psi_{j+3}(x), \cdots, \psi_{j+p}(x)$ is called the *degeneracy* (or *index*) of λ_j. In case an eigenvalue λ_k is a multiple root with degree m in the equation $|I - \lambda A| = 0$, i.e., a factor $(\lambda - \lambda_k)^m$ appears in this equation, then m is called the *multiplicity* of the eigenvalue. For the typically well behaved square integrable kernels, it can be shown that the index p never exceeds the multiplicity m of the root or eigenvalue of the kernel, i.e., $0 < p \leq m$, and for symmetric kernels $p = m$. For $p = 1$, the eigenvalue λ_j is termed "simple".

Example 2 Homogeneous Fredholm Equation with Degenerate Kernel. Solve the integral equation

$$u(x) = \lambda \int_0^{\pi} (\cos^2 x \cos 2t + \cos 3x \cos^3 t) u(t) dt. \tag{E.1}$$

This is a homogeneous Fredholm equation with degenerate kernel

$$\begin{aligned} K(x,t) &= \cos^2 x \cos 2t + \cos 3x \cos^3 t \\ &= \sum_{k=1}^{2} a_k(x) b_k(t) \end{aligned} \tag{E.2}$$

hence $a_1(x) = \cos^2 x$, $a_2(x) = \cos 3x$, $b_1(t) = \cos 2t$, and $b_2(t) = \cos^3 t$. We follow the method of Example 1 to find c_1 and c_2 from

$$c_m = \lambda \sum_{k=1}^{2} a_{mk} c_k \tag{5.18}$$

or the matrix equation

$$(I - \lambda A)C = 0. \tag{5.19}$$

The solution $u(x)$ of (E.1) is

$$u(x) = \lambda \sum_{k=1}^{2} c_k a_k(x) = \lambda c_1 \cos^2 x + \lambda c_2 \cos 3x. \tag{E.3}$$

To evaluate c_1 and c_2 from (5.18) we must evaluate a_{11}, a_{12}, a_{21}, and a_{22}, the elements of the matrix A:

$$\begin{aligned} a_{11} &= \int_0^{\pi} b_1(t) a_1(t) dt = \int_0^{\pi} \cos 2t \cos^2 t dt = \frac{\pi}{4} \\ a_{12} &= \int_0^{\pi} b_1(t) a_2(t) dt = \int_0^{\pi} \cos 2t \cos 3t dt = 0 \\ a_{21} &= \int_0^{\pi} b_2(t) a_1(t) dt = \int_0^{\pi} \cos^3 t \cos^2 t dt = 0 \\ a_{22} &= \int_0^{\pi} b_2(t) a_2(t) dt = \int_0^{\pi} \cos^3 t \cos 3t dt = \frac{\pi}{8}. \end{aligned}$$

Hence (5.19) becomes

$$\begin{bmatrix} 1 - \dfrac{\lambda\pi}{4} & 0 \\ 0 & 1 - \dfrac{\lambda\pi}{8} \end{bmatrix} \begin{bmatrix} c_1 \\ c_2 \end{bmatrix} = 0 \tag{E.4}$$

and if we multiply the matrices in (E.4), we obtain

$$\left(1 - \lambda\frac{\pi}{4}\right)c_1 = 0 \tag{E.5}$$

$$\left(1 - \lambda\frac{\pi}{8}\right)c_2 = 0. \tag{E.6}$$

For

$$C = \begin{bmatrix} c_1 \\ c_2 \end{bmatrix}$$

to be the nontrivial solution, we must have a zero determinant for $I - \lambda A$ in (E.4),

$$\begin{vmatrix} 1 - \dfrac{\lambda\pi}{4} & 0 \\ 0 & 1 - \dfrac{\lambda\pi}{8} \end{vmatrix} = \left(1 - \lambda\frac{\pi}{4}\right)\left(1 - \lambda\frac{\pi}{8}\right) = 0 \tag{E.7}$$

which is a quadratic equation in λ whose solutions are $\lambda_1 = 4/\pi$ and $\lambda_2 = 8/\pi$, which in turn are the eigenvalues of (E.1). As a solution to (E.1), we have two eigenfunctions $u_1(x)$ and $u_2(x)$, corresponding to the two eigenvalues $\lambda_1 = 4/\pi$ and $\lambda_2 = 8/\pi$, respectively. Next we consider the two eigenvalues separately and find their corresponding eigenfunctions.

(a) $\lambda_1 = \dfrac{4}{\pi}$

If we substitute $\lambda_1 = 4/\pi$ in (E.5) and (E.6) to solve for c_1 and c_2, we have

$$\left(1 - \frac{4}{\pi}\cdot\frac{\pi}{4}\right)c_1 = 0c_1 = 0, \qquad c_1 = c_1$$

from (E.5) and

$$\left(1 - \frac{4}{\pi}\cdot\frac{\pi}{8}\right)c_2 = \left(1 - \frac{1}{2}\right)c_2 = \frac{1}{2}c_2 = 0, \qquad c_2 = 0$$

from (E.6). Hence with $c_1 = c_1$ and $c_2 = 0$ in (E.3) the eigenfunction $u_1(x)$ corresponding to $\lambda_1 = 4/\pi$ is

$$u_1(x) = \frac{4}{\pi}c_1\cos^2 x + \frac{4}{\pi}(0)\cos 3x = \frac{4}{\pi}c_1\cos^2 x.$$

This means that the eigenfunction is known except for the arbitrary constant c_1, which determines its amplitude; we may arbitrarily let $(4/\pi)c_1 = 1$ to have

$$u_1(x) = \cos^2 x.$$

Now we consider the case of the second eigenvalue:

(b) $\lambda_2 = \dfrac{8}{\pi}$.

We again substitute $\lambda_2 = 8/\pi$ in (E.5) and (E.6) to obtain

$$\left(1-\frac{8}{\pi}\left(\frac{\pi}{4}\right)\right)c_1=(1-2)c_1=-c_1=0 \quad c_1=0$$
$$\left(1-\frac{8}{\pi}\left(\frac{\pi}{8}\right)\right)c_2=0c_2=0, \quad c_2=c_2.$$

Hence from $c_1 = 0$ and $c_2 = c_2$ in (E.3) we obtain $u_2(x)$, the eigenfunction corresponding to the eigenvalue $\lambda_2 = 8/\pi$,

$$u_2(x)=\frac{8}{\pi}(0)\cos^2 x+\frac{8}{\pi}c_2\cos 3x=\frac{8}{\pi}c_2\cos 3x.$$

Now we may let $(8/\pi)c_2 = 1$ to have

$$u_2(x)=\cos 3x.$$

Fredholm alternative

The statements regarding the existence of the solutions of the nonhomogeneous and homogeneous system of n linear equations (5.15) and (5.19), respectively, and how they relate to the existence of solutions of the nonhomogeneous and homogeneous Fredholm integral equations are valid even when the kernel is not degenerate, and are summarized in the following statement (Theorems 1 and 2) of the Fredholm alternative.

Theorem 1 Fredholm Alternative—The Main Part

If the homogeneous integral equation (5.20)

$$u(x)=\lambda\int_a^b K(x,\xi)u(\xi)d\xi \tag{5.20}$$

has only the trivial solution $u(x) \equiv 0$, then the corresponding nonhomogeneous equation,

$$u(x)=f(x)+\lambda\int_a^b K(x,\xi)u(\xi)d\xi \tag{5.21}$$

always has one and only one solution. On the contrary, if the homogeneous equation has some nontrivial solutions, then the nonhomogeneous integral equation has either no solution or an infinity of solutions depending on the given function $f(x)$ (see Theorem 2).

As mentioned before, nontrivial solutions $\{u_n(x)\}$ of the homogeneous equation (5.20) are called the eigenfunctions $u_n(x)$ of (5.20) corresponding to the eigenvalues $\{\lambda_n\}$,

$$u_n(x)=\lambda_n\int_a^b K(x,t)u_n(t)dt. \tag{5.22}$$

To complement the second part of the Fredholm alternative, when the homogeneous equation (5.20) has a nontrivial solution, we state the following theorem without proof, which gives us the necessary and sufficient condition for the existence of solutions of its associated nonhomogeneous equation (5.21) for the important special case of *symmetric kernels*, i.e., $K(x,t) = K(t,x)$. We make this choice of special kernels to facilitate a more clear initial presentation of the main idea behind conditions for the existence of the solution.

Theorem 2 If the homogeneous equation with *symmetric* kernel,

$$u(x) = \lambda \int_a^b K(x,\xi)u(\xi)d\xi, \quad K(x,t) = K(t,x) \tag{5.20}$$

has a nontrivial solution or solutions, $\{u_j(x)\}$ (corresponding to $\lambda = \lambda_j$ in (5.20)), then the associated nonhomogeneous equation (with the same fixed parameter λ)

$$u(x) = f(x) + \lambda \int_a^b K(x,\xi)u(\xi)d\xi, \quad K(x,t) = K(t,x) \tag{5.21s}$$

will have a solution if and only if the nonhomogeneous term $f(x)$ in (5.21s) is *orthogonal* to every solution $u_j(x)$ (corresponding to λ_j) of the homogeneous equation (5.20).

Of course, other theorems are available to accommodate equations with nonsymmetric kernels, but we chose the above very important special case of symmetric kernels to simplify a clear presentation of the main features of Fredholm alternative for the existence of solutions to Fredholm integral equations of the second kind.

For completeness, we present such theorems after this initial discussion, in Theorems 3 and 4, and illustrate them in detail in Example 5.

We may mention here that in comparison to the conditions of the above two theorems, for the existence of the solutions of Fredholm equations of the second kind, the theory for the equations of the first kind is much more restrictive, as we shall discuss and illustrate in Section 5.4.

We should note that in contrast to the last Theorem 1 of the Fredholm alternative, Theorem 2 is for symmetric kernels, and its statement also assumes the same value for λ in both (5.20) and (5.21). This means that we are considering the usually special case when the fixed parameter λ of the nonhomogeneous equation (5.21) is equal to λ_n, the eigenvalue of the homogeneous equation (5.20). In Section 5.2.2 we will consider the general solution in (5.47) for λ of (5.21) not equal to λ_n of (5.20), then treat the problem of $\lambda = \lambda_n$ as a special case in (5.57). The next Example 3 illustrates Theorem 1 when $\lambda \neq \lambda_n$ for the nonsymmetric kernel of Example 2, while Example 4 illustrates both Theorems 1 and 2 as it covers the two cases of $\lambda \neq \lambda_n$ and $\lambda = \lambda_n$ for a symmetric kernel.

Example 3 Existence of the Fredholm Equation Solution-Nonsymmetric Kernel

In light of the Fredholm alternative, let us discuss the possibility of a solution to the nonhomogeneous Fredholm equation

$$u(x) = f(x) + \lambda \int_0^\pi \left(\cos^2 x \cos 2t + \cos 3x \cos^3 t\right) u(t)dt \tag{E.1}$$

which is associated with the homogeneous equation of Example 2 with its nonsymmetric kernel. Hence, with Theorems 1 and 2 at our disposal, we can only use Theorem 1 of the Fredholm alternative for such a nonsymmetric kernel.

(a) From Example 2, the homogeneous equation associated with (E.1),

$$u(x) = \lambda \int_0^{\pi} \left(\cos^2 x \cos 2t + \cos 3x \cos^3 t\right) u(t)dt \qquad (E.2)$$

has two nontrivial solutions (eigenfunctions),

$$u_1(x) = \cos^2 x \ \ and \ \ u_2(x) = \cos 3x$$

corresponding to the eigenvalue $\lambda_1 = 4/\pi$ and $\lambda_2 = 8/\pi$, respectively. So according to Fredholm alternative in Theorem 1, the Fredholm integral equation in (E.1) will definitely have a unique solution if the parameter λ in (E.1) is not equal to either one of the above two eigenvalues of its kernel, i.e., when $\lambda \neq 4/\pi, 8/\pi$. For the special two cases of $\lambda = 4/\pi, 8/\pi$ in (E.1), the equation may have either *no* solution or *an infinity* of solutions depending on $f(x)$ as it is clearly stated in Theorem 1 of the Fredholm alternative. However, since the kernel is not symmetric in (E.1), we cannot appeal to the only tool available to us in this book (at this stage without having Theorems 3 and 4), i.e., Theorem 2 to do a follow-up and determine which of these two possibilities may occur. For $\lambda \neq 4/\pi$ or $8/\pi$ we can also construct the unique solution, since, fortunately, the kernel in (E.1) is *degenerate* with only two terms, where the method of Section 5.1.1 can be followed as illustrated in Example 1 (see also Example 2). The Fredholm alternative for nonsymmetric kernels is complemented in Theorems 3 and 4, and fully illustrated in Example 5.

In the following Example 4 we will consider a Fredholm integral equation of the second kind with *symmetric* kernel, where we can employ Theorem 1 and Theorem 2 as the complete statement of Fredholm alternative. We choose a symmetric degenerate kernel to allow us the method of Section 5.1.1 for constructing the unique solution or an infinity of solutions when they exist according to Fredholm alternative.

The case of symmetric but nondegenerate kernels is discussed in Section 5.2, with detailed illustration for the construction of the solution, which is done in Example 7, and where the conditions of the Fredholm alternative in Theorems 1 and 2 will be very evident in Examples 9 and 10 of the same section.

Example 4 Existence of Fredholm Equation's Solution-Symmetric Kernel

Consider the Fredholm integral equation of the second kind

$$u(x) = f(x) + \lambda \int_0^{2\pi} \sin(x+t)u(t)dt. \qquad (E.1)$$

We shall discuss the possible existence of the solution (or solutions) for the three particular cases:

(i) $\lambda = 3,\ f(x) = x$

(ii) $\lambda = 1/\pi,\ f(x) = \sin 2x$

(iii) $\lambda = 1/\pi,\ f(x) = \sin x.$

Here we have a symmetric kernel, $K(x,t) = \sin(x+t) = \sin(t+x) = K(t,x)$, so we can rely on Theorems 1 and 2 for the Fredholm alternative to determine the possible existence of a unique solution, an infinity of solutions, or no solution. In case a unique solution does exist, we can appeal to the method of Section 5.1.1 for constructing such solution, since the kernel here is also degenerate, $\sin(x+t) = \sin x \cos t + \cos x \sin t$. Moreover, at this stage, and as the Fredholm alternative states, we must know whether or not the parameter λ in (E.1) is one of the eigenvalues $\{\lambda_k\}$ of the kernel $K(x,t) = \sin(x+t)$. This says that before doing anything, we must resort to the method of Section 5.1.1 to determine the eigenvalues and eigenfunctions of the (associated) homogeneous case of (E.1),

$$\phi_k(x) = \lambda_k \int_0^{2\pi} \sin(x+t)\phi_k(t)dt \qquad (E.2)$$

as was done in Example 2 of Section 5.1.1.

In order not to lose track of the analysis of Fredholm alternative for the three above cases, we shall first quote the results of the eigenvalues and eigenfunctions as shall be determined by the method of Section 5.1.1 and delay the details of the method to the end of this example. Such details will also be very much needed for the construction of any possible solution of (E.1). The method described gives two distinct eigenvalues $\lambda_1 = 1/\pi$ and $\lambda_2 = -1/\pi$ corresponding to their eigenfunctions $\phi_1(x) = \sin x + \cos x$ and $\phi_2(x) = \sin x - \cos x$, respectively.

With these results we have for case (i) $\lambda = 3 \neq \mp 1/\pi$, i.e., the parameter λ in (E.1) is not equal to any of the eigenvalues of the symmetric kernel. Thus according to Theorem 1 of the Fredholm alternative, the nonhomogeneous Fredholm equation (E.1) has a *unique* solution for arbitrary $f(x)$, and, of course, this includes the $f(x) = x$ given in case (i).

For case (ii) we have $\lambda = 1/\pi = \lambda_1$, i.e., the parameter λ in (E.1) is equal to one of the eigenvalues of the symmetric kernel, which should be a sign of some alarm. However, since the kernel $K(x,t) = \sin(x+t)$ is symmetric, we can appeal to Theorem 2, which states that the existence of (infinite—not unique!) number of solutions depends on whether the given nonhomogeneous term $f(x) = \sin 2x$ in (E.1) is *orthogonal* to the eigenfunction $\phi_1(x) = \sin x + \cos x$, corresponding to the eigenvalue $\lambda_1 = 1/\pi$, on the interval $(0, 2\pi)$. To show this we will evaluate the following integral, whose vanishing proves the orthogonality of these two functions on $(0, 2\pi)$,

$$\begin{aligned}\int_0^{2\pi} f(x)\phi_1(x)dx &= \int_0^{2\pi} \sin 2x[\sin x + \cos x]dx \\ &= \frac{1}{2}\int_0^{2\pi} [\cos x - \cos 3x + \sin x + \sin 3x]dx \\ &= \frac{1}{2}[\sin x - \sin 3x/3 - \cos x - \cos 3x/3]\Big|_0^{2\pi} \\ &= \frac{1}{2}[(-4/3) - (-4/3)] = 0,\end{aligned}$$

(after using trigonometric identities for the first integral) which says that $f(x) = \sin 2x$ is orthogonal to $\phi_1(x) = \sin x + \cos x$ on $(0, 2\pi)$, whence the problem (E.1) for case (ii) has infinite number of solutions. These solutions will be constructed at the end of this analysis, where the method of Section 5.1.1 is employed.

For case (iii) with $\lambda = 1/\pi = \lambda_1$ and $f(x) = \sin x$, we will show next that $f(x) = \sin x$ is not orthogonal to $\phi_1(x) = \sin x + \cos x$, the eigenfunction corresponding to $\lambda_1 = \frac{1}{\pi}$,

$$\begin{aligned}\int_0^{2\pi} f(x)\phi_1(x)dx &= \int_0^{2\pi} \sin x[\sin x + \cos x]dx \\ &= \frac{1}{2}\int_0^{2\pi} [1 - \cos 2x + \sin 2x]dx\end{aligned}$$

$$\begin{aligned}&= \frac{1}{2}\left[x - \sin 2x/2 - \cos 2x/2\right]\Big|_0^{2\pi} \\ &= \frac{1}{2}\left[\{2\pi - 0 - (1/2)\} - \{0 - 0 - (1/2)\}\right] = \pi \neq 0\end{aligned}$$

as the integral does not vanish. Hence, according to Theorem 2 there exists no solution for (E.1) in case (iii) of $\lambda = \frac{1}{\pi}$ and $f(x) = \sin x$.

Now we follow the method described in Section 5.1.1 to find the eigenvalues λ_n and eigenfunctions $\phi_n(x)$ of the (associated) homogeneous problem in (E.2),

$$\phi_n(x) = \lambda \int_0^{2x} \sin(x+t)\phi_n(t)dt \tag{E.3}$$

as was illustrated in Example 2. Then we use the same method, that was illustrated in Example 1, for constructing the unique solution of the nonhomogeneous equation (E.1) for case (i), and the infinite number of solutions of (E.1) for case (ii).

Here we have a degenerate kernel,

$$\sin(x+t) = \sin x \cos t + \cos x \sin t = \sum_{k=1}^{2} a_k(x) b_k(t)$$

with $a_1(x) = \sin x$, $a_2(x) = \cos x$, $b_1(t) = \cos t$ and $b_2(t) = \sin t$.

We will follow the procedure from (5.7) to (5.14) with all the necessary details except for evaluating the simple integrations involved. We let

$$\begin{aligned}c_1 &= \int_0^{2\pi} b_1(t)\phi(t)dt = \int_0^{2\pi} \cos t\phi(t)dt, \\ c_2 &= \int_0^{2\pi} b_2(t)\phi(t)dt = \int_0^{2\pi} \sin t\phi(t)dt,\end{aligned}$$

which are to be determined from solving the homogeneous case of the matrix equation (5.14), i.e., with $F = 0$. We now need to compute a_{11}, a_{12}, a_{21}, and a_{22}, the entries for the matrix A,

$$\begin{aligned} a_{11} &= \int_0^{2\pi} b_1(t)a_1(t)dt = \int_0^{2\pi} \cos t \sin t dt = 0, \\ a_{12} &= \int_0^{2\pi} b_1(t)a_2(t)dt = \int_0^{2\pi} \cos t \cos t dt = \pi, \\ a_{21} &= \int_0^{2\pi} b_2(t)a_1(t)dt = \int_0^{2\pi} \sin t \sin t dt = \pi, \\ a_{22} &= \int_0^{2\pi} b_2(t)a_2(t)dt = \int_0^{2\pi} \sin t \cos t dt = 0. \end{aligned} \tag{E.4}$$

Also for the solution of case (i) we will need, as shown in (5.14),

$$\begin{aligned} f_1 &= \int_0^{2\pi} b_1(t)f(t)dt = \int_0^{2\pi} t \cos dt = 0, \\ f_2 &= \int_0^{2\pi} b_2(t)f(t)dt = \int_0^{2\pi} t \sin dt = -2\pi. \end{aligned} \tag{E.5}$$

So the homogeneous case of the matrix equation (5.14) now becomes

$$(I - \lambda A)C = \begin{bmatrix} 1 & 0 \\ 0 & 1 \end{bmatrix} \begin{bmatrix} c_1 \\ c_2 \end{bmatrix} - \lambda \begin{bmatrix} 0 & \pi \\ \pi & 0 \end{bmatrix} \begin{bmatrix} c_1 \\ c_2 \end{bmatrix} = \begin{bmatrix} 0 \\ 0 \end{bmatrix},$$

$$= (I - \lambda A)C = \begin{bmatrix} 1 & -\lambda\pi \\ -\lambda\pi & 1 \end{bmatrix} \begin{bmatrix} c_1 \\ c_2 \end{bmatrix} = \begin{bmatrix} 0 \\ 0 \end{bmatrix}, \tag{E.6}$$

which will have solution if the following determinant vanishes,

$$|I - \lambda A| = \begin{vmatrix} 1 & -\lambda\pi \\ -\lambda\pi & 1 \end{vmatrix} = 1 - \lambda^2\pi^2 = 0,$$

hence $\lambda_1 = \dfrac{1}{\pi}$ and $\lambda_2 = -\dfrac{1}{\pi}$ as the two distinct eigenvalues of the symmetric (and degenerate) kernel $\sin(x + t)$ of (E.2), as we quoted them at the beginning of this example.

To find the eigenfunctions $\phi_1(x)$ and $\phi_2(x)$ corresponding to the eigenvalues $\lambda_1 = 1/\pi$ and $\lambda_2 = -1/\pi$ we substitute for each case in (E.6) to find c_1 and c_2, which are to be substituted in (5.9) with $f(x) = 0$,

$$u(x) = \lambda \sum_{k=1}^{n} c_k a_k(x) \tag{5.9h}$$

to have the corresponding eigenfunction $\phi_1(x)$.

For $\lambda_1 = 1/\pi$ in (E.6) we have

$$\begin{bmatrix} 1 & -1 \\ -1 & 1 \end{bmatrix} \begin{bmatrix} c_1 \\ c_2 \end{bmatrix} = \begin{bmatrix} 0 \\ 0 \end{bmatrix},$$

$$c_1 - c_2 = 0,$$
$$-c_1 + c_2 = 0$$

Hence $c_1 = c_2$, and from (5.9h) above,

$$u(x) = \lambda \sum_{k=1}^{n} c_k a_k(x) \qquad (5.9h), (E.7)$$

we have

$$\phi_1(x) = 1/\pi \sum_{k=1}^{2} c_k a_k(x) = 1/\pi[c_1 \sin x + c_1 \cos x] = \frac{c_1}{\pi}(\sin x + \cos x). \quad (E.8)$$

If we normalize by choosing $c_1 = \pi$, we have $\phi_1 = \sin x + \cos x$, which is what we quoted at the beginning of this example. However, the more common way is to normalize with a unit norm, i.e., $||\phi_1(x)|| = [\int_0^{2x}(\sin x + \cos x)^2 dx]^{\frac{1}{2}} = \sqrt{2\pi}$, whence the normalized eigenfunction becomes $\phi_1(x) = (1/\sqrt{2\pi})(\sin x + \cos x)$.

For completeness we determine the second eigenfunction $\phi_2(x)$ corresponding to the second eigenvalue $\lambda_2 = -1/\pi$, where (E.4) with $\lambda = -1/\pi$ becomes

$$\begin{bmatrix} 1 & 1 \\ 1 & 1 \end{bmatrix} \begin{bmatrix} c_1 \\ c_2 \end{bmatrix} = \begin{bmatrix} 0 \\ 0 \end{bmatrix},$$

$$c_1 + c_2 = 0, \qquad c_2 = -c_1,$$
$$c_1 + c_2 = 0, \qquad c_2 = -c_1.$$

Hence $c_2 = -c_1$, and from (5.9) with $f(x) = 0$ (or (5.9h)) we have

$$\begin{aligned} \phi_2(x) = -\frac{1}{\pi}\sum_{k=1}^{2} c_k a_k(x) &= -\frac{1}{\pi}[c_1 \sin x - c_1 \cos x] \\ &= -\frac{c_1}{\pi}[\sin x - \cos x], \end{aligned} \qquad (E.9)$$

and if we normalize by choosing $(-c_1/\pi) = 1$, we have $\phi_2(x)= \sin x - \cos x$ which is what we quoted at the beginning of this example. Again, if we normalize with a unit norm, i.e.; $||\phi_2||= [\int_0^{2x}(\sin x - \cos x)^2 dx]^{\frac{1}{2}} = \sqrt{2\pi}$, we have the more commonly normalized eigenfunction $\phi_2(x) = (1/\sqrt{2\pi})(\sin x - \cos x)$.

We will leave it as an exercise to verify that the above $\phi_1(x)$ and $\phi_2(x)$ are the eigenfunctions corresponding to the eigenvalues $\lambda_1 = 1/\pi$ and $\lambda_2 = -1/\pi$, i.e., each of the corresponding pairs satisfies the homogeneous Fredholm equation (E.2) (see Exercise 4).

To construct the *unique* solution of (E.1) for case (i) of $\lambda = 3$ ($\neq \pm\frac{1}{\pi}$) and $f(x) = x$, we use (5.14) or (5.15) with the entries of the known matrices A and F as computed above in (E.4) and (E.5) to find c_1 and c_2 of the unknown matrix C, that are needed for the solution of (E.1) as given by (5.9),

$$u(x) = f(x) + \lambda \sum_{k=1}^{n} c_k a_k(x) \tag{5.9}$$

$$u(x) = x + 3\left[c_1 \sin x + c_2 \cos x\right]. \tag{E.10}$$

Of course, these c_1, c_2 are not to be confused with those of the homogeneous matrix equation (E.6). If we substitute the known values of $\lambda = 3$ and the matrices A from (E.4) and F from (E.5) in (5.15) we have

$$C - \lambda AC = F = \begin{bmatrix} c_1 \\ c_2 \end{bmatrix} - 3\begin{bmatrix} 0 & \pi \\ \pi & 0 \end{bmatrix}\begin{bmatrix} c_1 \\ c_2 \end{bmatrix} = \begin{bmatrix} 0 \\ -2\pi \end{bmatrix},$$

$$\begin{bmatrix} 1 & -3\pi \\ -3\pi & 1 \end{bmatrix}\begin{bmatrix} c_1 \\ c_2 \end{bmatrix} = \begin{bmatrix} 0 \\ -2\pi \end{bmatrix},$$

$$c_1 - 3\pi c_2 = 0,$$

$$-3\pi c_1 + c_2 = -2\pi,$$

$$c_1 = 6\pi^2/(9\pi^2 - 1), \qquad c_2 = 2\pi/(9\pi^2 - 1).$$

If we now substitute these values in (E.10) we obtain the *unique* solution to the problem in (E.1) with $\lambda = 3$,

$$u(x) = x + 3c_1 \sin x + 3c_2 \cos x = x + \frac{18\pi^2 \sin x + 6\pi \cos x}{9\pi^2 - 1}. \tag{E.11}$$

To construct the *infinite* number of solutions for case (ii) with $\lambda = 1/\pi$ in (E.1) (for $\lambda = \lambda_1 = \frac{1}{\pi}$), we can use (5.15) with the same matrix A, but the matrix F is different, since here we have a different function $f(x) = \sin 2x$ from that of $f(x) = x$ in case (i). So we need to evaluate

$$f_1 = \int_0^{2\pi} \sin 2t \cos t dt = 0,$$

$$f_2 = \int_0^{2\pi} \sin 2t \sin t dt = 0,$$

to be substituted with those entries of matrix A in (5.15) to have

$$\begin{bmatrix} c_1 \\ c_2 \end{bmatrix} - (1/\pi)\begin{bmatrix} 0 & \pi \\ \pi & 0 \end{bmatrix}\begin{bmatrix} c_1 \\ c_2 \end{bmatrix} = \begin{bmatrix} 0 \\ 0 \end{bmatrix},$$

$$\begin{bmatrix} 1 & -1 \\ -1 & 1 \end{bmatrix}\begin{bmatrix} c_1 \\ c_2 \end{bmatrix} = \begin{bmatrix} 0 \\ 0 \end{bmatrix},$$

$$c_1 - c_2 = 0, \qquad c_1 = c_2$$

$$-c_1 + c_2 = 0, \qquad c_1 = c_2.$$

With such result of the arbitrary $c_1 = c_2$ we have no unique solution, as when we substitute these values in (5.9) we have

$$u(x) = \sin 2x + c_1(\sin x + \cos x) \tag{E.12}$$

which represents an $infinite$ number of solutions because of the solution dependence on the arbitrary value c_1 in (E.12).

Fredholm Alternative-Nonsymmetric Kernels

For completeness, we may now return to the Fredholm alternative for the case of *nonsymmetric* kernels, state Theorem 4 that complements Theorem 1 for such kernels, then illustrate it with a simple nonsymmetric kernel in Example 5. In Theorem 2 we stated the complement to the Fredholm alternative (Theorem 1) for the case of symmetric kernels, i.e., $K(x,t) = K(t,x)$ in equation (5.21s). Here, we state a more general Theorem that complements Theorem 1 for the case of not necessarily symmetric kernels. Before we state such a theorem we present a simple statement of Theorem 3 that relates to a Fredholm integral equation of the second kind in $\psi(x)$, associated with (5.21), but with kernel $K(t,x)$ instead of $K(x,t)$ of (5.21),

$$\psi(x) = g(x) + \lambda \int_a^b K(t,x)\psi(t)dt \tag{5.23}$$

with its homogeneous case as

$$\psi(x) = \lambda \int_a^b K(t,x)\psi(t)dt. \tag{5.23h}$$

Theorem 3 Fredholm Equation with Kernel $K(t,x)$ instead of $K(x,t)$

"Consider the Fredholm integral equation of the second kind (5.21)

$$u(x) = f(x) + \lambda \int_a^b K(x,t)u(t)dt \tag{5.21}$$

and its homogeneous case (5.20)

$$u(x) = \lambda \int_a^b K(x,\xi)u(\xi)d\xi. \tag{5.20}$$

If the Fredholm alternative of Theorem 1 holds for equation (5.21) in $u(x)$ with kernel $K(x,t)$ for a given fixed λ, then it also holds for its associated equation (5.23) in $\psi(x)$ with kernel $K(t,x)$. Moreover, the homogeneous equation (5.20) (with kernel $K(x,t)$) and its associated equation (5.23h) (with kernel $K(t,x)$) have one and the same finite number of linearly independent solutions."

Now, that we introduced the associated equations to (5.21) and (5.20) with kernel $K(t,x)$ as in (5.23) and (5.23h), we will present Theorem 4 that complements the Fredholm alternative of Theorem 1 for nonsymmetric kernels instead of the special case of Theorem 2 of the symmetric kernels.

Theorem 4 Complement to Fredholm Alternative-Nonsymmetric Kernels

"If the associated homogeneous equation (5.23h) (with not necessarily symmetric kernel) has a nontrivial solution or solutions $\{\psi_j(x)\}$,

$$\psi_j(x) = \lambda_j \int_a^b K(t,x)\psi_j(t)dt \tag{5.24}$$

then the nonhomogeneous Fredholm equation (5.21)

$$u(x) = f(x) + \lambda \int_a^b K(x,t)u(t)dt \tag{5.21}$$

(with the same fixed parameters $\lambda = \lambda_j$) will have a solution if and only if the nonhomogeneous term $f(x)$ in (5.21) is orthogonal to every solution (eigenfunction) $\psi_j(x)$ of the associated equation (5.24) (with kernel $K(t,x)$), i.e.,

$$\int_a^b f(x)\psi_j(x)dx = 0."$$

Clearly this theorem becomes Theorem 2 when $K(x,t)$ is symmetric, since the associated equations (5.23h) and (5.24) become (5.20) and (5.22), respectively, and the above orthogonality condition $\int_a^b f(x)\psi_j(x)dx = 0$ becomes $\int_a^b f(x)\phi_j(x)dx = 0$ of Theorem 2, where $\{\phi_j(x)\}$ are the eigenfunctions of (5.20) (or (5.22)) for symmetric kernels.

To illustrate Theorem 4, as it complements the Fredholm alternative for nonsymmetric kernels, we choose the following example with a very simple (nonsymmetric kernel) to avoid lengthy details. The main emphasis will be directed towards solving the two homogeneous equations (5.20) with $K(x,t)$ and (5.23h) with $K(t,x)$ as they supply the basic ingredients to our analysis of both parts of the Fredholm alternative for nonsymmetric kernels (Theorems 1 and 4).

Example 5 Fredholm Alternative-Nonsymmetric Kernels

Consider the following problem with its very simple nonsymmetric kernel $K(x,t) = \sin(\ln x)$, where $K(t,x) = \sin(\ln t)$,

$$u(x) = 2x + \lambda \int_0^1 \sin(\ln x)u(t)dt. \tag{$E.1$}$$

This kernel is also degenerate with one term where $a_1(x) = \sin(\ln x)$ and $b_1(t) = 1$, so we use the method of this section to first find the eigenvalue λ_1, of the homogeneous equation (5.20) (with $K(x,t) = \sin(\ln x)$),

$$\phi_1(x) = \lambda_1 \int_0^1 \sin(\ln x)\phi_1(t)dt \tag{$E.2$}$$

and in case λ of (E.1) is not equal to this eigenvalue λ_1, we construct the unique solution of (E.1) as guaranteed by the first part of the Fredholm alternative (Theorem 1). If we let $c_1 = \int_0^1 u(t)dt$ in (E.1), we have

$$u(x) = 2x + c_1\lambda \sin(\ln x), \tag{E.3}$$

and if we substitute this value of $u(x)$ in the above integral defining c_1, we obtain

$$\begin{aligned} c_1 &= \int_0^1 u(t)dt = \int_0^1 [c_1\lambda \sin(\ln t) + 2t]dt, \\ c_1 &= c_1\lambda \int_0^1 \sin(\ln t)dt + t^2\Big|_0^1 . \end{aligned} \tag{E.4}$$

The integral $\int_0^1 \sin(\ln t)dt$ can be done with one substitution $u = \ln t$, which will reduce it to $\int_{-\infty}^0 e^u \sin u du$, that can be evaluated by two (careful) integrations by parts to give a value of $-\frac{1}{2}$,

$$\begin{aligned} c_1 &= -\frac{1}{2}c_1\lambda + 1, \quad c_1\left(1 + \frac{\lambda}{2}\right) = 1, \\ c_1 &= \frac{2}{\lambda + 2}, \qquad\qquad \lambda \neq -2. \end{aligned} \tag{E.5}$$

So with this result of c_1 and the (important) condition $\lambda \neq -2$ in (E.1), the unique solution to (E.1) becomes

$$u(x) = 2x + \frac{2\lambda}{\lambda + 2}\sin(\ln x), \qquad \lambda \neq -2. \tag{E.6}$$

Of course, the condition $\lambda \neq -2$ for the above unique solution in (E.6) should, in the spirit of the Fredholm alternative (Theorem 1), be transparent to us as $\lambda \neq \lambda_1 = -2$, where $\lambda_1 = -2$ should be the eigenvalue of the homogeneous equation (E.2). This can be verified easily from (E.4) or (E.5) without the nonhomogeneous term $t^2|_0^1$ in (E.4) (as if we do (E.3) without the $2x$ term, which amounts to doing (E.2)),

$$c_1 = -\frac{1}{2}c_1\lambda, \qquad c_1(\lambda + 2) = 0. \tag{E.7}$$

So unless $\lambda = -2$, the arbitrary constant c_1 in (E.7) must be zero, which results in a trivial solution $u(x) \equiv 0$ for the homogeneous equation (E.2) as can be obtained from $u(x)$ in (E.3) without the nonhomogeneous term $2x$. This means that for (E.2) to have an eigenvalue $\lambda_1 = -2$ in (E.7), the constant c_1 is allowed to be arbitrary. Thus, the corresponding nontrivial solution, i.e., the eigenfunction corresponding to $\lambda_1 = -2$ is obtained from (E.3) (without the term $2x$) as

$$\phi_1(x) = c_1\lambda_1 \sin(\ln x) = -2c_1 \sin(\ln x) = C\sin(\ln x). \tag{E.8}$$

The second part of this illustration is to address that case for the nonhomogeneous problem (E.1) (with its nonsymmetric kernel) when its parameters λ is equal to $\lambda_1 = -2$ of this kernel. Since the kernel $K(x,t) = \sin(\ln x)$ is not symmetric, we must resort to Theorem 4 for the second part of the Fredholm alternative, which means we have to find the eigenvalue and eigenfunction $\psi_1(x)$ of the homogeneous equation (5.23h), associated with the kernel $K(t,x) = \sin(\ln t)$,

$$\psi_1(x) = \mu_1 \int_0^1 \sin(\ln t)\psi_1(t)dt. \tag{E.9}$$

We use here μ_1 instead of λ_1 just to emphasize solving a homogeneous problem as new because of its different kernel $K(t,x) = \sin(\ln t)$. So we set out to find the eigenfunction $\psi_1(x)$ in a similar way to what we did for (E.2) and (E.1), except here in (E.9) we have the kernel $\sin(\ln t)$ instead of $\sin(\ln x)$ in (E.2). In this case we have a degenerate kernel $K(x,t) = \sin(\ln t)$ with one term, where $a_1(x) = 1$ and $b_1(t) = \sin(\ln t)$. As was done before, we let

$$c_1 = \int_0^1 b_1(t)\psi(t)dt = \int_0^1 \sin(\ln t)\psi(t)dt, \tag{E.10}$$

which when substituted in (E.9) we obtain $\psi_1(x) = \mu_1 c_1$, and if this $\psi_1(x)$ is substituted inside the integral of (E.10), we obtain

$$\begin{aligned} c_1 &= \mu_1 c_1 \int_0^1 \sin(\ln t)dt = -\frac{1}{2}\mu_1 c_1, \\ c_1 &\left(\frac{2+\mu_1}{2}\right) = 0. \end{aligned} \tag{E.11}$$

For $\psi_1(x) = \mu_1 c_1$ not to be the trivial solution $\psi_1(x) \equiv 0$, i.e., to have it as an eigenfunction, we cannot assign $c_1 = 0$. Thus from (E.11), the eigenvalue to equation (E.9) is $\mu_1 = -2$ corresponding to the eigenfunction $\psi_1(x) = \mu_1 c_1 = -2c_1 = c$, an arbitrary constant. This $\psi_1(x) = c$ represents an infinity of solutions, but it can be normalized, and we have the eigenfunction needed for the following important orthogonality condition of Theorem 4:

$$\int_a^b f(x)\psi_j(x)dx = 0 \ for \ all \ j. \tag{E.12}$$

So according to Theorem 4, when $\lambda = -2$ in the nonhomogeneous equation (E.1), where $\lambda_1 = -2$ is the eigenvalue of its *(nonsymmetric)* kernel $K(x,t) = \sin(\ln x)$, the only way for it to have a solution is for its nonhomogeneous term $f(x) = 2x$ to be orthogonal to $\psi_1(x) = c$ of (E.9) as spelled out in (E.12), which, unfortunately is not the case since,

$$\int_a^b f(x)\psi_1(x)dx = c\int_0^1 2xdx = c \neq 0 \tag{E.13}$$

and c is assumed not to be zero for $\psi_1(x) = c$ to be an eigenfunction (i.e., not a trivial solution).

To summarize, the final result of this example is that the Fredholm integral equation (E.1) has a unique solution when $\lambda \neq -2$ and no solution when $\lambda = -2$.

5.1.3 Approximating a Kernel by a Degenerate One

In many cases a nondegenerate kernel $K(x,t)$ may be approximated by a degenerate kernel as a partial sum of the Maclaurin (or other) series expansion of $K(x,t)$. For example, the nondegenerate kernel $K(x,t) = \cos xt$ may be approximated by a degenerate kernel

$$M(x,t) = 1 - \frac{x^2t^2}{2!} + \frac{x^4t^4}{4!} \tag{5.25}$$

which consists of the first three terms of the Maclaurin series expansion of $\cos xt$. Let us consider the Fredholm equation with kernel $K(x,t)$,

$$u(x) = f(x) + \lambda \int_a^b K(x,t)u(t)dt \tag{5.21}$$

and its associated equation,

$$v(x) = f(x) + \lambda \int_a^b M(x,t)v(t)dt \tag{5.26}$$

with kernel $M(x,t)$ as the degenerate kernel approximation of $K(x,t)$. In principle, we may use this section method to solve for $v(x)$, which is considered as an approximate to the solution $u(x)$ of (5.21). Of course, there will be an error involved in such an approximation, which is defined as $\varepsilon = |u(x) - v(x)|$, and we may attempt to estimate this error to give us a measure of how good this approximation is.

Example 6 Approximating a One by a Degenerate Kernel Find the approximate solutions to the Fredholm integral equation

$$u(x) = \sin x + \int_0^1 (1 - x\cos xt)u(t)dt \tag{E.1}$$

by considering its associated Fredholm equation with degenerate kernel.

We note here that the kernel $1 - x\cos xt$ is not degenerate, but a finite number of terms of its Maclaurin series

$$1 - x\left(1 - \frac{x^2t^2}{2!} + \frac{x^4t^4}{4!} - \cdots\right) = 1 - x + \frac{x^3t^2}{2!} - \frac{x^5t^4}{4!} + \cdots \tag{E.2}$$

is degenerate or, in other words, separable in x and t. So if we consider only three terms of the series in (E.2), we have a degenerate kernel

$$M(x,t) = 1 - x + \frac{x^3t^2}{2!} \tag{E.3}$$

as an approximation to $K(x,t) = 1 - x\cos xt$ of (E.1). The associated equation in $v(x)$,

$$v(x) = \sin x + \int_0^1 \left(1 - x + \frac{x^3t^2}{2!}\right) v(t)dt \tag{E.4}$$

has degenerate kernel and can be solved by the method we discussed in this section and illustrated as in Example 1. Hence from (5.6) we have

$$M(x,t) = (1-x) + \frac{x^3t^2}{2!} = \sum_{k=1}^{2} a_k(x)b_k(t) \tag{E.5}$$

where $a_1(x) = (1-x)$, $a_2(x) = x^3$ and $b_1(t) = 1$, $b_2(t) = t^2/2$. Now we can employ the method of solving nonhomogeneous Fredholm equations with degenerate kernel as illustrated in Example 1 (leaving the detailed steps for Exercise 11(a)) to find the elements of the matrix C in (5.15) are

$$c_1 = 1.00308, \qquad c_2 = 0.16736.$$

Therefore, the solution to (E.4), according to (5.9), is

$$\begin{aligned} v(x) &= \sin x + c_1a_1(x) + c_2a_2(x) \\ &= \sin x + 1.00308(1-x) + 0.16736x^3 \end{aligned} \tag{E.6}$$

which is the approximate solution to (E.1). In this special case it happens that we also know the exact solution of (E.1) as $u(x) = 1$, which can be verified easily.

$$\begin{aligned} \sin x + \int_0^1 (1 - x\cos xt)(1)dt &= \sin x + 1 - x\int_0^1 \cos xt dt \\ &= \sin x + 1 - x\left.\frac{\sin xt}{x}\right|_0^1 = \sin x + 1 - \sin x \\ &= 1. \end{aligned}$$

The approximate solution $v(x)$ in (E.6) and the exact solution $u(x) = 1$ of (E.1) are presented in Table 5.1 to show how $v(x)$ corresponding to the degenerate kernel (E.5) approximates $u(x) = 1$ with nondegenerate kernel $1 - x\cos xt$. It is of interest to observe how close $v(x)$ will be to $u(x)$ when we consider more terms of the Maclaurin series expansion for $M(x,t)$ (see Exercise 11).

Exercises 5.1

1. Solve the following Fredholm equations in $u(x)$, then verify your answer.

   (a) $u(x) = \sin x + \lambda \int_0^{\pi/2} \sin x \cos t u(t)dt$

   Hint: Write $C = \int_0^{\pi/2} \cos t u(t)dt$, where the above equation becomes: $u(x) = \sin x + \lambda C \sin x$, use this $u(x)$ in the integral of C, then solve for

Table 5.1 Approximate (Kernel Replaced by a Degenerate One) and Exact Solutions of Fredholm Equation (E.1)

x	0	0.25	0.5	0.75	1.0
Approximate values, $v(x) = \sin x + 1.00308(1-x)$ $+0.16736x^3$	1.0031	1.0023	1.0019	1.0030	1.0088
Exact values, $u(x) = 1$	1.0000	1.0000	1.0000	1.0000	1.0000

C. Aso, note that all the Fredholm equations in this Exercise 1 are nonhomogeneous with degenerate kernel.

(b) $u(x) = x + \lambda \int_{-\pi}^{\pi} (\cos x \sin t + x \cos t + t^2 \sin x)u(t)dt$

Hint: Here we will end up with a rather long 3×3 system of equations, however, many of the entries of the coefficient matrix A vanish due to integrating odd functions on the symmetric interval $(-\pi, \pi)$.

(c) $u(x) = 2x - \pi + 4 \int_0^{\pi/2} \sin^2 xu(t)dt$

(d) $u(x) = \cos x + \lambda \int_0^{\pi} \sin(x-t)u(t)dt$

(e) $u(x) = e^x + \lambda \int_0^1 (5x^2 - 3)t^2 u(t)dt$

(f) $u(x) = 2x - 6 - 2 \int_0^1 u(t)dt$

(g) $u(x) = 201x^2 - 80x + 52 + \int_0^1 (4xt^2 - 3x^2 t - t^3)u(t)dt$

2. Solve the following *homogeneous* Fredholm equations by finding the eigenvalues and eigenfunctions of the equations.

(a) $u(x) = \lambda \int_0^{2\pi} \sin x \sin tu(t)dt$

Hint: See the hint to problem 1(a).

(b) $u(x) = \lambda \int_0^{\pi/2} \sin x \cos tu(t)dt$

(c) $u(x) = \lambda \int_0^{\pi} \cos(x+t)u(t)dt$

(d) $u(x) = 2\lambda \int_0^1 x(t-2x)u(t)dt$

(e) $u(x) = \lambda \int_0^1 (5x^2 - 3)t^2 u(t)dt$

3. **(a)** Discuss the existence of the solution to the integral equation

$$u(x) = f(x) + \lambda \int_0^{2\pi} \sin(x+t)u(t)dt \tag{E.1}$$

for the two cases.

(i) $\lambda = 1/\sqrt{\pi}$, $f(x) = x^2$

(ii) $\lambda = 1/\pi$, $f(x) = \sin 3x$

Hint: Consult the Fredholm alternative in Theorems 1 and 2, and its detailed illustration in Example 4 for the present symmetric kernel.

(b) Find the solution in the case (or cases) where it exists.

Hint: Consult part (a) and Example 4 for the existence of a unique solution or infinity of solutions, then the method of Section 5.1.1 and its illustration at the end of Example 4 for constructing the solution (or solutions).

4. In Example 4, verify that $\phi_1(x) = (\sin x + \cos x)$ and $\phi_2(x) = \sin x - \cos x$ are the two eigenfunctions corresponding, respectively, to the two eigenvalues $\lambda_1 = 1/\pi$ and $\lambda_2 = -1/\pi$ of the kernel $K(x,t) = \sin(x+t)$, i.e., show that each pair of an eigenfunction and its corresponding eigenvalue satisfies the homogeneous Fredholm equation (E.2) in Example 4.

5. In problem 1(a), and its associated homogeneous case in 2(b), use their results to illustrate the Fredholm alternative.

   Hint: Compare the parameter λ in the Fredholm equation of problem 1(a) with the one eigenvalue λ_1 in problem 2(b).

6. Solve the following homogeneous Fredholm integral equations

(a) $$u(x) = 6\int_0^1 (2xt - x^2)u(t)dt \tag{E.1}$$

(b) $$u(x) = 2\int_0^{\frac{\pi}{2}} \frac{u(t)}{1+\cos 2t}dt \tag{E.2}$$

Hint: See the hint to Exercise 1(a), but watch for $u(x) = C \neq 0$, since this will give a divergent integral on the right side of (E.2).

(c) $$u(x) = \frac{1}{4}\int_{-2}^{2} |x|u(t)dt. \tag{E.3}$$

Hint: See the hint to Exercise 1(a).

7. In light of the Fredholm alternative, how do you explain the validity of the solution to problem 1(d) for all real values of its parameter λ?

8. Compare the results of problem 1(e) and its associated homogeneous case in problem 2(e), how do you explain the validity of the solution in 1(e) for all values of its parameter λ?

9. Consider the following problem with degenerate kernel $K(x,t) = (1 - 3xt)$ and a general nonhomogeneous term $f(x)$,

$$u(x) = f(x) + \lambda \int_0^1 (1 - 3xt)u(t)dt \qquad (E.1)$$

(a) Find the (two) eigenvalues of the kernel in (E.1).

(b) Consider the resulting system of equations from (E.1), what is the condition for a unique solution? Also what about the existence of a solution to (E.1)?

(c) Consider the cases when $\lambda = \lambda_1 = 2$ and $\lambda = \lambda_2 = -2$ in the answer of part (a), what can be said about the system in part (b) for the given function $f(x)$?

(d) Find the corresponding eigenfunctions to the eigenvalues of the kernel in part (a).

(e) Write the general solution to (E.1) when $\lambda = 2$.

10. **(a)** Use the method of degenerate kernels to solve the nonlinear integral equation

$$u(x) = b + \lambda \int_0^1 u^2(t)dt$$

Hint: $K(x,t) = 1$, let $C = \int_0^1 u^2(t)dt$, where the above equation becomes: $u(x) = b + \lambda C$; then use this $u(x)$ in the integral of C, which results in a single equation in C.

(b) Use the same method to solve the homogeneous equation

$$u(x) = \lambda \int_0^1 u^2(t)dt$$

11. Consider the Fredholm integral equation of Example 6

$$u(x) = \sin x + \int_0^1 (1 - x \cos xt)u(t)dt. \qquad (E.1)$$

(a) As in Example 6, approximate the kernel by a degenerate one by considering only the first two terms of the Maclaurin series expansion of $\cos xt$. Use the method of this section to find the approximate solution $v(x)$ as given for Example 6.

(b) Compare the result of part (a) with the exact solution $u(x) = 1$ and another approximate of Example 6 [by considering the first three terms of the Maclaurin series expansion of $\cos xt$ in (E.1)].

12. **(a)** Use the first three terms of the Maclaurin series of the kernel in the integral equation

$$u(x) = e^x - x - \int_0^1 x(e^{xt} - 1)u(t)dt \qquad (E.1)$$

to reduce it to another approximate equation with degenerate kernel, then solve this problem.

(b) Verify that $u(x) = 1$ is the exact solution of (E.1).

(c) Tabulate the approximate solution in part (a) and compare with the exact solution in part (b).

13. **(a)** Reduce the Fredholm equation

$$u(x) = x + \int_{-1}^1 e^{xt}u(t)dt$$

to an approximate one (in $v(x)$) with degenerate kernel.

(b) Consider only the first two terms of the Maclaurin series of e^{xt} in part (a) and solve the resulting integral equation, using the method of Example 1 for a degenerate kernel.

14. Assume that the kernel $K(x,t)$ is not symmetric, i.e., $K(x,y) \neq \overline{K(y,x)}$, show that the following two kernels, associated with $K(x,t)$

$$K_1(x,y) = \int K(s,x)\overline{K(s,y)}ds$$

and

$$K_2(x,y) = \int K(x,s)\overline{K(y,s)}ds$$

are symmetric.

Hint: Take the complex conjugate $\overline{K_1(x,y)}$ of $K_1(x,y)$, noting that $\overline{\int f_1(x)f_2(x)dx} = \int \overline{f_1(x)}\ \overline{f_2(x)}dx$, and $\overline{\overline{f_1(x)}} = f_1(x)$.

15. Consider the problem of Example 5 for the Fredholm integral equation with *nonsymmetric* kernel (as in Example 5) but with more general nonhomogeneous term $f(x)$.

$$u(x) = f(x) + \lambda \int_0^1 \sin(\ln x)u(t)dt. \qquad (E.1)$$

(a) In light of the Fredholm alternative for Fredholm integral equation with *nonsymmetric* kernels, as stated in Theorems 1 and 4, discuss the existence of the solution (or solutions) to (E.1) when

(i) $\lambda = 3$, $f(x) = x^3$

(ii) $\lambda = -2$, $f(x) = \cos \pi x$

(b) Find the unique solution of (E.1) in part (a) when it exists.

(c) Find an infinity of solutions of (E.1) in part (a) when they exist.

5.2 FREDHOLM INTEGRAL EQUATIONS WITH SYMMETRIC KERNEL

In the preceding section we discussed methods of solving nonhomogeneous and homogeneous Fredholm equations for the special case of degenerate kernels and showed how the method reduced essentially to solving systems of linear equations. In this section we consider the Fredholm equation for another important special case of the symmetric kernel[2] [i.e., $K(x,t) = K(t,x)$ in (5.21)]:

$$u(x) = f(x) + \lambda \int_a^b K(x,t)u(t)dt, \quad K(x,t) = K(t,x). \qquad (5.21s)(5.27)$$

Similar to the case of (3.1), the Volterra integral equation, it turns out that the resolvent kernel $\Gamma(x,t;\lambda)$, for (5.27), can be expressed as an infinite series in terms of the orthonormal eigenfunctions of the homogeneous equation with symmetric kernel,

$$u(x) = \lambda \int_a^b K(x,t)u(t)dt, \quad K(x,t) = K(t,x) \qquad (5.20s)(5.28)$$

which is discussed next.

As in the case of (3.2) for the solution to the Volterra equation in terms of its resolvent kernel in (3.3), the resolvent kernel $\Gamma(x,t;\lambda)$ of (5.27) will be shown, in the analysis leading to (5.55), to give

$$u(x) = f(x) + \lambda \int_a^b \Gamma(x,t;\lambda)f(t)dt \qquad (5.55)$$

as the solution of (5.27), where $\Gamma(x,t;\lambda)$ is given in (5.46). This will be derived in Section 5.2.2. We stress here the difference between λ, the eigenvalue of the homogeneous Fredholm equation (5.28) and the parameter λ of the nonhomogeneous Fredholm equation (5.27). In most of our treatment we will assume that the parameter λ of (5.27) is different from all the eigenvalues $\{\lambda_n\}$ of the homogeneous Fredholm equation (5.28).

5.2.1 Homogeneous Fredholm Equations with Symmetric Kernel

There are many interesting results concerning the eigenvalues $\{\lambda_n\}$ and the eigenfunctions $\{u_n(x)\}$ of the symmetric kernel of (5.28),

[2]If the kernel $K(x,t)$ is a complex-valued function, then the definition of the *symmetric* kernel is $K(x,t) = \overline{K(t,x)}$, where $\overline{K}$ is the complex conjugate of K.

$$u_n(x) = \lambda_n \int_a^b K(x,t)u_n(t)dt, \qquad K(x,t) = K(t,x) \tag{5.29}$$

where $K(x,t)$ is expanded in terms of the eigenfunctions.

In the following we state, then illustrate or prove, the most important of the results needed for the development of the series expansion of the symmetric kernel in (5.28) and the resolvent kernel of the nonhomogeneous equation (5.27).

(a) The eigenvalues of the symmetric kernel in (5.28) are real.

This can be proved easily but one needs to consider complex-valued functions; that is left for an exercise.

(b) The eigenfunctions $u_n(x)$ and $u_m(x)$ of the symmetric kernel corresponding to two distinct eigenvalues $\lambda_n \neq \lambda_m$ are orthogonal [i.e., $\int_a^b u_n(x)u_m(x)dx = 0$, $\lambda_n \neq \lambda_m$].

To simplify the orthogonal expansion or the Fourier series in terms of the orthogonal eigenfunctions $u_n(x)$ of the symmetric kernel,

$$K(x,t) = \sum_n c_n u_n(x), \tag{5.30}$$

$$c_n = \frac{\int_a^b K(x,t)u_n(x)dx}{\int_a^b u_n^2(x)dx} \tag{5.31}$$

we will normalize them by redefining them as an *orthonormal* eigenfunction (as we did in Section 4.1),

$$\phi_n(x) = \frac{u_n(x)}{\sqrt{\int_a^b u_n^2(x)dx}}, \qquad \int_a^b \phi_n^2(x)dx = 1. \tag{5.32}$$

(c) The degeneracy or multiplicity p of the eigenvalue $\lambda \neq 0$ is finite for every symmetric kernel that is square integrable on the square $\{(x,t) : a \leq x \leq b, a \leq t \leq b\}$, that is,

$$\int_a^b \int_a^b K^2(x,t)dxdt = B^2 < \infty. \tag{5.33}$$

This simply means that for such a kernel there can be only a finite number p of eigenfunctions $u_{j+1}, u_{j+2}, \cdots, u_{j+p}$ (sometimes, for clarity, are written as $u_j^{(0)}, u_j^{(1)}, \cdots, u_j^{(p-1)}$) that corresponds to an eigenvalue λ_j of $K(x,t)$.

From the above and other results, a very important theorem, the *Hilbert-Schmidt theorem*, can be developed. This theorem expands $f(x)$ of the Fredholm equation of *the first kind* with symmetric kernel

$$f(x) = \lambda \int_a^b K(x,t)u(t)dt, \qquad K(x,t) = K(t,x) \tag{5.34}$$

in a Fourier series in terms of the orthonormal eigenfunctions $\{\phi_k(x)\}$ of the symmetric kernel,

$$\phi_k(x) = \lambda_k \int_a^b K(x,t)\phi_k(t)dt, \qquad K(x,t) = K(t,x) \tag{5.35}$$

Before we state the Hilbert-Schmidt theorem we must note that there is a limitation on the class of functions $f(x)$ that can be expressed as in (5.34), since thinking of a solution $u(x)$ for (5.34) means the existence of such solution $u(x)$ for the Fredholm integral equation of the first kind (5.34) for the given function $f(x)$. However this, in general, as was illustrated earlier with a very basic problem in Example 8 of Section 1.3, cannot be (easily) assured. Indeed the conditions for the existence of a solution to Fredholm integral equation of the first kind is much more restrictive when compared with those of the second kind. Such an important topic will be discussed briefly after the next Example 7, illustrated in Example 8, then it will be discussed in more detail in Section 5.4.

The following is a simple version of Hilbert-Schmidt theorem.

Hilbert-Schmidt Theorem-Fredholm Equation of the First Kind Let $f(x)$ be expressed as in the form of the Fredholm equation of the first kind (5.34), let $K(x,t)$ be symmetric and square integrable in the square $\{(x,t) : a \le x \le b,\ a \le t \le b)\}$ for the square integrable $u(x)$, then $f(x)$ can be expanded in a Fourier series,

$$f(x) = \sum_{k=1}^{\infty} a_k \phi_k(x) \tag{5.36}$$

$$a_k = \int_a^b f(x)\phi_k(x)dx \tag{5.37}$$

in terms of the orthonormal eigenfunctions $\{\phi_k(x)\}$ of the symmetric kernel $K(x,t)$ and the series (5.36) converges to $f(x)$ *in the mean* (as defined in (4.52) (see (4.51)). The series is also convergent absolutely and uniformly.

As we shall see in the next section this theorem is essential for developing the resolvent for the nonhomogeneous Fredholm equation with symmetric kernel (5.27). The following Mercer's theorem is of importance as it expresses the symmetric kernel as an infinite series of a product of its orthonormal eigenfunctions.

Mercer's Theorem

If the kernel $K(x,t)$ is symmetric and square integrable on the square $\{(x,t) : a \le x \le b, a \le t \le b\}$, continuous, and has only *positive* eigenvalues (or at most a finite number of negative eigenvalues), then the series

$$\sum_{k=1}^{n} \frac{\phi_k(x)\phi_k(t)}{\lambda_k}$$

converges absolutely and uniformly and gives the following *bilinear* form for the symmetric kernel:

$$K(x,t) = \sum_{k=1}^{\infty} \frac{\phi_k(x)\phi_k(t)}{\lambda_k}. \tag{5.38}$$

The conditions and results of Mercer's theorem and Hilbert-Schmidt theorem are illustrated in detail in the following example.

Example 7 Eigenfunctions for a Homogeneous Equation with a (Nondegenerate) Symmetric Kernel

The eigenfunctions $\{u_k(x)\}$,

$$u_k(x) = \lambda_k \int_0^1 K(x,t)u_k(t)dt \tag{5.39}$$

of the symmetric kernel

$$K(x,t) = \begin{cases} x(1-t), & 0 \le x \le t \\ t(1-x), & t \le x \le 1 \end{cases} \tag{5.40}$$

can be obtained by reducing (E.1) to its equivalent eigenvalue problem of (E.3) and (E.4) as was done in Example 6 of Section 2.5:

$$\frac{d^2u}{dx^2} + \lambda u = 0, \qquad 0 < x < 1 \qquad (E.3), (5.41)$$

$$u(0) = u(1) = 0. \qquad (E.4), (5.42)$$

The eigenfunctions of (E.3) and (E.4) are clearly $u_k(x) = \sin k\pi x$ and the eigenvalues are $\lambda_k = k^2\pi^2$. These eigenvalues $\lambda_k = \pi^2 k^2$ of the symmetric kernel in (5.40) are real, and the eigenfunctions $\{\sin k\pi x\}$ are orthogonal. From the definition of the norm square in (4.49) we have

$$||u_k||^2 = \int_a^b u_k^2(x)dx = \int_0^1 \sin^2 k\pi x dx = \frac{1}{2}.$$

Hence the orthonormal eigenfunctions are

$$\phi_k(x) = \frac{\sin k\pi x}{||u_k||} = \frac{\sin k\pi x}{\sqrt{\frac{1}{2}}} = \sqrt{2}\sin k\pi x.$$

The symmetric kernel in (5.40) is square integrable on the square $\{(x,t) : 0 \le x \le 1, 0 \le t \le 1\}$ since it is bounded in x and t there. The non-zero eigenvalues here are *simple* since for every $\lambda_k = k^2\pi^2$ there corresponds only one eigenfunction $\sin k\pi x$. Therefore, the conditions of the Hilbert-Schmidt theorem are satisfied, for the square integrable $u(x)$ of (5.34), as $K(x,t)$ in (E.1) is symmetric and square integrable. Also the conditions of Mercer's theorem are clearly met, thus $K(x,t)$ of (5.40) can be expressed in the bilinear series (5.38) with $\phi_k(x) = \sqrt{2}\sin k\pi x$ and $\lambda_k = k^2\pi^2$.

Next we will present a discussion concerning the difficulty in securing the existence of a solution to Fredholm integral equation of the *first kind* which is concluded by a detailed illustration in Example 8. The more detailed treatment with precise (practical) theorems is done in Section 5.4. This topic was touched upon very briefly in Section 1.3 and was illustrated with Example 8 there.

On the Existence of a Solution to Fredholm Equation of the First Kind

Before presenting any illustration for the Hilbert-Schmidt theorem in relation to the Fredholm integral equation of the first kind (5.34) we may first inquire into whether a solution, at all, does exist (let alone be unique) for the equation of such an illustration. We emphasize this point, since the theory is rather restrictive for such existence of solutions if compared to that of Fredholm equations of the second kind. Indeed, in Example 8 that follows shortly with a Fredholm equation of the first kind, we will illustrate this point in detail. In essence, the comparison will lead us to think of the generous theory for the existence of the solution to Fredholm equation of the second kind (5.21) and its associated homogeneous one (5.20), as given by Theorems 1 to 4 of Fredholm alternative in Section 5.1. In contrast, the theory for the existence of the solution for Fredholm integral equation of the first kind (5.34) (even with symmetric kernel),

$$f(x) = \lambda \int_a^b K(x,t)u(t)dt, \quad K(x,t) = K(t,x) \tag{5.34}$$

is much more limited.

This difficulty of finding a solution to the Fredholm integral equation (5.34) stems from the fact that the integration operation over the input as the sought solution $u(t)$, is a smoothing process, especially, when combined with a nicely behaved kernel $K(x,t)$. This means, for example, that if the solution $u(x)$ is piecewise continuous, then the above integration operation on the right hand side of (5.34), with continuous kernel $K(x,t)$, would result in a smoother, i.e., continuous output $f(x)$ of (5.34) as the given function at hand. So, if we are given a continuous function $f(x)$ in (5.34), we cannot, in general guarantee an answer in the search for a solution $u(x)$ among the class of continuous functions! In other words, if we look at the right hand side of (5.34) as an integral transform of *piecewise* continuous functions, then this transform maps such class of functions to a more restrictive one, in this case *continuous* functions. Indeed, for more smooth kernels, i.e., $K(x,t)$ differentiable, the class of piecewise or even integrable functions $u(t)$ is mapped into differentiable functions $f(x)$. Hence, for a continuous kernel $K(x,t)$ (in both x and t) and a continuous output $f(x)$, the integral equation of the first kind (5.34) cannot, in general, be solved by a continuous function $u(t)$. Of course, if $K(x,t)$ is not very regular, then it is possible that this irregularity is combined with the smoothness of the integration operation and a continuous solution $u(t)$, to produce a continuous output $f(x)$ in (5.34).

A very basic theorem, which addresses searching for a solution $u(x)$ among continuous functions, would even be more demanding on $f(x)$. Such demand of the theory translates in requiring that the given function $f(x)$ must be expressible in a Fourier series of the eigenfunctions of the *continuous, real* and *symmetric* kernel $K(x,t)$ of (5.34). It states that:

Theorem 5 "For the continuous real and symmetric kernel, and continuous $f(x)$, a solution to (5.34) exists only if the given function $f(x)$ can be expressed in a series of the eigenfunctions of the kernel $K(x,t)$, i.e., only if

$$f(x) = \sum_{k=1}^{\infty} a_k \phi_k(x), \tag{5.36}$$

$$a_k = \int_a^b f(x)\phi_k(x)dx \tag{5.37}$$

where we are using here the orthonormal set of eigenfunctions $\{\phi_k(x)\}_{k=1}^{\infty}$ on (a,b)."
With the condition of this theorem, the solution takes a similar form

$$u(x) = \sum_{k=1}^{\infty} b_k \phi_k(x) = \sum_{k=1}^{\infty} \lambda_k a_k \phi_k(x), \tag{5.43}$$

$$b_k = \lambda_k a_k. \tag{5.43a}$$

This form satisfies the condition (5.36) as we substitute the expression (5.43) for $u(t)$ inside the integral of (5.34)

$$\begin{aligned} f(x) &= \int_a^b K(x,t)\left[\sum_{k=1}^{\infty} \lambda_k a_k \phi_k(t)\right] dt \\ &= \sum_{k=1}^{\infty} \lambda_k a_k \int_a^b K(x,t)\phi_k(t)dt \\ &= \sum_{k=1}^{\infty} \lambda_k a_k \frac{1}{\lambda_k}\phi_k(x) = \sum_{k=1}^{\infty} a_k \phi_k(x) \end{aligned} \tag{5.36}$$

after using the fact that $\phi_k(x)$ and λ_k are the eigenfunctions and eigenvalues, as seen in (5.35), of the *symmetric* kernel $K(x,t)$ of the integral inside the sum.

We may note here that, although we are guaranteed the existence of the (continuous) solution of (5.34) in the form of the series (5.43), it is by no means a unique solution. This is the case, since if we add to the series in (5.43) a function $\Psi(x)$ that is *orthogonal* to the kernel $K(x,t)$, i.e.,

$$\int_a^b K(x,t)\Psi(t)dt = 0, \tag{5.44}$$

and substitute in (5.34), we obtain the same output $f(x)$ as in (5.36). So, for a *unique* solution $u(x)$ in (5.43), we must insist that there are no functions $\Psi(x)$ that are orthogonal to the *symmetric* kernel $K(x,t)$.

Perhaps, at this level of discussion, the safest way to come up with an example which has a solution for the Fredholm equation of the first kind, (5.34), is to assume a form of (continuous) solution $u(t)$ and find the resulting $f(x)$. This $f(x)$ may then be used as a given function in (5.34) to safely illustrate the Hilbert-Schmidt theorem, and the impor ıt condition (5.36) and (5.37) for the existence of the solution to (5.34). Understandably, we can start with the simplest form $u(t) = 1$ on $(0, 1)$ in the integral of the special case of (5.34) with the symmetric kernel,

$$K(x,t) = \begin{cases} x(1-t), & 0 < x < t \\ t(1-x), & t < x < 1 \end{cases} \tag{5.45}$$

and after paying attention to the two branches of the kernel $K(x,t)$ in (5.45), we can easily integrate to have the result as $f(x) = \dfrac{\lambda}{2}(x - x^2)$, which we have as a simple exercise (see Exercise 4). In the following example we illustrate the conditions for the existence of a solution to such resulting integral equation. We will then illustrate the related aspects discussed above.

Example 8

(a) Now we can write a reasonable Fredholm integral equation of the first kind (5.34) with $f(x) = \frac{1}{2}(x - x^2)$, $\lambda = 1$ where we know for sure that the solution does exist as $u(x) = 1, 0 < x < 1$,

$$\frac{1}{2}(x - x^2) = \int_0^1 K(x,t)u(t)dt \tag{E.1}$$

where $K(x,t)$ is the symmetric kernel

$$K(x,t) = \begin{cases} x(1-t), & 0 \le x \le t \\ t(1-x), & t \le x \le 1 \end{cases} \tag{E.2}$$

From Example 7 we have the orthonormal eigenfunctions of the kernel $K(x,t)$ as $\sqrt{2} \sin k\pi x$. So we will express $f(x)$ and $u(x)$ in a Fourier sine series (1.116), (1.115) of these functions. The Fourier (sine) series for these two functions ($f(x) = \dfrac{1}{2}(x - x^2)$ and $u(x) = 1$ on $(0, 1)$) can be easily written, respectively as (see Exercise 4(b))

$$f(x) = \frac{1}{2}(x - x^2) = \sum_{k=0}^{\infty} \frac{2\sqrt{2} \cdot \sqrt{2} \sin(2k+1)\pi x}{\pi^3 (2k+1)^3}, \quad 0 < x < 1 \tag{E.3}$$

$$u(x) = 1 = \sum_{k=0}^{\infty} \frac{2\sqrt{2} \cdot \sqrt{2} \sin(2k+1)\pi x}{\pi(2k+1)}, \quad 0 < x < 1. \tag{E.4}$$

Now if we compare the Fourier coefficients $a_k = \frac{2\sqrt{2}}{\pi^3(2k+1)^3}$ for $f(x) = \frac{1}{2}(x - x^2)$ and $b_k = \frac{2\sqrt{2}}{\pi(2k+1)}$ for $u(x) = 1$, we find that the condition (5.43a) for the existence of the solution to the specific (and well prepared in advance!) Fredholm integral equation of the first kind (E.1), is satisfied,

$$b_k = \lambda_{2k+1} a_k = \pi^2 (2k+1)^2 \cdot \frac{2\sqrt{2}}{\pi^3 (2k+1)^3} = \frac{2\sqrt{2}}{\pi(2k+1)}. \tag{E.5}$$

It is clear that this given $f(x) = \frac{1}{2}(x - x^2)$ in (E.1) and $K(x,t)$ in (E.2) satisfy Theorem 5, which we shall leave for an exercise (see Exercise 4). So the series expansion (5.36) of $f(x)$, as required by Theorem 5, is justified, thus in turn the existence of a solution to the special case (E.6) of the Fredholm equation of the first kind with symmetric kernel (5.34).

(b) With these words of caution about the rather restrictive conditions for the existence of the solution of Fredholm integral equation of the first kind, we leave this important subject for now, and we shall return to it in Section 5.4 with a more general theorem and a rather relaxed condition on the solution $u(t)$ of (5.34). It may be instructive to give here the spirit of such a theorem compared to the above rather restrictive Theorem 5.

As we had explained following (5.45) and in Exercise 4(a) of this section, that the simple continuous function $u(x) = 1,\ 0 < x < 1$ is a solution to the Fredholm integral equation of the first kind

$$\frac{1}{2}(x - x^2) = \int_0^1 K(x,t) u(t) dt, \tag{E.6}$$

$$K(x,t) = \begin{cases} x(1-t), & 0 < x < t \\ t(1-x), & t < x < 1 \end{cases} \tag{E.7}$$

which can be verified here easily. On the other hand, the equation

$$x = \int_0^1 K(x,t) u(t) dt \tag{E.8}$$

with the same kernel as in (E.7) has no solution. These results will be supported by a limited version of Picard's Theorem 7 with *necessary and sufficient* conditions for the existence of a not necessarily continuous, but *square integrable* solutions.

An important dividend of the more relaxed existence Theorem 7, that we shall present in Section 5.4, is that as it assures us of the solution, it also offers a *method for constructing* such a solution. This is a great relief when we know that integral equations of the first kind are denied the usual simple *iterative* method. The latter difficulty, of course, is due to the absence of the unknown function $u(x)$ as a separate term outside the integral of the Fredholm integral equation of the first kind in (5.34) as compared to that of the second kind in (5.21).

(c) To also illustrate Mercer's Theorem for the series expansion (5.38) of the above kernel $K(x,t)$ in (5.45) , we see that the theorem is satisfied since the kernel is continuous and all its eigenvalues $\{\lambda_k\} = \{k^2\pi^2\}$ are positive; therefore such a kernel can be expanded in terms of the (orthonormal) eigenfunctions $\{\phi_k(x)\} = \{\sqrt{2}\sin k\pi x\}$ as

$$K(x,t) = 2\sum_{k=1}^{\infty} \frac{\sin k\pi x \sin k\pi t}{k^2\pi^2}. \tag{E.9}$$

In the following section we will develop the resolvent kernel $\Gamma(x,t;\lambda)$ for the nonhomogeneous Fredholm equation (5.27) with symmetric kernel.

5.2.2 Solution of Fredholm Equations of the Second Kind with Symmetric Kernel

With the aid of the foregoing important development of the Fredholm homogeneous equation with symmetric kernel (5.28), we show here, at least formally, that the resolvent kernel $\Gamma(x,t;\lambda)$ of the nonhomogeneous equation with symmetric kernel (5.27) is expressed as an infinite series of the orthonormal eigenfunctions $\{\phi_k(x)\}$ of $K(x,t)$,

$$\Gamma(x,t;\lambda) = \sum_{k=1}^{\infty} \frac{\phi_k(x)\phi_k(t)}{\lambda_k - \lambda}, \qquad \lambda \neq \lambda_k \tag{5.46}$$

and hence the solution to (5.27) is

$$u(x) = f(x) + \lambda \sum_{k=1}^{\infty} \frac{a_k\phi_k(x)}{\lambda_k - \lambda}, \qquad \lambda \neq \lambda_k \tag{5.47}$$

$$a_k = \int_a^b f(x)\phi_k(x)dx. \tag{5.37}$$

To prove (5.47), we write (5.27) in the form

$$h(x) = u(x) - f(x) = \lambda \int_a^b K(x,t)u(t)dt \tag{5.48}$$

which is suitable for the Hilbert-Schmidt theorem with the function $h(x) = u(x) - f(x)$ in (5.48) instead of $f(x)$ in (5.34). According to the Hilbert-Schmidt theorem, remembering its important conditions here on $u(x)(= h(x) + f(x))$ being square integrable on $a \leq x \leq b$ and $K(x,t)$ symmetric and square integrable on the square $\{(x,t) : a \leq x \leq b, a \leq t \leq b\}$, we can expand $h(x)$ in a Fourier series (5.36) and (5.37) of the orthonormal eigenfunctions $\{\phi_k(x)\}$ of the symmetric kernel $K(x,t)$,

$$h(x) = u(x) - f(x) = \sum_{k=1}^{\infty} b_k \phi_k(x) \tag{5.49}$$

$$\begin{aligned} b_k &= \int_a^b h(x)\phi_k(x)dx = \int_a^b u(x)\phi_k(x)dx - \int_a^b f(x)\phi_k(x)dx \\ &= d_k - a_k. \end{aligned} \tag{5.50}$$

Here d_k is the Fourier coefficient of the unknown function $u(x)$,

$$d_k = \int_a^b u(x)\phi_k(x)dx \tag{5.51}$$

and a_k is the Fourier coefficient of the given function $f(x)$. In (5.50) we have now a relation between b_k, d_k, and a_k. It is clear that we need to express b_k of (5.49) in terms of a_k to arrive at the final solution (5.47). To do this we need another relation, $b_k = \lambda d_k/\lambda_k$, which we can easily show, since

$$\begin{aligned} b_k &= \int_a^b [u(x) - f(x)]\phi_k(x)dx \\ &= \int_a^b \lambda \int_a^b K(x,t)u(t)dt\phi_k(x)dx \\ &= \lambda \int_a^b u(t) \int_a^b K(t,x)\phi_k(x)dxdt \end{aligned} \tag{5.52}$$

after using the integral of (5.48) for $h(x)$, interchanging the two integrals, and using the fact that the kernel is symmetric [i.e., $K(x,t) = K(t,x)$]. Now according to (5.35), the inside integral is $\dfrac{\phi_k(t)}{\lambda_k}$,

$$b_k = \lambda \int_a^b u(t)\frac{1}{\lambda_k}\phi_k(t)dt = \frac{\lambda}{\lambda_k}\int_a^b u(t)\phi_k(t)dt = \frac{\lambda}{\lambda_k}d_k \tag{5.53}$$

after using the definition of d_k in (5.51). If we substitute from (5.53) for d_k in (5.50), we obtain b_k in terms of a_k,

$$b_k\frac{\lambda_k - \lambda}{\lambda} = a_k, \qquad b_k = \frac{\lambda}{\lambda_k - \lambda}a_k \tag{5.54}$$

and if we now substitute b_k from (5.54) in (5.49) , we obtain

$$u(x) = f(x) + \lambda \sum_{k=1}^{\infty} \frac{a_k \phi_k(x)}{\lambda_k - \lambda}, \qquad \lambda \neq \lambda_k \tag{5.47}$$

which is (5.47), the solution of (5.27) with symmetric kernel. This solution (5.47) can be rewritten using a_k from (5.37) as

$$
\begin{aligned}
u(x) &= f(x) + \lambda \sum_{k=1}^{\infty} \frac{\phi_k(x)}{\lambda_k - \lambda} \int_a^b f(t)\phi_k(t)dt \\
&= f(x) + \lambda \int_a^b f(t) \left[\sum_{k=1}^{\infty} \frac{\phi_k(x)\phi_k(t)}{\lambda_k - \lambda} \right] dt \qquad (5.55) \\
u(x) &= f(x) + \lambda \int_a^b \Gamma(x,t;\lambda) f(t)dt, \qquad \lambda \neq \lambda_k
\end{aligned}
$$

after exchanging the infinite summation with the integration and defining $\Gamma(x,t;\lambda)$, the resolvent as in (5.46),

$$
\Gamma(x,t;\lambda) = \sum_{k=1}^{\infty} \frac{\phi_k(x)\phi_k(t)}{\lambda_k - \lambda}, \qquad \lambda \neq \lambda_k. \tag{5.46}
$$

The very clear condition $\lambda \neq \lambda_k$ in (5.47) on the parameter λ, in the Fredholm integral equation of the second kind (5.27), not equal to any of the eigenvalues $\{\lambda_k\}$ of its symmetric kernel is consistent with the Fredholm alternative in Theorem 1. In case $\lambda = \lambda_k$, as we shall illustrate in the next Example 9 for a symmetric kernel, we will use the second part of the Fredholm alternative as stated in Theorem 2.

The Gibbs Phenomenon in (the Truncated) Fourier Series (Eigenfunctions) Expansion

In the above development we had the Hilbert-Schmidt conditions for $h(x) = u(x) - f(x)$ represented as in $f(x)$ of (5.34), whence $h(x)$ was expressed in terms of the Fourier series (5.49) of the eigenfunctions $\{\phi_k(x)\}$ of the symmetric kernel.

We may remark here that in the practical applications, we can compute only a finite sum of N terms to approximate the infinite series in (5.49), which will, of course, incur a *truncation error*. With a fast convergence of the above series, such an error may be reduced by increasing N. There is also another "stubborn" error that may appear, which cannot be reduced by the mere increase of the series' N terms. Such an error will appear when we write the solution as in (5.49) where $f(x)$ may be sectionally continuous. For such input function $f(x)$ with a jump discontinuity we must watch for the very well known error, namely, *the Gibbs phenomenon*,[3] which manifests itself as overshoots and undershoots in the neighborhood of the jump discontinuity of the approximated function $h(x) = u(x) - f(x)$ in (5.49). Indeed the size of the first overshoot (in the truncated Fourier series) is about 8.95% of the size of the jump discontinuity. The next is an undershoot of about 4.86% of the size of the jump discontinuity. This phenomenon also appears in the general orthogonal (eigenfunctions) expansion, such as the Fourier-Bessel series and the Fourier-Legendre polynomial series, to give two familiar examples. The sizes of the first overshoot and undershoot are about the same as that of the above (truncated) Fourier-trigonometric series.

[3]See Jerri [1998] for the first comprehensive book treatment of the Gibbs phenomenon that covers the basic elements of the subject and its research development since its discovery in 1848.

Example 9 Nonhomogeneous Fredholm Equation with Symmetric Kernel
Solve the Fredholm equation of the second kind

$$u(x) = x + \lambda \int_0^1 K(x,t)u(t)dt \tag{E.1}$$

with the symmetric kernel given as in Example 7:

$$K(x,t) = \begin{cases} x(1-t), & 0 \le x \le t \\ t(1-x), & t \le x \le 1 \end{cases} \tag{E.2}$$

In Example 7 we showed that this symmetric kernel satisfies all the conditions required for deriving the solution (5.47) for (E.1), and that the eigenvalues and the orthonormal eigenfunctions of $K(x,t)$ in (E.2) are $\lambda_k = \pi^2 k^2$ and $\phi_k(x) = \sqrt{2}\sin k\pi x$, respectively. If we use these results, the solution to (E.1) according to (5.47) is

$$\begin{aligned} u(x) &= x + \lambda \sum_{k=1}^{\infty} \frac{a_k\sqrt{2}\sin k\pi x}{\pi^2k^2 - \lambda}, \qquad \lambda \neq \pi^2 k^2 \\ &= x + \frac{2\lambda}{\pi}\sum_{k=1}^{\infty}\frac{(-1)^{k+1}\sin k\pi x}{k(\pi^2k^2-\lambda)} \end{aligned} \tag{E.3}$$

since

$$a_k = \int_0^1 x\sqrt{2}\sin k\pi x dx = \frac{(-1)^{k+1}\sqrt{2}}{k\pi}. \tag{E.4}$$

The resolvent kernel of (E.1) according to (5.46) is

$$\Gamma(x,t;\lambda) = 2\sum_{k=1}^{\infty}\frac{\sin k\pi x \sin k\pi t}{\pi^2k^2 - \lambda}, \qquad \lambda \neq \pi^2k^2. \tag{E.5}$$

The foregoing treatment for constructing the solution (5.47) of the nonhomogeneous equation (5.27) was based on the fact that the parameter λ of (5.27) is not equal to any of the eigenvalues $\{\lambda_k\}$ of the symmetric kernel $K(x,t)$ in (5.35). When λ is equal to one eigenvalue λ_{j+1}, with degeneracy p, then for $\lambda = \lambda_k = \lambda_{j+1}$, the coefficient $a_k/(\lambda - \lambda_k)$ in (5.47) is not defined unless $a_k = 0$, which makes the $a_k/(\lambda - \lambda_k)$ indeterminate and hence arbitrary. From the definition of a_k,

$$a_k = \int_a^b f(x)\phi_k(x)dx$$

the condition $a_k = 0$ would mean that $f(x)$ must be orthogonal to $\phi_k(x)$, and hence a solution to the integral equation (5.27) in the form (5.47) does not exist unless $f(x)$ is orthogonal to all the eigenfunctions $\phi_{j+1}, \phi_{j+2}, \cdots, \phi_{j+p}$ that correspond to the (degenerate) eigenvalue $\lambda_{j+1} = \lambda_{j+2} = \cdots = \lambda_{j+p}$.

$$a_k = \int_a^b f(x)\phi_k(x)dx = 0, \qquad k = j+1, \cdots, j+p. \tag{5.56}$$

We may remark here that this condition on the nonhomogeneous term $f(x)$ of (5.27) is consistent with Theorem 2, the second part of the Fredholm alternative (Theorem 1) for symmetric kernels.

In the case that this condition (5.56) is satisfied, the series will include arbitrary constants $B_1, B_2, \cdots, B_p$ resulting from the p indeterminate forms

$$\frac{a_k}{\lambda - \lambda_k} = B_k, \qquad \lambda = \lambda_k, \qquad a_k = 0, \qquad k = j+1, j+2, \cdots, j+p$$

and (5.47) becomes

$$\begin{aligned} u(x) &= f(x) + \lambda \bullet \sum_{\substack{k=1 \\ k \neq j+1, j+2, \cdots, j+p}}^{\infty} \frac{a_k \phi_k(x)}{\lambda_k - \lambda} \\ &\quad + B_1 \phi_{j+1}(x) + B_2 \phi_{j+2}(x) + \cdots + B_p \phi_{j+p}(x) \end{aligned} \tag{5.57}$$

which represents infinity of solutions because of the arbitrary constants $B_1, B_2, \cdots, B_p$. (For the degeneracy in (5.57), its last line may be (for more clarity) written as $+B_1\phi_{j+1}^{(0)} + B_2\phi_{j+2}^{(1)} + \cdots + B_p\phi_{j+p}^{(p-1)}$.) Thus, and according to Theorem 2, for $\lambda = \lambda_k$ the Fredholm integral equation with symmetric kernel (E.1) has an infinity of solutions in (5.57) provided that condition (5.56), of $f(x)$ being orthogonal to all the eigenfunctions corresponding to the (degenerate, index p) eigenvalue $\lambda_{j+1} = \lambda$, where λ is the given parameter in (E.1).

Example 10

The Fredholm integral equation

$$u(x) = x + 4\pi^2 \int_0^1 K(x,t)u(t)dt \tag{E.1}$$

with the symmetric kernel of Example 9,

$$K(x,t) = \begin{cases} x(1-t), & 0 \le x \le t \\ t(1-x), & t \le x \le 1 \end{cases} \tag{E.2}$$

is not solvable since here we have that the parameter λ of (E.1) is equal to $\lambda_2 = 4\pi^2$, the eigenvalue of $K(x,t)$ in (E.2) of Example 7, and $f(x) = x$ is not orthogonal to its corresponding eigenfunction $\phi_2(x) = \sqrt{2}\sin 2\pi x$ since

$$\int_0^1 x\sqrt{2}\sin 2\pi x dx = \frac{-\sqrt{2}}{2\pi} = \frac{-1}{\sqrt{2}\pi} \neq 0.$$

However, the integral equation with $f(x) = \sin 3x$ instead of the above $f(x) = x$ in (E.1),

$$u(x) = \sin 3\pi x + 4\pi^2 \int_0^1 K(x,t)u(t)dt \tag{E.3}$$

(with $K(x,t)$ as in (E.2)) does have solutions even though its $\lambda = 4\pi^2 = \lambda_2$, since $f(x) = \sin 3\pi x$ here is orthogonal to $\phi_2(x) = \sqrt{2}\sin 2\pi x$,

$$\sqrt{2}\int_0^1 \sin 3\pi x \sin 2\pi x dx = 0.$$

The solution is obtained from (5.57) after computing a_k for $f(x) = \sin 3\pi x$, where we note that this $f(x)$ is a very special case, as it is a member of the orthogonal set $\{\sin k\pi x\}$. Thus $a_k = 0$ except for $a_3 = \sqrt{2}\int_0^1 \sin^2 3\pi x dx = 1/\sqrt{2}$, where the sum in (5.57) becomes only one term, and we have

$$\begin{aligned} u(x) &= \sin 3\pi x + 4\pi^2 a_3 \frac{\sqrt{2}\sin 3\pi x}{(3\pi)^2 - 4\pi^2} + B_2 \sin 2\pi x \\ &= \frac{9}{5}\sin 3\pi x + B_2 \sin 2\pi x. \end{aligned} \tag{E.4}$$

This represents an infinite number of solutions for (E.3) because of the aribtrary constant B_2 in (E.4). We note here that the multiplicity is $p = 1$ for the eigenvalue $\lambda_2 = 4\pi^2$.

Comments on the Numerical Evaluation of the Eigenvalues: Rayleigh-Ritz Method

In solving the homogeneous Fredholm equation (5.20),

$$u(x) = \lambda \int_a^b K(x,t)u(t)dt \tag{5.20}$$

in order to find its (nontrivial solutions) eigenfunctions and eigenvalues, we had up till now the chance of the special case of degenerate kernels in Section 5.1.2, where the method of Section 5.1.1 was employed as it was illustrated in Example 2 and at the end of Example 4. From the start of Section 5.2 we needed the eigenvalues and eigenfunctions of the homogeneous equation (5.20) with nondegenerate (but symmetric) kernel. This was essential for the statement of the Fredholm alternative (Theorems 1, 2) concerning the existence of a solution (or solutions) to the Fredholm integral equation of the second kind (5.27) (with the same kernel). To find such eigenvlaues and eigenfunctions we resorted to reducing the homogeneous Fredholm equation (5.20) to its equivalent boundary value problem, associated with a differential equation and boundary conditions, as discussed and illustrated in Section 2.5. We used this method in Example 7, and we even repeated using the same kernel and its eigenvalues and eigenfunctions in the rest of the Examples 8, 9 and 10 concerning the existence of a solution to the Fredholm integral equation of the first kind in Example 8, and the construction of the solution to the Fredholm equation of the second kind (5.27) in Examples 9 and 10, with the same nondegenerate (but symmetric) kernel.

We emphasize here that in most of the illustrations that we have presented, the resulting boundary vlaue problem was a very familiar one and hence its eigenfunctions and eigenvalues were obtained with minimum effort. However, in general the resulting boundary value problem may be a general Sturm-Liouville problem, and hence we cannot expect such easily obtained familiar solutions, so we may resort to the approximate or numerical methods to find the eigenfunctions and eigenvalues. One of the most familiar numerical methods for estimating the eigenvalues is called the *Rayleigh-Ritz method*, whose derivation is based on variational principles, which we shall not pursue here, and be satisfied with another method that we will cover briefly at the end of Section 5.3.2. For the Rayleigh-Ritz method simple presentation and detailed illustration, we refer the reader to the first edition of this book.[4] Also a summary of the method with a number of detailed illustrations are found in the "Student's Solutions Manual" to accompany this book[5] (see the end of the preface for more information).

Exercises 5.2

1. Consider the homogeneous Fredholm equation with symmetric kernel

$$u(x) = \lambda \int_0^\pi \cos(x+t)u(t)dt \qquad (E.1)$$

of Exercise 2(c), Section 5.1.

(a) Use the results of Exercise 2(c), Section 5.1, to verify that for this symmetric kernel ($\cos(x+t) = \cos(t+x)$) the eigenvalues are real and the corresponding eigenfunctions are orthogonal.

(b) Use differentiation to reduce the integral equation to an ordinary differential equation from which you determine the eigenfunctions, then the eigenvalues. Compare those with the results of Exercise 2(c), Section 5.1.

(c) Find the orthonormal eigenfunctions.

(d) Use (E.1) to find the eigenvalues. *Hint:* Substitute each eigenfunction of part (c) in (E.1) to find their corresponding eigenvalues.

(e) Show that the symmetric kernel is square integrable on $\{(x,t) : 0 \le x \le \pi, 0 \le t \le \pi\}$.

(f) Determine whether Mercer's theorem applies to this problem and if so, write the Kernel's bilinear expansion of (5.38).

2. Use the results in problem 1 to solve the nonhomogeneous integral equation

$$u(x) = x + \lambda \int_0^\pi \cos(x+t)u(t)dt \qquad (E.2)$$

[4] Jerri [1985, pp. 146–151]. See also Kanwal [1971, 1997 (2nd ed.)] and Green [1969].
[5] Jerri [1999].

by finding the resolvent kernel.

3. Consider the nonhomogeneous Fredholm equation

$$u(x) = \cos 2x + 2\int_0^{\pi/2} K(x,t)u(t)dt \tag{E.1}$$

with the kernel

$$K(x,t) = \begin{cases} \sin x \cos t, & 0 \le x \le t \\ \sin t \cos x, & t \le x \le \dfrac{\pi}{2} \end{cases} \tag{E.2}$$

(a) Verify that the kernel is symmetric and is square integrable on the square $\{(x,t) : 0 \le x \le \pi/2, 0 \le t \le \pi/2\}$.

(b) Reduce the homogeneous equation

$$u(x) = \lambda \int_0^{\pi/2} K(x,t)u(t)dt \tag{E.3}$$

with $K(x,t)$ as in (E.2) to a differential equation to obtain the eigenvalues and eigenfunctions.

(c) Use the information in part (b) to solve the nonhomogeneous equation (E.1).

(d) Just as in Exercise 2, use the Fredholm alternative (Theorems 1, 2) to show that the Fredholm integral equation in (E.1) above does indeed have a unique solution.

4. Consider the Fredholm integral equation of the first kind in (5.34) with $\lambda = 1$, (with the given particular $f(x)$ and the symmetric kernel) as it was considered in Example 8. (This is the same problem as Exercise 2 in Section 5.4.)

(a) Show that the solution $u(t) = 1$ corresponds to the nonhomogeneous term $f(x) = \frac{1}{2}(x - x^2)$. *Hint:* Watch for the two branches of the kernel $K(x,t)$, write the integral on the two subintervals $(0, x)$ and $(x, 1)$.

(b) As needed for (E.3) and (E.4) of Example 8, write the Fourier sine series for both the solution $u(x) = 1$, and the nonhomogeneous term $f(x) = \frac{1}{2}(x - x^2)$ on the interval (0,1).

(c) Show that the nonhomogeneous term $f(x) = \frac{1}{2}(x - x^2)$ in (E.6) and $K(x,t)$ in (E.7) of Example 8 satisfy Theorem 5. *Hint:* Note that $f(x) = \dfrac{1}{2}(x - x^2)$ is continuous on $(0, 1)$, and that the clearly symmetric kernel $K(x,t)$ in (E.2) is square integrable on the square $\{x\epsilon(0,1), t\epsilon(0,1)\}$. (See the hint to part (a).)

5. For Example 10, verify that $u(x)$ in (E.4) satisfies the Fredholm equation in (E.1).

5.3 FREDHOLM INTEGRAL EQUATIONS OF THE SECOND KIND

5.3.1 Method of Fredholm Resolvent Kernel

One of the methods of solving the general Fredholm integral equations of the second kind (5.21),

$$u(x) = f(x) + \lambda \int_a^b K(x,t)u(t)dt \tag{5.21}$$

is the method of evaluating the Fredholm resolvent kernel $\Gamma(x,t;\lambda)$,

$$u(x) = f(x) + \lambda \int_a^b \Gamma(x,t;\lambda) f(t)dt \tag{5.58}$$

$$\Gamma(x,t;\lambda) = \frac{D(x,t;\lambda)}{D(\lambda)}, \quad D(\lambda) \neq 0 \tag{5.59}$$

where $\Gamma(x,t;\lambda), D(x,t;\lambda)$, and $D(\lambda)$ are called the *Fredholm resolvent kernel of (5.21), the Fredholm minor, and the Fredholm determinant, respectively.* The $D(x,t;\lambda)$ is defined as

$$D(x,t;\lambda) = \sum_{n=0}^{\infty} \frac{(-\lambda)^n}{n!} B_n(x,t), \tag{5.60}$$

where $B_0(x,t) = K(x,t)$, and

$$B_n(x,t) = C_n K(x,t) - n \int_a^b K(x,s) B_{n-1}(s,t)ds, \quad n = 1, 2, \cdots, \tag{5.61}$$

where

$$C_n = \int_a^b B_{n-1}(t,t)dt, \qquad n = 1, 2, \cdots, C_0 = 1 \tag{5.62}$$

and $D(\lambda)$ is defined as

$$D(\lambda) = \sum_{n=0}^{\infty} \frac{(-\lambda)^n}{n!} C_n. \tag{5.63}$$

Note also that from (5.60) $D(x,t;0) = B_0(x,t) = K(x,t)$ as seen in (5.61) with $C_0 = 1$ in (5.62).

Before we present an illustration for this method in the following Example 11 we may comment on the importance of the above results (5.58) to (5.63) with $D(\lambda) \neq 0$ as they encompass the following *Fredholm's first theorem* (Theorem 6). Without assuming complex analysis we may give a simple version of this important theorem with our attention being fixed, primarily, on the basic conditions for the existence of the unique solution to Fredholm integral equation of the second kind (5.21) as presented in (5.58). This is besides good qualities of the Fredholm resolvent kernel $\Gamma(x,t;\lambda)$ in (5.59), and the convergence for all λ of the series (5.60) and (5.63) for the Fredholm minor $D(x,t;\lambda)$ and the Fredholm determinant $D(\lambda)$, respectively.

Theorem 6 Fredholm's First Theorem: A Simple Version

"The Fredholm integral equation of the second kind (5.21) with $f(x)$ and $K(x,t)$ integrable, has for $D(\lambda) \neq 0$, a unique solution of the form given in (5.58) via the Fredholm resolvent kernel $\Gamma(x,t;\lambda)$. Moreover, this resolvent kernel, as seen in (5.59) is a ratio of two infinitely differentiable functions of λ, namely $D(x,t;\lambda)$ and $D(\lambda)$."

Example 11 (The Fredholm Resolvent Kernel Method)

Solve the Fredholm integral equation of the second kind

$$u(x) = f(x) + \lambda \int_0^1 xe^t u(t)dt \tag{E.1}$$

According to (5.58), the solution to this equation is

$$u(x) = f(x) + \lambda \int_0^1 \Gamma(x,t;\lambda) f(t)dt.$$

To evaluate the resolvent kernel $\Gamma(x,t;\lambda)$ we should start evaluating the functions required for it in (5.59) which are found in (5.60)–(5.63).

Here $B_0(x,t) = K(x,t) = xe^t$, $C_0 = 1$, and hence from (5.62) we have

$$C_1 = \int_0^1 B_0(t,t)dt = \int_0^1 te^t dt = 1. \tag{E.2}$$

For C_2 we need $B_1(t,t)$, which we can evaluate from (5.61),

$$\begin{aligned} B_1(x,t) &= C_1 K(x,t) - \int_0^1 K(x,s)B_0(s,t)ds \\ &= xe^t - \int_0^1 xe^s se^t ds = xe^t - xe^t \int_0^1 se^s ds \\ &= xe^t - xe^t = 0. \end{aligned} \tag{E.3}$$

From (5.62) we have

$$C_2 = \int_0^1 B_1(t,t)dt = 0. \tag{E.4}$$

If we use $C_2 = 0$ and $B_1 = 0$ in (5.61) for $B_2(x,t)$ we obtain $B_2(x,t) = 0$ and this can be used again in (5.62) for $C_3 = 0$. This can be continued to obtain

$$\begin{aligned} &C_1 = 1, \quad C_n = 0, \qquad n = 2, 3, \cdots \\ &B_n = 0, \quad n = 1, 2, \cdots \end{aligned}$$

It is clear from (5.63) and the values of $C_0 = C_1 = 1$, $C_n = 0$, $n = 2, 3, \cdots$ above that

$$D(\lambda) = C_0 - \lambda C_1 - 0 \cdots = 1 - \lambda \neq 0 \tag{E.5}$$

and from (5.60) and the values of $B_0 = xe^t$, $B_n = 0, n = 1, 2, \cdots$, we have

$$D(x,t;\lambda) = xe^t - 0 = xe^t. \tag{E.6}$$

From (5.59), (E.5), and (E.6) the (Fredholm) resolvent kernel becomes

$$\Gamma(x,t;\lambda) = \frac{D(x,t;\lambda)}{D(\lambda)} = \frac{xe^t}{1-\lambda} \tag{E.7}$$

and the solution to (E.1) is

$$u(x) = f(x) + \lambda \int_0^1 \frac{xe^t}{1-\lambda} f(t)dt. \tag{E.8}$$

We may remark here that the kernel $K(x,t) = xe^t$ of (E.1) is degenerate with one term, and hence it is much easier to solve (E.1) using the method for solving Fredholm equations of the second kind with degenerate kernel which we discussed in Section 5.1 and illustrated in Example 1 for a degenerate kernel with two terms,

$$K(x,t) = xt^2 + t^2 x.$$

Another way of expressing the repeated integral expressions for $B_n(x,t)$ in (5.61) and C_n in (5.62) is in the form of n repeated integrals whose integrand is a determinant of order n and whose entries are determined by the kernel $K(x,t)$. These new expressions are

$$B_n(x,t) = \int_a^b \int_a^b \cdots \int_a^b \begin{vmatrix} K(x,t) & K(x,t_1) & \cdots & K(x,t_n) \\ K(t_1,t) & K(t_1,t_1) & \cdots & \\ \vdots & & & \\ K(t_n,t) & \cdots & & K(t_n,t_n) \end{vmatrix} dt_1 dt_2 \cdots dt_n \tag{5.64}$$

$$C_n = \int_a^b \int_a^b \cdots \int_a^b \begin{vmatrix} K(t_1,t_1) & K(t_1,t_2) & \cdots & K(t_1,t_n) \\ K(t_2,t_1) & \cdots & & \\ \vdots & & & \\ K(t_n,t_1) & \cdots & & K(t_n,t_n) \end{vmatrix} dt_1 dt_2 \cdots dt_n. \tag{5.65}$$

These kinds of expressions for $B_n(x,t)$ and C_n may have a great advantage for those who are familiar with determinants and their properties and manipulations, which are used for more efficient computations. For example, we needed two steps of substitution to compute $B_2(x,t)$ from (5.61) in Example 11, whereas if we use (5.64) we can immediately write

$$B_2(x,t) = \int_0^1 \int_0^1 \begin{vmatrix} xe^t & xe^{t_1} & xe^{t_2} \\ t_1 e^t & t_1 e^{t_1} & t_1 e^{t_2} \\ t_2 e^t & t_2 e^{t_1} & t_2 e^{t_2} \end{vmatrix} dt_1 dt_2. \tag{5.66}$$

It is advisable here to exhaust the properties of determinants, which may produce a simple result for the integrand, before embarking on doing the double integration. For example, the result of $C_2 = 0$ in (E.4) of Example 11, can be obtained easily, since from (5.65) we have

$$C_2 = \int_0^1 \int_0^1 \begin{vmatrix} t_1 e^{t_1} & t_1 e^{t_2} \\ t_2 e^{t_1} & t_2 e^{t_2} \end{vmatrix} dt_1 dt_2 \tag{5.67}$$

where it is obvious that the determinant is zero,

$$\begin{vmatrix} t_1 e^{t_1} & t_1 e^{t_2} \\ t_2 e^{t_1} & t_2 e^{t_2} \end{vmatrix} = t_1 t_2 e^{t_1+t_2} - t_1 t_2 e^{t_1+t_2} = 0$$

and hence $C_2 = 0$. Also the integral in (5.66) can be shown to vanish after noting that the first and second columns of the determinant are proportional, which results in the vanishing of the determinant [see Exercise 2(a)].

5.3.2 Method of Iterated Kernels

Another method of solving Fredholm integral equation of the second kind

$$u(x) = f(x) + \lambda \int_a^b K(x,t)u(t)dt \tag{5.21}$$

is the method of iterated kernels. This method starts, as in the case of the Volterra equation of the second kind (of Section 3.1) by the zeroth approximation $u_0(x) = f(x)$ for the solution $u(x)$ in the integral of (5.21), to obtain the first approximation $u_1(x)$,

$$u_1(x) = f(x) + \lambda \int_a^b K(x,t)f(t)dt = f(x) + \lambda\phi_1(x) \tag{5.68}$$

where

$$\phi_1(x) \equiv \int_a^b K(x,t)f(t)dt. \tag{5.69}$$

This $u_1(x)$ of (5.68) is substituted again in the integral of (5.21) to obtain the second aproximation $u_2(x)$,

$$\begin{aligned} u_2(x) &= f(x) + \lambda \int_a^b K(x,t)u_1(t)dt = f(x) \\ &\quad + \lambda \int_a^b K(x,t)\left[f(t) + \lambda \int_a^b K(t,y)f(y)dy\right] dt \\ &= f(x) + \lambda \int_a^b K(x,t)f(t)dt + \lambda^2 \int_a^b \left[\int_a^b K(x,t)K(t,y)dt\right] f(y)dy \\ &= f(x) + \lambda\phi_1(x) + \lambda^2 \int_a^b K_2(x,y)f(y)dy \end{aligned} \tag{5.70}$$

after using (5.69) for the first integral and defining the *iterated kernel*

$$K_2(x,y) = \int_a^b K(x,t)K_1(t,y)dt \tag{5.71}$$

with $K_1(t,y) \equiv K(t,y)$.

If we define

$$\phi_2(x) \equiv \int_a^b K_2(x,y)f(y)dy \tag{5.72}$$

then $u_2(x)$ in (5.70) becomes

$$u_2(x) = f(x) + \lambda\phi_1(x) + \lambda^2\phi_2(x). \tag{5.73}$$

This second approximation is then substituted in (5.21) following the same steps as those used above to obtain $u_3(x)$:

$$\begin{aligned} u_3(x) &= f(x) + \lambda\phi_1(x) + \lambda^2\phi_2(x) + \lambda^3 \int_a^b K_3(x,y)f(y)dy \\ &= f(x) + \lambda\phi_1(x) + \lambda^2\phi_2(x) + \lambda^3\phi_3(x) \end{aligned} \tag{5.74}$$

where

$$K_3(x,y) \equiv \int_a^b K(x,t)K_2(t,y)dt$$

$$\phi_3(x) \equiv \int_a^b K_3(x,y)f(y)dy \tag{5.75}$$

and $K_2(t,y)$ is given by (5.71).

If this process is continued n times, we obtain $u_n(x)$, the nth approximation for the solution of (5.21), as

$$\begin{aligned} u_n(x) &= f(x) + \lambda\phi_1(x) + \lambda^2\phi_2(x) + \cdots + \lambda^n\phi_n(x) \\ &= f(x) + \sum_{i=1}^{n} \lambda^i\phi_i(x), \end{aligned} \tag{5.76}$$

$$\phi_i(x) = \int_a^b K_i(x,y)f(y)dy \tag{5.77}$$

$$K_i(x,y) = \int_a^b K(x,t)K_{i-1}(t,y)dt, \qquad i = 2, 3, \cdots, n. \tag{5.78}$$

$K_i(x,y)$ is called the ith *iterated kernel*. It remains to find under what condition the series (5.76) converges to $u(x)$, the solution of (5.21). It turns out that the series (5.76) converges for $|\lambda B| < 1$,[6] $|\lambda| < 1/B$, where

[6]See Pogorzelski [1966].

$$B = \sqrt{\int_a^b \int_a^b K^2(x,t)dxdt.} \tag{5.79}$$

The convergent series

$$u(x) = f(x) + \sum_{i=1}^{\infty} \lambda^i \phi_i(x) \tag{5.80}$$

is called the *Neumann series* and can be rewritten, after substituting for $\phi_i(x)$ from (5.77), as

$$\begin{aligned} u(x) &= f(x) + \sum_{i=1}^{\infty} \lambda^i \int_a^b K_i(x,t) f(t)dt \\ &= f(x) + \int_a^b \left[\sum_{i=1}^{\infty} \lambda^i K_i(x,t) \right] f(t)dt \\ &= f(x) + \lambda \int_a^b \Gamma(x,t;\lambda) f(t)dt \end{aligned} \tag{5.81}$$

and hence we find a new resolvent kernel for (5.21), which is

$$\Gamma(x,t;\lambda) = \sum_{i=1}^{\infty} \lambda^{i-1} K_i(x,t) \tag{5.82}$$

in addition to the Fredholm resolvent kernel of (5.59). In Example 13 we will present a simple proof for showing that the above Fredholm resolvent kernel $\Gamma(x;t;\lambda)$ is *unique*.

We may remark here that assuming the uniqueness of the solution to (5.21), we can show that this resolvent kernel (5.82) associated with the Neumann series solution is unique, a result that we will relegate its' proof as a simple illustration in Example 13. Next we illustrate the foregoing iterated kernel-Neumann series method for the same problem as that of Example 11, where we used the Fredholm resolvent kernel method of (5.58)–(5.63) to solve it. We may remind of the other (special) resolvent kernel of (5.46) that we used in (5.55) of Seciton 5.1 for the solution of the Fredholm integral equation with *symmetric* kernel (5.21s).

Example 12 Iterated Kernels: Neumann Series Method

Solve the integral equation of Example 11,

$$u(x) = f(x) + \lambda \int_0^1 xe^t u(t)dt. \tag{E.1}$$

To arrive at the Neumann series solution (5.81) for this problem we must prepare $K_i(x,t)$, the ith iterate of the kernel $K(x,t) = xe^t$. Here we have $K_1(x,t) = K(x,t) = xe^t$. For $i = 2$ we obtain the second iterate $K_2(x,y)$ from (5.78),

$$\begin{aligned} K_2(x,y) &= \int_0^1 K(x,t)K_1(t,y)dt = \int_0^1 xe^t te^y dt \\ &= xe^y \int_0^1 te^t dt = xe^y. \end{aligned} \tag{E.2}$$

Now we use this result again in (5.78) for $i = 3$ to obtain

$$\begin{aligned} K_3(x,y) &= \int_0^1 K(x,t)K_2(t,y)dt = \int_0^1 xe^t te^y dt \\ &= xe^y \int_0^1 te^t dt = xe^y. \end{aligned} \tag{E.3}$$

and it is obvious from (E.2), (E.3) and (5.78), that if these calculations are repeated, we obtain the general expression for the ith iterate of the kernel as

$$K_i(x,y) = xe^y. \tag{E.4}$$

This is now substituted in the Neumann series (5.81) to obtain the final solution to (E.1),

$$u(x) = f(x) + x\sum_{i=1}^{\infty} \lambda^i \int_0^1 e^y f(y)dy. \tag{E.5}$$

We note that this series converges for $|\lambda| < \frac{1}{|B|} = \sqrt{6/(e^2-1)} \approx 0.97$ since according to (5.79) with $K(x,t) = xe^t$, we have

$$\begin{aligned} B^2 &= \int_0^1 \int_0^1 K^2(x,t)dxdt = \int_0^1 \int_0^1 x^2 e^{2t} dxdt = \int_0^1 e^{2t} \int_0^1 x^2 dxdt \\ &= \int_0^1 e^{2t} \left(\frac{1}{3}\right) dt = \frac{1}{3}\frac{e^2-1}{2} = \frac{e^2-1}{6}. \end{aligned} \tag{E.6}$$

So $B = \sqrt{(e^2-1)/6}$, and for the series in (E.5) to converge we must have $|\lambda B| < 1$, which means that $|\lambda| < 1/|B| = \sqrt{6/(E^2-1)} \approx 0.97$. If we write (E.5) in terms of resolvent kernel, we obtain the same answer as in Example 11,

$$\begin{aligned} u(x) &= f(x) + \int_0^1 \left(\sum_{i=1}^{\infty} \lambda^i xe^y\right) f(y)dy \\ &= f(x) + x\int_0^1 \left(\sum_{i=1}^{\infty} \lambda^i\right) e^y f(y)dy \\ &= f(x) + \lambda \int_0^1 \frac{xe^y}{1-\lambda} f(y)dy \end{aligned} \tag{E.7}$$

after recognizing that the geometric series $\sum_{i=1}^{\infty} \lambda^{i-1}$ converges to $1/(1-\lambda)$.

We note here that Example 12 was carefully chosen with simple kernel to facilitate the illustration of the method of iterated kernels by keeping the effort of performing the repeated integrations at a minimum. As a consequence is the special feature of Example 12 of the simple form of the ith iterate $K_i(x,y) = xe^y$ of (E.4), which in more general problems will be of a more complicated form.

A special class of kernels that may result in a finite (instead of an infinite) Neumann series is that of the *orthogonal kernels*. Two kernels $K(x,t)$ and $L(x,t)$ are called orthogonal on $\{(x,t) : a \le x \le b, a \le t \le b\}$ if the following two integrals vanish (see Exercise 25, Section 4.1):

$$\int_a^b K(x,\tau)L(\tau,t)d\tau = 0 \tag{5.83}$$

$$\int_a^b L(x,\tau)K(\tau,t)d\tau = 0. \tag{5.84}$$

As a special case, if it turns out that the kernel $K(x,t)$ is orthogonal to all kernel iterates $K_i(x,t)$, $i = n+1, n+2, \cdots$, then according to (5.78), all the iterates with order above n will vanish and the Neumann series (5.80) will have n terms only. In the very special case when the kernel $K(x,t)$ is orthogonal to itself, then according to (5.78), we have

$$K_2 = K_3 = \cdots = 0 \tag{5.85}$$

and the Neumann series (5.80) becomes a one-term series with the resolvent kernel of (5.82) as a λ multiple of the kernel itself.

In the next example we prove that the Fredholm resolvent kernel $\Gamma(x;t;\lambda)$ of (5.82) is unique.

Example 13 Proof of Uniqueness of the Resolvent Kernel $\Gamma(x,t;\lambda)$ in (5.82)

For a fixed $\lambda = \lambda_0$ let there be two resolvent kernels $\Gamma_1(x,t;\lambda_0)$ and $\Gamma_2(x,t;\lambda_0)$ for the solution in (5.81). Substituting these two values of the resolvent kernel in (5.81) (assuming it is a unique solution to (5.21)), we obtain

$$f(x) + \lambda_0 \int_a^b \Gamma_1(x,t;\lambda_0)f(t)dt \equiv f(x) + \lambda_0 \int_a^b \Gamma_2(x,t;\lambda_0)f(t)dt \tag{E.1}$$

$$\int_a^b \Gamma_1(x,t;\lambda_0)f(t)dt \equiv \int_u^b \Gamma_2(x,t;\lambda_0)f(t)dt \tag{E.2}$$

which can be written as

$$\int_a^b [\Gamma_1(x,t;\lambda_0)f(t)dt - \int_a^b \Gamma_2(x,t;\lambda_0)f(t)dt \equiv 0. \tag{E.3}$$

We note that (E.3) is valid for arbitrary function $f(t)$; hence if we set $\Gamma_1(x,t;\lambda_0) - \Gamma_2(x,t;\lambda_0) \equiv \Phi(x,t;\lambda_0)$ and let $f(t) = \Phi(x,t;\lambda_0)$ in (E.3), we obtain

$$\int_a^b |\Phi(x,t;\lambda_0)|^2 dt \equiv 0$$

which implies that $\Phi(x, t; \lambda_0) \equiv 0$; hence

$$\Gamma_1(x, t; \lambda_0) - \Gamma_2(x, t; \lambda_0) \equiv 0$$

and

$$\Gamma_1(x, t; \lambda_0) = \Gamma_2(x, t; \lambda_0)$$

which says that the resolvent kernel of (5.81) is unique.

Numerical Evaluation of the Eigenvalues: Method of Traces

In Section 5.2.3 we mentioned the Rayleigh-Ritz method for estimating the eigenvalues λ for the homogeneous Fredholm integral equation

$$u(x) = \lambda \int_a^b K(x, t)u(t)dt. \tag{5.86}$$

We present here another method for estimating the eigenvalues since it essentially makes use of the iterated kernels $K_i(x, t)$ in (5.78) of this section. Here we will only state the results of the method and illustrate it with a detailed example. This method gives the following formula for estimating the *smallest* eigenvalue λ_1:

$$\lambda_1 \sim \sqrt{\frac{A_{2i}}{A_{2i+2}}} \tag{5.87}$$

where A_j is defined in terms of $K_j(x, t)$, the jth iterate of the kernel $K(x, t)$; as

$$A_j = \int_a^b K_j(t, t)dt \tag{5.88}$$

which is called the jth trace of the kernel $K(x, t)$ and hence the name for the method of traces. For the symmetric kernel $K(x, t)$ we can show that the even-indexed trace A_{2i} can be expressed in terms of $K_i^2(x, t)$ as

$$A_{2i} = \int_a^b \int_a^b K_i^2(x, t)dxdt. \tag{5.89}$$

Example 14 Method of Traces for Estimating Eigenvalues

In this example we consider the following problem

$$u(x) = \lambda \int_{-1}^1 xtu(t)dt. \tag{E.1}$$

Here we have $K(x, t) = K_1(x, t) = xt$ and for simplicity we will seek an estimate for the lowest eigenvalue λ_1 from (5.87) as

$$\lambda_1 \sim \sqrt{\frac{A_2}{A_4}} \tag{E.2}$$

which corresponds to $i = 1$. According to (5.89), since we have a symmetric kernel $K(x,t) = xt$ here we must evaluate the first and second iterated kernels of $K(x,t) = xt$ for A_2 and A_4, respectively, to be used for a (rough!) approximation of λ_1 in (5.87) with $i = 1$. So we start with $K_1(x,t) = K(x,t) = xt$, to obtain

$$\begin{aligned} A_2 &= \int_{-1}^{1}\int_{-1}^{1} K_1^2(x,t)dxdt = \int_{-1}^{1}\int_{-1}^{1} x^2t^2dxdt \\ &= \int_{-1}^{1} \frac{x^3}{3}\bigg|_{-1}^{1} t^2dt = \frac{2}{3}\int_{-1}^{1} t^2dt = \frac{2}{3}\cdot\frac{t^3}{3}\bigg|_{-1}^{1} = \frac{4}{9}. \end{aligned} \tag{E.3}$$

For A_4 we must have $K_2(x,t)$, which can be evaluated from (5.78) with

$$\begin{aligned} K_1(x,t) &= K(x,t) = xt, \\ K_2(x,t) &= \int_{-1}^{1} K(x,y)K_1(y,t)dy \\ &= \int_{-1}^{1} xyytdy = xt\int_{-1}^{1} y^2dy = xt\,\frac{y^3}{3}\bigg|_{-1}^{1} = \frac{2}{3}xt. \end{aligned} \tag{E.4}$$

Now we substitute $K_2(x,t) = (2/3)xt$ from (E.4) in (5.89) to obtain the value for A_4,

$$\begin{aligned} A_4 &= \int_{-1}^{1}\int_{-1}^{1} K_2^2(x,t)dxdt = \int_{-1}^{1}\int_{-1}^{1} \frac{4}{9}x^2t^2dxdt \\ &= \frac{4}{9}\int_{-1}^{1} \frac{x^3}{3}\bigg|_{-1}^{1} t^2dt = \frac{4}{9}\left(\frac{2}{3}\right)\int_{-1}^{1} t^2dt \\ &= \frac{8}{27}\int_{-1}^{1} t^2dt = \left(\frac{8}{27}\right)\frac{t^3}{3}\bigg|_{-1}^{1} = \frac{8}{27}\left(\frac{2}{3}\right)\bigg| = \frac{16}{81}. \end{aligned} \tag{E.5}$$

When we substitute in (E.2) the values of $A_2 = 4/9$ from (E.3) and $A_4 = 16/81$ from (E.5), we obtain the estimate for the lowest eigenvalue,

$$\lambda_1 \sim \sqrt{\frac{A_2}{A_4}} = \sqrt{\frac{4/9}{16/81}} = \frac{3}{2} \tag{E.6}$$

and hence $\lambda_1 \sim 3/2$.

5.3.3 Some Basic Approximate Methods

The approximate methods that we will present here for solving the Fredholm equation of the second kind

$$u(x) = f(x) + \int_a^b K(x,t)u(t)dt \tag{5.90}$$

are based on approximating the solution $u(x)$ of (5.90) by a partial sum.

$$S_N(x) = \sum_{k=1}^{N} c_k \phi_k(x) \tag{5.91}$$

of N *linearly independent* functions $\phi_1, \phi_2, \cdots, \phi_N$ on the interval (a, b). Of course, if this approximate solution (5.91) is to be substituted in (5.90) for $u(x)$, there will be an error $\epsilon(x, c_1, c_2, \cdots, c_N)$ involved, which depends on x and on the way the coefficients c_k, $k = 1, 2, \cdots, N$ are chosen,

$$S_N(x) = f(x) + \int_a^b K(x,t) S_N(t) dt + \varepsilon(x, c_1, c_2, \cdots, c_N). \tag{5.92}$$

The main point here is how we can find or impose N conditions to give us the N equations required for determining the N coefficients $c_1, c_2, \cdots, c_N$ of the approximate solution (5.91). The methods employed will differ by the way these conditions are set, and of course the better method will be the one that keeps the error in (5.92) to a minimum.

Collocation Method

This method presents the N conditions by insisting that the error in (5.92) vanishes at N points $x_1, x_2, \cdots, x_N$. This reduces (5.92) to the N equations

$$S_N(x_i) = f(x_i) + \int_a^b K(x_i, t) S_N(t) dt, \qquad i = 1, 2, \cdots, N \tag{5.93}$$

for determining the coefficients $c_1, c_2, \cdots, c_N$ of the approximate solution $S_N(x)$ in (5.91). To determine these coefficients in (5.93) we first substitute for $S_N(x)$ from (5.91) in terms of the given N linearly independent functions $\phi_1, \phi_2, \cdots, \phi_N(x)$, perform the integration, then substitute $x = x_1, x_2, \cdots, x_N$ for which the error $\varepsilon(x, c_1, c_2, \cdots, c_N)$ vanishes.

Example 15 The Collocation Approximate Method

We illustrate this approximate method with the following simple Fredholm equation of the second kind:

$$u(x) = x + \int_{-1}^{1} xtu(t) dt \tag{E.1}$$

which of course can be solved by using any of the exact methods discussed in the preceding sections as the kernel $K(x,t) = xt$ is degenerate and symmetric. We choose here three linearly independent functions $\phi_1(x) = 1$, $\phi_2(x) = x$, and $\phi_3(x) = x^2$, and so the approximate solution from (5.91) is

$$S_3(x) = \sum_{k=1}^{3} c_k \phi_k(x) = c_1 + c_2 x + c_3 x^2. \tag{E.2}$$

If we substitute this in (5.92), we obtain

$$\begin{aligned} S_3(x) &= c_1 + c_2 x + c_3 x^2 = x + \int_{-1}^{1} xt(c_1 + c_2 t + c_3 t^2)dt + \epsilon(x, c_1, c_2, c_3) \\ &= c_1 + c_2 x + c_3 x^2 = x + x\int_{-1}^{1} (c_1 t + c_2 t^2 + c_3 t^3)dt + \epsilon(x, c_1, c_2, c_3) \end{aligned} \tag{E.3}$$

and after performing the integration,

$$\begin{aligned} \int_{-1}^{1} (c_1 t + c^2 t^2 + c_3 t^3)dt &= c_1 \frac{t^2}{2} + c_2 \frac{t^3}{3} + c_3 \frac{t^4}{4}\bigg|_{-1}^{1} \\ &= \frac{1}{2}c_1 + \frac{1}{3}c_2 + \frac{1}{4}c_3 - \left(\frac{1}{2}c_1 - \frac{1}{3}c_2 + \frac{1}{4}c_3\right) \\ &= \frac{2}{3}c_2 \end{aligned} \tag{E.4}$$

(E.3) becomes

$$\begin{aligned} c_1 + c_2 x + c_3 x^2 &= x + x\left(\frac{2}{3}\right) c_2 + \varepsilon(x, c_1, c_2, c_3) \\ &= x\left(1 + \frac{2}{3}c_2\right) + \varepsilon(x, c_1, c_2, c_3). \end{aligned} \tag{E.5}$$

To find c_1, c_2, and c_3 we need three equations, which we provide (via the collocation method) by insisting that the error $\varepsilon(x, c_1, c_2, c_3)$ in (E.5) vanishes at three points (among other choices) $x_1 = 1$, $x_2 = 0$, and $x_3 = -1$, which gives, respectively,

$$c_1 + c_2 + c_3 = 1 + \frac{2}{3}c_2, \qquad c_1 + \frac{1}{3}c_2 + c_3 = 1 \tag{E.6}$$

$$c_1 + 0 + 0 = 0, \qquad c_1 = 0 \tag{E.7}$$

$$c_1 - c_2 + c_3 = -1 - \frac{2}{3}c_2, \qquad c_1 - \frac{1}{3}c_2 + c_3 = -1. \tag{E.8}$$

It is simple to solve for c_1, c_2, and c_3 from (E.6)–(E.8), which gives $c_1 = c_3 = 0$ and $c_2 = 3$. The approximate solution to (E.1) is $S_3(x) = 3x$. For this example it happens that we can easily verify that the exact solution to (E.1) is also $u(x) = 3x$. However, the perfect agreement between the approximate and exact solutions should not be surprising since the particular form of the approximate solution $S_3(x) = c_1 + c_2 x + c_3 x^2$ included the exact solution as a very special case, $u(x) = c_2 x = 3x$. It should be clear, however, that such agreement is not possible when we consider another form for the approximate solution of (E.1), say, $S_3(x) = c_1 + c_2 \sin x + c_3 \cos x$ in terms of the three linearly independent functions 1, $\sin x$, and $\cos x$, which we leave as an exercise. We may remark again that we have chosen this very particular problem to minimize the detailed computations in favor of clarifying the main steps of the method. In the following example we consider a more general problem with a known exact solution with which to compare our approximate solution.

Example 16 The Collocation Approximate Method

The Fredholm equation

$$u(x) = e^{-x} - \int_0^1 xe^t u(t)dt \tag{E.1}$$

is a special case of problem (E.1) of Example 12 with $f(x) = e^{-x}$ and $\lambda = -1$; hence its exact solution is easily obtained as

$$u(x) = e^{-x} - \frac{x}{2} \tag{E.2}$$

using (E.7) of Example 12.

Now we will use the collocation method to find an approximate solution to (E.1) which we will compare with the exact solution (E.2). We again choose the three linearly independent simple functions $1, x, x^2$, so the approximate solution is $S_3(x) = c_1 + c_2 x + c_3 x^2$. If we substitute in (5.92) with $f(x) = e^{-x}$ and $K(x,t) = -xe^t$, we obtain

$$c_1 + c_2 x + c_3 x^2 = e^{-x} - x\int_0^1 e^t(c_1 + c_2 t + c_3 t^2)dt + \epsilon(x, c_1, c_2, c_3). \tag{E.3}$$

We perform the integration on the right side of (E.3) to obtain

$$c_1 + c_2 x + c_3 x^2 = e^{-x} - x(c_1 e - c_1 + c_2 + c_3 e - 2c_3) + \varepsilon(x, c_1, c_2, c_3). \tag{E.4}$$

To determine c_1, c_2, and c_3, we insist that the error $\varepsilon(x, c_1, c_2, c_3)$ in (E.4) vanishes at three points. In this case we take $x = 0,\ 1/2$, and 1, which gives us the desired three equations in c_1, c_2, and c_3.

$$x = 0: \qquad c_1 + 0 + 0 = 1 - 0, \qquad c_1 = 1 \tag{E.5}$$

$$\begin{aligned} x = \frac{1}{2} : c_1 + \frac{1}{2}c_2 + \frac{1}{4}c_3 &= e^{-1/2} - \frac{1}{2}(c_1 e - c_1 + c_2 + c_3 e - 2c_3) \\ &= 1 + \frac{1}{2}c_2 + \frac{1}{4}c_3 = e^{-1/2} - \frac{1}{2}(e - 1 + c_2 + c_3 e - 2c_3) \end{aligned} \tag{E.6}$$

$$\begin{aligned} x = 1 : c_1 + c_2 + c_3 &= e^{-1} - (c_1 e - c_1 + c_2 + c_3 e - 2c_3) \\ &= 1 + c_2 + c_3 = e^{-1} - (e - 1 + c_2 + c_3 e - 2c_3). \end{aligned} \tag{E.7}$$

From (E.5)–(E.7) it is easy to solve for c_1, c_2, and c_3 as $c_1 = 1$, $c_2 = -1.441$, and $c_3 = 0.310$, which makes the approximate solution

$$S_3(x) = 1 - 1.441x + 0.310x^2. \tag{E.8}$$

In Table 5.2 and Figure 5.1 we present a comparison between this approximate solution (E.8) and the exact solution (E.2) of the problem (E.1).

Table 5.2 Comparison of Approximate (Collocation Method) and Exact Solutions of Fredholm Integral Equation (E.1)

x	0	0.25	0.5	0.75	1.0
Approximate values					
$u(x) \sim S_3(x) = 1 - 1.441x + 0.310x^2$	1	0.6590	0.3568	0.0933	-0.1310
Exact values, $u(x) = e^{-x} - \frac{x}{2}$	1	0.6538	0.3565	0.0974	-0.1321

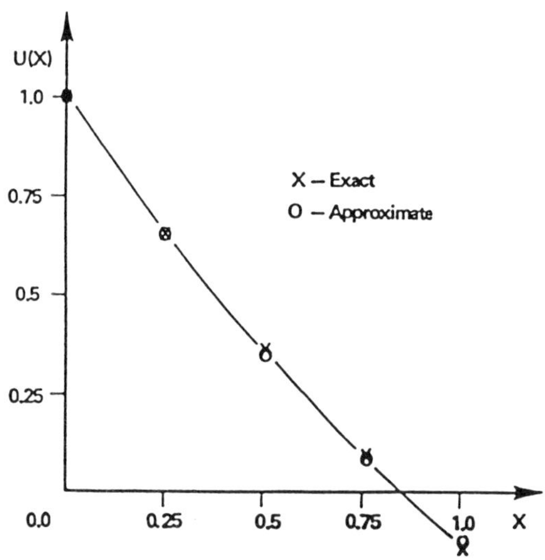

Fig. 5.1 Comparison of approximate (collocation method) and exact solutions of Fredholm equation (E.1) of Example 16.

Galerkin (or the Weighted Functions) Approximate Method

This method establishes the N conditions necessary for the determination of the N coefficients in (5.91) by making the error $\varepsilon(x, c_1, c_2, \cdots, c_N)$ of (5.92), as a function of x, orthogonal to N given linearly independent functions $\psi_1(x), \psi_2(x), \cdots, \psi_N(x)$ on the interval (a, b). We will use the definition of orthogonality in (4.46) on the error $\varepsilon(x, c_1, c_2, \cdots, c_N)$ in (5.92), where these N conditions become

$$\int_a^b \psi_j(x)\varepsilon(x, c_1, c_2, \cdots, c_N)dx$$
$$= \int_a^b \psi_j(x)\left[S_N(x) - f(x) - \int_a^b K(x,t)S_N(t)dt\right]dx = 0, \quad j = 1, 2, \cdots, N \tag{5.94}$$

which can be rewritten as the following N equations in the N unknowns $c_1, c_2, \cdots, c_N$,

$$\int_a^b \psi_j(x) \left[S_N(x) - \int_a^b K(x,t) S_N(t) dt \right] dx = \int_a^b \psi_j(x) f(x) dx, \quad j = 1, 2, \cdots, N$$

or

$$\begin{aligned} \int_a^b \psi_j(x) \left\{ \sum_{k=1}^N c_k \phi_k(x) - \int_a^b K(x,t) \left[\sum_{k=1}^N c_k \phi_k(t) \right] dt \right\} dx \\ = \int_a^b \psi_j(x) f(x) dx, \qquad j = 1, 2, \cdots, N \end{aligned} \tag{5.95}$$

after substituting for $S_N(x)$ from (5.91). We remark here that in general the linearly independent functions $\psi_j(x)$ are different from $\phi_j(x)$ used for the approximation, but sometimes it is convenient to use the same functions.

Example 17 The Galerkin Approximate Method

For simplicity we will illustrate this method for the same problem of Example 15,

$$u(x) = x + \int_{-1}^1 xtu(t) dt \tag{E.1}$$

and we choose the same linearly independent functions $\phi_1(x) = 1$, $\phi_2(x) = x$, and $\phi_3(x) = x^2$ to approximate the solution $u(x)$ by

$$S_3(x) = c_1 + c_2 x + c_3 x^2. \tag{E.2}$$

If we substitute this in (E.1), then according to (5.92), the error is

$$\varepsilon(x, c_1, c_2, c_3) = c_1 + c_2 x + c_3 x^2 - x - \int_{-1}^1 xt(c_1 + c_2 t + c_3 t^2) dt. \tag{E.3}$$

To find the three equations necessary for determining c_1, c_2, and c_3, the Galerkin method requires this error to be orthogonal to three linearly independent functions $\psi_1(x)$, $\psi_2(x)$, and $\psi_3(x)$, which again for simplicity we choose as $1, x$, and x^2, respectively. The orthogonality condition (5.95) gives the three desired equations in c_1, c_2, and c_3:

$$\int_{-1}^1 1 \left[c_1 + c_2 x + c_3 x^2 - \int_{-1}^1 xt(c_1 + c_2 t + c_3 t^2) dt \right] dx = \int_{-1}^1 1(x) dx \tag{E.4}$$

$$\int_{-1}^1 x \left[c_1 + c_2 x + c_3 x^2 - \int_{-1}^1 xt(c_1 + c_2 t + c_3 t^2) dt \right] dx = \int_{-1}^1 x(x) dx \tag{E.5}$$

$$\int_{-1}^1 x^2 \left[c_1 + c_2 x + c_3 x^2 - \int_{-1}^1 xt(c_1 + c_2 t + c_3 t^2) dt \right] dx = \int_{-1}^1 x^2(x) dx \tag{E.6}$$

We note from (E.4) in Example 15 that the inside integral in the equations above,

$$\int_{-1}^{1} t(c_1 + c_2 t + c_3 t^2)dt = \frac{2}{3}c_2.$$

We use this result and perform the rest of the simple integrations to obtain the three equations in c_1, c_2, and c_3:

$$\begin{aligned}\int_{-1}^{1}\left(c_1 + \frac{1}{3}c_2 x \quad + c_3 x^2\right)dx &= \int_{-1}^{1} x dx = \left.\frac{x^2}{2}\right|_{-1}^{1} = 0\\ &= c_1 x + \frac{1}{6}\, c_2 x^2 + \frac{1}{3}c_3 x^3\Big|_{-1}^{1} = 2c_1 + \frac{2}{3}c_3 = 0\end{aligned} \tag{E.7}$$

$$\begin{aligned}\int_{-1}^{1}\left(c_1 x \quad + \frac{1}{3}c_2 x^2 + c_3 x^3\right)dx &= \int_{-1}^{1} x^2 dx = \left.\frac{x^3}{3}\right|_{-1}^{1} = \frac{2}{3}\\ &= \frac{1}{2}c_1 x^2 + \frac{1}{9}\, c_2 x^3 + c_3 x \frac{x^4}{4}\Big|_{-1}^{1} = \frac{2}{9}c_2 = \frac{2}{3},\quad c_2 = 3\end{aligned} \tag{E.8}$$

$$\begin{aligned}\int_{-1}^{1}\left(c_1 x^2 \quad + \frac{1}{3}c_2 x^3 + c_3 x^4\right)dx &= \int_{-1}^{1} x^2 \cdot x dx = \left.\frac{x^4}{4}\right|_{-1}^{1} = 0\\ &= \frac{1}{3}\, c_1 x^3 + \frac{1}{12}c_2 x^4 + \frac{1}{5}c_3 x^5\Big|_{-1}^{1} = \frac{2}{3}c_1 + \frac{2}{5}c_3 = 0.\end{aligned} \tag{E.9}$$

From (E.7), (E.8), and (E.9) we solve for c_1, c_2, and c_3, to find that $c_1 = 0$, $c_2 = 3$, and $c_3 = 0$, which gives $S_3(x) = 3x$ as the approximate solution. But as we pointed out in Example 15, this is also the exact solution. This is because of our choice of the three linearly independent functions $1, x, x^2$ for $S_N(x)$, where the exact solution happened to be $3x$, only a constant multiple of one of them, namely x. We leave it as an exercise $[9(a, i)]$ to illustrate this Galerkin method with the problem of Example 15 where we take the three linearly independent functions 1, $\sin x$, and $\cos x$ instead of $1, x$, and x^2.

Other approximate methods for solving Fredholm integral equations include that of the least squares method, which in summary insists on the integral of the square of the error,

$$\int_a^b \varepsilon^2(x, c_1, c_2, \cdots, c_N)dx = minimum \tag{5.96}$$

on the interval (a, b) being a minimum. We shall not discuss this or other approximate methods here due to their somewhat lengthy computations; we refer the reader to their more complete treatment in other texts that cover approximate methods of solving integral equations.[7]

[7]See Green [1969, p.96], Baker and Miller [1977], Delves and Mohammed [1988].

Exercises 5.3

1. Use the method of the Fredholm resolvent kernel (5.58) and (5.59) to solve the following Fredholm equations of the second kind, then verify your answer.

   (a) $u(x) = x^2 + \lambda \int_0^1 (x - 2t)u(t)dt$

   Hint: We have $C_0 = 1$, $B_0 = K(x,t) = x - 2t$, so start with C_1 from (5.62), then $B_1(x,t)$ from (5.61), and we continue as in Example 11 to obtain the resolvent kernel $\Gamma(x,t;\lambda)$ for the solution $u(x)$ in (5.58).

   (b) $u(x) = e^x - \int_0^1 e^{x-t}u(t)dt$

   (c) $u(x) = x + \lambda \int_0^1 (4xt - x^2)u(t)dt$

   (d) $u(x) = 1 + \lambda \int_0^\pi \sin(x+t)u(t)dt.$

2. **(a)** Use the properties of determinants to show that the double integral in (5.66) vanishes. *Hint:* See that the first column is proportional to the second column in the determinant of (5.66) after writing it as

$$\begin{bmatrix} xe^t \\ t_1 e^t \\ t_2 e^t \end{bmatrix} = e^{t-t_1} \begin{bmatrix} xe^{t_1} \\ t_1 e^{t_1} \\ t_2 e^{t_1} \end{bmatrix}$$

   (We may note that all the three columns in (5.66) are proportional to each other, so are the three rows!)

   (b) Solve the problem of Example 11, by using (5.64) and (5.65) for $B_n(x,t)$ and C_n instead of (5.61) and (5.62), respectively.

3. Use the iterated kernels method to solve the integral equation

$$u(x) = 2 + \lambda \int_0^{2\pi} \sin(x - 2t)u(t)dt.$$

   Hint: Note that the kernel $K(x,t) = \sin(x - 2t)$ is orthogonal to itself (see Exercise 22, Section 4.1).

4. Solve the Fredholm equation

$$u(x) = 2 + \lambda \int_0^\pi \sin(x+t)u(t)dt.$$

   Hint: Use the Neumann series (5.81). (Also, you can use the result of problem 5 with very minor changes!)

5. Use the iterated kernels-Neumann series method to solve the following integral equation. Verify your answer.

$$u(x) = 3 + \lambda \int_0^{\pi} \sin(x+t)u(t)dt.$$

6. Use the method of traces (Section 5.3.2) to find an estimate for λ_1, the smallest eigenvalue of the kernels.

   (a) $K(x,t) = x^2t^2$

   (b) $K(x,t) = \begin{cases} t, & x \geq t \\ x, & x \leq t \end{cases}$

   of the homogeneous Fredholm equation

$$u(x) = \lambda \int_0^1 K(x,t)u(t)dt$$

   Hint: See Example 14.

   In the following problems 7 and 8, where the collocation method is to be used, you may choose your own convenient collocation points. (In problem 8, for example, you may try the collocation points $x_1 = 0$, $x_2 = \frac{1}{3}$, $x_3 = \frac{2}{3}$ and $x_4 = 1$.)

7. **(a)** Use the collocation method to find an approximate solution for the equation of Example 15

$$u(x) = x + \int_{-1}^{1} xtu(t)dt$$

   in terms of

   (i) The three linearly independent functions $\phi_1(x) = 1$, $\phi_2(x) = \sin x$, and $\phi_3(x) = \cos x$.

   (ii) The eight linearly independent functions 1, $\sin x$, $\cos x$, $\sin 2x$, $\cos 2x$, $\sin 3x$, $\cos 3x$, and $\sin 4x$.
   handle the lengthy computations of solving the linear equations.

   (b) Tabulate the two approximate solutions in part (a) and compare them with the exact solution $u(x) = 3x$ of Example 15.

8. Use the collocation method to find an approximate solution for the equation of Example 16,

$$u(x) = e^{-x} - \int_0^1 xe^t u(t)dt$$

   in terms of the linearly independent functions

   (a) $\phi_1(x) = 1$, $\phi_2(x) = x$, $\phi_3(x) = x^2$, $\phi_4(x) = x^3$

   (b) $\phi_1(x) = \sin x$, $\phi_2(x) = \cos x$

(c) $\phi_1(x) = e^{-x}$, $\phi_2(x) = x$

and compare the results of part (a), (b), and (c) with the approximate and exact solutions in Example 16.

9. **(a)** Use the Galerkin method to find an approximate solution for the equation of Example 15,

$$u(x) = x + \int_{-1}^{1} xtu(t)dt$$

in terms of

(i) The three linearly independent functions $\phi_1(x) = 1$, $\phi_2(x) = \sin x$, and $\phi_3(x) = \cos x$.

(ii) The eight linearly independent functions 1, $\sin x$, $\cos x$, $\sin 2x$, $\cos 2x$, $\sin 3x$, $\cos 3x$, and $\sin 4x$. You may use $\psi_i(x) = \phi_i(x)$.

(b) Tabulate the two approximate solutions in part (a) and compare them with the exact solution $u(x) = 3x$ and the approximate solution obtained by the collocation method in exercise 7(a,i,ii).

(c) Use the least squares criterion (5.96) to compare how good the approximations in exercises 7(a,i) and 9(a,i) are.

(d) Do part (c) for exercises 7(a,ii) and 9(a,ii) and show how they in turn compare with 7(a,i) and 9(a,i), respectively.

10. Do Exercise 8 using the Galerkin method instead of the collocation method and compare your results.

5.4 FREDHOLM INTEGRAL EQUATIONS OF THE FIRST KIND

Towards the end of Section 5.2.1, and in relation to the Hilbert-Schmidt theorem, we discussed then illustrated in Example 8 the difficulty of insuring the existence of the solution $u(x)$ to Fredholm integral equations of the *first* kind,

$$f(x) = \int_{a}^{b} K(x,t)u(t)dt \tag{5.97}$$

and how the given function $f(x)$ must be restricted to have such a solution. Moreover, even when, perhaps on other grounds, we know that there is a solution, we lack the usual *iterative* method to construct it. This is due to the absence of the solution $u(x)$ outside the integral of (5.97), which is in contrast to integral equations of the second kind, where the iterative (or successive approximations) method plays an important role, as we had discussed in Sections 5.3.2 and 3.1, respectively, for Fredholm and Volterra equations of the second kind.

At the level of this book, the simplest statement on the *existence* of *a unique* solution for the Fredholm integral equation of the first kind (5.97) is found in (the following) Theorem 7, which is limited to a special class of symmetric kernels

$(K(x,t) = \overline{K(t,x)})$ that we shall describe in the following simple Definition 1. This Theorem 7 is a restricted version of Picard's theorem. For the general theory, the kernel $K(x,t)$ can be complex-valued, the reason for using the complex conjugation in the definition of the symmetric kernel as $K(x,t) = \overline{K(t,x)}$; it is dropped when we deal with only real-valued kernels, and we write, $K(x,t) = K(t,x)$ for symmetric real kernels as we did in (5.27). For the definitions needed for Theorem 7, we shall rely on the basic elements of Fourier series, that we have introduced and used for the theory of homogeneous Fredholm integral equations with symmetric kernels in Section 5.2.1. So, here we will limit ourselves to symmetric kernels, but we may have the chance later (or in the exercises) to briefly discuss cases or examples of non-symmetric kernels.

5.4.1 Fredholm Equations of the First Kind with Symmetric Kernels

In Example 8 of Section 5.2, and the first basic Theorem 5 for the existence of a solution to Fredholm integral equation of the first kind with symmetric kernel (5.34), we showed how conditions for such an existence are rather demanding on the given function $f(x)$ in (5.34). We now present another very basic theorem, which is aimed at the existence of not necessarily continuous solutions to (5.34), namely, *square integrable* solutions. Also the condition of this theorem guarantees a unique solution to (5.34). For this theorem, we need to present a few definitions, which describe the particular symmetric kernel that allows the existence of a unique solution to Fredholm integral equation of the first kind (5.34). Such special symmetric kernels are called *closed* symmetric kernels, which we shall describe in the following two definitions. This will enable us to give a precise statement of the simplest possible theorem on the existence of the solutions without the need for more abstract development that is necessary for most of the other theorems. The theorem will be illustrated very clearly in Example 18.

While the first part of this section deals with the rather demanding conditions for the existence of the solution, the second part of the section deals with another difficulty that such a solution may have. Briefly, Fredholm integral equations of the first kind are termed *ill-posed*, a rather advanced subject which we shall attempt to explain on the level of this book, and where we complement our discussion with a number of examples for various applied problems.

Definition 1: A function $f(t)$ is termed *orthogonal* to a symmetric kernel $K(x,t)$ on (a,b), if

$$\int_a^b K(x,t)f(t)dt = 0. \tag{5.98}$$

We will need the following basic result, where it can be shown that "a square integrable function $f(x)$ on (a,b) is orthogonal to a symmetric kernel $K(x,t)$ if and only if it is orthogonal to all eigenfunctions $\{\phi_n(x)\}$ of the kernel as defined in (5.35),

$$\phi_n(x) = \lambda_n \int_a^b K(x,t)\phi_n(t)dt, \quad K(x,t) = \overline{K(t,x)}\text{''}. \tag{5.99}$$

Also we may repeat the definition of the *null function n(x)*, which is the function that has its (square) norm vanish on the indicated interval (a, b),

$$\int_a^b n^2(x)dx = 0. \tag{5.100}$$

Now we define the special class of symmetric kernels that would allow the simple statement of Theorem 7 for the existence of the unique solution of Fredholm integral equations of the first kind (5.97). This is the class of *closed symmetric* kernels.

Definition 2: Closed Symmetric Kernels

The symmetric kernel $(K(x,t) = \overline{K(t,x)})$ that is orthogonal to no other function but the null function $n(x)$, is called *closed.* With the result following (5.98), this definition says that, for a closed symmetric kernel, there is no function other than the null function which can be orthogonal to all the eigenfunctions of such a kernel. This statement is another definition for such a set of eigenfunctions to be called *complete*, and we remind that we are using the square norm, as we did in (5.32) of Section 5.2.1. The following is a limited version of *Picard's theorem*, which is stated without proof, but will be clearly illustrated in Example 18.

Theorem 7: Existence of a Unique Solution to Fredholm Equation of the First Kind – with Closed Symmetric Kernel

"The Fredholm integral equation of the first kind (5.97) with a *closed symmetric* kernel has a unique (square integrable) solution if and only if the following series

$$\sum_{n=1}^{\infty} |\lambda_n a_n|^2 \tag{5.101}$$

converges, where $\{\lambda_n\}$ are the eigenvalues of the kernel $K(x,t)$ (as indicated in (5.99), and the a_n are the Fourier coefficients of the given function $f(x)$ on the interval (a, b) in terms of the *orthonormal* eigenfunctions of the kernel as given in (5.31), (5.32) and (5.29),

$$a_n = \int_a^b f(x)\overline{\phi_n(x)}dx, \tag{5.102}$$

$$f(x) = \sum_{n=1}^{\infty} a_n\phi_n(x).\text{''} \tag{5.103}$$

Also, as it shall become clear from the illustration in Example 18, the important condition of the convergence of the series in (5.101) is necessary for the class of square integrable solutions $u(x)$ of (5.97) to have the Fourier series representation

$$u(x) = \sum_{n=1}^{\infty} \lambda_n a_n \phi_n(x) \tag{5.104}$$

in terms of the orthonormal eigenfunctions of the kernel of (5.97). If we compare (5.104) and (5.103) for the Fourier coefficients $b_n = \lambda_n a_n$ and a_n of $u(x)$ and $f(x)$, respectively, we find the condition $b_n = \lambda_n a_n$, which is what we used in (5.43) for illustrating the theorem (Theorem 5) presented in Section 5.2 for the existence of the solution to the Fredholm integral equation of the first kind, which was illustrated in Example 8. Thus, while we have to know λ_n and a_n to check the condition (5.101) for the existence of the solution $u(x)$, the same product $\lambda_n a_n$ provides us with a method of constructing such solution in a form of the Fourier series (5.104) for the Fredholm integral equation of the first kind (5.97). This is a relief for having a method of solving (5.97), and it is a main advantage, of having the *complete* orthonormal set of eigenfunctions of the (special) *closed symmetric* kernel, for constructing such solution (5.104). This is especially true when we know that, in general, integral equations of the first kind are denied the well known *simple iterative* method used for the equations of the second kind. Next, we will make some general comments and give an illustration of the theorem in Example 18. As we remarked earlier, having a closed symmetric kernel in (5.97) means that we can work with its complete orthonormal set of eigenfunctions $\{\phi_n(x)\}$ on (a, b). With such completeness, we feel at ease writing a Fourier series expansion in terms of such functions for any square integrable function on the interval (a, b). In that vein of constructing the square integrable solution (5.104) in terms of such Fourier series (with special condition on its coefficients), we need the following Riesz-Fisher theorem, which we state without a proof.

Riesz-Fisher Theorem: "If $\{u_n(x)\}$ is a given orthonormal set of functions that are defined and integrable along with their square $|u_n(x)|^2$ on (a, b), and if $\{c_n\}$ is a given sequence such that $\sum_{n=1}^{\infty} |c_n|^2$ converges. Then there exists a unique function $f(x)$, integrable together with its square $|f(x)|^2$ on (a, b) for which $\{c_n\}$ are the Fourier coefficients of its Fourier series in terms of the complete set of eigenfunctions $\{u_n(x)\}$,

$$f(x) = \sum_{n=1}^{\infty} c_n u_n(x), \tag{5.105}$$

$$c_n = \int_a^b f(x) u_n(x) dx \tag{5.106}$$

and to which the Fourier series (5.105) converges *in the mean*, i.e.

$$\lim_{N\to\infty} \int_a^b \left| f(x) - \sum_{n=1}^{N} c_n u_n(x) \right|^2 dx = 0.'' \tag{5.107}$$

A very relevant comment on the present discussion is how condition (5.101) restricts the class of functions $f(x)$ (for a given symmetric kernel) in (5.97) for this equation of the first kind to have a solution. This is especially when we know that the eigenvalues λ_n are increasing, as was illustrated in Example 7 with $\lambda_n = n^2\pi^2$.

So, $f(x)$ must have coefficients a_n that are decaying fast enough to make the series in (5.101) with its nth term $\lambda_n a_n$ converge. Such restriction should be borne in the mind of anyone that wants to give a simple example of a Fredholm integral equation of the first kind. This is so true, since for a casually given function $f(x)$ in (5.97) the solution $u(x)$ may not exist! This will be illustrated in the following Example 18 for the two simple functions used in Example 8, namely, $f(x) = x$ and $f(x) = \frac{1}{2}(x - x^2)$ on the interval $(0, 1)$. We will show that, according to the condition (5.101), a solution to (5.97) does not exist for the first case with the function $f(x) = x$, while it does exist for the second case with the function $\frac{1}{2}(x - x^2)$. Moreover we can construct this latter solution via its Fourier series as in (5.104).

Example 18 On the Existence of a Unique Solution to Fredholm Equation of the First Kind

(a) To illustrate how difficult it is to satisfy the (necessary and sufficient) condition (5.101) we may consider the simple example with $f(x) = x$, $0 < x < 1$,

$$x = \int_0^1 K(x,t)u(t)dt \tag{E.1}$$

where $K(x,t)$ is a symmetric kernel which we used in Example 8,

$$K(x,t) = \begin{cases} x(1-t), & 0 \le x \le t \\ t(1-x), & t \le x \le 1 \end{cases} \tag{E.2}$$

and where we can secure from that example its orthonormal eigenfunctions $\{\phi_k(x)\}_{K=1}^{\infty} = \{\sqrt{2}\sin k\pi x\}$ and note its (clearly increasing!) eigenvalues $\{\lambda_k\}_{k=1}^{\infty}$ $=\{k^2\pi^2\}_{k=1}^{\infty}$. We also note that this set of eigenfunctions is complete on the interval $(0, 1)$ of the closed (symmetric) kernel $K(x,t)$ of (E.2).

We will show here that a solution to (E.1) does not exist. This is so, since as we force it on (E.1), we may write the Fourier series for $f(x) = x$ on $(0, 1)$ in terms of the above eigenfunctions, according to (5.104), (5.103), as

$$x = \sum_{k=1}^{\infty} a_k \sqrt{2} \sin k\pi x, \qquad 0 < x < 1 \tag{E.3}$$

$$a_k = \int_0^1 x\sqrt{2}\sin k\pi x dx = (-1)^{k+1}\frac{\sqrt{2}}{k\pi} \tag{E.4}$$

where the above integral for the Fourier coefficients a_k is done with one simple integration by parts;

$$x = \frac{\sqrt{2}}{\pi}\sum_{k=1}^{\infty}\frac{(-1)^{k+1}}{k}\sin k\pi x, \qquad 0 < x < 1. \tag{E.5}$$

We note here that (E.3) to (E.5) are all fine since the eigenfunctions are complete[8] on $(0, 1)$, and $f(x) = x$ is square integrable on $(0, 1)$, i.e., $\int_0^1 x^2 dx = \frac{1}{3}$, so this function

[8] See (4.47), (4.48), and (4.52) and the discussion immediately following (4.52).

is entitled to its Fourier sine series representation in (E.5), which does converge in the mean to $f(x) = x$ on $(0,1)$. The problem arises as soon as we look at (E.1), where we see clearly that we are forcing a solution $u(x)$ for it, which does not exist, as the violation of condition (5.101) will indicate.

For (5.101), we have now $a_k = \frac{(-1)^{k+1}\sqrt{2}}{k\pi}$ and $\lambda_k = k^2\pi^2$, so

$$\sum_{k=1}^{\infty} |\lambda_k a_k|^2 = \sum_{k=1}^{\infty} \left| k^2\pi^2 \frac{(-1)^{k+1}\sqrt{2}}{k\pi} \right|^2 = 2\pi^2 \sum_{k=1}^{\infty} k^2 \tag{E.6}$$

which is a divergent series. But since (5.101) is a necessary and sufficient condition for the existence of the solution to (5.97), we easily conclude the non-existence of such solution to (E.1). Another way of showing this negative result for (E.1) is to force a Fourier series representation for the (assumed) solution $u(x)$, then find that (E.1) implies that such series diverges, which we leave for an exercise (see Exercise 1).

From this illustration for Theorem 7, we should learn that before embarking on solving a Fredholm integral equation of the first kind we must first have the eigenvalues $\{\lambda_n\}$ of its symmetric kernel, then we proceed to find the Fourier coefficients $\{a_n\}$ of the Fourier series expansion of the given function $f(x)$ in terms of the orthonormal eigenfunctions of the kernel. Then it is a matter of the condition on the product

$$|\lambda_n a_n| = O\left(\frac{1}{n^k}\right) \quad \text{i.e., } of\ the\ order\ \frac{1}{n^k}, \quad k > \frac{1}{2} \tag{E.7}$$

for the series (5.101) to converge. In the above example we can see that it is not the case since

$$|\lambda_n a_n| = \sqrt{2}\pi n = O(n) \tag{E.8}$$

where in this case $k = -1$, and its corresponding series (5.101) clearly diverges.

(b) In the following we will consider the Fredholm integral equation of the first kind (5.97) for the above problem (E.1) with the same symmetric kernel (E.2) of (E.1), except that we have here $f(x) = \frac{1}{2}(x - x^2)$.

$$\frac{1}{2}(x - x^2) = \int_0^1 K(x,t)u(t)dt. \tag{E.9}$$

As we did in Example 8, we first write the Fourier sine series for $f(x) = \frac{1}{2}(x - x^2)$, on $(0,1)$,

$$\frac{1}{2}(x - x^2) = \sum_{n=0}^{\infty} \frac{2\sqrt{2}}{\pi^3(2n+1)^3}\sqrt{2}\sin(2n+1)\pi x, \quad 0 < x < 1 \tag{E.10}$$

where the Fourier coefficients are easily computed, using integration by parts, from its Fourier coefficients integral as given in (5.103) with $\phi_n(x) = \sqrt{2}\sin n\pi x$,

$$a_n \equiv \begin{cases} a_{2n+1} = \int_0^1 \frac{1}{2}(x - x^2)\sqrt{2}\sin(2n+1)\pi x dx = \dfrac{2\sqrt{2}}{\pi^3(2n+1)^3} \\ a_{2n} = 0 \end{cases} . \tag{E.11}$$

Recalling that the eigenvalues are $\lambda_n = n^2\pi^2$, we have for condition (5.101)

$$|a_{2n+1}\lambda_{2n+1}| = \left| \frac{2\sqrt{2}}{\pi^3(2n+1)^3}\pi^2(2n+1)^2 \right| = O\left(\frac{1}{n}\right) \tag{E.12}$$

and the series in (5.101) converges since $k = 1 > \frac{1}{2}$ in (E.7). Indeed, the sought solution to (E.9) is $u(x) = 1, 0 < x < 1$, as can be verified after simple integration (see Exercise 2(a)). As a matter of fact, and as we did for Example 8, a practical way of making an example, for a Fredholm integral equation of the first kind that does have a solution, is to plug in a known function as a solution $u(x)$ inside the integral, and find the result of the integral as $f(x)$ to be used for the example as a sure thing to guarantee the solution to the problem. On the other hand, once we have λ_n for the kernel and a_n for $f(x)$ of the equation of the first kind (5.97), we first use $\lambda_n a_n$ in (5.101) to see whether a solution does exist, and if so we use the same $\lambda_n a_n$ in (5.104) to construct that solution as a Fourier series in terms of the eigenfunctions $\{\phi_n(x)\}$ of the kernel with coefficients $b_n = a_n\lambda_n$.

5.4.2 Ill-Posed Problems and the Fredholm Equation of the First Kind

As we remarked in Section 1.5, in the practical applications we often resort to approximate or numerical methods for solving linear systems, and in particular integral equations. For such systems, it is desirable that a small error in the given data of the system causes a correspondingly small error in the output as the desired solution. In other words we would like to see that the solution (output) depends in a *continuous* way on the input (given data), and such a system is termed a *stable* system. So it is of utmost importance for a stated problem to represent a stable system, especially when very complex (expensive) computations are to be involved. In most introductory treatments with some theoretical touch, we usually emphasize the fact that we should insure the *existence* of the solution before we go after it, moreover such a solution better be *unique* for us to focus on it as the only useful one. These two concepts, of *existence* and *uniqueness* of the solution, were the most important to guarantee *the classical solution*, via very well known theorems, where powerful analytical methods are used. With the advent of contemporary complex problems and the urgency for their solutions, powerful approximate and numerical methods had to be utilized, of course, with an awareness for the inevitable practical error in the data. As a consequence, the above mentioned *stability* condition is now added to the previous *existence* and *uniqueness* of the solution. Indeed such three conditions were postulated by Hadamard for initial and boundary value problems. The stability condition is motivated by the fact that in a physical system, the input is a measured data, and we want to make sure that a small inaccuracy in this data (error in the input) will cause only a small error in the output as the solution of the problem.

A problem stated with the assurance of the *existence, uniqueness and stability* of its solution is termed a *well-posed problem*, otherwise it is *ill-posed.* A typical example of a well-posed problem is that of the potential distribution u in a disc due to given input potential $u = f$ on its rim that we presented in (1.24), where we can prove the existence and uniqueness of the solution (potential) u in the interior of the disc. For now our physical intuition suggests that such solution u depends continuously on the data f at the boundary, i.e., it is a stable problem. This example is to be differentiated from the one that we shall present in Example 19, which is due to Hadamard, where we give the potential as well as its gradient on the boundary, and which illustrates the earliest analytical example of an ill-posed problem.

Another example is the solution of the temperature distribution in a bar with given initial temperature (data), and boundary conditions. Again it can be proved that a solution in the interior (temperature $u(x,t)$ for, $t > 0$; $0 < x < l$), exists, and it is unique. Also, on physical grounds we can see that a small change in the initial temperature causes only a small change in the temperature in the interior. Definitions and theorems are introduced to prove these results but they are beyond the scope of this book, the interested reader may consult the available references on the subject[9] For us, we may look at the input-output problem symbolically as with operator notation, without going in depth to the theorems in the above references. However, we may give a descriptive, though not so precise, notion of some of their results, which will be followed by a specific clear illustration in Example 20, and a discussion of the ill-posedness of Fredholm integral equations of the first kind. For example, consider the operator equation,

$$A\phi = f \tag{5.108}$$

where A is an operator, say the integral operator in the Fredholm equation of the first kind,

$$A\phi \equiv \int_a^b K(x,t)\phi(t) = f(t) \tag{5.109}$$

mapping the desired solution ϕ as an element of (an acceptable) space of functions X into f, as an element of another space Y of the same type functions,

$$A : X \to Y. \tag{5.110}$$

The idea of well-posedness will depend on the existence of an inverse operator A^{-1} that will return $f \epsilon Y$ to $\phi \epsilon X$,

$$A^{-1}f = \phi, \tag{5.111}$$

and thus obtaining the solution of the integral equation (5.109).

For a small change in the input f to cause only a small change in the output (solution) ϕ, or in other words, the continuous dependence of ϕ on the data f, means

[9]See Kress [1989] and Weinberger [1965].

that this inverse operator should be continuous. Unfortunately, in general, and for a large class of such operators, this may not be the case. This would mean that, according to (5.111), a small change in f may cause a very large change in ϕ, and the problem becomes ill-posed. Such a situation is familiar to us, where we may have the linear system of n equations in the n unknowns of the (column) matrix U,

$$AU = F \tag{5.112}$$

where A is the known n by n matrix of the coefficients, and F is the known column matrix. However, when it comes to solving for U in (5.112) A may not have an inverse, whose simple check is when its determinant $|A|$ vanishes. Another situation is that of the heat equation (see Exercise 2) which is stable as the heat is a diffusive process, and a small change in the initial temperature will not cause big a change in the (diffusing) output, which can be described as "forgetting its past". However, the inverse heat problem of knowing the temperature now, and we are to find the initial temperature, which is called the *inverse*, or *backward heat equation* is an ill-posed problem. In physical terms this means that the heat diffusion is an irreversible physical process. In the following Example 19 we will use a very well known example due to Hadamard to illustrate what we mean by an ill-posed problem. It will be followed by a discussion of the ill-posedness of Fredholm integral equations of the first kind.

Example 19 An Ill-Posed Problem-Hadamard's Example[10]

Here we will illustrate that the following potential distribution problem is ill-posed in the sense of Hadamard. Consider the boundary value problem for the potential distribution $u(x, y)$ in two dimensional space (upper half plane), which is free of charge, it is governed by the *Laplace equation* in the interior,

$$\nabla^2 u(x,y) = \frac{\partial^2 u}{\partial x^2} + \frac{\partial^2 u}{\partial y^2} = 0, \quad -\infty < x < \infty, \quad y > 0. \tag{E.1}$$

The given boundary conditions are a grounded lower edge $y = 0$,

$$u(x, 0) = 0, \qquad -\infty < x < \infty \tag{E.2}$$

and where the gradient $\frac{\partial u}{\partial y}$ of the potential is given at the same edge $y = 0$,

$$\frac{\partial u(x,0)}{\partial y} = f(x), \qquad -\infty < x < \infty \tag{E.3}$$

where $f(x)$ is a continuous function. Hadamard's example is for the choice of the data $f(x)$ as the particular sequence

$$f_n(x) = \frac{\sin nx}{n}, \qquad -\infty < x < \infty. \tag{E.4}$$

[10]Optional

We can easily show that the sequence

$$u_n(x,y) = \frac{1}{n^2}\sin nx \sinh ny \tag{E.5}$$

is a solution to the boundary value problem (E.1), (E.2) and (E.3) with $f(x) = f_n(x)$ as in (E.4). Also the input $f_n(x) = \frac{\sin nx}{n}$ of (E.4) is convergent to zero as $n \to \infty$, i.e., for large n there could be only small changes in the input data of (E.4). However, the solution (output) in (E.5) (with its factor $\sinh ny = \frac{e^{ny}-e^{-ny}}{2}, y > 0$) will sustain a very large change due to the e^{ny} term for the same large n. Hence the solution $u_n(x,y)$ in (E.5) to the boundary value problem (E.1)–(E.3) and (E.4) is not stable, and the problem is ill-posed. To show that the inverse, or backward heat equation is also ill-posed, we refer the reader to Kress (1989).

The treatments and methods for a stable approximate solution of ill-posed problems are called *regularization methods*. Briefly, and to use operator notation, the operator A of the ill-posed problem $A\phi = f$ is replaced by one (or a family) of a bounded operator R_α such that for the perturbed data $f^\delta \equiv f + \delta f$ of f with a known error $|f^\delta - f| \le \delta$, the (resulting perturbed) solution ϕ^δ, corresponding to this perturbed data, is a reasonable approximation of the actual solution ϕ i.e. ϕ^δ depends continuously on f^δ. A detailed treatment with powerful theorems, that describe such regularization methods, is found in Kress (1990).

Ill-Posedness of Fredholm Integral Equations of the First Kind

What concerns us in this section is that Fredholm integral equations of the first kind can easily show the signs of ill-posedness. In Theorem 7, as a special case of Picard's theorem, we established a unique solution for Fredholm equations of the first kind,

$$f(x) = \int_a^b K(x,t)u(t)dt, \quad K(x,t) = \overline{K(t,x)} \tag{5.97a}$$

for the *closed* symmetric kernel $K(x,t)$ as

$$u(x) = \sum_{n=1}^{\infty} a_n \lambda_n \phi_n(x) \tag{5.104}$$

where λ_n and ϕ_n are the eigenvalues and eigenfunctions, respectively, of the symmetric kernel, and a_n is the Fourier coefficients of $f(x)$ in terms of such a (complete) set of orthonormal eigenfunctions,

$$a_n = \int_a^b f(x)\overline{\phi_n(x)}dx, \tag{5.102}$$

$$f(x) = \sum_{n=1}^{\infty} a_n \phi_n(x). \tag{5.103}$$

The necessary and sufficient condition of Theorem 7 for the existence of such a solution is that the series $\sum_{n=1}^{\infty} |\lambda_n a_n|^2$ converges. This sounds very fine as far as the two desired qualities of *existence* and *uniqueness* of the solution to our problem of the Fredholm equation of the first kind. What remains, for the present discussion, is the third quality of the *stability* of the solution for the problem to be the desired and acceptable well-posed problem. Unfortunately, from the solution $u(x)$ in (5.104) we can show that the problem is not stable. This is the case, since if we perturb the given data $f(x)$ by a small $\delta f(x)$, the solution $u(x)$ in its Fourier series representation in (5.104) will not be perturbed by what we wish, a Fourier series representation of $\delta f(x)$ (or a constant multiple of it), but some *magnification*, i.e., a much larger corresponding change δu in $u(x)$. This, as we shall see shortly, is due to the eigenvalues λ_n factor in (5.104), where they are increasing. If we write (5.104), using a_n as in (5.102), we have

$$u(x) = \sum_{n=1}^{\infty} \lambda_n \phi_n(x) \int_a^b f(y)\overline{\phi_n(y)}dy. \tag{5.113}$$

Now if we perturb $f(x)$ by $\delta f(x)$, we substitute $f(x) + \delta f(x)$ inside the integral of (5.113) to have $u(x) + \delta u(x)$ on the left hand side,

$$\begin{aligned} u(x) + \delta u(x) &= \sum_{n=1}^{\infty} \lambda_n \phi_n(x) \int_a^b [f(y) + \delta f(y)]\overline{\phi_n(y)}dy \\ &= \sum_{n=1}^{\infty} \lambda_n \phi_n(x) \int_a^b f(y)\overline{\phi_n(y)}dy \\ &\quad + \sum_{n=1}^{\infty} \lambda_n \phi_n(x) \int_a^b \delta f(y)\overline{\phi_n(y)}dy, \end{aligned} \tag{5.114}$$

$$\delta u(x) = \sum_{n=1}^{\infty} \lambda_n \epsilon_n \phi_n(x), \quad \epsilon_n = \int_a^b \delta f(y)\overline{\phi_n(y)}dy \tag{5.115}$$

where ϵ_n is the above Fourier coefficient of the small perturbation $\delta f(x)$, and where we recognize the first series in (5.114) to represent $u(x)$ as in (5.104) to cancel $u(x)$ from both sides of the equation (5.114). In (5.115) we see that while ϵ_n is the Fourier coefficient of the small perturbation of the input $\delta f(x)$, the Fourier coefficients of the corresponding perturbation for the output $\delta u(x)$, is *magnified* by the multiplicative *increasing* factor λ_n, which clearly will cause $\delta u(x)$ to be a larger change compared to the given small change $\delta f(x)$. Hence, there seems to be no continuous dependence of the solution $u(x)$ on the given data $f(x)$ in the Fredholm equation of the first kind (5.97). This possible ill-posedness of the Fredholm equation of the first kind adds to other difficulties for the existence of solutions for problems with more general kernel than the above symmetric one that we limited ourselves to in all our discussions up to now. Then, it is no wonder that Fredholm integral equations of the first kind are in the forefront with regard to the research priority in integral equations.

Exercises 5.4

1. Consider the Fredholm integral equation of the first kind

$$x = \int_0^1 K(x,t)u(t)dt \qquad (E.1)$$

with a symmetric kernel as was considered in Example 18. Follow steps (i)–(iii) to show, as in Example 18, that a solution does not exist for this equation.

(i) Assume a Fourier series representation for the (not so sure!) solution $u(x)$ in terms of the eigenfunctions of the kernel,

$$u(x) = \sum_{k=1}^{\infty} b_k \sqrt{2} \sin k\pi x. \qquad (E.2)$$

(ii) Substitute this $u(x)$ in the integral of (E.1), interchange the summation with integration as though the quality of the convergence of the series (E.2) allows that.

Hint: For the integration inside the series involving the kernel, use the fact that $\sqrt{2}\sin k\pi x$ are the eigenfunctions of the kernel as described in (5.35) or (5.29).

(iii) Write a similar Fourier series for $f(x) = x$ on (0,1) and use in (E.1), then compare coefficients, where you find that $b_k = \pi\sqrt{2}(-1)^{k+1}k$ which makes the (assumed) Fourier series for the solution in (E.2) divergent. Thus, there exists no solution to (E.1).

2. Consider the Fredholm integral equation of the first kind (E.1) of Example 8 in Section 5.2. (This is the same problem as in Exercise 4 of Section 5.2.)

(a) Verify that $u(x) = 1$, $0 < x < 1$ is a solution to this problem. *Hint:* Watch for the two branches of the kernel $K(x,t)$; write the integral on the two subintervals $(0, x)$ and $(x, 1)$.

(b) Write the Fourier series for the solution $u(x) = 1, 0 < x < 1$ (of part (a)) and the given function $f(x) = \frac{1}{2}(x - x^2)$, $0 < x < 1$ in terms of the eigenfunctions of the kernel, to verify $b_k = \lambda_k a_k$ in (5.43) (and (5.43a)).

(c) Verify that for the function $f(x) = \frac{1}{2}(x - x^2)$, $0 < x < 1$ in (E.10), the Hilbert-Schmidt theorem is satisfied.

Hint: Note that $f(x) = \frac{1}{2}(x - x^2)$ is continuous on $(0, 1)$, and that the clearly symmetric kernel $K(x,t)$ in (E.2) is square integrable on the square $\{x\epsilon(0,1), t\epsilon(0,1)\}$. (See the hint to part (a).)

3. **(a)** Show that a Fredholm integral equation of the first kind,

$$f(x) = \int_a^b K(x,t)u(t)dt \qquad (E.1)$$

with degenerate kernel,

$$K(x,t) = \sum_{k=1}^{n} a_k(x) b_k(t) \tag{E.2}$$

does not have a solution unless the given function $f(x)$ is restricted to a linear combination of the functions $a_k(x)$,

$$f(x) = \sum_{k=1}^{n} c_k a_k(x).$$

(b) Consider a Fredholm integral equation of the first kind,

$$f(x) = \int_a^b K(x,t) u(t) dt \tag{E.1}$$

with continuous kernel $K(x,t)$ and continuous $f(x)$. Would we necessarily search for a continuous solution $u(t)$?

(c) Assume that $f(x)$ and the solution $u(x)$ of (E.1) have each a Fourier series expansion in terms of $\{\phi_n(x)\}$, the set of eigenfunctions of the kernel $K(x,t)$ in (E.1). What restriction on the Fourier coefficients of $f(x)$, and hence $f(x)$, would that entail?

4. Consider the Fredholm integral equation of the first kind,

$$f(x) = \int_a^b K(x,t) u(t) dt \tag{E.1}$$

where $K(x,t)$ is continuous, real and symmetric.

Assume that $K(x,t)$ has only finite number of eigenfunctions, $\{\phi_1(x), \phi_2(x), \cdots, \phi_n(x)\}$, show that the equation (E.1) then becomes solvable only for a restricted class of functions regardless of $u(x)$. (See Exercise 3, which is very similar).

Hint: Substitute $K(x,t) = \sum_{i=1}^{n} c_i(t)\phi_i(t)$ in (E.1) and integrate with respect to t.

5. **(a)** Illustrate problem 4 for the example,

$$f(x) = \int_0^{2\pi} \sin(x+t) u(t) dt \tag{E.1}$$

according to the following instructions in (i) - (ii).

(i) Solve for the eigenvalues and eigenfunctions of $K(x,t) = \sin(x+t)$, and

(ii) Show that any functions $f(x)$ of the form in (E.1) is restricted to the linear combination of the (two) eigenfunctions found in part (a).

(b) Show that the solution to (E.1) is also not unique.

Hint: If you add to the solution a function $g(x)$, which is orthogonal to the eigenfunctions $\phi_1(x)$ and $\phi_2(x)$ of (E.1), it will still be a solution.

(c) In general, when can the solution of a Fredholm integral equation of the first kind with symmetric kernel, be unique?

6. **(a)** Consider the integral equation of the first kind in $K(\alpha, x)$,

$$\int_0^\pi K(\alpha, x) \sin \omega x dx = F(\omega) \tag{E.1}$$

and let $K(\alpha, x)$ be square integrable on $(0, \pi)$ for particular values of α. Assume the very well known Riemann-Lebesgue lemma

$$\lim_{\omega \to \infty} \int_0^\pi K(\alpha, x) \sin \omega x dx = 0 \tag{E.2}$$

and use it to show that this result (E.2) illustrates the $ill - posedness$ of the equation of the first kind (E.1).

Hint: See that for large values of ω, $F(\omega)$ and so is its change $\delta F(\omega)$ will be *small*, however the solution $K(\alpha, x)$ maybe piecewise continuous in x with $large$ jump discontinuities.

(b) Consider the Laplace transform, of the piecewise continuous and of exponential order $f(t)$, $(f(t) = o(e^{\alpha t}))$,

$$F(s) = \int_0^\infty e^{-st} f(t) dt, \quad s > \alpha. \tag{E.3}$$

Show that $\lim_{s \to \infty} F(s) = 0$, and use this result to comment about the well-posedness of the singular Fredholm equation of the first kind (E.3) in $f(t)$. *Hint:* See part (a).

7. Show that for the integral transform $f(x)$ of $u(t)$ (or the integral equation of the first kind in $u(t)$), with continuous kernel $K(x, t)$,

$$f(x) = \int_a^b K(x, t) u(t) dt. \tag{E.1}$$

(a) If $u(t)$ is piecewise continuous, then $f(x)$ is continuous.

(b) Based on the result in (a), can we guess at only a continuous solution $u(t)$ for the problem of the first kind (E.1), when $f(x)$ is continuous?

(c) If the solution $u(t)$ in (E.1) is considered as the output corresponding to the input (or data) $f(x)$ of the system represented by the equation (E.1), what is the consequence of the result in part (b) on the well-posedness of the solution for such Fredholm integral equation of the first kind?

8. Consider the Green's function of the loaded string $G(x,t)$ as in the hanging chain problem (2.28). On physical grounds show that

$$\int_0^l G(x,t)f(t)dt = 0$$

has only the trivial solution $f(t) \equiv 0$.

5.5 NUMERICAL SOLUTION OF FREDHOLM INTEGRAL EQUATIONS

In the preceding section we illustrated the many different exact and approximate methods for solving integral equations using special examples that needed moderate amounts of work. For more general cases we sometimes resorted to approximate methods where one integral equation is approximated by another which can be handled by the usual methods illustrated. When both approaches do not apply, we may have to resort to the numerical method of approximating the integral by a finite sum, and hence the integral equation is approximated by a set of simultaneous equations whose number is determined by the number of values or samples of the approximate solution $u(x_i)$ on the desired interval.

In this section we will first remind of the most basic numerical integration formulas such as the trapezoidal and Simpson's rule that we have already discussed in Section 1.5 in (1.141) and (1.144), respectively. Then we will prepare for the numerical approximation setting of Fredholm integral equations of the second kind, and where both the trapezoidal rule and Simpson's rule will be used for approximating the integration term in the equation. Such an approximation setting becomes a (square) set of $n+1$ linear equations in the $n+1$ (approximate) samples of the solution $u(x_i), i = 0, 1, 2, \cdots, n$. This preparation will be concluded by an example where the approximate numerical values are compared with the exact solution of a simple Fredholm integral equation (see Example 20 and Exercises 1,2 and 3.) In this section we will concentrate on using only the very basic integration formulas such as the trapezoidal rule and the Simpson's rule. As we emphasized in Section 1.5, the higher quadrature rules, and their use in approximating the integral, for the numerical solutions of Fredholm integral equations, is relegated to Section 7.3 of Chapter 7. The treatment there is supported with the necessary tables, and a good number of very detailed examples and exercises.

We will also have a chance to make some comments concerning the numerical solution of a particular class of singular Fredholm integral equations. These are the ones characterized by their infinite limit (or limits) of integration.

5.5.1 Numerical Approximation Setting of Fredholm Integral Equations

After introducing the basic numerical integration rules in Section 1.5, we are now in a position to discuss the numerical setting of Fredholm integral equations. We will first consider the Fredholm equation of the second kind.

Consider the Fredholm integral equation of the second kind, as we used it in (1.148) in Section 1.5.1,

$$u(x) = f(x) + \int_a^b K(x,t)u(t)dt, \qquad (5.116), (1.148)$$

where we approximated its integral by a sum as in (1.149),

$$S_n(x) = \sum_{j=0}^{n} K(x,t_j)u(t_j)\Delta_j t. \qquad (1.149)$$

As was indicated in Section 1.5, we usually use equal increment Δt instead of the above more general $\Delta_j t$. Here j as the index in $\Delta_j t$ may indicate a *weight* D_j assigned to the ordinates $K(x,t_j)u(t_j)$ (of the integrand) by the particular numerical integration rule that we discussed and illustrated for the trapezoidal rule (1.141) and Simpson's rule (1.144).

With the approximation to the integral in (1.149), we have the approximate result to the Fredholm integral equation (1.148)

$$u(x) \approx f(x) + \sum_{j=0}^{n} K(x,t_j)u(t_j)\Delta_j t. \qquad (5.117)$$

Now, it becomes clear that if we are to solve for approximate sample values $u(x_j)$ of the solution $u(x)$, we may require (5.117) to be an equality at the $n+1$ locations x_i, $i = 0,1,2,\cdots,n$ of the (approximate) sample values $u(x_i)(= u(t_i))$, $i = 0,1,2,\cdots,n$,

$$u(x_i) = f(x_i) + \sum_{j=0}^{n} K(x_i,t_j)u(t_j)\Delta_j t, \quad i = 0,1,2,3,\cdots,n. \qquad (5.118)$$

With such "forcing" of the approximation (5.117) to the equality (5.118), it should be clear that the $\{u(x_i)\}$ in (5.117) are only approximations to the solution $u(x)$ of the integral equation (5.116) at $\{x_i\}$, and they should really be designated differently. In (5.118) we see that the (linear) Fredholm integral equation (5.116) is approximated by a system of $n+1$ *linear* equations in the (approximate) samples of its solution $u_i \equiv u(x_i)$, $i = 0,1,2,3,\cdots,n$. This should definitely remind us of a matrix equation, whereby we can rely on our knowledge of solving systems of linear algebraic equations with the help of matrix analysis, and more importantly our dependence on its theory for the existence of such sought solution. Indeed, the strong relation between matrix theory and the theory of linear Fredholm integral equations goes a long way to Fredholm's original work on linear integral equations, as it became abundantly clear in the first few sections of this Chapter, where such theory is developed. If we use the notation $u_i = u(x_i)$, $f_i = f(x_i)$, $K_{ij} = K(x_i, t_j)$, where clearly $U = [u_i]$, $F = [x_i]$ are column matrices while $\mathcal{K} = [K_{ij}]$ is an $n + 1$ by $n + 1$ square matrix, we can rewrite (5.118) as a *matrix equation*,

$$U = F + D\mathcal{K}U. \tag{5.119}$$

where $D = [D_i \delta_{ij}]$ is a diagonal matrix of order $n + 1$, and δ_{ij} is the Kronecker delta. So in matrix notation we are after the unknown column matrix U,

$$IU - D\mathcal{K}U = F,$$

$$[I - D\mathcal{K}]U = F \tag{5.120}$$

where $I = [\delta_{ij}]$ is the unit (square) matrix of order $n + 1$. If the inverse $[I - D\mathcal{K}]^{-1}$ of the matrix $[I - D\mathcal{K}]$ on the left of (5.120) exists, we have

$$U = [I - D\mathcal{K}]^{-1}F \tag{5.121}$$

as the solution of the *approximate* sample values $u(x_i)$, $i = 0, 1, 2, 3, \cdots, n$ of the Fredholm integral equation (5.116). From matrix theory we know that the inverse of a square matrix A exists if its determinant $|A|$ does not vanish. So a unique solution to our system of equations in (5.120) exists if $|I - D\mathcal{K}| \neq 0$. On the other hand if $|I - D\mathcal{K}| = 0$, the system in (5.120) has infinite solutions or no solutions. To be more explicit, we will attempt in the following illustrations to set up the numerical approximation of the Fredholm equation of the second kind, the Volterra equation of the second kind case was covered in Section 3.3. For the convenience of the reader we have included in Section 1.5.4 a brief review of Cramer's rule to be used for solving the above system of equations in (5.120).

Nonhomogeneous Fredholm Equations of the Second Kind

Let us consider again the Fredholm equation of the second kind

$$u(x) = f(x) + \int_a^b K(x,t)u(t)dt. \tag{5.116}$$

We subdivide the interval (a, b) into n equal increments $\Delta t = (b - a)/n$ and we call $t_0 = a$, $t_j = a + j\Delta t = t_0 + j\Delta t$; since we will be using either t or x as our variable, we will call $x_0 = t_0 = a$, $x_n = t_n = b$, and $x_i = x_0 + i\Delta t$ (or in short $x_i = t_i$). We will refer to the known function values at x_i as $f(x_i) = f_i$, the value of the kernel $K(x, t)$ at (x_i, t_j) as $K(x_i, t_j) = K_{ij}$, and the (approximate!) values of the unknown function $u(x)$ at x_i or t_i as $u(x_i) = u_i$ or $u(t_i) = u_i$.

Numerical Integration with the Trapezoidal Rule

So if we use the trapezoidal rule (1.141) to approximate the integral of (5.116), we have

$$\begin{aligned} u(x) &= f(x) + \int_a^b K(x,t)u(t)dt \approx f(x) + \Delta t \left[\frac{1}{2}K(x,t_0)u(t_0) \right. \\ &\left. + K(x,t_1)u(t_1) + \cdots + K(x,t_{n-1})u(t_{n-1}) + \frac{1}{2}K(x,t_n)u(t_n) \right] \end{aligned} \tag{5.122}$$

or

$$\begin{aligned} u(x) &\approx f(x) + \Delta t \left[\frac{1}{2}K(x,t_0)u_0 + K(x,t_1)u_1 \right. \\ &\left. + \cdots + K(x,t_{n-1})u_{n-1} + \frac{1}{2}K(x,t_n)u_n \right] \end{aligned} \tag{5.123}$$

where the solutions of (5.123) are approximate solutions of (5.116) since there is an error involved in replacing the integral in (5.116) by the $n+1$ sum of the trapezoidal rule. With this note, we shall from now on use the equal = sign instead of the approximate $\approx$ sign in (5.123).

If we consider $n+1$ values of $u_i = u(x_i) = u(t_i)$, $i = 0, 1, 2, 3, \cdots, n$, then (5.123) becomes

$$u_i = f_i + \Delta t \left[\frac{1}{2}K_{i0}u_0 + K_{i1}u_1 + \cdots + K_{i,n-1}u_{n-1} + \frac{1}{2}K_{in}u_n \right],$$

$$i = 0, 1, 2, \cdots, n \tag{5.124}$$

which are $n+1$ equations in u_i, the approximate solution to $u(x)$ at $x = x_i = a + i\Delta t$, $i = 0, 1, \cdots, n$.

If we transform all the terms involving the solution u_i to the left side of (5.124) leaving only the nonhomogeneous part f_i on the right side, then write all the $n+1$ equations for u_i, $i = 0, 1, 2, \cdots, n$ explicitly, we have the following $n+1$ system of equations in $u_0, u_1, \cdots, u_n$ to be solved:

$$\begin{aligned} \left(1 - \frac{\Delta t}{2}K_{00}\right) u_0 - \Delta t K_{01} u_1 \quad & -\Delta t K_{02} u_2 - \cdots - \Delta t K_{0,n-1} u_{n-1} \\ & -\frac{\Delta t}{2} K_{0,n} u_n = f_0 \\ -\frac{\Delta t}{2} K_{10} u_0 + (1 - \Delta t K_{11}) u_1 \quad & -\Delta t K_{12} u_2 - \cdots - \Delta t K_{1,n-1} u_{n-1} \\ & -\frac{\Delta t}{2} K_{1,n} u_n = f_1 \end{aligned}$$

$$\begin{aligned}
&\vdots\\
-\frac{\Delta t}{2}K_{n-1,0}u_0 - \Delta t K_{n-1,1}u_1 \quad &-\Delta t K_{n-1,2}u_2 + \cdots\\
+(1-\Delta t K_{n-1,n-1})u_{n-1} \ &- \tfrac{\Delta t}{2}K_{n-1,n}u_n = f_{n-1}\\
-\frac{\Delta t}{2}K_{n,0}u_0 - \Delta t K_{n1}u_1 \quad &-\Delta t K_{n,2}u_2 - \cdots - \Delta t K_{n,n-1}u_{n-1}\\
&+\left(1-\frac{\Delta t}{2}K_{nn}\right)u_n = f_n
\end{aligned} \tag{5.125}$$

which can be written in a matrix form as

$$[I - D_T\mathcal{K}]U = F \tag{5.126}$$

where $I = [\delta_{ij}]$, the identity matrix, $D_T = [D_j\delta_{ij}]$, the diagonal matrix representing the weights of the quadrature rule used, which is here the trapezoidal rule, as they appear in (1.141) (with $D_0 = \frac{1}{2}\Delta t$, $D_1 = \Delta t$, $D_2 = \Delta t$, $\cdots, D_n = \frac{1}{2}\Delta t$). $\mathcal{K}$ is the matrix for the kernel $\mathcal{K} = [K_{ij}]$, thus the matrix of the coefficients of the linear system in (5.125) or (5.126) is

$$A = I - D_T\mathcal{K} =$$

$$\begin{bmatrix}
1-\frac{\Delta t}{2}K_{00} & -\Delta t K_{01} & \cdots & & -\frac{\Delta t}{2}K_{0n}\\
-\frac{\Delta t}{2}K_{10} & 1-\Delta t K_{11} & \cdots & & -\frac{\Delta t}{2}K_{1n}\\
\vdots & & & \vdots & \\
-\frac{\Delta t}{2}K_{n-1,0} & \cdots & & (1-\Delta t K_{n-1,n-1}) & -\frac{\Delta t}{2}K_{n-1,n}\\
-\frac{\Delta t}{2}K_{n0} - \Delta t K_{n1} & \cdots & & & 1-\frac{\Delta t}{2}K_{nn}
\end{bmatrix} \tag{5.127}$$

U is the matrix of the solutions,

$$U = \begin{bmatrix} u_0\\ u_1\\ \vdots\\ u_n \end{bmatrix} \tag{5.128}$$

and F is the matrix of the nonhomogeneous part,

$$F = \begin{bmatrix} f_0\\ f_1\\ \vdots\\ f_n \end{bmatrix}. \tag{5.129}$$

So now we may summarize that in approximating the integral in the (linear) Fredholm integral equation by the $n+1$ terms of the trapezoidal rule, we have reduced the integral equation to a set of $n+1$ (linear) equations (5.125) in $u_0, u_1, \cdots, u_n$, or to the matrix equation (5.126) to be solved for the unknown matrix U whose elements $u_0, u_1, \cdots, u_n$ are the $n+1$ approximate samples of the solution to the integral equation (5.116) (or (1.148)). As we mentioned earlier, an obvious result from the theory of linear systems of equations regarding the solution of the matrix equation (5.126) is that there is a unique solution U to (5.126) when $|A| = |I - D_T\mathcal{K}|$, the determinant of the coefficients matrix $I - D_T\mathcal{K}$, does not vanish, and that (5.126) has infinite solutions or no solution when the determinant $|I - D_T\mathcal{K}|$ vanishes. To this end, then, it is a matter of how efficient we are in solving matrix equations and how prepared in choosing a more suitable method of numerical integration instead of the trapezoidal rule. Since this book assumes preparation only in elementary calculus and differential equations, we will not attempt to seek efficiency in our present illustrations, as our main purpose here is to introduce the subject in the clearest way possible. It is left to the readers to choose their own method of solving the resulting system of linear equations (5.125). This, however, does not prevent us from noting some special features, such as the symmetry of the kernel, which will simplify the computations. For our illustrations, and for the purpose of a more self-contained treatment, we felt it helpful to have a brief presentation in Section 1.5.4 of Cramer's rule for solving system of linear equations. Of course, one may consult other efficient methods, for example, the Gauss elimination method. The illustration of the numerical approximation of the Fredholm integral equation (5.116) when Simpson's rule (1.144) is used for approximating its integral, (and where, n is an *even* number) is left for an exercise (see Exercise 5.). In Section 7.3, of the (optional) Chapter 7, we will use higher quadrature rules for approximating the integral in (5.116), the trapezoidal rule and the Simpson's rule, used here, are only two special cases of such rules.

Example 20

Use the trapezoidal rule with $n = 2$ to set up the approximate numerical representation of a 3×3 system of linear equations in (the approximate values) $u(x_i)$, $i = 0, 1, 2$ of the following Fredholm integral equation,

$$u(x) = \sin x + \int_0^1 (1 - x\cos xt)u(t)dt. \tag{E.1}$$

With $n = 2$ we have $\Delta t = (1-0)/2 = 1/2$, so $t_i = i\Delta t = (1/2)i = x_i$. If we use the trapezoidal rule for the integral in (E.1) (with the weights D_j of (5.118) corresponding to the trapezoidal rule as in (1.141)) we have

$$u_i = f_i + \frac{1}{2}\left(\frac{1}{2}K_{i0}u_0 + K_{i1}u_1 + \frac{1}{2}K_{i2}u_2\right), \quad i = 0, 1, 2 \tag{E.2}$$

or in matrix form.

$$\begin{bmatrix} 1-\frac{1}{4}K_{00} & -\frac{1}{2}K_{01} & -\frac{1}{4}K_{02} \\ -\frac{1}{4}K_{10} & 1-\frac{1}{2}K_{11} & -\frac{1}{4}K_{12} \\ -\frac{1}{4}K_{20} & -\frac{1}{2}K_{21} & 1-\frac{1}{4}K_{22} \end{bmatrix} \begin{bmatrix} u_0 \\ u_1 \\ u_2 \end{bmatrix} = \begin{bmatrix} \sin 0 \\ \sin \frac{1}{2} \\ \sin 1 \end{bmatrix} \qquad (E.3)$$

Now if we substitute for $f_i = f(x_i) = \sin(i/2)$ and $K_{ij} = K(x_i, t_j) = 1 - (i/2)\cos(ij/4)$ in (E.3), we obtain

$$\begin{bmatrix} 0.75 & -0.5 & -0.25 \\ -0.125 & 0.741 & -0.140 \\ 0 & -0.061 & 0.885 \end{bmatrix} \begin{bmatrix} u_0 \\ u_1 \\ u_2 \end{bmatrix} = \begin{bmatrix} 0 \\ 0.479 \\ 0.842 \end{bmatrix} \qquad (E.4)$$

as a simple matrix equation $AU = F$ to be solved for the approximate value u_0, u_1 and u_2 of the integral equation (E.1). It should be on our mind to check that the determinant $|A|$ does not vanish for the system (E.4) to have a unique solution. We shall leave (the rather lengthy!) details of finding the final numerical solution of (E.4) to Exercise 2. The result of such approximate values are $u_0 = 1.013$, $u_1 = 1.009, u_2 = 1.021$, which compare very well with the exact solution $u(x) = 1$. Also they compare well with the approximate values of Example 6 in Section 5.1 as shown in Table 5.1, $u_0 = 1.003$, $u_1 = 1.002$, $u_2 = 1.009$. We shall return to the numerical methods of solving Fredholm integral equations in Section 7.3 of the (optional) Chapter 7 where the more efficient Gauss quadrature rules are used.

In the next section we will discuss and illustrate the numerical approximate solution of the very important case of the *homogeneous* Fredholm integral equations. We should remind here of the special feature of these equations, where they are associated with an eigenvalue problem, as we had discussed in Section 5.1.2.

5.5.2 Homogeneous Fredholm Equations

In Section 5.5.1, we considered the nonhomogeneous Fredholm integral equation of the second kind

$$u(x) = f(x) + \int_a^b K(x,t)u(t)dt \qquad (5.116)$$

then used the trapezoidal rule for approximately the integral that resulted in a set of $n+1$ nonhomogeneous algebraic equations in $n+1$ unknowns $\{u_i\}_{i=0}^{n}$, and which we wrote in the following matrix form (5.126) as follows from (5.125)

$$[I - D_T\mathcal{K}]U = F \qquad (5.126)$$

where I, D_T, $\mathcal{K}$ and F are clearly defined after (5.126) as in (5.127)–(5.129).

In this section, we consider the numerical method of solving a homogeneous Fredholm equation

$$u(x) = \lambda \int_a^b K(x,t)u(t)dt, \qquad (5.130)$$

which can be developed in the same way as we did for the nonhomogeneous Fredholm integral equation (5.116). We will again use the trapezoidal rule, with n subintervals to approximate the integral above, and reduce (5.130) to $n+1$ linear homogeneous equations, in the $n+1$ (approximate) unknowns u_i, $i = 0, 1, \cdots, n$.

$$u_i = \lambda\Delta t \left[\frac{1}{2}K_{i0}u_0 + K_{i1}u_1 + \cdots + K_{i,n-1}u_{n-1} + \frac{1}{2}K_{in}u_n\right],$$

$$i = 0, 1, 2, \cdots, n. \tag{5.131}$$

Here it looks that such numerical approximation setting, as a system of *homogeneous* linear equations for the homogeneous Fredholm integral equation (5.130), should follow as a special case of (5.116) with $f(x) \equiv 0$. However, any discussion of the results will need what might be new concepts of the *eigenvalues* and *eigenfunctions*, which we have already discussed in detail in Section 5.1.2, (and earlier at the end of Section 4.1.3). So, attention should be made to the parameter λ of the homogeneous integral equation (5.130) and its numerical approximation (5.131). In summary, the values of this λ in (5.130) (or (5.131)) that results in *nontrivial* solutions for these equations are called the *eigenvalues*, while the corresponding (nontrivial) solutions are called the *eigenfunctions.*

If we bring all the terms to the left side of (5.131) and write the $n+1$ homogeneous equations for $i = 0, 1, 2, \cdots, n$, we have

$$\begin{aligned}
&\left(1 - \frac{\lambda\Delta t}{2}K_{00}\right)u_0 - \lambda\Delta t K_{01}u_1 - \lambda\Delta t K_{02}u_2 - \cdots - \frac{\lambda\Delta t}{2}K_{0,n}u_n = 0\\
&-\lambda\frac{\Delta t}{2}K_{10}u_0 + (1 - \lambda\Delta t K_{11})u_1 - \lambda\Delta t K_{12}u_2 - \cdots - \frac{\lambda\Delta t}{2}K_{1,n}u_n = 0\\
&\vdots\\
&-\frac{\lambda\Delta t}{2}K_{n0}u_0 - \lambda\Delta t K_{n1}u_1 - \lambda\Delta t K_{n2}u_2 + \cdots + \left(1 - \frac{\lambda\Delta t}{2}K_{nn}\right)u_n = 0.
\end{aligned} \tag{5.132}$$

There is one simplification that can be attained by letting $\lambda = 1/\mu$ and hence μ will appear only in one term of each equation instead of appearing in every term; that is, (5.132) reduces to

$$\begin{aligned}
&\left(\mu - \frac{\Delta t}{2}K_{00}\right)u_0 - \Delta t K_{01}u_1 - \Delta t K_{02}u_2 - \cdots - \frac{\Delta t}{2}K_{0,n}u_n = 0\\
&-\frac{\Delta t}{2}K_{10}u_0 + (\mu - \Delta t K_{11})u_1 - \Delta t K_{12}u_2 - \cdots - \frac{\Delta t}{2}K_{1,n}u_n = 0\\
&\vdots\\
&-\frac{\Delta t}{2}K_{n0}u_0 - \Delta t K_{n1}u_1 + \cdots + \left(\mu - \frac{\Delta t}{2}K_{nn}\right)u_n = 0
\end{aligned} \tag{5.133}$$

which is in the same form now as (5.125) except for that $f_i = 0$, $i = 0, 1, 2, \cdots, n$, and the 1 in parentheses on the diagonal (of (5.125) is replaced by μ. So if we write this set of $n+1$ homogeneous equations in matrix form, we have

$$K_H U = 0$$

where

$$0 = \begin{bmatrix} 0 \\ \vdots \\ 0 \end{bmatrix}$$

is the zero matrix, U is the same matrix as in (5.128).

$$U = \begin{bmatrix} u_0 \\ u_1 \\ \vdots \\ u_n \end{bmatrix} \tag{5.134}$$

and K_H signifies the coefficient matrix for the homogeneous equation (5.133):

$$K_H = \begin{bmatrix} \mu - \frac{\Delta t}{2}K_{00} & -\Delta t K_{01} & \cdots & & -\frac{\Delta t}{2}K_{0n} \\ -\Delta t K_{10} & \mu - \Delta t K_{11} & \cdots & & -\frac{\Delta t}{2}K_{1n} \\ \vdots & & & \vdots & \\ -\frac{\Delta t}{2}K_{n-1,0} & -\Delta t K_{n-1,1} & \cdots & \mu - \Delta t K_{n-1,n-1} & -\frac{\Delta t}{2}K_{n-1,n} \\ -\frac{\Delta t}{2}K_{n0} & -\Delta t K_{n1} & \cdots & & \mu - \frac{\Delta t}{2}K_{nn} \end{bmatrix}. \tag{5.135}$$

We must recall here that a nontrivial solution to this system of $n+1$ linear homogeneous equations exists if and only if the determinant $|K_H|$ of the coefficients matrix K_H in (5.135) vanishes. This condition is used to find the (approximate) eigenvalues λ of (5.130) through finding $\mu = 1/\lambda$ as the zeros of $|K_H| = 0$.

We may recall that while $|A| = |I - D_T \mathcal{K}| \neq 0$ guarantees a unique (approximate) solution for (the nonhomogeneous equation) (5.127), the foregoing condition $|K_H| = 0$ guarantees a nontrivial but not a unique solution to the homogeneous system (5.135), which means that we may have to determine the values $u_0, u_1, \cdots, u_n$ in terms of one of them as an arbitrary value. Such an arbitrary constant can be evaluated in practice when we normalize the approximated solution. This will become clear in the following illustration.

Example 21 Numerical Solution of Homogeneous Fredholm Equations

For illustrating the numerical method of solving homogeneous Fredholm equations, we consider the following equation of Example 7:

$$u(x) = \lambda \int_0^1 K(x,t)u(t)dt \tag{E.1}$$

with the symmetric kernel

$$K(x,t) = \begin{cases} x(1-t), & 0 \le x \le t \\ t(1-x), & t \le x \le 1 \end{cases} \tag{E.2}$$

where we found that the normalized eigenfunctions are

$$u_k(x) = \sqrt{2}\sin k\pi x, \qquad k = 1, 2, 3, \cdots \tag{E.3}$$

corresponding to the eigenvalues $\lambda_k = \pi^2 k^2$. To simplify the computations we will attempt to find an approximate solution at $x = 0,\ 1/2$, and 1, hence $n = 2$ and $\Delta t = \frac{1}{2}$. So we proceed to evaluate $K_{ij} = K(i/2, j/2)$, $i, j = 0, 1, 2$, where we have $K_{00} = K_{01} = K_{02} = K_{12} = 0$ and $K_{11} = \frac{1}{4}$ from the first branch of $K(x,t)$ in (E.1) and $K_{10} = K_{20} = K_{21} = K_{22} = 0$ from the second branch of $K(x,t)$ in (E.1). Hence if we substitute these values in (5.133), we obtain

$$\begin{aligned} \mu u_0 + 0 + 0 &= 0 \\ 0 + \left(\mu - \frac{1}{8}\right) u_1 + 0 &= 0 \\ 0 + 0 + \mu u_2 &= 0. \end{aligned} \tag{E.4}$$

For this system of homogeneous equations to have a nontrivial solution, the determinant of the coefficients must vanish,

$$\begin{vmatrix} \mu & 0 & 0 \\ 0 & \mu - \frac{1}{8} & 0 \\ 0 & 0 & \mu \end{vmatrix} = \mu^2 \left(\mu - \frac{1}{8}\right) = 0,$$

$$\mu = 0, \qquad \mu = \frac{1}{8}. \tag{E.5}$$

If we consider $\mu = 1/8$, this will give $\lambda = 1/\mu = 8$ and if we substitute this value of μ in (E.4), we obtain $u_0 = u_2 = 0$ and $u_1 = u_1$ as an arbitrary constant. Hence we have the two zero values at $x = 0, 1$ but an arbitrary value u_1 at $x = 1/2$. What we did here is, of course, a very rough approximation to the integral in (E.1), where we used only three points, but it can be improved by considering more points. It remains to find the arbitrary value u_1. For this we may approximate the solution function by two straight lines connecting the three points (0,0), $(1/2, u_1)$, and (1,0) as

$$u(x) = \begin{cases} 2u_1 x, & 0 \le x \le \dfrac{1}{2} \\ -2u_1(x-1), & \dfrac{1}{2} \le x \le 1 \end{cases} \tag{E.6}$$

then make its norm be unity,

$$\int_0^1 u^2(x)dx = 1 \tag{E.7}$$

to find u_1 and then compare $u(x)$ with an orthonormal solution from (E.3). If we substitute $u(x)$ from (E.6) in (E.7), we obtain

$$4u_1^2 \int_0^{\frac{1}{2}} x^2 dx + 4u_1^2 \int_{\frac{1}{2}}^1 (x-1)^2 dx = \frac{1}{6}u_1^2 + \frac{1}{6}u_1^2 = \frac{u_1^2}{3} = 1, \quad u_1 = \sqrt{3} \sim 1.73.$$

So the approximate numerical values are

$$u(0) \approx u_0 = 0, \qquad u\left(\frac{1}{2}\right) \approx u_1 = 1.73, \qquad u(1) \approx u_2 = 0$$

corresponding to an approximate eigenvalue of $\lambda = 8$. Now if we want to compare these values to an exact orthonormal eigenfunction from (E.3), we must choose $u_1(x) = \sqrt{2}\sin \pi x$, since this corresponds to the eigenvalue $\lambda_1 = \pi^2 \sim 10$, which is the closest to $\lambda = 8$. This exact solution gives

$$u_0 = u(0) = \sqrt{2}\sin 0 = 0, \;\; u_1 = u\left(\frac{1}{2}\right) = \sqrt{2}\sin\frac{\pi}{2} = 1.41, \;\; u_2 = u(1) = 0.$$

As we have indicated at the beginning of this section, we have included here only the most basic numerical integration rules to approximate the integral of the Fredholm integral equations. The higher order quadrature rules of approximating the integral, their tables, and the numerical setting of the Fredholm integral equations using such rules, are covered in Section 7.3. There we support the use of such different rules with a good number of detailed examples and exercises.

Exercises 5.5

1. **(a)** Use a numerical method (trapezoidal rule) to solve for the approximate values of the solution of the Fredholm equation of Example 20

$$u(x) = \sin x + \int_0^1 (1 - x\cos xt)u(t)dt$$

at

(i) $x = 0, \frac{1}{3}, \frac{2}{3}, 1$

(ii) $x = 0, \frac{1}{10}, \frac{2}{10}, \frac{3}{10}, \cdots, 1$

Hint: Note that for (5.125) with a 3×3 system in Exercise 1, a 4×4 system of part (i) and an 11×11 system of part (ii), you need to use a computer to handle the lengthy computations for solving the resulting linear equations.

(b) Tabulate the two approximate results in part (a) and compare them with the approximate solution

$$v(x) = \sin x + 1.003(1-x) + 0.1674x^3$$

of Example 6 in Section 5.1.

2. **(a)** In problem 1(a)(i) use Simpson's rule instead of the trapezoidal rule.

   (b) Compare the approximate results of part (a) with the exact answer $u(x) = 1$.

3. **(a)** Use a numerical method (trapezoidal rule) to solve for the approximate values of the solution of the equation of Example 16 in Section 5.3

$$u(x) = e^{-x} - \int_0^1 xe^t u(t)dt$$

   at

   (i) $x = 0, \frac{1}{2}, 1$

   (ii) $x = 0, \frac{1}{10}, \frac{2}{10}, \frac{3}{10}, \cdots, 1$. See the hint for Exercise 1(a).

   (b) Compare the two approximate results in part (a) with the exact and approximate results of Example 16 as presented in Table 5.2 and Figure 5.1 (of Section 5.3).

4. **(a)** Use a numerical method (trapezoidal rule) to solve for the approximate values of the solution of the Fredholm equation

$$u(x) = 2 - \frac{1}{\pi}\int_{-1}^1 \frac{1}{1+(x-t)^2}u(t)dt \qquad (E.1)$$

   at $x = -1, 0, 1$.

   (b) Attempt to verify such a crude approximate solution.

   Hint: Try to integrate numerically with the three approximate values of $u(x)$ and see how the two sides of (E.1) compare for each value of $x = -1, 0$, and 1.

   (c) Repeat parts (a) and (b) for the approximate values of the solution at $x = -1, -9/10, -8/10, \cdots, 0, 1/10, 2/10, \cdots, 1$, then graph and compare with the results in part (a). See the hint for Exercise 1(a,ii).

5. **(a)** Use Simpson's rule of integration (1.144) instead of the trapezoidal rule to reduce the Fredholm integral equation (5.116) to a system of $2n+1$ linear equations similar to that of (5.124).

   Hint: Note that n must be even in (1.144) of the Simpson's rule.

   (b) Use the result in part (a) to solve for the equation of Exercise 1(a).

$$u(x) = \sin x + \int_0^1 (1 - x\cos xt)u(t)dt \qquad (E.1)$$

   at $x = 0, \frac{1}{2}$, and 1.

(c) Compare the results in part (b) with those of Exercise 1(a) and the approximate solution of Example 6,

$$v(x) = \sin x + 1.0031(1 - x) + 0.1674x^3.$$

6. **(a)** Use a numerical method (trapezoidal rule) to solve for the approximate values at $x = 0, 1/2$, and 1 of the homogeneous Fredholm equation

$$u(x) = \lambda \int_0^1 K(x,t)u(t)dt \tag{E.1}$$

$$K(x,t) = \begin{cases} t(1-x)(2x - t^2 - x^2), & 0 \le t \le x \\ x(1-t)(2t - x^2 - t^2), & x \le t \le 1 \end{cases} \tag{E.2}$$

This problem represents the deflection $u(x)$ of a rotating shaft (1.19) with unit length and constant density, where λ combines most of the shaft physical properties.

Hint: Note that the kernel is symmetric.

(b) Repeat part (a) for approximate eigenvalues and the solution values at $x = 0, 1/4, 1/2, 3/4$, and 1. See the hint for Exercise 6(c).

7. **(a)** Use a numerical method (trapezoidal rule) to solve for the approximate values at $x = 0, \frac{1}{3}, \frac{2}{3}, 1$ of the homogeneous Fredholm equation of Example 21.

$$u(x) = \lambda \int_0^1 K(x,t)u(t)dt \tag{E.1}$$

$$K(x,t) = \begin{cases} x(1-t), & 0 \le x \le t \\ t(1-x), & t \le x \le 1 \end{cases} \tag{E.2}$$

Hint: Use $n = 3$, search for the approximate solution that corresponds to the largest finite eigenvalue, and follow Exmaple 21.

(b) Compare the results of part (a) with an exact eigenfunction of $u_k(x) = \sqrt{2}\sin k\pi x$ corresponding to the exact eigenvalue $\lambda_k = \pi^2 k^2$.

Hint: Try to approximate the function by three straight lines between the four approximate values, then make it with a norm of 1 as we did in Example 21.

8. Use a numerical method (trapezoidal rule) to solve the Fredholm equation

$$u(x) = x + \int_0^1 K(x,t)u(t)dt \tag{E.1}$$

$$K(x,t) = \begin{cases} t, & 0 \le t \le x \\ x, & x \le t \le 1 \end{cases} \tag{E.2}$$

at $x = 0, \frac{1}{4}, \frac{1}{2}, \frac{3}{4}$, and 1.

9. For the three samples u_1, u_2, and u_3 of problem 3a(i), use the Lagrange interpolation formula (1.153) and (1.154) to interpolate the approximate solution, then compare with the exact answer of $u(x) = e^{-x} - \frac{x}{2}$, $0 < x < 1$ and the answer of problem 3a(ii) at $x = 0, \frac{1}{10}, \frac{2}{10}, \cdots, 1$.

6

Existence of the Solutions: Basic Fixed Point Theorems

With the main emphasis of this edition on a simple introductory and applicable course in integral equations, this chapter must definitely be considered as an *optional* one. Indeed we could have relegated it to an appendix, but since its simple and descriptive presentation[1] relates to basic topics in Chapters 3 and 5, we opted to retain it in this edition. Of course the introductory course depends, primarily, on good parts of the first five chapters as we described it in our "suggestions for course adoption" at the end of the preface. For a more advanced applied course, parts of this chapter may prove helpful to the reader with a desire to look into more basic theory, besides the methods of solutions in Chapters 3 and 5.

6.1 PRELIMINARIES: TOWARD A CONTRACTIVE MAPPING

Our treatment in Chapters 3 and 5 for the Volterra and Fredholm integral equations centered mainly on illustrations of the known methods of finding exact, approximate, or numerical solutions. In so doing we either had to assume the existence of a unique solution or stated some conditions to secure it.

In this chapter we present and prove a few basic theorems that are necessary for establishing the existence and uniqueness of the solutions of integral equations. We start with a descriptive presentation to motivate the basic mathematical concepts

[1]For more information on the existence of solutions to linear as well as nonlinear integral equations, see Kress [1989], Hochstadt [1973], Pogorzelski [1966] (greater depth), Cochran [1972], and Collatz [1966] (numerical methods).

needed for an accurate and clear statement of the principal theorem: the *fixed point theorem* of Banach. It is our intention first to give a clear presentation of several applications of the fixed point theorem, which have been selected with the goal of keeping this chapter at the same level as, and in harmony with, the remainder of the text.

The very basic iterative method that we employed in Chapters 3 and 5,

$$u_{n+1}(x) = f(x) + \lambda \int K(x,t)u_n(t)dt \tag{6.1}$$

was instrumental in constructing the solutions, and in many instances we even showed the convergence of the sequence $u_n(x)$ to $u(x)$, the solution of the original integral equation

$$u(x) = f(x) + \lambda \int K(x,t)u(t)dt. \tag{6.2}$$

Even when we accept such practical constructive proofs, we still should inquire about their applicability to other, more general problems that cannot be solved in closed forms. In particular, all our treatment in this text has been directed toward solving only *linear* integral equations as in (6.2), with no method or illustration given of how to proceed when we have *nonlinear* integral equations. The reason for this is that while the existence of a unique solution may be assumed or established by direct computations for the linear problem (6.2), it is a very different matter to tackle that of the much more complicated nonlinear integral equation

$$u(x) = \int F(x,t,u(t))dt \tag{6.3}$$

whose successive approximations (iterations) are

$$u_{n+1}(x) = \int F(x,t,u_n(t))dt. \tag{6.4}$$

In this section we motivate the preparations necessary for accurate statements and proofs of the few very basic theorems on the existence and uniqueness of the solutions for such general problems. The iterative method, which we have used so extensively in this text, will be a principal vehicle for the proofs of these theorems.

Compared to the constructive-type proofs that we have employed until now, the theorems of the present section and their proofs will have more of a geometric approach. For example, the integral equation (6.3) is looked at in the following way: The right-hand side is considered as a mapping or transformation T on u denoted by $T(u)$, while the left-hand side indicates that such transformation had left this one element u unchanged,

$$u = T(u) \tag{6.5}$$

This means that the solution u which we seek for the integral equation (6.3) represents a very special element in the domain of the operator T, namely, that which remains unaltered or fixed under the T transformation. Such an element u as in (6.5) is called a *fixed point* of the transformation or mapping T, which is the solution sought for the integral equation (6.3). In this sense the successive approximations (iterations) of (6.4) can be written as

$$u_{n+1} = T(u_n). \tag{6.6}$$

The question still remains as to whether the general mapping T has a fixed point, and if so whether such a point is unique. This, as we expect, will depend on the function $F(x, t, u(t))$ in (6.3) or $K(x, t)$ in the linear case in (6.2). However, there are other very important factors that enter into play, including the nature of the iterative process, the measure we use for the distance (*metric*) in determining how close the members of the sequence u_n are clustering together toward a limit point, and most important, the quality of the set or space from which we select such sequences. A very familiar space to us is R, the set of real numbers, with the distance between u_n and u (Euclidean distance) defined by

$$d(u_n, u) = |u_n - u|. \tag{6.7}$$

This, as we shall see, is but one of a variety of measures of distance (or metric) that we may choose to adopt in order to facilitate the proofs of the desired theorems.

For the n-dimensional Euclidean space $R^n = \{x = (x_1, x_2, \cdots, x_n);\ x_i \in R\}$, the distance above is easily generalized to

$$d(x, y) = \sqrt{\sum_{i=1}^{n} (x_i - y_i)^2}, \qquad x, y \in R^n.$$

Another different measure of distance between two elements x and y of R^n is defined as

$$d_1(x, y) = \max_{i=1,2,\cdots,n} |x_i - y_i|, \qquad x, y \in R^n. \tag{6.8}$$

This type of distance d_1 of (6.8) proves very useful when modified to give a measure of the difference between continuous functions. For $f(x)$ and $g(x)$ as two elements of the set $C[a, b]$ of continuous functions on the closed interval $[a, b]$, we define the distance between them as

$$d(f(x), g(x)) = \max_{x \in [a,b]} |f(x) - g(x)|, \qquad f, g \in C[a, b] \tag{6.9}$$

which is graphically the largest distance between the two functions on the closed interval $[a, b]$, as in Figure 6.1. A simple example of this type of distance is that between $f(x) = \cos x$ and $g(x) = \sin x$ on $[0, \pi/2]$, which is the maximum distance of 1 occurring at $x = 0$ and $x = \pi/2$.

In practice, if $g(x)$ is the approximation to the solution $u(x)$ we are seeking, then the (maximum) metric $d(u, g)$ in (6.9) measures the maximum deviation of $g(x)$ from the desired solution $u(x)$. So if we are to require an accuracy of 10^{-6}, for example, the way to express it is via the maximum of the metric as $\epsilon = d(u, g) = 10^{-6}$.

We shall soon present the formal definition for the distance or metric $d(x, y)$ between two elements x, y of a given set X, but first we would like to motivate the type of convergence that is more suitable for describing the clustering or closeness of the members of the sequence u_n. In our construction of the solution via the iterative process we were after the sequence u_n approaching the limit u as n approaches infinity, which is the usual type of convergence

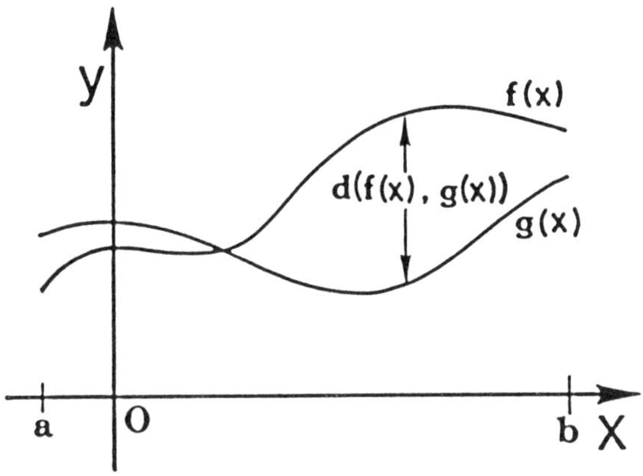

Fig. 6.1 The distance $d(f, g)$ of (6.9) between two continuous functions.

$$\lim_{n\to\infty} |u_n - u| = 0 \tag{6.10}$$

that one encounters in the basic calculus course. However, in practice we very often do not have a way of knowing the *limit point u*, but instead we know merely that as n increases, u_{n+1} gets closer to u_n (i.e., the sequence is clustering). It is even a better sign when not only the consecutive members u_{n+1}, u_n but the members of the sequence u_{n+p}, u_n become close, that is, when their distance $|u_{n+p} - u_n|$ becomes very small as n, the number of iterations, increases, that is,

$$\lim_{n\to\infty} |u_{n+p} - u_n| = 0. \tag{6.11a}$$

This would be a very good sign for the convergence of the sequence, but without specifying the particular limit point. We should note that in (6.11a) we may use m instead of $n + p$, and write

$$\lim_{n,m\to\infty} |u_m - u_n| = 0. \tag{6.11b}$$

Such a practical concept of convergence is called *Cauchy convergence* as opposed to the usual convergence in (6.10), which we will refer to as "convergence to the limit point u." There seems to be a drawback to Cauchy-type convergence in that it does not specify the limit point that we are after. In other words, we are concerned about whether Cauchy convergence (6.11) would ever imply the convergence (6.10), which spells out the limit point. To answer this question in the affirmative will depend on *the particular space* that contains the sequence and on *the type of metric* we use to measure the distance between the elements of this space. A space with its assigned metric (distance) is called a *metric space*. We will soon show that in a metric space, convergence (6.10) to a limit u always implies Cauchy convergence (6.11), but the converse, which is what we are after, is not always true.

A metric space in which Cauchy convergence implies convergence to a limit is a very special one termed *complete metric space*. This is the metric space we shall work with and in which we state and prove the fixed point theorem.

Before we begin the formal definitions necessary for the accurate statements of the fixed point theorems, there is still an extremely desirable property of the transformation or mapping $T(u)$ of (6.5). This property can be described as a kind of *focusing* effect of T as it maps the input estimate u_n to its output u_{n+1} as in (6.6). By this we mean that the distance between the images $u' = T(u)$ and $v' = T(v)$ would be closer than the distance between their objects u and v in the domain of T, which can be expressed as

$$d(u', v') = d(T(u), T(v)) \leq \alpha d(u, v), \qquad 0 \leq \alpha < 1 \tag{6.12}$$

and is illustrated in Figure 6.2. A transformation T with this property (6.12) is called *contractive*, as indeed it results in a contracted or closer distance between its outputs (images). It is just this contractive property which is responsible for clustering the sequence $\{u_n\}$ of the iterative process (6.6),

$$u_{n+1} = T(u_n) \tag{6.6}$$

toward a limit point.

It has been our attempt to give, in a very descriptive way, an idea of the main concepts that are needed for the statement of a fixed point theorem, namely, a complete metric space and a contractive mapping. With this informal introduction, a very basic fixed point theorem of Banach states that "for a *contractive* mapping T on a *complete metric space*, there exists a *unique* solution u to $u = T(u)$." The detailed statement and proof of this important theorem are given in Section 6.2.

For our purposes this would ensure the existence of a unique solution to the linear integral equation used in earlier chapters,

$$u(x) = f(x) + \lambda \int K(x, t)u(t)dt \tag{6.2}$$

as well as the (generally) nonlinear integral equation

$$u(x) = \int F(x, t, u(t))dt \tag{6.3}$$

as special cases of a contractive mapping $u = T(u)$.

Although our principal concern in this text is with solutions of integral equations, this should not distract us from other possible applications of the fixed point theorem

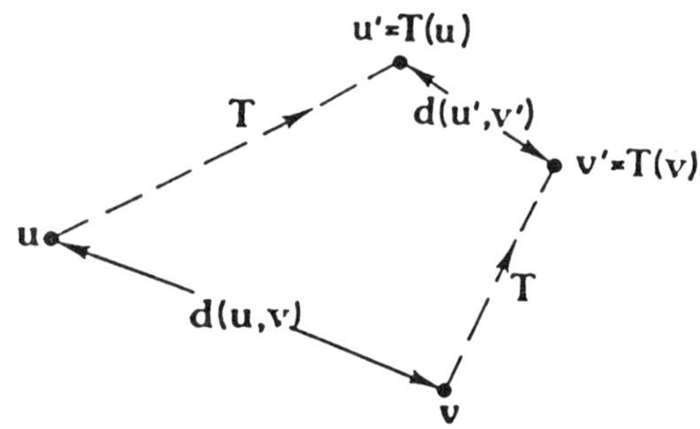

Fig. 6.2 Contractive mapping.

in proving the existence of solutions for various types of equations that can be described by the mapping

$$u = T(u). \tag{6.5}$$

For example, instead of T being the integral operator in the integral equations above, it can represent a *differential operator* in the case of a differential equation like

$$u = \frac{1}{\lambda}\frac{d^2u}{dx^2} = T(u). \tag{6.13}$$

Another example is that of algebraic equations, in particular the matrix equation

$$\begin{aligned} x = Ax, \quad x &= (x_1, x_2, \cdots, x_n) \\ A &= [a_{ij}]_{i,j=1,2,\cdots,n} \end{aligned} \tag{6.14}$$

Up to this point in the discussion we have not mentioned that the mapping T is linear-hence the role of the fixed point theorem in assuring the existence and uniqueness of solutions to a class of "usually" intractable nonlinear integral and differential equations. We refer here to a certain class, as it remains for us to show that the particular equation has a contraction operator T.

Even though the fixed point theorem can be applied to integral equations as well as differential equations, the successive approximations (iterative) process (6.1) favors the integral equation representation of the problem, since in practice we watch the approach of the sequence u_{n+1} of (6.6) toward the desired solution of the integral equation (6.3). This means that in order to apply the fixed point theorem to differential equations, we may first change the differential equation to an integral equation to make it suitable for the iterative process (6.6). We will illustrate this application for initial value problems associated with differential equations after reducing them to Volterra integral equations.

With the foregoing intuitive and very descriptive introduction of the basic concepts necessary for stating the fixed point theorem, we turn now to the formal definitions of these concepts. It is our intention to keep the treatment brief, but clear and mostly self-contained.

6.1.1 Basic Definitions: Complete Metric Spaces

Metric Space

A *metric space*, designated as $(\mathcal{M}, d)$, is a set $\mathcal{M}$ with a mapping $(d : \mathcal{M} \times \mathcal{M} \to R)$ that associates a real number (distance) $d(x, y) = r \epsilon R$ to every ordered pair (x, y) in the domain of d and such that this distance (or metric) $d(x, y)$ satisfies the following three conditions:

(a) $d(x, y) \geq 0$ for $x, y \epsilon \mathcal{M}$, and $d(x, y) = 0 \Leftrightarrow x = y$ (6.15)

This means that the distance between any two elements is always *nonnegative*, and the distance being zero is equivalent to the two elements being identical.

(b) $d(x, y) = d(y, x)$ for $x, y \epsilon \mathcal{M}$. (6.16)

That is, the distance is *symmetric* in x and y.

(c) $d(x, z) \leq d(x, y) + d(y, z)$ for $x, y, z \epsilon \mathcal{M}$. (6.17)

This means that the present general definition of distance still satisfies an inequality that parallels the usual *triangle inequality*,

$$|x - z| \leq |x - y| + |y - z| \tag{6.18}$$

see (Figure 6.3).

The triangle inequality (6.17) will be used very often in proofs of the basic theorems. We note that the present mapping which defines the distance $d(x, y)$ is to be distinguished from $T(u)$ in (6.5).

A familiar example of a metric space is (R, d), the set of real numbers R with the distance (metric) $d(x, y) = |x - y|$, which can easily be shown to satisfy the three properties of a metric listed above. The set $C[a, b]$ of *continuous functions* on the closed interval $[a, b]$, together with the metric

$$d(f(x), g(x)) = \max_{x \epsilon [a,b]} |f(x) - g(x)|, \qquad f, g \epsilon C[a, b] \tag{6.19}$$

also constitutes an important metric space, especially for the present development, where we are dealing with continuous functions of the integral equation (6.3), such as $F(x, t, u(t))$.

It is instructive to show that this new type of distance $d(f(x), g(x))$ in (6.19) does satisfy the triangle inequality (6.17),

$$d(f, g) \leq d(f, h) + d(h, g).$$

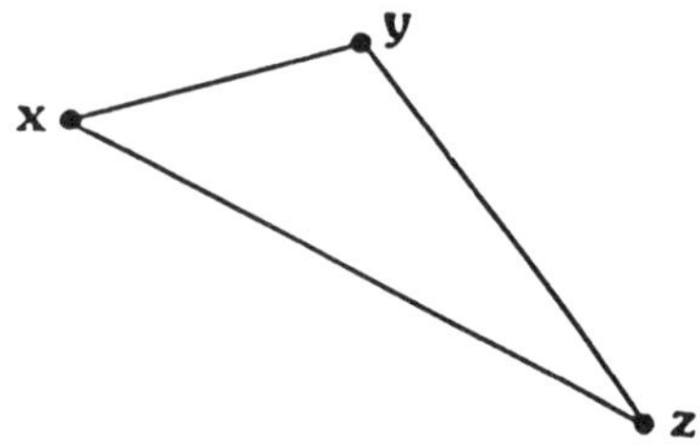

Fig. 6.3 Triangle for (6.18).

To prove this we observe

$$\begin{aligned} d(f,g) &= \max_x |f(x)-g(x)| = \max_x |f(x)-h(x)+h(x)-g(x)| \\ &\le \max_x\{|f(x)-h(x)|+|h(x)-g(x)|\} \\ &\le \max_x |f(x)-h(x)| + \max_x |h(x)-g(x)| \\ &\le d(f,h)+d(h,g) \end{aligned}$$

where the maximum is taken over all $x\epsilon[a,b]$.

Limit of a Sequence in Metric Space

Let $\{u_n\}_{n=1}^{\infty}$ be a sequence of points $u_n\epsilon\mathcal{M}$. The point u is called a *limit point* of the sequence, that is,

$$u = \lim_{n\to\infty} u_n \tag{6.20}$$

if for each $\epsilon > 0$ there is a number $n_0 = n_0(\epsilon)$ such that for $n > n_0(\epsilon)$ the element u_n is within the distance ϵ from u [i.e., $d(u,u_n) < \epsilon$]. In this case we say that the sequence u_n converges to u.

As we mentioned earlier, especially for the iterative process (6.4) or (6.6), it is sometimes the case that the elements u_n of the sequence get very close to each other but no limit u is known [i.e., $d(u_n,u_m) \to 0$ as $n,m \to \infty$]. This brings us to the Cauchy-type convergence. The sequence $\{u_n\}_{n=1}^{\infty}$ in $\mathcal{M}$ is called *Cauchy* if for each $\epsilon > 0$ there is $n_0 = n_0(\epsilon)$ such that for $n,m > n_0(\epsilon)$ we have $d(u_n,u_m) < \epsilon$.

We will prove here that in a metric space every convergent sequence (6.20) is Cauchy convergent. From the definition of the sequence $\{u_n\}$ being convergent we have

$$d(u,u_n) < \epsilon' \quad \mathit{for} \quad n > n_0(\epsilon'). \tag{6.21}$$

We want to show that this implies indexCauchy convergence Cauchy convergence, that is,

$$d(u_n, u_m) < \epsilon \quad for \quad n, m > N(\epsilon). \tag{6.22}$$

From the definition of the metric we have the triangle inequality (6.18),

$$d(u_n, u_m) \leq d(u_n, u) + d(u, u_m) \tag{6.23}$$

where for the right side we can use the assumed convergence to have

$$d(u_n, u) < \epsilon' \quad for \quad n > n_0(\epsilon') \tag{6.24}$$

and in the same way,

$$d(u, u_m) < \epsilon' \quad for \quad m > m_0(\epsilon'). \tag{6.25}$$

So if we take $n > \max(n_0, m_0) = N_0(\epsilon')$, we can use this $N_0(\epsilon')$ to satisfy both (6.24) and (6.25), which are then used in (6.23) to yield

$$d(u_n, u_m) < \epsilon' + \epsilon', \qquad n, m > N_0(\epsilon'). \tag{6.26}$$

If we let $\epsilon' = \epsilon/2$, we have

$$d(u_n, u_m) < \epsilon \quad for \quad n, m > N_0\left(\frac{\epsilon}{2}\right) = N(\epsilon) \tag{6.27}$$

which, according to (6.22), constitutes the Cauchy convergence.

Although *convergence* to a *limit implies Cauchy convergence*, we should bear in mind that the converse is not always true. As mentioned earlier, only in very special metric spaces called *complete metric spaces* do we also have *Cauchy convergence implying convergence to a limit*. For example, in the metric space $(\mathcal{M}, d)$ with $\mathcal{M} = \mathcal{Q}$ as the set of rational numbers, a sequence of rational numbers may converge to a limit point $u = \sqrt{2}$, but this limit point is not a member of the set of rational numbers.

It can be shown that the metric space $(\mathcal{M}, d)$, with $\mathcal{M} = R$ the set of real numbers and $d(x, y) = |x - y|$ for $x, y \epsilon R$, is complete. Also, the set $C[a, b]$ of continuous functions on the closed interval $[a, b]$, together with the metric

$$d(f(x), g(x)) = \max_{x \in [a,b]} |f(x) - g(x)|, \qquad f(x), g(x) \in C[a, b] \tag{6.9}$$

can be shown as another very important complete metric space. Because of space limitations we will not pursue proofs of these results; instead, we concentrate our efforts on proving the main result, the fixed point theorem.

Fixed Point of a Mapping

Let $(\mathcal{M}, d)$ be a metric space and $\mathcal{M}'$ a subset of $\mathcal{M}$ with the mapping

$$T : \mathcal{M}' \to \mathcal{M}, \qquad T(u) = v.$$

An element $u_0 \in \mathcal{M}'$ is a *fixed point* of the mapping T if $T(u_0) = u_0$. For all elements $u \in \mathcal{M}'$ we are looking for those particular elements u_0 that remain unchanged under the transformation T, that is,

$$u_0 = T(u_0). \tag{6.28}$$

Contractive Mapping

The mapping T in (6.28) is called *contractive* if there is a nonnegative real number α less than $1, 0 \leq \alpha < 1$, such that for each $u_1, u_2 \in \mathcal{M}'$ we have

$$d(T(u_1), T(u_2)) \leq \alpha d(u_1, u_2). \tag{6.29a}$$

In other words, a contractive mapping brings the images $T(u_1)$ and $T(u_2)$ closer in the range of the operator T than their corresponding objects u_2 and u_1 in the domain, as illustrated in Figure 6.2. In terms of our iterative process

$$u_{n+1} = T(u_n) \tag{6.6}$$

we have, for example, u_2 as the image of u_1 and u_3 as the image of u_2, so with a contractive mapping $d(T(u_2), T(u_1)) \leq \alpha d(u_2, u_1)$, but $T(u_2) = u_3, T(u_1) = u_2$, hence

$$d(u_3, u_2) \leq \alpha d(u_2, u_1). \tag{6.29b}$$

In the same way we can show that

$$d(u_4, u_3) = d(T(u_3), T(u_2)) \leq \alpha d(u_3, u_2) \leq \alpha^2 d(u_2, u_1) \tag{6.30}$$

after using (6.29a) and (6.29b).

This process can be continued to obtain the general case,

$$d(u_{n+1}, u_n) \leq \alpha d(u_n, u_{n-1}) \leq \alpha^2 d(u_{n-1}, u_{n-2}) \cdots \leq \alpha^{n-1} d(u_2, u_1) \tag{6.31}$$

which says that the sequence is clustering since the outputs u_{n+1} and u_n are closer than the inputs u_2 and u_1 by a *geometric factor* of $\alpha^{n-1}, 0 \leq \alpha < 1$.

6.1.2 Contractive Mapping for Linear Fredholm Equations

In the following example we illustrate conditions for the linear Fredholm integral equations of Chapter 5 to represent a contractive mapping.

Consider the Fredholm integral equation of the second kind (5.7a),

$$u(x) = g(x) + \lambda \int_a^b K(x,t)u(t)dt \equiv T(u). \qquad (6.32), (5.7a)$$

We assume that $g(x)$ is continuous on the interval $[a, b]$ and $K(x, t)$ is continuous on the square $D = \{(x, t) : x \in [a, b], t \in [a, b]\}$, as indicated in Figure 6.4. For such functions we shall work with the complete metric space $C[a, b]$ of continuous functions and its metric $d(x, y)$ as in (6.9).

To find a sufficient condition for the mapping $T(u)$ of (6.32) to be contractive, we first indicate that the kernel $K(x, t)$ here is bounded [i.e., $|K(x, t)| \leq M$] since it is continuous on the bounded domain of the square in Figure 6.4. To show the

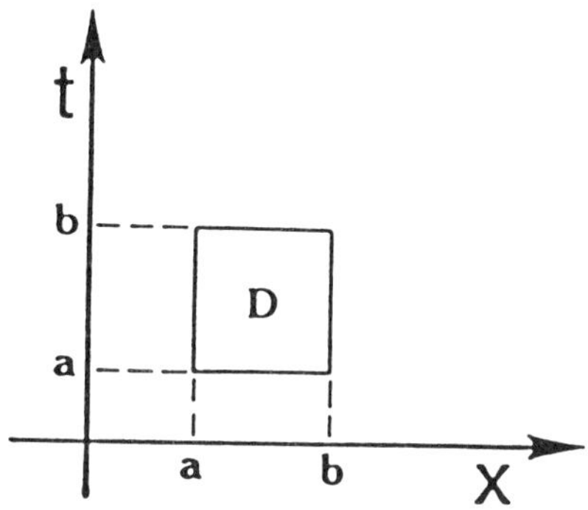

Fig. 6.4 Domain D of (6.32).

contraction property of T, we use the metric of (6.9) on the images $T(\beta(x))$ and $T(\gamma(x))$ of the two continuous functions $\beta(x)$, $\gamma(x)$ in $C[a,b]$,

$$\begin{aligned} d(T(\beta(x)), T(\gamma(x))) &= \max_{x\in[a,b]} \left| g(x) + \lambda \int_a^b K(x,t)\beta(t)dt - [g(x) \right. \\ &\qquad \left. + \lambda \int_a^b K(x,t)\gamma(t)dt] \right| \\ &= \max_{x\in[a,b]} \left| \lambda \int_a^b K(x,t)[\beta(t) - \gamma(t)]dt \right| \\ &\leq \max_{x\in[a,b]} \int_a^b |\lambda K(x,t)[\beta(t) - \gamma(t)]|dt \\ &\leq |\lambda| M \max_{x\in[a,b]} \int_a^b |[\beta(t) - \gamma(t)]|dt \\ &\leq |\lambda| M \max_{x\in[a,b]} |\beta(x) - \gamma(x)| \int_a^b dt \\ &\leq |\lambda| M(b-a) d(\beta(x), \gamma(x)) = \alpha d(\beta(x), \gamma(x)) \end{aligned} \tag{6.33}$$

after using the upper bound M for $|K(x,t)|$. Hence with

$$\alpha = |\lambda M(b-a)| < 1 \tag{6.34}$$

or

$$|\lambda| < \frac{1}{M(b-a)} \tag{6.35}$$

the mapping of the linear Fredholm equation (6.32) becomes contractive, since this ensures $d(T(\beta(x), T(\gamma(x)) < d(\beta(x), \gamma(x))$ in (6.33).

6.1.3 Contractive Mapping for Linear Volterra Equations

Next we illustrate that to ensure a contractive mapping $T(u)$ for the linear Volterra integral equation (3.1),

$$u(x) = f(x) + \lambda \int_0^x K(x,t)u(t)dt \tag{3.1}$$

$$= T(u) \tag{6.36}$$

we need much less restrictive conditions than those for the Fredholm equation (6.32).

In Section 3.1.2 we considered the successive approximation method (3.25) of solving (3.1),

$$u_n(x) = f(x) + \int_0^x K(x,t)u_{n-1}(t)dt \tag{3.25}$$

which we write here as

$$u_{n+1}(x) = f(x) + \lambda \int_0^x K(x,t)u_n(t)dt. \tag{6.37}$$

To assure the convergence of this approximation, we stated the result (without proof) that "if $f(x)$ is continuous on $[0,a]$ and $K(x,t)$ is also continuous for $0 \le x \le a$, $0 \le t \le x$, then the sequence $u_n(x)$ converges to the solution $u(x)$ of (3.1)." In terms of our present development, where we are working in $C[a,b]$, the space of continuous functions, we should be able to reach a conclusion of convergence without any extra conditions.

This indeed is possible but needs a number of preliminary results. The most important of these results is to show that for large enough n, the nth-order mapping $T^n(u)$ of the Volterra equation is a contractive one. We will limit our efforts in the following example to showing this result, which we feel captures the main idea of the contraction for $T(u)$, and we leave it to Example 3 at the end of Section 6.2.1, after we already have the fixed point theorem, to show that if $T^n(u)$ is contractive, then $T(u) = u$ has a unique solution.

$T^n(u)$ is easily illustrated when applied in (6.37),

$$\begin{aligned} u_{n+1} = T(u_n) = T(T(u_{n-1})) = T^2(u_{n-1}) &= T^2(T(u_{n-2})) = T^3(u_{n-2}) \\ = \cdots &= T^{n-1}(T(u_1)) = T^n(u_1). \end{aligned} \tag{6.38}$$

Example 1 A Contractive Mapping for the Volterra Equation

We will show here that $T^n(u)$ of (3.1) is contractive when n is large. Following what we did for the successive approximations and the iterated kernels method in Section 3.1, we write

$$T(u_1) = u_2(x) = f(x) + \lambda \int_a^x K(x,\xi)u_1(\xi)d\xi \tag{E.1}$$

$$
\begin{aligned}
T^2(u_1) &= u_3(x) = T(u_2(x)) = f(x) + \lambda \int_a^x K(x,\xi)u_2(\xi)d\xi \\
&= f(x) + \lambda \int_a^x K(x,\xi)[f(\xi) + \lambda \int_a^\xi K(\xi,t)u_1(t)dt]d\xi \\
&= f(x) + \lambda \int_a^x K(x,\xi)f(\xi)d\xi + \lambda^2 \int_a^x \int_a^\xi K(x,\xi)K(\xi,t)u_1(t)dtd\xi \\
&= f(x) + \lambda \int_a^x K(x,\xi)f(\xi)d\xi + \lambda^2 \int_a^x K_2(x,\xi)u_1(\xi)d\xi \\
&= T(f) + \lambda^2 \int_a^x K_2(x,\xi)u_1(\xi)d\xi
\end{aligned}
\tag{E.2}
$$

since the last double integral reduces to the single integral $\int_\xi^x K(x,y)K(y,\xi)dy$ that defines the iterated kernel $K_2(x,\xi)$ as defined in (3.4). Note how the first two terms in (E.2) are known operations on the known function $f(x)$, where we consider them fixed $T(f)$ as far as the mapping of the (variable) estimate u_1 is concerned, and where we are seeking a contractive mapping on u_1. If we repeat this successive process to $u_{n+1} = T(u_n) = T^n(u_1)$, we have

$$
T^n(u_1) = T(u_n) = f(x) + \lambda \int_a^x K(x,\xi)f(\xi)d\xi + \lambda^2 \int_a^x K_2(x,\xi)f(\xi)d\xi
$$

$$
+ \cdots + \lambda^{n-1} \int_a^x K_{n-1}(x,\xi)f(\xi)d\xi + \lambda^n \int_a^x K_n(x,\xi)u_1(\xi)d\xi \tag{E.3}
$$

where $K_n(x,\xi)$ is the nth iterated kernel of (3.4),

$$
K_{n+1}(x,\xi) = \int_\xi^x K(x,t)K_n(t,\xi)dt, \quad K_1(x,\xi) = K(x,\xi) \qquad (3.4)(E.4)
$$

and we note that all the terms, except the last one with u_1, are considered known. We emphasize this point since as we write $|T^n(u_1) - T^n(v_1)|$ next in preparation to show $T^n(u)$ as a contractive mapping, all these known terms will cancel out, leaving us with only the last term, which will involve the desired $|u_1 - v_1|$ difference of the first estimates,

$$
|T^n(u_1) - T^n(v_1)| = |T(u_n) - T(v_n)| \le |\lambda|^n \int_a^x |K_n(x,\xi)||u_1(\xi) - v_1(\xi)|d\xi \tag{E.5}
$$

where, of course, $T^n(v_1)$ is obtained as in (E.3) of $T^n(u_1)$..

Before we use the metric (6.9) on (E.5) in (E.9), we should prepare for an upper bound of the iterated kernel $K_n(x,\xi)$ on the square indicated. Since $K(x,\xi) = K_1(x,\xi)$ is assumed continuous on this bounded square domain, we can conclude that $K_1(x,\xi)$ is bounded by some positive number M, $|K_1(x,\xi)| \le M$. It remains to show the following bound for the iterated kernel $K_n(x,\xi)$:

$$|K_n(x,\xi)| \le \frac{M^n}{(n-1)!}(x-\xi)^{n-1}, \qquad a \le \xi \le x \tag{E.6}$$

which can be established by mathematical induction. We illustrate the cases for $K_2(x,\xi)$ and $K_3(x,\xi)$:

$$\begin{aligned} |K_2(x,\xi)| &= \left| \int_\xi^x K(x,t)K(t,\xi)dt \right| \le \int_\xi^x |K(x,t)||K(t,\xi)|d\xi \le M \int_\xi^x \\ &\quad \cdot |K(t,\xi)|dt \\ &\le M \left[M \max \left| \int_\xi^x dt \right| \right] \le M^2(x-\xi) \end{aligned} \tag{E.7}$$

$$\begin{aligned} K_3(x,\xi) &= \left| \int_\xi^x K(x,t)K_2(t,\xi)dt \right| \le M \int_\xi^x |K_2(t,\xi)|dt \\ &\le M \int_\xi^x M^2(t-\xi)dt \le M^3 \int_\xi^x (t-\xi)dt \\ &\le M^3 \frac{(x-\xi)^2}{2!}, \end{aligned} \tag{E.8}$$

after using the result of (E.7) for the bound on $|K_2(t,\xi)|$ in (E.8). With this result (E.6) and the result (E.5), we write

$$\begin{aligned} d(T^n(u_1), T^n(v_1)) &= \max_x |T^n(u_1) - T^n(v_1)| \\ &= \max_x |\lambda^n| \left| \int_a^x K_n(x,\xi)[u_1(\xi) - v_1(\xi)]d\xi \right| \\ &\le |\lambda|^n \max_x \int_a^x |K_n(x,\xi)||u_1(\xi) - v_1(\xi)|d\xi \\ &\le |\lambda|^n \max_x \int_a^x M^n \frac{(x-\xi)^{n-1}}{(n-1)!}|u_1(\xi) - v_1(\xi)|d\xi \\ &\le |\lambda|^n M^n \max_x |u_1(x) - v_1(x)| \int_a^x \frac{(x-\xi)^{n-1}}{(n-1)!}d\xi \\ &\le |\lambda|^n M^n d(u_1,v_1) \left(-\frac{(x-\xi)^n}{n!} \right) \Bigg|_{\xi=a}^x \\ &= |\lambda|^n M^n \frac{(x-a)^n}{n!} d(u_1,v_1) \le |\lambda|^n M^n \frac{(x-a)^n}{n!} d(u_1,v_1), \\ & d(T^n(u_1), T^n(v_1)) \le \alpha d(u_1,v_1) \end{aligned} \tag{E.9}$$

where

$$\alpha = |\lambda|^n M^n \frac{(x-a)^n}{n!}. \tag{E.10}$$

Hence $T^n(u)$ is contractive if $\alpha < 1$, which, with the help of the $n!$ in the denominator, is the case when n is sufficiently large (i.e., if we wait for more iterations). Of course, if we have our problem on the unit square, $0 \le x \le 1$, then $|x-a| = |x| < 1$ in the factor $|x-a|^n$ of (E.10) will help even more in speeding α of (E.10) toward being less than 1.

If we consult Example 1 of Chapter 3,

$$u(x) = f(x) + \lambda \int_0^x e^{x-t} u(t)dt \tag{E.11}$$

we had a closed form for the iterated kernel,

$$K_n(x,t) = \frac{(x-t)^{n-1}}{(n-1)!} e^{x-t}. \tag{E.12}$$

So on the unit square we have

$$|K_n(x,t)| \le \frac{x^{n-1}}{(n-1)!} e, \qquad M = e \tag{E.13}$$

which is within the (more conservative) bound we obtain from (E.6),

$$|K_n(x,t)| \le \frac{M^n x^{n-1}}{(n-1)!} = \frac{x^{n-1}}{(n-1)!} e^n. \tag{E.14}$$

In this case we use (E.13) in the second line above that of (E.9) to obtain

$$\alpha = |\lambda|^n e \frac{x^n}{n!} \tag{E.15}$$

instead of

$$\alpha = |\lambda|^n e^n \frac{x^n}{n!} \tag{E.16}$$

which corresponds to using (E.14) in (E.9) and (E.10).

6.2 FIXED POINT THEOREM OF BANACH

With the definitions of *metric space, fixed point* of the mapping, and *contractive mapping*, we are now in a position to state and prove a very basic fixed point theorem, the Banach (or *Banach-Cacciopoli*) theorem.

Fixed Point Theorem

Let $(\mathcal{M}, d)$ be a complete metric space and let the mapping $T : \mathcal{M} \to \mathcal{M}$ be a contraction; then T has exactly one fixed point.

Proof: We are to prove the following for the mapping

$$u = T(u) \tag{6.39}$$

(a) The uniqueness of the fixed point when it exists.

(b) The existence of the fixed point, where we show first that the sequence of the successive approximations

$$u_{n+1} = T(u_n) \tag{6.40}$$

is Cauchy convergent, hence convergent since it is in a complete metric space. More important, we show that the limit point for this convergent sequence $u = \lim_{n\to\infty} u_n$ is indeed the fixed point of the actual problem $u = T(u)$.

(a) To prove the uniqueness of the fixed point, suppose that there are two distinct fixed points u and v, $u \neq v$ [i.e., $u = T(u)$ and $v = T(v)$, $u \neq v$]. Since $u \neq v$, the distance between them is not zero: $d(u, v) \neq 0$. Because u and v are fixed points of T, we also have

$$d(T(u), T(v)) = d(u, v) \neq 0. \tag{6.41}$$

But since the mapping T is also contractive, we have, according to (6.12),

$$d(T(u), T(v)) \leq \alpha d(u, v), \qquad 0 \leq \alpha < 1. \tag{6.42}$$

If we combine (6.41) and (6.42), we see clearly that there is a contradiction

$$d(u, v) = d(T(u), T(v)) \leq \alpha d(u, v),$$

$$(1 - \alpha) d(u, v) \leq 0$$

where since $d(u, v) > 0$ by assumption, then $1 - \alpha \leq 0$, $\alpha \geq 1$, which contradicts the assumption of contractive mapping whose α is strictly less than 1. Hence the distance $d(u, v)$ must be identically zero, which is equivalent to u being equal to v, and which proves the uniqueness of the fixed point when it exists.

(b) To prove the existence of a limit point as a fixed point for $u = T(u)$, we will first prove that the sequence u_n of the iterative process

$$u_{n+1} = T(u_n) \tag{6.40}$$

is a Cauchy sequence.

With the help of the contraction property, we will find the distance $d(u_n, u_{n+1})$ between two consecutive approximations in terms of the distance $d(u_2, u_1)$ between the first two approximations (input estimates) u_1 and u_2. The next step is to find the distance $d(u_n, u_{n+p})$, that we need to use in proving the Cauchy convergence (6.11b).

From (6.4) and the assumption of contractive mapping we have

$$d(u_2, u_3) = d(T(u_1), T(u_2)) \leq \alpha d(u_1, u_2). \tag{6.43}$$

By the same reasoning

$$d(u_3, u_4) = d(T(u_2), T(u_3)) \leq \alpha d(u_2, u_3) \tag{6.44}$$

and if we invoke on the right side the previous result for $d(u_2, u_3)$, we have

$$d(u_3, u_4) \leq \alpha d(u_2, u_3) \leq \alpha^2 d(u_1, u_2).$$

If we continue this to u_n and u_{n+1}, we have, as we did for (6.30) and (6.31),

$$d(u_n, u_{n+1}) \leq \alpha^{n-1} d(u_1, u_2) \tag{6.45}$$

where clearly the higher order consecutive iterates $u_n, u_{n+1} (n >> 1)$ are much closer together than the first ones, u_1 and u_2, due to the geometric factor α^{n-1}; $0 \leq \alpha < 1$. Still we have to show the Cauchy convergence, which will entail the use of the important result (6.45) and the triangle inequality of the metric $d(u_n, u_{n+p})$. Observe that

$$\begin{aligned} d(u_n, u_{n+p}) \leq d(u_n, u_{n+1}) + d(u_{n+1}, u_{n+2}) + \\ + d(u_{n+2}, u_{n+3}) + \cdots + d(u_{n+p-1}, u_{n+p}) \end{aligned} \tag{6.46}$$

after repeated use of the triangle inequality (6.17). Now we use property (6.45) on each of the terms on the right side, which are distances for consecutive sequences. Therefore,

$$\begin{aligned} d(u_n, u_{n+p}) &\leq d(u_n, u_{n+1}) + d(u_{n+1}, u_{n+2}) + d(u_{n+2}, u_{n+3}) \\ &\quad + \cdots + d(u_{n+p-1}, u_{n+p}) \\ &\leq \alpha^{n-1} d(u_1, u_2) + \alpha^n d(u_1, u_2) + \alpha^{n+1} d(u_1, u_2) \\ &\quad + \cdots + \alpha^{n+p-2} d(u_1, u_2) \\ &= \left[\alpha^{n-1} + \alpha^n + \alpha^{n+1} + \cdots + \alpha^{n+p-2}\right] d(u_1, u_2) \\ &= \alpha^{n-1}(1 + \alpha + \alpha^2 + \cdots + \alpha^{p-1}) d(u_1, u_2) \\ &= \alpha^{n-1} \frac{1 - \alpha^p}{1 - \alpha} d(u_1, u_2), \end{aligned} \tag{6.47}$$

after realizing that we have a geometric series in the parentheses above. Since $0 \leq \alpha < 1$, the right side would clearly go to zero as $n \to \infty$, which makes $d(u_n, u_{n+p}) \to 0$ as $n \to \infty$ (i.e., the sequence converges in the Cauchy sense).

Since this sequence u_n is an element of a complete metric space, it will converge to a limit u in this space (i.e., $\lim_{n\to\infty} u_n = u$).

What remains is to show that this limit point u is indeed the fixed point of our equation; that is, it must satisfy $u = T(u)$, or in other words, $d(u, T(u)) = 0$. From (6.6) we have

$$u_{n+1} = T(u_n)$$

and from the proof of the existence of the limit point above we can say that

$$\lim_{n\to\infty} u_{n+1} = \lim_{n\to\infty} u_n = u$$

or

$$d(u, T(u_n)) = d(u, u_{n+1}) \to 0, \qquad d(u, u_n) \to 0$$

as $n \to \infty$. With these results we will use the triangle inequality to have

$$\begin{aligned} d(u,T(u)) &\leq d(u,T(u_n)) + d(T(u_n),T(u)) \leq d(u,T(u_n)) + \alpha d(u_n,u) \\ &\leq d(u,u_{n+1}) + \alpha d(u_n,u) \end{aligned} \tag{6.48}$$

after using the contraction property of the operator T in the last term.

As $n \to \infty$ each of the two terms on the right would approach zero, which makes $d(u,T(u)) \leq 0$; but since the metric d is nonnegative by definition, we must have $d(u,T(u)) = 0$, which means that $u = T(u)$, the desired result of the fixed point theorem.

Next we will illustrate this important Banach fixed point theorem to prove the existence of unique solutions to linear and nonlinear Fredholm and Volterra integral equations that exhibit contraction.

As we mentioned earlier, the importance of the foregoing type of proof for the fixed point theorem is that it presents us with a method of constructing the solution. Moreover, it gives an upper bound on the error $\epsilon_n = |u - u_n|$ or in general $d(u,u_n)$, incurred in approximating the solution u by the nth successive approximation u_n, in terms of the difference between the first two estimates, u_1 and u_2,

$$\epsilon_n = d(u,u_n) \leq \frac{\alpha^{n-1}}{1-\alpha} d(u_1,u_2). \tag{6.49a}$$

This is obtained easily from the last line in (6.47); we take the limit as $p \to \infty$, where $\lim_{p\to\infty} \alpha^p = 0$ for $0 \leq \alpha < 1$ on the right side and $\lim_{p\to\infty} u_{n+p} = u$ (since $n + p = m \to \infty$, $\lim_{m\to\infty} u_m = u$) on the left side, to give

$$\lim_{p\to\infty} d(u_n,u_{n+p}) = d(u_n,u) \leq \frac{\alpha^{n-1}}{1-\alpha} d(u_1,u_2) \tag{6.49b}$$

6.2.1 Existence of the Solution for Linear Integral Equations

Linear Fredholm Equations

For our illustration of the linear Fredholm integral equation

$$u(x) = g(x) + \lambda \int_a^b K(x,t)u(t)dt \qquad (6.32), (5.7a)$$

we found in (6.34) that $\alpha = \lambda M(b-a)$, which gives a contractive mapping if we insist that $\alpha < 1$ [i.e., $\lambda < 1/M(b-a)$, where M is the upper bound of $|K(x,t)|$ on the square of Figure 6.4]. In this case the u_n estimate as an input would produce an output u_{n+1} that has a maximum error bounded as in (6.49a),

$$\epsilon_n = \max_{x\in[a,b]} |u - u_{n+1}| \leq \frac{|\lambda M(b-a)|^n}{1 - |\lambda M(b-a)|} \max_{x\in[a,b]} |u_2 - u_1|, \quad |\lambda M(b-a)| < 1 \tag{6.50}$$

in terms of the maximum difference between the first two estimates, u_2 and u_1.

Next we consider Example 12 of Chapter 5, where we solved the Fredholm integral equation

$$u(x) = f(x) + \lambda \int_0^1 xe^t u(t)dt$$

by the successive approximation (iterated kernels–Neumann series) method. We will compare the contraction condition for the existence of the solution, which we have established here, with the one we merely stated above equation (5.79) in Chapter 5 for the convergence of the Neumann series solution (5.80).[2]

Example 2 Existence of the Unique Solution for Fredholm Linear Equations

Consider the following Fredholm integral equation; let us find a condition that assures the existence of the solution.

$$u(x) = f(x) + \lambda \int_0^1 xe^t u(t)dt. \tag{E.1}$$

Assume that $f(x)$ is continuous on $[0, 1]$. $K(x,t) = xe^t$ is obviously continuous on the square $x \in [0,1], t \in [0,1]$ (see Figure 6.4); hence it is bounded there and we can easily see that a bound M is e, that is,

$$M = \max_{\substack{x\in[0,1] \\ t\in[0,1]}} |xe^t| = e. \tag{E.2}$$

So according to (6.35), this Fredholm equation represents a contractive mapping, and hence assures the existence of a unique solution if

$$\alpha = \lambda M(b-a) = \lambda e(1-0) = \lambda e < 1$$

that is, if $|\lambda| < 1/e \sim 0.37$.

Now we would like to compare this condition with that which we gave for the linear Fredholm integral equation

$$u(x) = f(x) + \lambda \int_a^b K(x,t)u(t)dt \tag{5.21}$$

to be $|\lambda| < 1/B$, where B is given in (5.79) as

$$B = \sqrt{\int_a^b \int_a^b K^2(x,t)dxdt} \tag{5.79}$$

B was calculated in Example 12 of Chapter 5 to be $B = \sqrt{(e^2-1)/6}$, where the condition

$$|\lambda| < \frac{1}{B} \quad \textit{becomes} \quad |\lambda| < \frac{1}{B} = \sqrt{\frac{6}{e^2-1}} \sim 0.97.$$

[2]See Pogorzelski [1966].

This $|\lambda| < 0.97$, when compared with ours of $\lambda < 0.37$, makes the contraction approach to convergence appear more conservative. The reason lies in the nature of a special complete metric space of *square integrable* functions on (a, b) in which the condition $|\lambda| < 1/B$ was obtained. These square integrable functions $f(x)$ on (a, b) were discussed in Section 4.1.3 in relation to their Fourier series representation in (4.47) and (4.48). They are defined such that $\int_a^b |f(x)|^2 dx < \infty$. For the space of these functions we define the metric $d(f(x), g(x)) = \sqrt{\int_a^b |f(x) - g(x)|^2 dx}$, whence they constitute a complete metric space. (See (4.52) and some of the references given in the first page of this Chapter.)

Linear Volterra Equations

The following example illustrates the existence of a unique solution for Volterra equations.

Example 3 Existence of the Unique Solution for Linear Volterra Integral Equation

In Example 1 we showed that the nth-order mapping $T^n(u)$ for the Volterra equation

$$u(x) = f(x) + \lambda \int_a^x K(x,t)u(t)dt \tag{3.1}$$

$$= T(u) \tag{E.1}$$

is contractive. We show here that this implies the existence of a solution u to $u = T(u)$ above, and that this solution u is unique.

Let $T^n(u) = S(u) = u$. Since $T^n(u)$ is contractive, then by the Banach fixed point theorem it has a unique solution u,

$$T^n(u) = S(u) = u. \tag{E.2}$$

This means that with the first estimate u_1, we have the sequence $u_{k+1} = S(u_k) = S^k(u_1)$ converging to u, that is,

$$u = \lim_{k\to\infty} u_{k+1} = \lim_{k\to\infty} S^k(u_1) = \lim_{k\to\infty} (T^n)^k(u_1) = \lim_{k\to\infty} T^{nk}(u_1) = u. \tag{E.3}$$

Recall from Example 1 that $T^n(u)$ was proved contractive for large enough n, so with the unique solution u for $T^n(u) = u$, it should be clear that for the even larger kn, $T^{kn}(u) = u$ has the same solution u of $T^n(u) = u$, n large.

In (E.3) we have the first estimate u_1, being arbitrary, so we may choose it to be $u_1 = T(u)$,

$$\begin{aligned} u &= \lim_{k\to\infty} T^{nk}(u_1) = \lim_{k\to\infty} T^{nk}(T(u)) \\ &= \lim_{k\to\infty} T^{nk+1}(u) = \lim_{k\to\infty} T[T^{kn}(u)] = \lim_{k\to\infty} T(u) = T(u), \\ u &= T(u) \end{aligned} \tag{E.4}$$

after using $T^{nk}(u) = u$ in the second line for large n.

Hence (E.4) represents the existence of the solution u to $T(u) = u$. To prove that u is unique, let $\gamma, \beta, \gamma \neq \beta$, be two different solutions to $u = T(u)$ [i.e., $\gamma = T(\gamma)$, $\beta = T(\beta)$]. But since $\gamma = T(\gamma)$, then

$$T^n(\gamma) = T^{n-1}[T(\gamma)] = T^{n-1}(\gamma) = \cdots = T(\gamma) = \gamma,$$

$$T^n(\gamma) = \gamma. \tag{E.5}$$

The same can be shown for β,

$$T^n(\beta) = \beta. \tag{E.6}$$

But since T^n is known to be contractive, it must have a unique solution which forces $\gamma = \beta$. Hence $T(u) = u$ has a unique solution.

The following section represents our only (brief) discussion on analysis of *nonlinear* integral equations. It deals first with applying the fixed point theorem to nonlinear integral equations. This is followed by a simple initial value problem associated with a first-order nonlinear differential equation to illustrate the importance of the integral representation of differential equations.

6.2.2 Existence of the Solution for Nonlinear Integral Equations

In the preceding section we limited our illustrations of the Banach fixed point theorem to the existence of unique solutions of the *linear* Fredholm and Volterra integral equations. In this section we apply the fixed point theorem to nonlinear Fredholm and Volterra equations. This is followed in Section 6.2.3 by an initial value problem associated with first-order nonlinear differential equation. The latter problem is added to indicate the importance of having to change to the integral representation in order to enjoy the method of successive approximations, and where proving the contraction property is greatly facilitated when working with an integral operator, as we illustrated for linear integral equations in Section 6.2.1.

Nonlinear Fredholm Equations

Consider the nonlinear Fredholm integral equation

$$u(x) = f(x) + \lambda \int_a^b F(x, t, u(t))dt = T(u) \tag{6.51}$$

where we assume $f(x)$ continuous on $[a, b]$ and that $F(x, t, u(t))$ is continuous, hence bounded on the square of Figure 6.4, $|F(x, t, u(t))| \leq M$ for bounded $u(t)$: $c < u(t) < d$. Consider also the successive approximation of (6.51),

$$u_{n+1}(x) = f(x) + \lambda \int_a^b F(x, t, u_n(t))dt \tag{6.52}$$

and the metric (6.9) with the metric space of continuous functions $C[a, b]$.

To show whether the mapping in (6.51) is a contractive one, we must first look at the distance between the images $T(\beta(x))$ and $T(\gamma(x))$ of the inputs $\beta(x)$ and $\gamma(x)$ in $C[a, b]$,

$$\begin{aligned} d(T(\beta),\ T(\gamma)) & \\ &= \max_x \left| f(x) + \lambda \int_a^b F(x, t, \beta(t)) - \left[f(x) + \lambda \int_a^b F(x, t, \gamma(t))dt \right] \right| \\ &= \max_x |\lambda| \int_a^b |F(x, t, \beta(t)) - F(x, t, \gamma(t))|dt. \end{aligned} \tag{6.53}$$

To relate this distance $d(T(\beta), T(\gamma))$ of the outputs to that of the inputs $d(\beta(t), \gamma(t))$ we need to have the maximum operation of the last line taken on $|\beta(t) - \gamma(t)|$, which clearly is not available in this form, as seen inside the integral of (6.53). To have $|\beta(t) - \gamma(t)|$ freed from the operation of the function F inside this integral, we impose a well-known condition on F that would satisfy our goal, called the *Lipschitz condition.*

The function $F(x, t, u(t))$ is called *Lipschitz* with respect to the variable $u(t)$ if there is a positive constant L such that

$$|F(x, t, \beta(t)) - F(x, t, \gamma(t))| \leq L|\beta(t) - \gamma(t)| \tag{6.54}$$

for $(x, t, \beta(t))$ and $(x, t, \gamma(t))$ in the domain of F.

If we impose this Lipschitz condition on F inside the integral of (6.53), we have

$$\begin{aligned} d(T(\beta), T(\gamma)) &\leq |\lambda| L \max_x \int_a^b |\beta(t) - \gamma(t)|dt \leq |\lambda| L(b - a) \max_x |\beta(x) - \gamma(x)| \\ &\leq |\lambda| L(b - a) d(\beta, \gamma) \end{aligned} \tag{6.55}$$

where we have $\alpha = |\lambda| L(b - a)$. Hence the mapping $T(u)$ of (6.51) becomes contractive when $\alpha = |\lambda| L(b - a) < 1$,

$$|\lambda| < \frac{1}{L(b - a)} \tag{6.56}$$

where L is the Lipschitz constant of $F(x, t, u(t))$ as in (6.54).

We note that if F is linear in $u(t)$, as in our illustrations in Section 6.2.1, then F is always Lipschitz, since for $F(x, t, u(t)) = K(x, t)u(t)$ we have

$$\begin{aligned} |F(x, t, \beta)) - F(x, t, \gamma(t))| &= |K(x, t)\beta(t) - K(x, t)\gamma(t)| = |K(x, t)[\beta(t) - \gamma(t)]| \\ &= |K(x, t)||\beta(t) - \gamma(t)| \leq M|(\beta(t) - \gamma(t)| \end{aligned} \tag{6.57}$$

where M, the upper bound of $|K(x,t)|$, can stand for L, the Lipschitz constant.

We also note that from the start we assumed that $F(x,t,u(t))$ is continuous in all three variables, but clearly the continuity of F in $u(t)$ does not imply that it is Lipschitz in $u(t)$. A simple counterexample is that of $F(x,t,u(t)) = xt\sqrt{u(t)}$, which is continuous but not Lipschitz in $u(t)$. However, if $F(x,t,u(t))$ has a continuous partial derivative $\partial F/\partial u$ in the domain D of F, then F is Lipschitz in u, as we will show next, and

$$L = \max_{(x,t)\epsilon D} \left| \frac{\partial F}{\partial u} \right|. \tag{6.58}$$

Since $F(x,t,u(t))$ is assumed to have continuous partial derivative $\partial F/\partial u$ in D, we can use the *mean value theorem*, which states that for any $u_1(t)$ and $u_2(t)$ in D there is an $\eta(t)$ between them, $u_1(t) < \eta(t) < u_2(t)$, such that

$$\frac{F(x,t,u_1(t)) - F(x,t,u_2(t))}{u_1(t) - u_2(t)} = \frac{\partial F}{\partial u}(x,t,\eta(t)). \tag{6.59}$$

From this result we have

$$|F(x,t,u_1(t)) - F(x,t,u_2(t))| = \left| \frac{\partial F}{\partial u}(x,t,\eta(t)) \right| |u_1(t) - u_2(t)| \le L|u_1(t) - u_2(t)| \tag{6.60}$$

where $L = \max_{(x,t)\epsilon D} |\partial F/\partial u|$, as given in (6.58).

To take a simple example consider $F(x,t,u(t)) = t^2 \sin u(t)$ on the rectangle $0 \le x \le 3, 0 \le t \le 1$. $\partial F/\partial u = t^2 \cos u(t)$, where according to (6.58), L is

$$L = \max_{(x,t)\epsilon D} \left| \frac{\partial F}{\partial u} \right| = \max_{0 \le t \le 1} |t^2 \cos u(t)| \le 1$$

since both t^2 and $\cos u(t)$ are bounded from above by 1.

Besides this important Lipschitz condition on $F(x,t,u(t))$ for ensuring the contractive property of $T(u)$ in (6.51) when $|\lambda| < 1/L(b-a)$ in (6.56), we should also ensure that the output $T(u)$ is bounded. This is accomplished by assuming that $F(x,t,u(t))$ is bounded by a constant k,

$$|F(x,t,u(t))| < k. \tag{6.61}$$

With this condition on F in (6.51) we have

$$|u(x) - f(x)| = \left| \lambda \int_a^b F(x,t,u(t))dt \right| \le |\lambda| \int_a^b |F(x,t,u(t))|dt \le |\lambda| k(b-a). \tag{6.62}$$

This means that if our input estimates $u_n(t)$ in (6.52) are bounded [i.e., $c \le u_n(t) \le d$], the outputs $u_{n+1}(t)$ can also be bounded within the *same* range by limiting the value of λ in (6.62) and taking into consideration the bounds on f, $m_1 \le f \le m_2$. This amounts to choosing λ as

$$|\lambda| \leq \min\left(\frac{m_1 - c}{k(b-a)}, \frac{d - m_2}{k(b-a)}\right). \tag{6.63}$$

So to have this condition of boundedness on the successive approximations as well as the important contractive property condition (6.56) on λ, we require that

$$|\lambda| < \min\left(\frac{m_1 - c}{k(b-a)}, \frac{d - m_2}{k(b-a)}, \frac{1}{L(b-a)}\right) \tag{6.64}$$

when $m_1 < f(x) < m_2$, $c < T(u_n) < d$, and L is the Lipschitz constant of F as in (6.54), (6.58), or (6.60).

In regards to the iterative process (6.52) and its mapping T, there can be different variations on it that may result in a better contraction property for its associated modified mapping T_m (see Jerri [1991] Jerri et. al. [1987], Jerri and Herman [1996]).

Nonlinear Volterra Equations

Consider the nonlinear Volterra integral equation

$$u(x) = f(x) + \lambda \int_a^x F(x, t, u(t))dt = T(u). \tag{6.65}$$

As for the nonlinear Fredholm equation (6.51), we will assume that $f(x)$ is continuous on $[a, b]$ and bounded: $m_1 < f(x) < m_2$; $F(x, t, u(t))$ is continuous with respect to the three variables x, t, and $u(t)$ on the domain D: $a \leq x \leq b$, $a \leq t \leq x$, $u(t)$ unbounded: $c < u(t) < d$; and $F(x, t, u(t))$ is Lipschitz with respect to $u(t)$. To ensure that the outputs $u_{n+1}(x)$,

$$u_{n+1}(x) = f(x) + \lambda \int_a^x F(x, t, u_n(t))dt = T(u_n) \tag{6.66}$$

are always bounded within the range $c < u_n(t) < d$ of the inputs, we follow the same steps for the Fredholm equation in getting the condition (6.63) on λ to come up with similar condition on λ of (6.66),

$$|\lambda| < \min\left(\frac{m_1 - c}{k(b-a)}, \frac{d - m_2}{k(b-a)}\right) \tag{6.67}$$

where k is the upper bound of F (i.e., $|F| \leq k$). As we have shown for the linear Volterra integral equations, the proof of a contractive mapping for the Volterra equations does not require an extra condition on λ as long as we take large enough n for $u_n(x)$. This is, of course, a welcome nicety of the Volterra equations which stems from the nature of its origin as an initial value problem.

For the linear Volterra equations, we showed in Example 1 that $T^n(u)$ is contractive, then concluded from that in Example 3 that $T(u) = u$ of the Volterra equation has a unique solution. This was accomplished with the aid of the iterated kernels, which are clearly exclusive for the linear case only. Here we will follow a slightly different procedure to get to the contraction of $T(u)$ in (6.65) and the convergence of the sequence $u_n(t)$ of (6.66) to the unique solution $u(t)$.

Since $F(x,t,u(t))$ in (6.65) and (6.66) is assumed Lipschitz, we can follow what we did in (6.54) and (6.55), for the nonlinear Fredholm equation, and write

$$|u_{n+1}(x) - u_n(x)| < L|\lambda| \int_a^x |u_n(t) - u_{n-1}(t)|dt \tag{6.68}$$

where $u_n(x)$ is given in (6.66).

Shortly we will show that [see (6.73)]

$$|u_{n+1}(x) - u_n(x)| < |c-d||\lambda L|^{n-1} \frac{|x-a|^{(n-1)}}{(n-1)!}. \tag{6.69}$$

This will allow us to conclude that the series

$$\sum_{n=1}^{\infty} [u_{n+1}(x) - u_n(x)] \tag{6.70}$$

is absolutely and uniformly convergent since it is dominated by an infinite series [of the sequence on the right of the inequality in (6.69)] which obviously does converge uniformly with no restriction on λ.

To show the convergence of the successive approximation (6.66) sequence $u_n(x)$ to the solution $u(x)$ of (6.65), we first note that $u_n(x)$ can be written, with the help of the telescoping terms, as

$$\begin{aligned} u_n(x) &= u_1(x) + u_2(x) - u_1(x) + u_3(x) - u_2(x) + \cdots + u_n(x) - u_{n-1}(x) \\ &= u_1(x) + \sum_{j=1}^{n-1} [u_{j+1}(x) - u_j(x)]. \end{aligned} \tag{6.71}$$

So with the uniform convergence of the series (6.70) which we have here on the right side of (6.71), we can take the limit of both sides of (6.71) as $n \to \infty$ to conclude that $\lim_{n\to\infty} u_n(x)$ exists for all $x \in [a,b]$. But this is exactly the sequence $u_n(x)$ in (6.66), and since we assumed that $F(x,t,u_n(t))$ is continuous in $u_n(t)$, we can take the limit as $n \to \infty$ on both sides of (6.66), allowing $\lim\limits_{n\to\infty} F(x,t,u_n(t)) = F(x,t,u(t))$, to conclude that $\lim_{n\to\infty} u_n(x) = u(t)$, the solution to the nonlinear Volterra equation (6.65).

What remains is to prove the result in (6.69). From (6.68) we have, for $n=2$,

$$|u_3(x) - u_2(x)| < L|\lambda| \int_a^x |u_2(t) - u_1(t)|dt < L|\lambda||c-d||x-a| \tag{6.72}$$

after using the bounds set on $u(t) : c < u(t) < d$. In the same way we show that

$$
\begin{aligned}
|u_4(x) - u_3(x)| &< L|\lambda| \int_a^x |u_3(t) - u_2(t)|dt < L\lambda \int_a^x L|\lambda||c-d||t-a|dt \\
&< |L\lambda|^2 |c-d| \frac{|x-a|^2}{2} \\
|u_5(x) - u_4(x)| &< L|\lambda| \int_a^x |u_4(t) - u_3(t)|dt < |L\lambda|^3 |c-d| \int_a^x \frac{|t-a|^2}{2} dt \\
&< |L\lambda|^3 |c-d| \frac{|x-a|^3}{3!}
\end{aligned}
\tag{6.73}
$$

where a simple mathematical induction establishes (6.69).

6.2.3 Existence of the Solution for Nonlinear Differential Equations

Here we consider an initial value problem associated with a first-order nonlinear differential equation in $y(x)$ on the domain $D(x, y)$ of a rectangle as indicated in Figure 6.5.

$$\frac{dy}{dx} = f(x, y), \quad |x - x_0| < a, \quad |y - y_0| < b \tag{6.74}$$

$$y(x_0) = y_0 \tag{6.75}$$

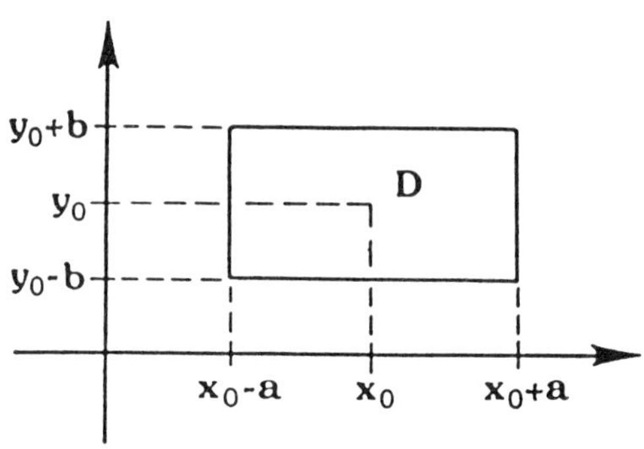

Fig. 6.5 Domain D of (6.74).

We will assume that $f(x, y)$ is continuous, and in anticipation of the use of the fixed point theorem for the integral representation of this problem we assume that $f(x, y)$ is also Lipschitz in $y(x)$,

$$|f(x, \beta(x)) - f(x, \gamma(x)| < L|\beta(x) - \gamma(x)| \tag{6.76}$$

If we integrate (6.74) from x_0 to x, and involve the initial condition $y(x_0) = y_0$, we have the nonlinear (special) Volterra integral equation

$$y(x) = y_0 + \int_{x_0}^{x} f(t, y(t))dt. \tag{6.77}$$

In light of the foregoing development, this represents a very special case of (6.65) with $F(x, t, u(t)) = f(t, u(t))$. To keep the outputs $u_{n+1}(x)$ bounded, we also assume that f is bounded by k [i.e., $|f(t, u(t))| < k$]. Thus the integral representation (6.77) would have a unique solution, which is obtained via the successive approximation of this representation,

$$y_{n+1}(x) = y_0 + \int_{x_0}^{x} f(t, y_n(t))dt. \tag{6.78}$$

Hence we observe the clear advantage of using this iterative process (6.78) once the initial value problem associated with the differential equations is changed to its equivalent Volterra integral equation representation.

7

Higher Quadrature Rules for the Numerical Solutions

As we emphasized at the end of Section 1.5, besides the very basic numerical integrations formulas of the trapezoidal rule (1.141) and Simpson's rule (1.144), there are many other numerical integration formulas (or numerical quadrature rules). They are, of course, used for more accurate way of approximating the integral as compared to their special cases (lower order quadrature rules) of the trapezoidal and Simpson's rules. These two rules correspond to using, respectively, first and second degree polynomials, while the higher quadrature rules, to be discussed here, use high degree polynomials.

7.1 HIGHER QUADRATURE RULES OF INTEGRATION WITH TABLES

To give a brief discussion of the quadrature rules we return to our first numerical approximation of integrals using weights (or quadratures) D_i in (1.140) with a minor modification

$$\int_a^b f(x)dx = \sum_{i=1}^{N} D_i f(x_i) + \epsilon = S_N + \epsilon \tag{7.1}$$

where ϵ is the error of the approximation, and the summation here is over $i = 1$ to N,

$$S_N = \sum_{i=1}^{N} D_i f(x_i) \tag{7.2}$$

Here N denotes the number of (in general) not necessarily equidistant sample locations $\{x_i\}_{i=1}^N$. With this note, we should mention that there are numerical methods that use the end points of the interval $x = a$ and b as x_1 and x_N, respectively, which are termed *closed* rules. Those rules which avoid the end points and use $x_1 = a + \epsilon_1$ and $x_N = b - \epsilon_2$ as first and last sample locations in the interior of the interval (a, b) are called *open* rules. The latter rules are useful in case there are *singularities* at the end points. Also since $b - a$ is constant, we may write the weight $D_i = (b - a)w_i$, where we list the values of the weights w_i in Tables 7.1 to 7.6 (in this Section) for the representative quadrature rules of interest in this presentation.

In (7.2) we observe that the approximation sum S_N has two variables; namely, the *locations* $\{x_i\}$ and the *weights* D_i. As we mentioned earlier, there are basically two groups of numerical integration methods, where the main difference depends on their use of these two variables. For lack of space, and instead of just presenting the formulas without derivation, we shall be satisfied with sketching the outline of the essential steps of such derivation. The details are left to the exercises with ample guiding hints, and can be found in the already cited references in Section 1.5. These two groups of methods start by expressing the function to be integrated $f(x)$ in terms of P known functions (*basis*) $h_i(x)$, $i = 1, 2, \cdots, P$,

$$f(x) = \sum_{i=1}^{P} \alpha_i h_i(x). \tag{7.3}$$

This is to be substituted in the integral of (7.1), and the criterion here is to have the error ϵ [as in (7.1)] of the approximation vanish for all coefficients α_i, $i = 1, 2, \cdots, P$. After evaluating the P integrations involved of $\int_a^b h_i(x)dx$, $i = 1, 2, \cdots, P$, this amounts to equating coefficients for each α_i in the resulting (7.1) with the condition that the error $\epsilon = 0$. The result is a system of P linear equations in the $2N$ unknowns of the locations $\{x_i\}_{i=1}^N$ and the weight $\{D_i\}_{i=1}^N$. The two different groups of numerical *quadrature rules* differ basically in their dealing with such $2N$ variables. We shall discuss the *(closed) Newton-Cotes rule*, and the *(open) Gauss* (or *Gauss-Legendre) rule* as representatives of the two groups.

A. The Newton-Cotes (NC) *(closed)*[1] Quadrature Type Rules

The first group of quadrature rules fixes the locations by considering *equidistant* samples at $x_i = a + (i - 1)h$, $i = 1, 2, \cdots, N$, where $h = \frac{b-a}{N-1}$, and takes $P = N$. This results in a convenient N by N *square* system of linear equations in the N unknown weights D_i, $i = 1, 2, \cdots, N$ of (7.2) to be determined. So, if the determinant of the coefficients matrix of these D_i is not zero, and if all the above computed integrals $\int_a^b h_i(x)dx$ did exist, then there is a unique solution for the weights D_i.

As a representative of this group, the Newton-Cotes (NC) rule uses a simple *monomial* $h_i(x) = x^{i-1}$ for the basis in the expansion (7.3) of $f(x)$. This results in

[1]Newton-Cotes rule of the *open* type are found in Table 7.1(b), while a list of most of the present closed type rules are found in Table 7.1(a).

simplifying the above needed integrations, and gives the N weights D_i for what is termed the $N-point$ *rule* of degree $N-1$. These weights are listed in Table 7.1(a) for approximating the integral $\int_a^b f(x)dx$,[2]

$$\int_a^b f(x)dx = \int_{x_1}^{x_N} f(x)dx \approx \sum_{i=1}^{N} hw_i f(x_i) \tag{7.4}$$

for the cases $N = 2, 3, 4$ and 5. The weights are tabulated as w_i for $D_i = hw_i$, where $h = \frac{b-a}{N-1}$.

For this table, it is important to note that if we write this rule for approximating the integral $\int_0^1 f(x)dx$ as

$$S_N(0,1) = \sum_{i=1}^{N} \frac{1}{N-1} w_i f\left(\frac{i-1}{N-1}\right) \tag{7.5}$$

then for the integral $\int_a^b f(x)dx$, the sample locations are scaled by $b-a$ and translated by a as $x_i = a + (i-1)h$, and that the weights are also scaled by $(b-a)$ as $w_i = hw_i, h = \dfrac{b-a}{N-1}$,

$$S_N(a, a+(N-1)h) = \sum_{i=1}^{N} hw_i f(a+(i-1)h). \tag{7.6}$$

For illustration we will write the first three cases explicitly, for completeness we will also give the error estimate of such approximations. For $N = 2$, we have $x_1 = a$, $x_2 = b$, and

$$\int_{x_1}^{x_2} f(x)dx = \frac{h}{2}[f(x_1) + f(x_2)] - \frac{h^3}{12} f''(\xi), \quad x_1 < \xi < x_2 \tag{7.7}$$

where $h = \frac{b-a}{N-1}$, which we recognize as a two-point *trapezoidal* rule. The trapezoidal rule for the N points is called the *extended, composite, or repeated trapezoidal* rule, which is clearly based on this two-point rule, where its first degree polynomial (straight line) is used for each two adjacent points to result in (1.141). Attention should be paid to the difference between the error bound given in (1.143) for the (extended) trapezoidal rule (1.141), and the error estimate given in (7.7) for the two-point (basic trapezoidal) rule.

For $N = 3$, we have, $x_1 = a$, $x_2 = \dfrac{a+b}{2}$, $x_3 = b$, and

[2]For the rest of the related quadrature formulas, and the more detailed tables with high accuracy, see Abramowitz and Stegun [1965, pp. 885-890 and pp. 916-924, respectively].

Table 7.1 Newton-Cotes Rules

(a) Newton-Cotes Rules of the *Closed* Type

Trapezoidal Rule

$$\begin{aligned}\int_{x_1}^{x_2} f(x)dx &= \frac{h}{2}(f_1 + f_2) - \frac{h^3}{12}f''(\xi),\\ &\quad x_1 < \xi < x_2,\\ h &= x_2 - x_1\end{aligned}$$

Extended (composite) Trapezoidal Rule

$$\begin{aligned}\int_{x_1}^{x_N} f(x)dx &= h\left[\frac{f_1}{2} + f_2 + \cdots + f_{N-1} + \frac{f_N}{2}\right] - \frac{(N-1)h^3}{12}f''(\xi),\\ f_n &\equiv f(x_n); \quad x_1 < \xi < x_N, \quad h = \frac{x_N - x_1}{N-1}\end{aligned}$$

Simpson's Rule

$$\begin{aligned}\int_{x_1}^{x_2} f(x)dx &= \frac{h}{3}[f_1 + 4f_2 + f_3] - \frac{h^5}{90}f^{(4)}(\xi),\\ &\quad x_1 < \xi < x_3, \qquad h = \frac{x_3 - x_1}{2}\end{aligned}$$

Extended (composite) Simpson's Rule

$$\begin{aligned}\int_{x_1}^{x_{2n+1}} f(x)dx &= \frac{h}{3}[f_1 + 4(f_2 + f_4 + \cdots f_{2n})\\ &\quad +2(f_3 + f_5 + \cdots + f_{2n-1}) + f_{2n+1}] - \frac{nh^5}{90}f^{(4)}(\xi),\\ h &= \frac{x_{2n+1} - x_1}{2n}\end{aligned}$$

(Simpson's $\frac{3}{8}$ Rule)

$$\int_{x_1}^{x_4} f(x)dx = \frac{3h}{8}(f_1 + 3f_2 + 3f_3 + f_4) - \frac{3f^{(4)}(\xi)h^5}{80}$$

(n-point Newton-Cotes Rules, $n = 5, 6, 7, 8, 9$)

$$\int_{x_1}^{x_5} f(x)dx = \frac{2h}{45}(7f_1 + 32f_2 + 12f_3 + 32f_4 + 7f_5) - \frac{8f^{(6)}(\xi)h^7}{945}$$

$$\vdots$$

(b) Newton-Cotes Rules of the *Open* Type

$$\int_{x_1}^{x_4} f(x)dx = \frac{3h}{2}(f_2 + f_3) + \frac{f^{(2)}(\xi)h^3}{4}$$

$$\int_{x_1}^{x_5} f(x)dx = \frac{4h}{3}(2f_2 - f_3 + 2f_4) + \frac{28f^{(4)}(\xi)h^5}{90}$$

$$\int_{x_1}^{x_3} f(x)dx = \frac{h}{3}[f(x_1) + 4f(x_2) + f(x_3)] - \frac{h^5}{90} f^{(4)}(\xi), \qquad x_1 < \xi < x_3 \tag{7.8}$$

where we use the notation $f^{(n)}(x) \equiv \frac{d^n f}{dx^n}$. This is the three-point and degree 2 rule, which is the basic Simpson's rule for three points. Again this represents the backbone for the repeated (or extended) Simpson's rule (1.144) with N points, where its derivation uses the first three samples $f(x_0)$, $f(x_1)$, and $f(x_2)$ and fits them to a polynomial of degree 2 (parabola), which in other words uses the above basic three-point rule (7.8) to approximate the integral on (x_0, x_2). This process is repeated for the three samples $f(x_2)$, $f(x_3)$, and $f(x_4)$ using the same rule in (7.8) and so on $\cdots$ to result in (1.144), where n is even, $N = n+1$. For the basic Simpson's rule (7.8) of three-point (and degree $m = 2$), if it is repeated M times, the total number of points is $N = mM + 1 = 2M + 1 = n + 1$, and it is termed Simpson's of M *panels* or the familiar composite (or repeated) Simpson's rule (1.144). The same is said about the composite trapezoidal rule (1.141) as the basic two-point Newton-Cotes rule of degree 1 with M panels, $N = M + 1 = n + 1$. After we present the other higher degree Newton-Cotes rules next, we will see that they can be extended in the same way as high degree NC rules with repeated M panels.

For $N = 4$, we have the four-point, third degree $(m = 3)$ (closed) rule,

$$\int_{x_1}^{x_4} f(x)dx = \frac{3h}{8}[f(x_1) + 3f(x_2) + 3f(x_3) + f(x_4)] - \frac{3h^5 f^{(4)}(\xi)}{80}, \qquad x_1 < \xi < x_4 \tag{7.9}$$

which is termed Simpson's 3/8-rule.

In addition to the tabulation of the rest of these (*closed*) Newton-Cotes rules in Table 7.1(a) for N up to 5, we refer the reader to the most basic reference of Abramowitz and Stegun (1965, pp. 885–924) for the rest of the above explicit high quadrature rules with N up to 11, along with estimates for their errors. This important reference of a "Handbook of Mathematical Functions with Formulas, Graphs and Tables" is a very valuable reference, which, for our purpose here, has all the different quadrature rules, along with their very accurate weights w_i and locations x_i for reasonable values of N, and very high N for the Gauss-Legendre quadrature rule.

The Newton-Cotes Repeated Rules: $NC(m, M)$ of Panel M and Degree m

As was discussed above, the familiar Simpson's rule (1.144) is a repeated three-point, degree $m = 2$ Newton-Cotes rule (7.8) with size panel M, where $N = mM + 1 = 2M + 1 = n + 1$. The size of the panel is obtained from $b - a = Mh$, where h is the size of the subinterval used repeatedly with the three-point rule (7.8). The same repeated process can be done for the higher order Newton-Cotes method with panel size M, and are called the *repeated* Newton-Cotes (here *closed*) rules. For an indication of the importance of such extension, one needs only to look at the basic three-point Simpson's rule (7.8), and how inefficient it would be for approximating integrals on $[a, b]$ with its mere three samples of the integrand. The same unsatisfactory approximation property is observed about the high degree Newton-Cotes N-point rules, even when the integrated function is well behaved.

The derivation of the repeated Newton-Cotes (closed) rules with M panels parallels exactly what we described, and is well known in calculus texts, for Simpson's (composite or repeated) rule (1.144) and the (composite or repeated) trapezoidal rule (1.141). Here we present the final result

$$\int_a^b f(x)dx \approx \sum_{k=1}^{M}\sum_{i=1}^{N} hw_i f(a+(k-1)h+(i-1)h) \tag{7.10}$$

and remind of the allowed translation and scaling that we discussed in (7.6) versus (7.5) for each subinterval of integration in the above summation, and where the weight for the rule, according to (7.6), is hw_i.

Example 1 Newton-Cotes Rule

(a) As an illustration we use the integral $\int_0^1 \frac{dx}{1+x^2}$ whose exact value is $\frac{\pi}{4} \simeq 0.785398$. We first use the three-point (second degree) Newton-Cotes rule [the nonrepeated Simpson's rule (7.8)] with $N = 3$, $h = \frac{1-0}{2} = \frac{1}{2}$ to have

$$\begin{aligned}\int_0^1 \frac{1}{1+x^2}dx &\approx S_3(0,1) = \frac{1}{6}\left[f(0) + 4f\left(\frac{1}{2}\right) + f(1)\right] \\ &= \frac{1}{6}\left[1 + 4\left(\frac{4}{5}\right) + \frac{1}{2}\right] = \frac{1}{6}\frac{47}{10} = \frac{47}{60} \\ &= 0.78333\end{aligned} \tag{E.1}$$

(b) Now we use the four-point (degree 3) Newton-Cotes (or the $\frac{3}{8}$ Simpson's) rule of (7.9) with $h = \frac{1}{3}$ to have

$$\begin{aligned}S_4(0,1) &= \frac{3}{8}\left(\frac{1}{3}\right)\left[f(0) + 3f\left(\frac{1}{3}\right) + 3f\left(\frac{2}{3}\right) + f(1)\right] \\ &= \frac{1}{8}\left[1 + 3\left(\frac{9}{10}\right) + 3\left(\frac{9}{13}\right) + \frac{1}{2}\right] = \frac{51}{65} \\ &= 0.7846\end{aligned} \tag{E.2}$$

which shows a very good improvement over the above (nonrepeated) Simpson's rule in (E.1). We will leave the details to an Exercise for comparing these results with the results of the other more accurate rules such as the 2-panels Simpson's rule, the six-point, degree 5 Newton-Cotes rule, and the repeated six-point, degree 5 Newton-Cotes rule with $M = 2$ panels. In Example 2(a) we will use a four-point Gauss-Legendre rule with its better approximation compared to the present result of the four-point Newton-Cotes rule.

The Maclaurin Rule

There is another (hybrid!) method, which sets equidistant samples locations x_i, but varies the weight w_i in order to minimize the error ϵ of S_N in (7.1), namely, *the Maclaurin Method*. We list its *fixed uniform* locations x_i and *variable* weights w_i in Table 7.2. This method, as seen from its locations in Table 7.2[3] (with $c = 1$) starting at $x_1 = -\frac{3}{8}$ for a four-point formula on the interval $(-\frac{1}{2}, \frac{1}{2})$ is of the *open* type, i.e., it does not involve the end points of the interval as samples points. It is also known to be good for Volterra integral equations, which will be illustrated in Example 6 of Section 7.2 for its use in the numerical solution of Volterra integral equations of the first kind.

Table 7.2 Locations x_i and Weights w_i for the *Maclaurin* Rule (equidistant samples)

$$\int_{-\frac{c}{2}}^{\frac{c}{2}} f(x)dx \approx c\sum_{i=1}^{N} w_i f(x_i)$$

where $\pm x_i$ are the samples' locations and w_i are the weight factors[3]

$\pm x_i$		w_i	$\pm x_i$		w_i
	$N=2$			$N=5$	
1/2		1/2	0		402/1152
	$N=3$		2/10		100/1152
0		2/8	4/10		275/1152
1/3		3/8		$N=6$	
	$N=4$		1/12		254/1280
1/8		11/48	3/12		139/1280
3/8		13/48	5/12		247/1280

Integrals with Infinite Limits (associated with singular equations)

At this point we may inquire about integrals with infinite limits, for example, $\int_0^\infty \frac{dx}{1+x^2}$, whose exact value is $\frac{\pi}{2} \approx 1.5708$. Here we must truncate the upper limit of integration to L that results in an approximate integral $\int_0^L \frac{dx}{1+x^2}$, which happens to have the exact value of $\tan^{-1} L$ to compare our numerical approximation with. Since the function $\frac{1}{1+x^2}$ is slowly varying, we must take large enough value of L,

[3]From Kondo [1991, p. 148], courtesy of Oxford University Press (Clarendon Press).

for example $L = 200$. We will illustrate next the use of a high degree four-point Newton-Cotes rule (7.9), where $N = 4$, $h = \frac{L}{N-1} = \frac{200}{3}$, and

$$\begin{aligned}\int_0^{200} \frac{1}{1+x^2}dx &\approx \frac{3}{8}\left(\frac{200}{3}\right)\left[f(0)+3f\left(\frac{200}{3}\right)+3f\left(\frac{400}{3}\right)+f(200)\right]\\ &= \frac{200}{8}\left[1+6.75\times10^{-4}+1.69\times10^{-4}+2.50\times10^{-5}\right]\\ &= 25.0217,\end{aligned}$$

which is a bad approximation when compared with the exact value $\tan^{-1} 200 = 1.56580$. This shows how inefficient such methods may be, and thus the need for more efficient rules like the following *Gauss quadrature* rules of the next section.

One may think that the Newton-Cotes rules can do well for the integral $\int_0^\infty e^{-x}$ $\sin 2x dx$ where the function $e^{-x}\sin 2x$ decays much faster than $\frac{1}{1+x^2}$ of the above example, but still with taking a limit $L = 200$, and using the four-point Newton-Cotes rule (7.9), we have

$$\begin{aligned}\int_0^{200} e^{-x}\sin 2x dx &\approx \frac{3}{8}\left(\frac{200}{3}\right)\left[f(0)+3f\left(\frac{200}{3}\right)+3f\left(\frac{400}{3}\right)+f(200)\right]\\ &= \frac{200}{3}\left[3.29\times10^{-29}+1.343\times10^{-58}-1.178\times10^{-87}\right]\\ &= 8.22\times10^{-28},\end{aligned}$$

which is not good when compared to the exact value $\frac{2}{5}$ of the infinite integral. As we shall see in the illustrations [Example 2(b)] of the next *Gauss-Legendre rule*, we will have a much more accurate result for the integral $\int_0^\infty e^{-x}\sin 2x dx$ with only $L = 10$, and an eight-point Gauss-Legendre rule. However, for the first integral $\int_0^\infty \frac{1}{1+x^2}dx$ of the slowly varying $f(x) = \frac{1}{1+x^2}$, the eight-point Gauss-Legendre rule with $L = 100$ will not be anywhere close in accuracy!. Last we will try in Example 3 the *Gauss-Laguerre* quadrature rule, which is (natural) for integration over the infinite interval $(0, \infty)$, and compare its (better) results with the above methods for the same two integrals.

B. Gauss-Legendre and Other Quadrature Rules

We present here the Gauss (or Gauss-Legendre) rule as a representative of the other principal group of quadrature rules. This, as was mentioned earlier, is characterized by *nonequidistant* locations of the samples x_i as well as variable weights D_i in (7.1). Thus, for the N-point rule of (7.2), and after substituting the expansion (7.3) for $f(x)$ in (7.1), as we did at the beginning of this section, we will end up with a P system of linear equations in the $2N$ variables of the locations x_i and the weights D_i. In this case we set $P = 2N$, but the assurance of a unique solution for the N weights is not so obvious! The analysis here needs some familiarity with the topic of orthonormal polynomials, which we have already discussed and used in Chapter 5 (see Section 5.2.1 for the orthonormal eigenfunctions and Section 4.1.3 and its related exercise 23 for the Legendre polynomials). For now we will supply, in a simple manner, a few very basic elements of this topic to allow us a general sketch of what is behind the Gauss quadrature rules. First, this second group of quadrature rules uses *orthonormal polynomials* $q_n(x)$ of degree n, $n = 1, 2, \cdots, N-1$ instead of the simple monomial x^{i-1} of the first group of Newton-Cotes rules. Also, the locations of the samples $\{x_i\}$ will be the zeros of the polynomial of the highest degree (of such polynomials) $q_N(x)$, i.e., $q_N(x_i) = 0, i = 1, 2, 3, \cdots, N$.

The special property of the polynomials is that they are *orthonormal* on the interval (a, b) of the integration considered, which means that

$$\int_a^b \rho(x) q_m(x) q_n(x) dx = \begin{cases} 0, & n \neq m \\ 1, & n = m \end{cases} \tag{7.11}$$

where $\rho(x)$ is called a *weight function*, $\rho(x) > 0$.

For the present case of the Gauss-Legendre polynomials, $q_n(x) = (\frac{2n+1}{2})^{\frac{1}{2}} P_n(x)$, $n = 1, 2, 3, \cdots, N-1$ are used on the interval $[-1, 1]$, where $P_n(x)$ is the Legendre polynomial of degree n. To give a few examples of the Legendre polynomials $P_n(x)$,

$$P_0(x) = 1, P_1(x) = x, \quad P_2(x) = \frac{1}{2}(3x^2 - 1),$$

$$P_3(x) = \frac{1}{2}(5x^3 - 3x), \quad P_4(x) = \frac{1}{8}(35x^4 - 30x^2 + 3).$$

More Legendre polynomials can be generated via the *Rodrigues formula*,

$$P_n(x) = \frac{1}{2^n n!} \frac{d^n (x^2 - 1)^n}{dx^n}, \qquad P_n(1) = 1. \tag{7.12}$$

It is easy to show that for $\rho(x) = 1$ in (7.11), the integral vanishes for $q_1(x) = (\frac{2+1}{2})^{\frac{1}{2}} P_1(x) = \sqrt{\frac{3}{2}} x$, $q_2(x) = (\frac{4+1}{2})^{\frac{1}{2}} P_2(x) = \sqrt{\frac{5}{2}} (\frac{3}{2}x^2 - \frac{1}{2})$ where for $n = 1 \neq m = 2$, we have

$$\frac{\sqrt{15}}{2} \int_{-1}^{1} x \left(\frac{3}{2} x^2 - 1 \right) dx = 0$$

since the integrand is an odd function over the symmetric interval $[-1, 1]$. However, when $n = m = 1$ for $q_1(x) = \sqrt{\frac{3}{2}} x$, the integral in (7.11) gives the value of 1,,

$$\int_{-1}^{1} \left[\sqrt{\frac{3}{2}} P_1(x) \right]^2 dx = \frac{3}{2} \int_{-1}^{1} x^2 dx = \frac{3}{2} \frac{x^3}{3} \Bigg|_{-1}^{1} = 1.$$

Also, it is easy to show that $P_2(x)$ has two real zeros of $\mp \frac{1}{\sqrt{3}}$ in the interval $[-1, 1]$, when we look at the equation $P_2(x) = (\frac{3}{2}x^2 - \frac{1}{2}) = 0$. The generalization of this result is that $P_k(x)$ has k *real distinct* zeros in the interval $[-1, 1]$. As seen in Table 7.3 these zeros are symmetric around the origin, so only half of them on $(0, 1)$ are tabulated with $\pm$ signs. For example with $N = 2$ we are using $P_2(x)$ with its two zeros on $(-1, 1)$ as $-\frac{1}{\sqrt{3}} = -0.57735$ and $\frac{1}{\sqrt{3}} = 0.57735$.

The derivation of the Gauss-Legendre rule

$$\int_{-1}^{1} f(x)dx = \sum_{i=1}^{N} w_i f(x_i) \tag{7.13}$$

takes advantage of the orthonormality (7.11) of the polynomials on $[-1, 1]$, and the special locations $\{x_i\}_{i=1}^{N}$ as the zeros of $P_N(x)$, i.e., $P_N(x_i) = 0, i = 1, 2, 3, \cdots, N$ to determine the N weights D_i of (7.2) from the $2N$ equations. The details may take us more away from our main line of sketching the idea, but the net result is that we end up with an N-point rule of degree $2N - 1$. We must note that since these (orthonormal polynomials) rules result in an approximation of degree $2N - 1$, then they clearly will give an *exact* approximation to the integral of any function which happens to be a polynomial of degree $\leq 2N - 1$. The weights w_i, and the locations (zeros of $P_N(x)$) are listed in the following Table 7.3, for N up to 8. For higher values of N, and extremely accurate value of w_i and x_i for this Gauss-Legendre rule as well as other orthonormal polynomials rules, see Abramowitz and Stegun [1965, pp. 916-924].

Table 7.3 Locations x_i and Weights w_i for the *Gauss-Legendre* Rule

$$\int_{-1}^{1} f(x)dx \approx \sum_{i=1}^{N} w_i f(x_i)$$

$\pm x_i$ (zeros of Legendre polynomials $P_N(x)$), w_i weight factors

$\pm x_i$	w_i	$\pm x_i$	w_i
$N = 2$		$N = 6$	
0.577350	1.000000	0.238619	0.467914
		0.661209	0.360762
$N = 3$		0.932469	0.171325
0.000000	0.888889		
0.774597	0.555556	$N = 7$	
		0.000000	0.417959
$N = 4$		0.405845	0.381830
0.339981	0.652145	0.741531	0.279705
0.861136	0.347855	0.949108	0.129485
$N = 5$		$N = 8$	
0.000000	0.568888	0.183435	0.362684
0.538469	0.478629	0.525532	0.313707
0.906180	0.236927	0.796666	0.222381
		0.960290	0.101229

From Table 7.3 we see that for $N = 2$ we have equal weights $w_1 = w_2 = 1$, and we have already found the two zeros $x_1 = -\frac{1}{\sqrt{3}} \approx -0.577350$ and $x_2 = \frac{1}{\sqrt{3}} \approx 0.577350$ of $P_2(x)$ as the two (symmetric) locations in the interval $[-1, 1]$. Hence, we have two-point Gauss-Legendre rule of degree $2N - 1 = 3$, i.e., we use a Legendre polynomial of degree 3 to have the Gauss-Legendre approximation for S_2 of (7.2),

$$S_2 \approx f\left(-\frac{1}{\sqrt{3}}\right) + f\left(\frac{1}{\sqrt{3}}\right).$$

To give a simple example, we consider the integral $\int_{-1}^{1} e^x dx$ with its exact value $e - \frac{1}{e} = 2.350402$. The two-point Gauss rule of degree 3 gives the following approximate value $S_2 = e^{-\frac{1}{\sqrt{3}}} + e^{\frac{1}{\sqrt{3}}} = 0.561384 + 1.781312 = 2.342696$, while the two-point Newton-Cotes rule of degree 1 in (7.7) gives $S_2 = \frac{1}{e} + e$ =0.367879+ 2.718282 = 3.086161, which is a bad approximation compared to the Gauss rule as it is evident from comparing these two results with the exact value of 2.350402. Although, this is a very simple example, it demonstrates what is well known about the power of the Gauss type quadrature rules. This is especially true in the treatment of the numerical solution of Fredholm integral equations, where we end up with an $N \times N$ system of linear equations, and for large N, the cost for such numerical computations becomes prohibitive! This is unless we have efficient methods like the Gauss quadrature rules. In contrast, the Newton-Cotes type N-point rules of degree $N - 1$ are known to be inefficient in this situation, and the accuracy is in much doubt for large N. However, they are adequate for the more simple *triangular* system of the resulting linear equations in the case of Volterra integral equations.

The setting up of the numerical approximation of Fredholm integral equations parallels that which we have illustrated already with the help of the simple (repeated) trapezoidal rule in (5.124) and (5.125) of Section 5.5.1, except for looking up the sample locations and weights from Tables 7.1-7.6, which we will return to after presenting another illustration, and few more useful and efficient quadrature rules.

Example 2 Gauss-Legendre Rule

(a) In this example we return to the integral $\int_0^1 \frac{1}{1+x^2} dx$ of Example 1, and use a four-point Gauss-Legendre rule to be compared with the result of the four-point Newton-Cotes rule of Example 1. For lack of space we shall present only the final results for the comparison.

We first note that the Gauss-Legendre rule, with its above Table 7.3 of locations and weights, is done for the integral $\int_{-1}^{1} f(x) dx$ on the symmetric interval $(-1, 1)$. So for the above integral on (0,1), we use a scaling and translation via the transformation $y = 2x - 1$, $dy = 2dx$,

$$\begin{aligned}\int_0^1 \frac{1}{1+x^2} dx &= \frac{1}{2} \int_{-1}^{1} \frac{dy}{1 + \frac{1}{4}(y+1)^2} = \frac{1}{2} \int_{-1}^{1} f(y) dy \\ &\approx \frac{1}{2}[0.652145 f(-0.339981) + 0.347855 f(-0.861136)\end{aligned}$$

$$+0.652145f(0.339981) + 0.347855f(0.861136)] = \frac{1}{2}[0.588098$$
$$+0.346186 + 0.450101 + 0.186422] = 0.785403.$$

This is a much better approximation to the exact value of $\frac{\pi}{4} \approx 0.785398$ than that of 0.78461, obtained with the use of the four-point Newton-Cotes rule that was done at the end of Example 1.

(b) Next we return to our two examples of integration over the infinite interval $(0, \infty)$, namely, $\int_0^\infty e^{-x} \sin 2x dx$ (with its exact value of 0.4) and that of the slowly varying function $\int_0^\infty \frac{1}{1+x^2} dx$ (with its exact value of $\frac{\pi}{2} \approx 1.5708$). We will truncate the infinite limit of integration in the first integral to $L = 10$, and use an eight-point Gauss-Legendre rule which results in a much better value of 0.40041 than that of 8.22×10^{-28} when using $L = 200$ with a four-point Newton-Cotes rule in the computations for integrals with infinite limits (following Example 1). The exact value of the infinite integral is 0.4. In the second integral, with its slowly varying function $f(x) = \frac{1}{1+x^2}$, this eight-point Gauss rule with $L = 100$ gives a good approximation of 1.17915 (compared to the exact value of 1.5608 of $\tan^{-1} 100$ of the truncated integral on (0,100)) which is far better than the very bad approximation of 25.0217 of the four-point Newton-Cotes rule with $L = 200$, that we did following Example 1.

For the Gauss-Legendre rule of approximating the truncated integral

$$\int_0^{10} e^{-x} \sin 2x dx,$$

we must use the transformation $y = \frac{1}{5}x - 1$ in order to have an integral over the symmetric integral $-1 < y < 1$,

$$\int_0^{10} e^{-x} \sin 2x dx = 5 \int_{-1}^{1} e^{-5(y+1)} \sin(10(y+1)) dy.$$

We must mention here that the above are illustrations to show some indication for the approximation of the two groups of numerical integration rules, namely, the Newton-Cotes rules versus the Gauss type rules. These illustrations are by no means exhaustive, since much analysis must be done regarding a suitable truncation limit L for the infinite integral. For example, with the choice $L = 25$ for the above integral, the eight-point Gauss-Legendre rule gave a much better result of 1.578364 to the exact result of $\tan^{-1} 25 = 1.5308176$ of the truncated integral $\int_0^{25} \frac{dx}{1+x^2}$. This may be explained in terms of investigating the eight points in the region $0 < x < 25$ where the integrated function $\frac{1}{1+x^2}$ counts the most instead of spreading those eight points in the region $0 < x < 100$, where beyond $x = 25$, the function changes little from zero.

In Example 3, we will use the Gauss-Laguerre (polynomial) quadrature rule, which is designed for integrals on the *semi-infinite* interval $(0, \infty)$, to show better results with $L = 10$ for the first integral $\int_0^{10} e^{-x} \sin 2x dx$ and $L = 25$ for the second integral $\int_0^{25} \frac{1}{1+x^2} dx$, when an eight-point Gauss-Laguerre rule is used. These results will be compared with those of another efficient rule, also designed for integrals with infinite limits of slowly varying functions, namely, the *Gauss-rational* rule.

Gauss-Tchebychev and Tchebychev Quadrature Rules

For other known orthonormal polynomials used for the Gauss quadrature rules, we list their locations and weights in Tables 7.4–7.6. To illustrate the power of using such orthogonal polynomials we will first discuss the Gauss-Tchebychev N-point rule of degree $2N - 1$ for approximating the integral $\int_{-1}^{1} \frac{1}{\sqrt{1-x^2}} f(x)dx$ where we use the orthonormal Tchebychev polynomials of the *first kind*,

$$q_n(x) = \begin{cases} \sqrt{\frac{2}{\pi}} T_n(x), & n = 1, 2, \cdots, N-1 \\ \sqrt{\frac{1}{\pi}} T_0(x), & n = 0 \end{cases} \tag{7.14a}$$

on the interval (-1,1) with respect to the *weight function* $\rho(x) = \frac{1}{\sqrt{1-x^2}}$ [see (7.11)]. The very special property that makes such polynomials extremely useful, is that they can be related, after a simple change of variable $x = \cos\theta$ for $T_n(x)$, to a cosine function,

$$T_n(x) = \begin{cases} \frac{1}{2^n - 1} \cos n(\arccos x), & n \neq 0 \\ 1, & n = 0 \end{cases} \tag{7.14b}$$

This enables us to arrive at the zeros of $T_N(x)$ in the N-point rule from the following very simple formula,

$$\begin{aligned} T_N(x_i) &= 0 = \cos N(\arccos x_i), \\ N(\arccos x_i) &= (2i-1)\frac{\pi}{2}, \\ x_i &= \cos\left(\frac{(2i-1)\pi}{2N}\right), \quad i = 1, 2, \cdots, N \end{aligned} \tag{7.15}$$

The above change of variable $x = \cos\theta$ is the reason for the weight $\rho(x) = \frac{1}{(1-\cos^2\theta)^{\frac{1}{2}}} = \frac{1}{\sin\theta}$, where the integral $\int_0^{\pi} f(\theta)d\theta$ becomes $\int_{-1}^{1} \frac{1}{\sqrt{1-x^2}} f(\cos^{-1} x)dx$ since $d\theta = d(\cos^{-1} x) = \frac{1}{\sqrt{1-x^2}} dx$. The Tchebychev polynomials also have the simple property of *constant* weights w_i for their N-point Gauss quadrature rule,

$$\int_{-1}^{1} \frac{1}{\sqrt{1-x^2}} f(\cos^{-1} x)dx \approx \frac{\pi}{N} \sum_{i=1}^{N} f\left(\frac{(2i-1)\pi}{2N}\right), \tag{7.16}$$

where the above-mentioned change of variable in $g(x) = f(\cos^{-1} x)$ is very apparent in the integral to be approximated by such a simple rule. Of course, if $g(x) = f(\cos^{-1} x)$ is a polynomial of degree $\leq 2N - 1$, then the approximation (7.16), as it is for Gauss quadrature rules in general, is exact. For example, the integral $\int_{-1}^{1} \frac{1}{\sqrt{1-x^2}} x^2 dx$ is approximated exactly by the N =two-point Gauss-Tchebychev rule (7.16) of degree $2N-1=3$, since $g(x)=x^2$ is of degree $2 \leq 2N-1=3$. The exact value of this integral can be obtained by simple trigonometric substitution for its value as $\frac{\pi}{2}$. The above Tchebychev rule with $N=2$ gives the same value, since with $x_1 = -\cos\frac{\pi}{4} = -\frac{1}{\sqrt{2}}$, $x_2 = \cos\frac{\pi}{4} = \frac{1}{\sqrt{2}}$ we have $S_2 = \frac{\pi}{2}[f(-\frac{1}{\sqrt{2}}) + f(\frac{1}{\sqrt{2}})] = \frac{\pi}{2}[(-\frac{1}{\sqrt{2}})^2 + (\frac{1}{\sqrt{2}})^2] = \frac{\pi}{2}$. We may note that for the rule (7.16) with its equal weights, we need no tables for the locations x_i,since they can be obtained very easily from (7.15).

The above discussion makes use of the important relationship (7.14b) between the Tchebychev polynomial and the trigonometric cosine function. The result is a constant weight quadrature formula (7.16), but it is for a weighted integral,

$$\int_{-1}^{1} \frac{1}{\sqrt{1-x^2}} f(\cos^{-1} x) dx = \int_{-1}^{1} \frac{1}{\sqrt{1-x^2}} g(x) dx$$

of $g(x)$ on $(-1,1)$ with weight function $w(x) = \dfrac{1}{\sqrt{1-x^2}}$. There is, however, a *simpler* Tchebychev quadrature rule for the integral $\displaystyle\int_{-1}^{1} f(x)dx$ (without weight), and with equal weights $\dfrac{2}{N}$ in the sum,

$$\int_{-1}^{1} f(x)dx = \frac{2}{N} \sum_{i=1}^{N} f(x_i). \tag{7.17}$$

However, the locations of the samples x_i are different from those used in (7.16) [as obtained from the simple formula in (7.15)]. The locations x_i for (7.17) are listed in Table 7.4, and they are, more difficult to obtain, as zeros of the polynomial part of[4]

$$G(x) = x^N \exp\left(\frac{-N}{2 \cdot 3x^2} - \frac{N}{4 \cdot 5x^3} - \frac{N}{6 \cdot 7x^4} - \cdots\right).$$

Note that the x_i here in the Tchebychev rule (E.1) are different from those for the Gauss-Tchebychev rule.

We may also note that both sums, of the Gauss-Tchebychev rule (7.16) and the Tchebychev rule (7.17), are with equal weights ($w_i = \frac{\pi}{N}$ and $\frac{2}{N}$, respectively), compared to the rest of the Gauss quadrature rules used here, such as that of Legendre in (7.13) and what will follow as the Laguerre and the Hermite quadrature rules in (7.18) and (7.26), respectively.

We may remind again that in Tables 7.2, 7.3, and 7.4 for the Maclaurin rule, the Gauss-Legendre rule, and the Tchebychev rule (with equal weights), respectively, the

[4] Abramowitz and Stegun [1965, p. 887]

Table 7.4 Locations x_i for the Tchebychev Rule with Equal Weights (7.17)

$$\int_{-1}^{1} f(x)dx \approx \frac{2}{N} \sum_{i=1}^{N} f(x_i) \tag{E.1}$$

$\pm x_i$ (zeros of Tchebychev polynomials $T_N(x)$)

N	$\pm x_i$	N	$\pm x_i$
2	0.577350	5	0.832497
			0.374541
3	0.707107		0.000000
	0.000000		
		6	0.866247
4	0.794654		0.422519
	0.187592		0.266635

locations of the samples x_i (and the weights for Tables 7.2 and 7.3) are given for the integral $\int_{-a}^{a} f(x)dx$ on the symmetric interval $(-a, a)$. Hence the integral at hand $\int_{c}^{d} f(x)dx$ must be adjusted with a change of variable to define it on $(-a, a)$ as we did for Example 2 where the integral was defined on $(0, 1)$. Sometimes the data of the tables is adjusted accordingly to suit the limits of the given integral, as we shall do in preparing for (7.41) and in (E.5) of Example 7.

Gauss-Laguerre Quadrature Rule

Other very useful polynomials are the Laguerre polynomials $L_i(x)$ which are orthonormal on the *semi-infinite* interval $(0, \infty)$ with respect to the weight $\rho(x) = e^{-x}$ in (7.11). The locations x_i, which are the zeros of the Laguerre polynomial $L_N(x)$, and the weights w_i of this N-point Gauss rule are listed in Table 7.5 for the approximation of the following integral,

$$\int_{0}^{\infty} e^{-x} f(x)dx \approx \sum_{i=1}^{N} w_i f(x_i) \tag{7.18}$$

or its equivalent

$$\int_{0}^{\infty} g(x)dx \approx \sum_{i=1}^{N} w_i e^{x_i} g(x_i), \quad g(x) = e^{-x} f(x) \tag{7.19}$$

In Table 7.5, the locations x_i, weights w_i for (7.18) as well as weights $w_i e^{x_i}$ for (7.19) are tabulated for N up to 8, the rest are found in Abramowitz and Stegun [1965, p. 923].

The negative numbers in parenthesis $(-n)$ to the left of the numerical values of w_i in Table 7.5 are the negative exponent of the factor 10^{-n} used for the indicated small numbers. For integrals of the form $\int_a^\infty e^{-\beta x} f(x)dx$, we can simply make the change of variable $\xi = \beta(x-a)$ where this integral becomes $\frac{1}{\beta}e^{-\beta a}\int_0^\infty e^{-\xi} f\left(\frac{\xi}{\beta}+a\right)d\xi$, which then can be approximated by (7.18)

$$\begin{aligned}\int_a^\infty e^{-\beta x} f(x)dx &= \frac{e^{-\beta a}}{\beta}\int_0^\infty e^{-\xi} f\left(\frac{\xi}{\beta}+a\right)d\xi \\ &= \frac{e^{-\beta a}}{\beta}\int_0^\infty e^{-\xi} h(\xi)d\xi \approx \frac{e^{-\beta a}}{\beta}\sum_{i=1}^N w_i h(\xi_i) \\ &= \frac{e^{-\beta a}}{\beta}\sum_{i=1}^N w_i f\left(\frac{\xi_i}{\beta}+a\right)\end{aligned} \tag{7.20}$$

where, as seen in the middle sum for $h(\xi_i)$, the locations ξ_i are the zeros of the Laguerre polynomial $L_N(x)$. This generalized version (7.20) of the Gauss-Laguerre rule (7.19) is called the *shifted Gauss-Laguerre* rule.

A good illustration of this efficient rule for integration over the semi-infinite interval $(0,\infty)$ is to return to our examples of $\int_0^\infty e^{-x}\sin 2x dx$ and $\int_0^\infty \frac{1}{1+x^2}dx$, which we have already approximated in Example 2, using an eight-point Gauss-Legendre rule with truncation limit $L = 10$ for the first integral and $L = 25$ (and 100) for the second integral, and also right after Example 1, where Newton-Cotes rule was used with even larger $L = 200$ (see also Exercise 2).

Table 7.5 Locations x_i and Weights w_i for the Gauss-Laguerre Rule

$$\int_0^\infty e^{-x} f(x)dx \approx \sum_{i=1}^N w_i f(x_i) \tag{E.1}$$

$$\int_0^\infty g(x)dx \approx \sum_{i=1}^N w_i e^{x_i} g(x_i) \tag{E.2}$$

x_i (zeros of Laguerre polynomials $L_N(x)$), w_i weight factors

x_i	w_i	$w_i e^{x_i}$
	$N=2$	
0.585786	(-1)8.535534	1.533326
3.414214	(-1)1.464466	4.450957
	$N=3$	
0.415775	(-1)7.110930	1.077693
2.294280	(-1)2.785177	2.762143
6.289945	(-2)1.038926	5.601095

Table 7.5 continued Locations x_i and Weights w_i for the Gauss-Laguerre Rule

x_i	w_i	$w_i e^{x_i}$
	$N = 4$	
0.322548	(-1)6.031541	0.832739
1.745761	(-1)3.574187	2.048102
4.536620	(-2)3.888790	3.631146
9.395071	(-4)5.392947	6.487145
	$N = 5$	
0.263560	(-1)5.217556	0.679094
1.413403	(-1)3.986668	1.638488
3.596426	(-2)7.594245	2.769443
7.085810	(-3)3.611759	4.315657
12.640801	(-5)2.336997	7.219184
	$N = 6$	
0.222847	(-1)4.589647	0.573536
1.188932	(-1)4.170008	1.369253
2.992736	(-1)1.133734	2.260685
5.775144	(-2)1.039920	3.350525
9.837462	(-4)2.610172	4.886827
15.982874	(-7)8.985479	7.849016
	$N = 7$	
0.193044	(-1)4.093190	0.496478
1.026665	(-1)4.218313	1.177643
2.567877	(-1)1.471263	1.918250
4.900353	(-2)2.063351	2.771850
8.182153	(-3)1.074010	3.841249
12.734180	(-5)1.586546	5.380678
19.395728	(-8)3.170315	8.405432
	$N = 8$	
0.170280	(-1)3.691886	0.437723
0.903702	(-1)4.187868	1.033869
2.251087	(-1)1.757950	1.669710
4.266700	(-2)3.334349	2.376925
7.045905	(-3)2.794536	3.208541
10.758516	(-5)9.076509	4.268576
15.740679	(-7)8.485748	5.818083
22.863132	(-9)1.048001	8.906226

Example 3

Here we will use the four-point and eight-point Gauss-Laguerre rule ($N = 4$ and 8 in Table 7.5) for both integrals with the limit of integration being truncated to about 10 and 23, respectively. For the integral $\int_0^L e^{-x} f(x)dx$ with $L = 10$, we note from Table 7.5 that the closest value to 10 is $x_4 = 9.395071$ for the four-point Laguerre rule. For $N = 8$, we can go as far as $x_8 = 22.863132$.

The four-point Gauss-Laguerre rule approximation of the first integral $\int_0^\infty e^{-x}$ sin$2x dx$ is

$$\begin{aligned} &\approx 0.603154 \sin(2(0.322548)) + 0.357419 \sin(2(1.745761)) \\ &+0.0388790 \sin(2(4.536620) + 0.000539295\,(2(9.395071)) \\ &= 0.362662 - 0.12253 + 0.013388 - 3.2 \times 10^{-5} = 0.253486 \end{aligned}$$

which is not such a good approximation to the exact value 0.4 of the infinite integral. However an eight-point Gauss-Laguerre rule gave a better result of 0.3872805 as we shall illustrate in Example 4(b).

Now if we use an eight-point Gauss-Laguerre rule to approximate the integral $\int_0^\infty \frac{1}{1+x^2} dx$, which, effectively, truncates it to $L = x_8 = 22.863132$, we find that

$$\begin{aligned} \int_0^\infty \frac{1}{1+x^2} dx \;&\approx 0.425389 + 0.569099 + 0.275194 \\ &+\; 0.123768 + 0.063354 + 0.036563 + 0.023387 + 0.017006 \\ &= 1.533759 \end{aligned}$$

which is a good approximation to the exact result of $\tan^{-1} 22.863132$ =1.527085. It is also a better approximation than 1.578364 of the eight-point Gauss-Legendre rule used in Example 2 for the truncated integral $\int_0^{25} \frac{dx}{1+x^2}$ (see also Newton-Cotes approximation right after Example 1).

The Gauss-Rational Rule

Our treatment for the two integrals with infinite limits was based on simply truncating the infinite limit to a finite limit L, then using the Gauss-Legendre rule for the interval $(0, L)$ as was done in Example 2. The second method for accomplishing such truncation of the integral was done through the use of the (finite sum) Laguerre quadrature rule on the interval (x_1, x_N), where $\{x_i\}$ are the zeros of the Laguerre polynomial $L_N(x)$, as we have illustrated for the same two integrals in Example 3.

Another way around the infinite interval (a, ∞) of the integral,

$$I = \int_a^\infty f(x)dx \tag{7.21}$$

is to use a change of variable (or mapping) $x = v(\xi)$. This reduces the limits of integration to finite ones in the new variable ξ. For example, the following change of variable

$$x = v(\xi) = \frac{2(a+\beta)}{\xi+1} - \beta \tag{7.22}$$

(as a *rational* function of ξ) reduces the integral in (7.21) with $x\epsilon(a, \infty)$ to the following integral with *finite limits* for $\xi\epsilon(-1, 1)$,

$$
\begin{aligned}
I = \int_a^\infty f(x)dx &= -2(a+\beta)\int_1^{-1} \frac{f(v(\xi))d\xi}{(1+\xi)^2} \\
&= 2(a+\beta)\int_{-1}^{1} \frac{f(v(\xi))d\xi}{(1+\xi)^2} \\
&= 2(a+\beta)\int_{-1}^{1} \frac{F(\xi)d\xi}{(1+\xi)^2},
\end{aligned}
\tag{7.23}
$$

where $F(\xi) \equiv f(v(\xi))$, and the weight $\frac{2(a+\beta)}{(1+\xi)^2}$ in the above integral is the result of $dx = -\frac{2(a+\beta)}{(1+\xi)^2}d\xi$ after using $x = \frac{2(a+\beta)}{\xi+1} - \beta$. So, a *Gauss-Legendre* rule can now be used for approximating the integral in (7.23) on the symmetric interval $(-1, 1)$,

$$
\begin{aligned}
\int_a^\infty f(x)dx &= 2(a+\beta)\int_{-1}^{1} \frac{F(\xi)d\xi}{(1+\xi)^2} \\
&\approx 2(a+\beta)\sum_{i=1}^{N} \frac{w_i F(\xi_i)}{(1+\xi_i)^2}
\end{aligned}
\tag{7.24}
$$

with $F(\xi) \equiv f\left(\frac{2(a+\beta)}{1+\xi_i} - \beta\right)$, and w_i, ξ_i are, respectively, the weights and locations for the Gauss-Legendre rule as given in Table 7.3. Of course, in the (very special) case of the integrand $\frac{F(\xi)}{(1+\xi)^2}$ in (7.23) being a polynomial of degree $\leq 2N-1$, the Gauss-Legendre rule of approximating its integral in (7.24) would give the exact value of the integral.

This method is called the "*Gauss-rational rule*," which we will illustrate in the next Example 4 for the two integrals of Examples 2 and 3, and then compare the results for the different methods.

This rule reduces integrals with infinite limits of integration to those of finite limits. So, for our purpose of this book, it will help us in reducing some *singular* integral equations, those with infinite limits of integration, to non-singular integral equations, i.e., with finite limits of integration.

We emphasize here the different numerical integration methods, since they are essential for the accurate numerical setting of such singular integral equations. This is so, especially when at the level of this book we are not covering the theory behind and the analytic methods for solving such singular equations.

We may stress the point that while other methods deal first with truncating the infinite limit of the integral, the present Gauss-rational rule ends up dealing with *finite* limit integral as in (7.24). So, there is no surprise if it succeeds in approximating the integral $\int_0^\infty \frac{1}{1+x^2}dx$, with its slowly varying integrand $\frac{1}{1+x^2}$, which was a source of trouble for the former methods (such as the Maclaurin method and Newton-Cotes method that we discussed following Example 1) that must start with truncating the infinite limit.

The success of the Gauss-Laguerre rule in approximating integrals of the form $\int_0^\infty e^{-x}\sin 2x dx$, is also understood because of the inherent decaying factor e^{-x} in the integrand, that acts, effectively, to truncate the infinite limit of integration.

There remains the role of the parameter β in (7.22)–(7.24) for the Gauss-rational rule. To compare this rule with the Gauss-Laguerre rule for the integral $\int_0^\infty \frac{1}{1+x^2}dx$, we must consider this integral in the form of the integral in (7.20)

$$\begin{aligned} \int_0^\infty \frac{1}{1+x^2}dx &= \int_0^\infty e^{-\beta x} \cdot \frac{e^{\beta x}}{1+x^2}dx \\ &= \int_0^\infty e^{-\beta x} f(x)dx \approx \frac{1}{\beta}\sum_{i=1}^{N} \frac{w_i e^{\xi_i}}{1+(\frac{\xi_i}{\beta})^2} \end{aligned} \tag{7.25}$$

after using the *shifted* Gauss-Laguerre rule (7.20), and where ξ_i are the zeros of the Laguerre polynomial $L_N(x)$, and w_i are the weights of the Laguerre rule as given in Table 7.5.

The observation that the *Gauss-rational* rule may benefit from different (Larger! in this example) values of β than the Gauss-Laguerre rule, which uses smaller values of β, will appear in the illustration of the two methods in the following example.

Example 4

We consider again the same two integrals with infinite limits that we used in Examples 2 and 3.

(a)

$$\int_0^\infty \frac{1}{1+x^2}dx = \frac{\pi}{2} = 1.5707963 \tag{$E.1$}$$

$$\int_0^\infty e^{-x}\sin 2x dx = \frac{2}{5}. \tag{$E.2$}$$

We will first use an eight-point Gauss-rational rule as in (7.22)–(7.24) for the integral in (E.1) with $\beta = 10$ and from (7.22) with $a = 0$ we have $x = v(\xi) = \frac{2\beta}{\xi+1} - \beta$ to use in (7.23),

$$\int_0^\infty \frac{1}{1+x^2}dx = 2\beta\int_{-1}^1 \frac{F(\xi)}{(\xi+1)^2}d\xi, \tag{$E.3$}$$

where $F(\xi) = \frac{1}{1+(\frac{2\beta}{\xi+1}-\beta)^2}$. If we use $\beta = 10$ and an eight-point Gauss-Legendre rule on the second integral in (E.3), using Table 7.3, we have the following approximation,

$$\begin{aligned} \int_0^\infty \frac{1}{1+x^2}dx &\approx 20[0.000263 + 0.000689 + 0.001347 + 0.002577 \\ &+0.005327 + 0.012629 + 0.030205 + 0.025304 = 1.56685 \end{aligned}$$

Next we use an eight-point *Gauss-Laguerre* rule as in (7.25) with (a small) $\beta = 0.2$ to have

$$\begin{aligned} \int_0^\infty \frac{dx}{1+x^2} &= \int_0^\infty e^{-\beta x}\frac{e^{\beta x}}{1+x^2}dx \\ &\approx \frac{1}{\beta}\sum_{i=1}^{N} w_i \frac{e^{\xi_i}}{1+(\frac{\xi_i}{\beta})^2} \end{aligned}$$

$$= 5[0.25377 + 0.048273 + 0.013077 + 0.005211 + 0.002583$$
$$+0.001475 + 0.000939 + 0.000681] = 1.63005$$

which is not as good approximation, to the exact value $\frac{\pi}{2}$ =1.5707963 of the integral, as that of the above Gauss-rational rule that gave 1.56409. In this case it may be verified that pushing N to larger values is not enough to catch up with the better approximation of the Gauss-rational rule.

(b) For the second integral with its rapidly decaying integrand the numerical results are a bit different from that of the first integral.

First we use the *Gauss-rational rule* as in (7.22)–(7.24) to have

$$\int_0^\infty e^{-x} \sin 2x dx = 2\beta \int_{-1}^{1} \frac{F(\xi)}{(\xi+1)^2} d\xi, \tag{E.4}$$

where $F(\xi) = e^{-\frac{2\beta}{\xi+1}+\beta} \sin 2(\frac{2\beta}{\xi+1} - \beta)$. Then we use an eight-point Gauss-Legendre rule on the second integral with the help of Table 7.3 to have

$$\begin{aligned}\int_0^\infty e^{-x} \sin 2x dx \quad &\approx 20[0.0 + 1.62 \times 10^{-38} + 1.51 \times 10^{-14} \\ &-1.8 \times 10^{-7} + 0.000246 - 3.77 \times 10^{-4} + 0.017096 \\ &+0.008479] = 0.50888.\end{aligned}$$

Next we use an eight-point (shifted) *Gauss-Laguerre* rule (7.20) in parallel to (7.25) with $\beta = 1$, and consult Table 7.5 to have

$$\int_0^\infty e^{-x} \sin 2x dx \approx \sum_{i=1}^{8} w_i \sin(2\xi_i)$$

$$\approx [0.123315 + 0.407119 - 0.17193 + 0.025939 + 0.002792$$
$$+4.14 \times 10^{-5} + 5.55 \times 10^{-8} + 1.03 \times 10^{-9}] = 0.387281,$$

which is more accurate approximation of the exact value 0.4 of the integral, than the above Gauss-rational rule result of 0.51507.

Gauss-Hermite Quadrature Rule

For integrals $\int_{-\infty}^{\infty} e^{-x^2} f(x)dx$ over the *infinite* interval $(-\infty, \infty)$, the (orthogonal) *Hermite* polynomials $H_n(x)$ are used, which are orthogonal on the interval $(-\infty, \infty)$ with respect to the weight $\rho(x) = e^{-x^2}$ in (7.11), to give the N-point $Gauss - Hermite$ rule as

$$\int_{-\infty}^{\infty} e^{-x^2} f(x)dx \approx \sum_{i=1}^{N} w_i f(x_i) \tag{7.26}$$

where, of course, x_i are the zeros of $H_N(x)$, i.e., $H_N(x_i) = 0, i = 1, 2, \cdots, N$.

These samples locations x_i and the weights w_i are listed in Table 7.6 for the above integral (7.26) and its equivalent

$$\int_{-\infty}^{\infty} g(x)dx \approx \sum_{i=1}^{N} w_i e^{x_i^2} g(x_i) \tag{7.27}$$

for N up to 4. The same note regarding the negative numbers in parenthesis $(-n)$ for Table 7.6 applies here, where it is for (the exponent of) the factor 10^{-n} used for representing small numbers w_i to be employed in (7.26).

Table 7.6 Locations x_i and Weights w_i for Gauss-Hermite Rule

$$\int_{-\infty}^{\infty} e^{-x^2} f(x)dx \approx \sum_{i=1}^{N} w_i f(x_i) \tag{E.1}$$

$$\int_{-\infty}^{\infty} g(x)dx \approx \sum_{i=1}^{N} w_i e^{x_i^2} g(x_i) \tag{E.1}$$

$\pm x_i$ (zeros of Hermite polynomials $H_N(x)$), w_i weight factors

$\pm x_i$	w_i	$w_i e^{x_i^2}$
	$N = 2$	
0.707108	(-1)8.862269	1.461141
	$N = 3$	
0.000000	(0)1.181636	1.181636
1.224745	(-1)2.954091	1.323931
	$N = 4$	
0.524648	(-1)8.049141	1.059965
1.650680	(-2)8.131284	1.240226

Exercises 7.1

1. Consider the integral

$$\int_0^{\infty} \frac{1}{100 + x^2} dx,$$

which is very similar to the first integral used in Examples 3 and 4. In parallel to what we did in Example 4 use an eight-point Gauss-Laguerre rule with $\beta = 0.2$, then an eight-point Gauss-rational rule with $\beta = 10$ to approximate the first integral in (E.1). Compare your answer with the exact value of $\frac{\pi}{20} = 0.1570796$.

2. Do the same as in problem 1 for the following integral, except that $\beta = 1$ and 10 for the Gauss-Laguerre rule and the Gauss-rational rule, respectively.

$$\int_0^\infty e^{-x} \sin x dx.$$

Compare your answer with the exact value of 0.5.

3. Use the Gauss-Laguerre rule to compute the exact value of the integral

$$\int_1^\infty e^{-x} x^3 dx = \frac{16}{e}. \tag{E.1}$$

Hint: In order to use the Gauss-Laguerre rule for the $\int_0^\infty e^{-x} f(x)dx$, we make a change of variable $y = x - 1$ in the given integral of (E.1) to reduce it to $\int_0^\infty e^{-y} e^{-1} (y+1)^3 dy$, then apply the rule of (7.18) to this last integral (you can also use (7.20) on (E.1) with $a = 1$ as the *shifted* Gauss-Laguerre rule),

$$\begin{aligned} \int_1^\infty e^{-x} x^3 dx &= \frac{1}{e} \int_0^\infty e^{-y} (y+1)^3 dy \\ &\approx \frac{1}{e} \sum_{i=1}^{N} w_i (y_i + 1)^3 \end{aligned} \tag{E.2}$$

4. Consider the integral $\int_0^\infty e^{-x} x^4 dx$ and its exact value of 24.

   Use a three-point Gauss-Laguerre rule to show that we obtain, aside from a round-off error, etc., the exact value (since the integrated function $f(x) = x^4$, with respect to the weight function $p(x) = e^{-x}$ on $(0, \infty)$, is a polynomial of degree $4 \leq 2N - 1 = 2(3) - 1 = 5$).

5. Consider the integral

$$\int_{-\infty}^\infty e^{-x^2} x^4 dx$$

   whose exact value is $\frac{3\sqrt{\pi}}{4}$.

   (a) Use a two-point Gauss-Hermite rule to approximate the integral and compare with the exact value. This approximation is not exact! Why?, see part (b).

   (b) Use a three-point Gauss-Hermite rule to obtain the exact value of $\frac{3\sqrt{\pi}}{4}$ for the integral, to show that this polynomial rule gives an exact value since the integrated function $f(x) = x^4$, with respect to the weight function $\rho(x) = e^{-x^2}$ on $(-\infty, \infty)$, is a polynomial of degree $4 \leq 2(N) - 1 = 2(3) - 1 = 5$.

7.2 HIGHER QUADRATURE RULES FOR VOLTERRA EQUATIONS

As we mentioned earlier, the numerical setting of Volterra integral equations is usually done with Newton-Cotes rules, or their repeated versions. The simplest is the trapezoidal rule (1.141) that was used for the Volterra equation,

$$u(x) = f(x) + \int_a^x K(x,t)u(t)dt \tag{3.42}$$

to result in the $N \times N$, $N = n+1$, *triangular* system of equations in (3.45) or (3.46) (in Section 3.3). We may note that if we rush to suggest the repeated (high order) Simpson's rule we see that it cannot be started easily, since the second equation with $i = 1$, in (3.46) has only two samples u_0 and u_1 instead of the three required for Simpson's rule (since n must be even). This suggests staying with the trapezoidal rule for $i = 1$, but go to higher degree Newton-Cotes rules for few values of $i \geq 2$, then follow the latter by a repeated Newton-Cotes rule. One such successful version is to use the two-point Newton-Cotes rule, i.e., the trapezoidal rule (7.7) followed by the three-point Newton-Cotes rule, i.e., Simpson's rule (7.8) for $i = 2$, followed by a $\frac{3}{8}$ (or the four-point Newton-Cotes) rule (7.9) for $i = 3$, then return to the repeated Simpson's rule (1.144) for $i \geq 4$. For this combination, the approximation for Volterra integral equation (3.1) with increment Δt becomes

$$\begin{aligned}
u_0 &= f_0 \\
u_1 &= f_1 + \Delta t \left[\frac{1}{2} K_{10} u_0 + \frac{1}{2} K_{11} u_1 \right] : \text{ Trapezoidal} \\
u_2 &= f_2 + \frac{2\Delta t}{6} \left[K_{20} u_0 + 4K_{21} u_1 + K_{22} u_2 \right] : \text{ Simpson's} \\
u_3 &= f_3 + \frac{3\Delta t}{8} \left[K_{30} u_0 + 3K_{31} u_1 + 3K_{32} u_2 + K_{33} u_3 \right] : \frac{3}{8} - \text{rule}
\end{aligned} \tag{7.28}$$

$$\begin{aligned}
u_4 = {} & f_4 + \frac{2\Delta t}{6} [K_{40} u_0 + 4K_{41} u_1 + 2K_{42} u_2 \\
& +4K_{43} u_3 + K_{44} u_4] : \text{ repeated Simpson's.}
\end{aligned}$$

While Simpson's rule (1.144) and the $\frac{3}{8}$-rule are high-degree efficient rules, this method will still suffer from the inefficiency of the lower degree trapezoidal rule, used in determining u_1 above at the starting point, which is then used for the more accurate rules in determining u_2, u_3, u_4 and so on $\cdots$ in the following equations of (7.28). Such inaccurate value of the input u_1, would, of course, ruin the good accuracy of the high degree rules used for u_2, u_3, $u_4, \cdots$. One way around this difficulty is to use smaller Δt, say $\frac{\Delta t}{2}$ or $\frac{\Delta t}{4}$, from $t = 0$ to Δt to have a more accurate value for the starting value of u_1, then follow it by the other rules for u_2, u_3, and u_4, which is illustrated in the following example.

Example 5 We will return to Example 9 of Section 3.3, where the trapezoidal rule was used,

$$u(x) = x - \int_0^x (x-t)u(t)dt,$$

$$u(x) = x + \int_0^x (t-x)u(t)dt. \tag{E.1}$$

We first try (7.28) with $\Delta t = \frac{1}{2}$, and to improve the accuracy of the trapezoidal rule for determining u_1 in (7.28), we use $\Delta' t = \frac{1}{4}$ for $t = 0$ to $t = \Delta t = \frac{1}{2}$, where we have three points at $t = 0$, $\frac{1}{4}$ and $\frac{1}{2}$ with sample values u_0, u_1' and u_2' to use the trapezoidal rule on as a "mini" problem. The result u_2' can now stand for a more accurate value u_1 at $t = \frac{1}{2}$ to be used as u_1 in the third equation of (7.28) for determining u_2 via Simpson's rule.

First we choose $\Delta t = \frac{1}{2}$, $\Delta' t = \frac{1}{4}$, and use the trapezoidal rule in the second equation of (7.28) with $f(0) = 0$, $f_1 = f(\frac{1}{4}) = \frac{1}{4}$, $K_{10} = -\frac{1}{4}$, $K_{11} = (\frac{1}{4} - \frac{1}{4}) = 0$, to find $u_1' = u(\frac{1}{4})$,

$$\begin{aligned} u_0 &= f_0 = f(0) = 0, \qquad u_0 = 0 \\ u_1' &= f_1 + \Delta' t \left[\frac{1}{2} K_{10} u_0 + \frac{1}{2} K_{11} u_1' \right] \\ &= \frac{1}{4} + \frac{1}{4} \left[\frac{1}{2} \left(-\frac{1}{4} \right) (0) + \frac{1}{2} (0) u_1' \right], \quad u_1' = \frac{1}{4} \end{aligned}$$

Then we use $u_0 = 0, u_1' = u(\frac{1}{4}) = \frac{1}{4}$ and $f_2 = f(\frac{1}{2}) = \frac{1}{2}$ in a three-point trapezoidal rule with $\Delta' t = \frac{1}{4}$ to find the more refined $u_2' = u(\frac{1}{2})$, to be used later for Simpson's rule,

$$\begin{aligned} u_2' &= f_2 + \Delta' t \left[\frac{1}{2} K_{20} u_0 + K_{21} u_1' + \frac{1}{2} K_{22} u_2' \right] \\ &= \frac{1}{2} + \frac{1}{4} \left[\frac{1}{2} \left(-\frac{1}{2} \right) (0) + \left(\frac{1}{4} - \frac{1}{2} \right) \left(\frac{1}{4} \right) + \frac{1}{2} \left(\frac{1}{2} - \frac{1}{2} \right) u_2' \right] \\ &= \frac{1}{2} - \frac{1}{4} \times \frac{1}{16} = \frac{1}{2} - \frac{1}{64} = \frac{31}{64} = 0.4844. \end{aligned}$$

Now we return to our original (over all) problem with the larger t increment of $\Delta t = \frac{1}{2}$, $u_0 = 0$, and the more refined $u_1 = u(\frac{1}{2}) = u_2' = \frac{31}{64}$ to be used for Simpson's rule in (7.28) with $\Delta t = \frac{1-0}{2} = \frac{1}{2}$ to find $u_2 = u(1)$,

$$u_2 = f_2 + \frac{2\Delta t}{6} \left[K_{20} u_0 + 4K_{21} u_1 + K_{22} u_2 \right],$$

$$u_2 = 1 + \frac{1}{6} \left[(0-1)(0) + 4 \left(\frac{1}{2} - 1 \right) \left(\frac{31}{64} \right) + (1-1) u_2 \right],$$

$$u_2 = 1 - \frac{1}{6} \left(\frac{31}{32} \right) = 1 - \frac{31}{192} = \frac{161}{192} = 0.8385.$$

Next we use the $\frac{3}{8}$-rule in (7.28) with $\Delta t = \frac{t_3-0}{3} = \frac{1.5-0}{3} = \frac{1}{2}$, $u_0 = 0$, $u_1 = u(\frac{1}{2}) = \frac{31}{64}$, and $u_2 = u(1) = \frac{161}{192}$ to find $u_3 = u(\frac{3}{2})$,

$$\begin{aligned}
u_3 &= f_3 + \frac{3\Delta t}{8}\left[K_{30}u_0 + 3K_{31}u_1 + 3K_{32} + K_{33}u_3\right],\\
u_3 &= \frac{3}{2} + \frac{3}{16}\left[(0) - (\frac{3}{2})(0) + 3\left(\frac{1}{2} - \frac{3}{2}\right)\left(\frac{31}{64}\right) + 3\left(1 - \frac{3}{2}\right)\left(\frac{161}{192}\right)\right.\\
&\quad \left. + \left(\frac{3}{2} - \frac{3}{2}\right)u_3\right],\\
u_3 &= \frac{3}{2} + \frac{3}{16}\left[-\frac{93}{64} - \frac{483}{384}\right] = \frac{3}{2} - \frac{3}{16}(2.71094) = 1.5 - 0.5083\\
u_3 &= 0.9917.
\end{aligned}$$

These numerical (approximate) results $u_0 = u(0)$, $u_1' = u(\frac{1}{4}) = 0.25$, $u_1 = u_2' = u(\frac{1}{2}) = 0.4844$, $u_2 = u(1) = 0.8385$, and $u_3 = u(\frac{3}{2}) = 0.9917$ are to be compared with the values of the exact solution $u(x) = \sin x$, where $u(0) = 0$, $u(\frac{1}{4}) = u_1'$ =0.2474, $u_2' = u_1 = u(\frac{1}{2}) = 0.4794$, $u_2 = u(1) = 0.8415$, and $u_3 = u(\frac{3}{2}) = 0.9975$, as shown with the corresponding errors in the following Table 7.7.

Table 7.7 Numerical and Exact Solutions of a Volterra Equation of the Second Kind (Example 5)

i	x_i	$\tilde{u}_i$(num.)	$u_i = \sin x_i$(exact)	error $\varepsilon_i = \tilde{u}_i - u_i$
0	0.0	0.0000	0.0000	0.00000
$1'$	$\frac{1}{4}$	0.2500	0.2474	2.6×10^{-3}
$1, 2'$	$\frac{1}{2}$	0.4844	0.4794	5×10^{-3}
2	1	0.8385	0.8415	-3×10^{-3}
3	$\frac{3}{2}$	0.9917	0.9975	-5.8×10^{-3}

In Example 9 of Section 3.3, we used only the typical (extended) trapezoidal rule with $\Delta t = 1$, where the results are reported in Table 3.1 and Figure 3.1 of Section 3.3. A look at this data shows the better accuracy of the present computations. For example, we now have $u_2 = u(1) = 0.8385$, while in Example 9, we have $u_2 = u(1) = 1.0$, where the latter is much further away from the exact answer of $u(1) = \sin 1 = 0.8415$, i.e., 16% error compared to 3.3% error of the present computations.

For another illustration of this method, see Exercise 4 with its very detailed answer.

Volterra Equations of the First Kind

For Volterra integral equations of the first kind

$$f(x) = \int_a^x K(x,t)u(t)dt \tag{7.29}$$

we write its numerical setting, in parallel to (3.45), with $N = n + 1$ as

$$f(x_i) = \sum_{j=1}^{j\leq i} K(x_i, t_j)D_j u(t_j), \quad i = 1, 2, 3, \cdots, N, \quad t_j \leq x_i.$$

For this equation we will use the Maclaurin rule (see also Exercise 2).

From Table 7.2, we note that the Maclaurin rule on $(-\frac{1}{2}, \frac{1}{2})$ uses fixed *odd* number of increments for approximating the integration on $(-\frac{1}{2}, \frac{1}{2})$. For example, with $N = 4$, it uses $t_1 = -\frac{1}{2} + \frac{1}{8} = -\frac{3}{8}$, $t_2 = -\frac{1}{2} + \frac{3}{8} = -\frac{1}{8}$, $t_3 = -\frac{1}{2} + \frac{5}{8} = \frac{1}{8}$, and $t_4 = -\frac{1}{2} + \frac{7}{8} = \frac{3}{8}$. So with $\Delta x = h = \frac{1}{8}$ we must use $t_1 = a + h$, $t_3 = a + 3h$, $t_5 = a + 5h$, $t_7 = a + 7h$, i.e., we use an *odd* number of increments h. But as in (3.45), the (variable) upper limit for the integral of Volterra integral equation requires $t_j \leq x_i$, while the Maclaurin method, as an *open* method, does not use the end point a and the upper one of the considered interval. Thus to avoid the upper limit of the integration we must have $t_j < x_i$. This means that we may take x_i with *even* increments of h, $x_i = a + 2ih$, and (lower) *odd* increments of h for t_j, $t_j = a + (2j - 1)h$. Hence, we take $x_2 = a + 2h$, $x_4 = a + 4h$, $x_6 = a + 6h$, and $x_8 = a + 8h, \cdots$ for the variable x of the term $f(x)$ on the left side of (7.29) and also for the x of the kernel $K(x,t)$ inside the integral of (7.29). In this way, we have $f(x_i) = f(a + 2ih) \equiv f_{2i}$, $u(t_j) = u(a + (2j - 1)h) \equiv u_{2j-1}$, and $K(x_i, t_j)$ $=K(a + 2ih, a + (2j - 1)h) \equiv K_{2i,2j-1}$; $i, j = 1, 2, \cdots, N$. With this notation, the Maclaurin setting of the Volterra integral equation of the first kind (7.29) (for $i = 1, 2, \cdots, 5$,) becomes

$$\begin{aligned}
&2hK_{21}u_1 = f_2 \\
&4h\left[\frac{1}{2}K_{41}u_1 + \frac{1}{2}K_{43}u_3\right] = f_4 \\
&6h\left[\frac{3}{8}K_{61}u_1 + \frac{2}{8}K_{63}u_3 + \frac{3}{8}K_{65}u_5\right] = f_6 \\
&8h\left[\frac{13}{48}K_{81}u_1 + \frac{11}{48}K_{83}u_3 + \frac{11}{48}K_{85}u_5 + \frac{13}{48}K_{87}u_7\right] = f_8 \\
&10h\left[\frac{275}{1152}K_{10,1}u_1 + \frac{100}{1152}K_{10,3}u_3 + \frac{402}{1152}K_{10,5}u_5\right. \\
&\qquad\left. + \frac{100}{1152}K_{10,7}u_7 + \frac{275}{1152}K_{10,9}u_9\right] = f_{10} \\
&\qquad\vdots
\end{aligned} \tag{7.30}$$

and so on. We note again that in this example with $N = 5$, the equations stop at the sample u_9 before the one at the end point, namely u_{10}.

The following example illustrates this use of the Maclaurin rule for solving Volterra integral equations of the first kind (see also Exercise 2 and its detailed answers).

Example 6 Maclaurin Rule for Volterra Equation of the First Kind

Here we consider the Volterra integral equation of the first kind

$$\sin x = \int_0^x e^{x-t} u(t)dt$$

to illustrate the above Maclaurin method (rule) with $N = 4$. The numerical results are compared with the exact solution $u(x) = \cos x - \sin x$, which can be obtained from the answer of Example 7 of Section 3.2 (with $\lambda = 1$),which we arrived at easily via the use of Laplace transform.

Here we take $h = 0.05$, and with $N = 4$ we have $x_{2i} = 2ih$, $t_{2j-1} = (2j-1)h$, $i, j = 1, 2, 3, 4$. So, $x_2 = 0.1$, $x_4 = 0.2$, $x_6 = 0.3$, $x_8 = 0.4$ (the end point), and $t_1 = 0.05$, $t_3 = 0.15$, $t_5 = 0.25$ and $t_7 = 0.35$. Hence our unknowns $u(t_j)$ are labeled as $u_1 = u(t_1) = u(0.05)$, $u_3 = u(t_3) = u(0.15)$, $u_5 = u(t_5) = u(0.25)$, and $u_7 = u(t_7) = u(0.35)$.

If we substitute the values $f(x_2) = \sin x_2 = \sin 0.1$ and $K_{21} = K(x_2, t_1) = e^{0.1-0.05}$ in the first equation of (7.30), we find $u_1 = u(0.05)$,

$$\begin{aligned} 0.1e^{0.1-0.05}u_1 &= \sin 0.1 = 0.09983 \\ &= (0.1)(1.0513)u_1 = 0.09983, \quad u_1 = 0.9496. \end{aligned}$$

Next we substitute this value of u_1, along with the values of $f_4 = f(x_4) = \sin 0.2$, $K_{41} = e^{0.2-0.05}$, $K_{43} = e^{0.2-0.15}$, in the second equation of (7.30) to find $u_3 = u(0.15)$,

$$\begin{aligned} 0.2\left[\frac{1}{2}e^{0.2-0.05}(0.9496) + \frac{1}{2}e^{0.2-0.15}u_3\right] = \sin 0.2 \quad &= 0.1987, \\ u_3 \quad &= 0.8393. \end{aligned}$$

The same can be done for $u_5 = u(0.25)$ and $u_7 = u(0.35)$ in the third and the fourth equations of (7.30), respectively, to have

$$\begin{aligned} 0.3\left[\frac{3}{8}e^{0.3-0.05}(0.9496) + \frac{2}{8}e^{0.3-0.15}(0.8403) + \frac{3}{8}e^{0.3-0.25}u_5\right] &= \sin 0.3, \\ u_5 &= 0.7197 \end{aligned}$$

$$\begin{aligned} 0.5\left[\frac{13}{48}e^{0.4-0.05}(0.9496)\right. \quad &+\frac{11}{48}e^{0.4-0.15}(0.8403) + \frac{11}{48}e^{0.4-0.25}(0.7197) \\ &\left.+\frac{13}{48}e^{0.4-0.35}u_7\right] = \sin 0.4, \quad u_7 = 0.5960. \end{aligned}$$

These approximate values of $\tilde{u}_i$ are compared in the following Table 7.8 with the exact values of the solution $u(x_i) = \cos x_i - \sin x_i \equiv u_i$, along with the error $\varepsilon_i = \tilde{u}_i - u_i$. We used $\tilde{u}_i$ for the above approximate values to distinguish them from the exact values $u(x_i) \equiv u_i$.

Table 7.8 The Numerical Solution of a Volterra Equation of the First Kind—The Maclaurin Method (Example 6)

i	x_i	$4\tilde{u}_i$ (num.)	$u_i = 1\cos x_i - \sin x_i$ (exact)	error $\varepsilon_i = \tilde{u}_i - u_i$
1	0.05	0.9496	0.9488	8×10^{-4}
2	0.15	0.8403	0.8393	10×10^{-4}
3	0.25	0.7197	0.7215	-1.8×10^{-3}
4	0.35	0.5960	0.5965	-5×10^{-4}

Comments Regarding Some Singular Volterra Integral Equations

As we did in Section 1.2 in the simple classification of integral equations, we termed an integral equation *singular* because of the kernel being singular in the domain of the integration, or that the integral involves an unbounded limit (or limits) of ∞ or $-\infty$. The earliest example of an integral equation, and the very familiar example in most integral equations books, is that of the *Abel's problem* (1.20) in $\phi(y)$,

$$-\sqrt{2g}f(y) = \int_0^y \frac{\phi(\eta)}{\sqrt{y-\eta}}d\eta \tag{1.20}$$

where the kernel $K(y,\eta) = \frac{1}{\sqrt{y-\eta}}$ is singular at the end point $\eta = y$.

Fortunately, this singular equation (1.20) is in the special form of Laplace transform *convolution product* integral, where the Laplace transform can be used to solve it, which was done with complete details in Example 8 in Section 3.2. At the level of this introductory text, this example represents the only singular Volterra equation—with its singularity due to its kernel—that we have covered and solved analytically.

Another example of a singular Volterra integral equation, due to its infinite (lower) limit of integration, is that of the *torsion of a wire* (1.15) in the torsion $\omega(t)$,

$$m(t) = h\omega(t) + \int_{-\infty}^t \phi(t,\tau)\omega(\tau)d\tau, \tag{1.15}$$

where the integral represents the dependence of the torsion on all torques applied to the wire in time $(-\infty, t)$ besides the present torque $m(t)$.

At the level of this book we are not covering singular integral equations. However, we feel that a brief comment regarding the numerical approach to the class of singular Volterra integral equations, such as that of (1.15) above, is in order.

We limit our very brief remark to the one type of singular Volterra integral equations that are characterized by having an *infinite limit* for its integration as in (1.15) of the torsion of the wire,

$$m(t) = h\omega(t) + \int_{-\infty}^t \phi(t,\tau)\omega(\tau)d\tau \tag{1.15}$$

Here we consider its following general case

$$u(x) = f(x) + \int_{-\infty}^{x} K(x,t)u(t)dt, \quad -\infty < x < \infty. \tag{7.31}$$

The solution $u(x)$ is defined for $x\epsilon(-\infty,\infty)$, however the integration is done on $(-\infty, x)$, which can be considered as due to the assumption that the kernel $K(x,t)$ vanishes for $t > x$, or

$$K(x,t) = \begin{cases} K(x,t), & -\infty < t < x \\ 0, & x < t < \infty. \end{cases} \tag{7.32}$$

According to this definition of $K(x,t)$, if we write $K(t,x)$,

$$K(t,x) = \begin{cases} 0, & -\infty < t < x \\ K(t,x), & x < t < \infty \end{cases} \tag{7.33}$$

instead of $K(x,t)$ in (7.31), the limits of integration become from $t = x$ to $t = \infty$, and (7.31) reduces to

$$u(x) = f(x) + \int_{x}^{\infty} K(t,x)u(t)dt. \tag{7.34}$$

Either of the forms (7.31) with kernel $K(x,t)$ in (7.32) and (7.34) with kernel $K(t,x)$ in (7.33) is a singular Volterra integral equation of the second kind, and they are singular because of the presence of an *infinite* limit for their integrals.

In the following we present simple transformations that reduce the above Volterra integral equations with infinite limit for their integral, to ones with finite limits. However, in the new resulting equations, the kernel now becomes singular.

Mapping onto a Finite Interval

For the singularity of the equation due to the infinite limit of integration for Volterra (as well as Fredholm) equations, we can use the simple mappings $\xi = \frac{1}{x}$ and $\tau = \frac{1}{t}$ to map the infinite limit to 0. In the case of (7.31), this mapping reduces it to the following integral equation in the new unknown $\phi(\xi) = u(\frac{1}{x})$ on a finite $(0, \xi)$ interval instead of $u(t)$ on $(-\infty, x)$, $(x < 0)$

$$\phi(\xi) = F(\xi) + \int_{0}^{\xi} H(\xi,\tau) \cdot \frac{\phi(\tau)}{\tau^2} d\tau \tag{7.35}$$

where $F(\xi) = f(\frac{1}{x})$, $H(\xi,\tau) = K(\frac{1}{\xi}, \frac{1}{\tau})$.

We note here that even for a nice continuous kernel $K(x,t)$ in (7.31), the above integral equation (7.35) may reduce to one with a *singular kernel* with singularity at $\tau = 0$. For example, the equation in $u(x)$,

$$u(x) = xe^{x} + \int_{-\infty}^{x} e^{-2(x-t)}(x-t)u(t)dt \tag{7.36}$$

becomes the following in $\phi(\xi)$

$$
\begin{aligned}
u\left(\frac{1}{\xi}\right) &= \frac{1}{\xi}e^{\frac{1}{\xi}} - \int_0^{\xi} e^{-2\left(\frac{1}{\xi}-\frac{1}{\tau}\right)} \cdot \frac{1}{\tau^2}\left(\frac{1}{\xi}-\frac{1}{\tau}\right) u\left(\frac{1}{\tau}\right) d\tau \\
\phi(\xi) &= F(\xi) - \int_0^{\xi} \frac{(\tau-\xi)}{\xi} \cdot e^{-\frac{2(\tau-\xi)}{\xi\tau}} \cdot \frac{1}{\tau^3}\phi(\tau) d\tau
\end{aligned}
\tag{7.37}
$$

where the kernel is singular at $\tau = 0_-$ due to the factor $\frac{1}{\tau^3}$. (The factor $e^{-\frac{2(\tau-\xi)}{\xi\tau}}$ is bounded since $\xi < \tau < 0$.)

Since we are not covering the analysis of singular integral equations due to their kernel being singular, we shall leave this subject for the interest of the reader. (See the general references given towards the end of Section 5.5 on the numerical methods of integral equations including the singular ones Delves and Mohammed [1988], Baker and Miller [1977].)

Exercises 7.2

1. Consider the Volterra integral equation of the second kind

$$u(x) = 1 - 2x + 4x^2 + \int_0^x [3 + 6(x-t) - 4(x-t)^2]u(t)dt \tag{E.1}$$

(a) Let $t_1 = \Delta t = 0.05$, $t_2 = 0.1$, $t_3 = 0.15$, and $t_4 = 0.2$, $u(t_i) \equiv u_i$. Write its numerical Newton-Cotes "*gradual*" or "*combination*" rules in the sense of using the trapezoidal rule for determining u_1 of u_1, Simpson's rule for that of u_2, the $\frac{3}{8}$-rule for u_3, then returning to the repeated Simpson's rule for u_3, $u_5, \cdots$ *Hint:* See (7.28) and Example 5.

(b) Since we have a simple triangular 5×5 system, solve it successively to find the approximate values of the solution u_1, u_2, u_3, u_4, and u_5, then compare with those of the exact solution $u(x) = e^x$.

2. Consider the Volterra integral equation of the first kind,

$$x = \int_0^x (1 - x + t)u(t)dt. \tag{E.1}$$

(a) Use the *Maclaurin* method (that we used for (7.29) to obtain its numerical setting (7.30)) to write the numerical setting of (E.1), using $\Delta t = h = 0.05$, $u_i \equiv u(x_i) \equiv u(0.05i)$, $i = 1, 3, 5, 7, 9$. (See Example 6.)

(b) With such simple triangular system of equations, solve it successively to find the samples of the approximate solution u_i, $i = 1, 3, \cdots, 9$. Compare your results with the samples of the exact solution $u(x) = e^x$.

3. For the Volterra integral equation of the second kind in Exercise 1, repeat the problem with $\Delta t = 0.02$ and compare with the approximate and exact answers.

4. Consider the Volterra integral equation of the second kind of Exercise 1,

$$u(x) = 1 - 2x + 4x^2 + \int_0^x [3 + 6(x-t) - 4(x-t)^2]u(t)dt$$

and its numerical setting with $\Delta t = 0.05$ in part **(a)** of that exercise starting with the trapezoidal rule for u_1, Simpson's rule for u_2, the $\frac{3}{8}$-rule for u_3, then returning to the repeated Simpson rule for $u_4, u_5, \cdots$, as was done in (7.28) and illustrated in Example 5.

(a) In the present exercise use larger $\Delta t = 0.5$ instead of $\Delta t = 0.05$ in the above exercise, to solve the problem, and compare your results with its exact answer $u(x) = e^x$.

(b) The discussion following (7.28) suggests that the value u_1, obtained above with the (not so accurate low degree) trapezoidal rule would be the most inaccurate among the other approximate values. Use your results to comment on this discussion. Also, why is it that such inaccuracy is not so apparent in Exercise 1 when a smaller $\Delta t = 0.05$ was used?

For additional detailed computations of this problem see Exercise A.1 of Section 7.2.A in "The Student's Solution Manual" to accompany this text.[5]

5. For Exercise 1, use the Lagrange interpolation formula (1.153) and (1.154) to interpolate its numerical values of the solution, then compare this approximate interpolated solution $\tilde{u}(x)$ with the exact solution $u(x) = e^x$.

6. Consider the singular Volterra integral equation in $u(x)$,

$$u(x) = \frac{1}{x} - \int_{-\infty}^{x} \frac{1}{t^2}\left(\frac{1}{x} - \frac{1}{t}\right) u(t)dt$$

(a) Use the change of variable $\xi = \frac{1}{x}$, $\tau = \frac{1}{t}$ to reduce the integral to that of finite limits $\tau = 0$ to $\tau = \xi$ in the new unknown $\phi(\xi) = u(\frac{1}{x})$.

(b) Solve the resulting equation in $\phi(\xi)$, thus find $u(x) = \phi(\frac{1}{\xi})$. *Hint:* The resulting equation in $\phi(\xi)$ is non-singular with a Laplace convolution product type integral.

7.3 HIGHER QUADRATURE RULES FOR FREDHOLM EQUATIONS

In this section we consider the numerical approximation setting for solving Fredholm integral equations using the more efficient high quadrature rules of Section 7.1 such as the Gauss-Legendre and other Gauss quadrature rules. The results will be compared with the, relatively inefficient, trapezoidal rule that we have already used in (5.124)

[5] Jerri [1999]. See the end of the preface for more information.

of Section 5.5.1. As we discussed in Section 7.1, all such quadrature rules will differ in the weights D_j and the sample locations t_j of the numerical approximation setting (5.118) with $N = n + 1$,

$$u(x_i) = f(x_i) + \sum_{j=1}^{N} K(x_i, t_j) D_j u(t_j), \quad i = 1, 2, \cdots, N \qquad (7.38), (5.118)$$

for the Fredholm integral equation (of the second kind) (1.148), (5.116)

$$u(x) = f(x) + \int_a^b K(x,t) u(t) dt. \qquad (1.148), (5.116)$$

Here, the need for efficient numerical methods is clear, since we have an $N \times N$ square system of linear equations, and the cost is very high if we stay with the trapezoidal rule. For the case of Volterra integral equations, with its much simpler *triangular* system of linear equations, Newton-Cotes rules, or a combination of them, can be used to solve such systems, *successively*, or what is termed as the *marching method*, which was discussed in the previous Section 7.2.

We will illustrate (7.38) (or (5.117)) using a four-point *Gauss-Legendre rule* for approximating the integral on the interval (0,1), which we have already presented in Section 7.3. We must first note that in Table 7.3, we have the locations of the samples x_i and the weights w_i given on the (symmetric) interval (-1,1), and where both values of x_i and w_i are symmetric. Here, we use four sample points on (0,1). So, according to (7.5) and (7.6), we must translate the locations adding 1, then scale them by a factor of $h = \frac{1}{2}$, since our interval (0,1) has half the width of (-1,1) in Table 7.3. So the needed locations are $x_i' = \frac{1}{2}(1 + x_i)$, $i = 1, 2, 3, 4$. Also the weights w_i of Table 7.3 must be scaled by a factor of $\frac{1}{2}$ as indicated in (7.6) and $w_i' = \frac{1}{2} w_i$. Thus, we obtain the four locations on (0,1) from the following (symmetric) locations (of Table 7.3).

$$x_1 = -0.861136, x_2 = -0.339981, x_3 = 0.339981, x_4 = 0.861136$$

as

$$x_1' = \frac{1}{2}(1 - 0.861136) = 0.069432, \quad x_2' = 0.330009, \quad x_3' = 0.669991,$$
$$x_4' = 0.930568.$$

The new weights $w_i' = \frac{1}{2} w_i$ of the (symmetric) values w_i of Table 7.1 are:

$$w_1' = 0.173927, w_2' = 0.326073, w_3' = 0.326073, w_4' = 0.173927.$$

Now we can write the numerical setting of the Fredholm equation,

$$u(x) = f(x) + \lambda \int_0^1 K(x,t) u(t) dt \qquad (7.39)$$

with the help of a four-point Gauss-Legendre rule (in parallel to using the trapezoidal rule in setting (5.124) for (5.116)) as

$$\begin{aligned}
u_1 &= f_1 + \lambda\left[w_1' K_{11} u_1 + w_2' K_{12} u_2 + w_3' K_{13} u_3 + w_4' K_{14} u_4\right] \\
u_1 &= f_1 + \lambda\left[w_1' K_{11} u_1 + w_2' K_{12} u_2 + w_2' K_{13} u_3 + w_1' K_{14} u_4\right] \\
u_1 &= f_1 + \lambda[0.173927 K_{11} u_1 + 0.326073 K_{12} u_2 + 0.326073 K_{13} u_3 \\
&\quad +0.173927 K_{14} u_4].
\end{aligned}$$

The same is done for $i = 2, 3$, and 4 to have the system of four equations,

$$u_i = f_i + \lambda\left[w_1' K_{i1} u_1 + w_2' K_{i2} u_2 + w_2' K_{i3} u_3 + w_1' K_{14} u_4\right], \quad i = 1, 2, 3, 4, \tag{7.40}$$

and where it is understood that

$$u_i \equiv u(x_i'), f_i \equiv f(x_i') \text{ and } K_{ij} \equiv K(x_i', t_j'); \quad i, j = 1, 2, 3, 4.$$

Next we illustrate the same problem with the help of a four-point *Tchebychev* quadrature rule with its equal weights $w_i = \frac{b-a}{N} = \frac{1-0}{4} = \frac{1}{4}$, and with its locations x_i on (-1,1) as given in Table 7.4. Again, since our problem is defined on (0,1) we translate the locations x_i of Table 7.4 by 1, then scale them by a factor of $\frac{1}{2}$, i.e., $x_i = \frac{1}{2}(1+x_i)$, $i = 1, 2, 3, 4$ (as we did with the above values for the Gauss-Legendre quadrature rule) to have

$$\begin{gathered}
x_1' = \frac{1}{2}(1 + x_1) = \frac{1}{2}(1 - 0.794654) = 0.102673, \\
x_2' = 0.406204, \quad x_3' = 0.593796, \quad x_4' = 0.897327.
\end{gathered}$$

With the equal weights of $w_i = \frac{1}{4}$, $i = 1, 2, 3, 4$, the numerical setting of (7.39) with the Tchebychev rule becomes

$$u_i = f_i + \lambda \frac{1}{4}[K_{i1} u_1 + K_{i2} u_2 + K_{i3} u_3 + K_{i4} u_2], \quad i = 1, 2, 3, 4, \tag{7.41}$$

where $t_i' = x_i'$ are as determined above.

We recognize here that in order to have better accuracy for the solution of such system of linear equations, we often need a large number of points N and a high quadrature rule. Even with such very efficient rules of Section 7.1, the main task still remains on the shoulders of the user, where a good familiarity with matrix analysis is of utmost benefit. In this book, and for the sake of having the material self contained, we have included only a brief review of *Cramer's rule* in Section 1.5.4 for solving a system of linear equations.

Next, we will use Cramer's rule for solving the system of four linear equations as the result of the numerical approximation of Fredholm integral equation in Example 20 of Section 5.5.1,

$$u(x) = \sin x + \int_0^1 (1 - x \cos xt) u(t) dt \tag{7.42}$$

where this equation will now be approximated with the help of a four-point Gauss-Legendre quadrature rule as in (7.40), then a Tchebychev rule as in (7.41).

Example 7 Fredholm Equation of the Second Kind

(a) The Gauss-Legendre Rule

First we approximate the Fredholm integral equation of the second kind

$$u(x) = \sin x + \int_0^1 (1 - x\cos xt)u(t)dt \tag{E.1}$$

using a four-point Gauss-Legendre rule as in (7.40), to have the system of four equations in the four unknown sample values u_1, u_2, u_3 and u_4.

Of course, we first use the change of variable $\xi = 2t - 1$, $t = \frac{\xi+1}{2}$, $dt = \frac{1}{2}d\xi$, to have the integral on the symmetric interval $-1 < \xi < 1$, ready for its numerical approximation by the Gauss-Legendre rule and its Table 7.3. However we still have $x\epsilon(0,1)$, where we shall adjust x_i, w_i, given for the symmetric interval (-1,1) in Table 7.3 for $N = 4$, to give us $x_i' = \dfrac{1}{2}(x_i + 1)$, $w_i' = \dfrac{1}{2}w_i$ for $u(x_i') \equiv u_i$, $x_i'\epsilon(0,1)$.

$$u(x) = \sin x + \frac{1}{2}\int_{-1}^{1}\left[1 - x\cos x\left(\frac{\xi+1}{2}\right)\right]u\left(\frac{\xi+1}{2}\right)d\xi. \tag{E.2}$$

Now we use the four-point Gauss-Legendre rule for the integral with the help of Table 7.3, and evaluate u_1 at $x = x_1' = \frac{1+x_i}{2} = \frac{1-0.86113}{2} = 0.069432$ to generate the first linear equation of (7.40) corresponding to $i = 1$,

$$\begin{aligned} u_1 &= \sin(0.06943) + \frac{1}{2}[(0.34786)\left\{1 - (0.06943)\cos(0.06943)^2\right\}u_1 \\ &+(0.65215)\{1 - (0.06943)\cos(0.06943)(0.330001)\}u_2 \\ &+(0.65215)\{1 - (0.06943)\cos(0.06943)(0.66999)\}u_3 \\ &+(0.34786)\{1 - (0.06943)\cos(0.06943)(0.93057)\}u_4]. \end{aligned}$$

In a similar way we obtain the results at x_2', x_3', and x_4' to have the four linear equations in the four unknowns u_1, u_2, u_3, and u_4; $u_i \equiv u(x_i')$

$$\begin{aligned} u_1 &= 0.06938 + 0.16185u_1 + 0.30345u_2 + 0.30346u_3 \\ &+0.16188u_4 \\ u_2 &= 0.32405 + 0.11655u_1 + 0.21910u_2 + 0.22109u_3 \\ &+0.11922u_4 \\ u_3 &= 0.62098 + 0.05752u_1 + 0.11292u_2 + 0.12925u_3 \\ &+0.07932u_4 \\ u_4 &= 0.80196 + 0.01241u_1 + 0.03684u_2 + 0.07973u_3 \\ &+0.06906u_4. \end{aligned} \tag{E.3}$$

Then we rearrange to have the system of linear equations ready for a matrix equation form, with its resulting coefficient matrix A (in $AU = B$ of (E.3)) as

$$A = \begin{bmatrix} 0.83815 & -0.30344 & -0.30346 & -0.16188 \\ -0.11655 & 0.78090 & -0.22109 & -0.11922 \\ -0.05752 & -0.11292 & 0.87075 & -0.07932 \\ -0.01241 & -0.03684 & -0.07932 & 0.93094 \end{bmatrix}. \tag{E.4}$$

Here we must find the determinant of the (square) matrix A, which is essential to the use of Cramer's rule for obtaining the final solution, then we report the final solution (u_1, u_2, u_3, u_4). The result of evaluating the determinant of the square matrix A in (E.4) is $|A| = 0.450361$.

Now we use Cramer's rule (as in Section 1.5.4) to solve the system of equations in (E.3) where we find that the values $\tilde{u}_1$, $\tilde{u}_2$, $\tilde{u}_3$ and $\tilde{u}_4$ are almost equal to the exact value of $u(x) = 1, 0 < x < 1$ (within an error of about order 10^{-10}).

In Table 7.9 we present a comparison of these approximate (numerical) solutions $\tilde{u}_i$, with the exact solution $u_i = u(x_i) = 1$, along with the error $\varepsilon_i = \tilde{u}_i - u_i$.

Table 7.9 Numerical (Gauss-Legendre) and Exact Solutions of a Fredholm Equation (Example 7)

i	x_i	$\tilde{u}_i$(num.)	$u_i = 1$ (exact)	error $\varepsilon_i = \tilde{u}_i - u_i$
1	0.06943	$1+ \sim 10^{-10}$	1.0	$\sim 10^{-10}$
2	0.33001	$1+ \sim 10^{-10}$	1.0	$\sim 10^{-10}$
3	0.66999	$1+ \sim 10^{-10}$	1.0	$\sim 10^{-10}$
4	0.93057	$1+ \sim 3 \times 10^{-10}$	1.0	$\sim 3 \times 10^{-10}$

We note here that the absolute error of about 10^{-10} represents an improvement for the Legendre rule, when compared with the trapezoidal rule that we used for the same problem in Example 20 of Section 5.5.1, though with $N = 3$, where the error ranged between 10^{-2} and 2×10^{-2}.

(b) The Tchebychev (Equal Weight) Rule

Here we use a four-point Tchebychev rule, with the help of Table 7.4, as in (7.41) to solve the same Fredholm integral equation (E.1); and to have the integral on the symmetric interval, we use its transformed version in (E.2). We also adjust the x_i given on the symmetric interval (-1,1) in Table 7.4 to give us $x'_i = \frac{1}{2}(x_i + 1)$ for $u(x'_i) \equiv u_i$, $x'_i \epsilon (0, 1)$, since our integral of (E.1) is defined on (0,1).

$$\begin{aligned} u_1 &= \sin(0.102673) + \frac{1}{4}[(1 - (0.102673)\cos(0.102673)^2)u_1 \\ &+(1 - (0.102673)\cos\{(0.102673)(0.406204)\})u_2 \\ &+(1 - (0.102673)\cos\{(0.102673)(0.593796)\})u_3 \\ &+(1 - (0.102673)\cos\{(0.102673)(0.897327)\})u_4]. \end{aligned} \tag{E.5}$$

In a similar way we obtain the results at x_2, x_3, and x_4 to have four linear equations in u_1, u_2, u_3 and u_4,

$$\begin{aligned} u_1 &= 0.102493 + 0.224333u_1 + 0.224354u_2 + 0.224379u_3 \\ &+0.224441u_4 \\ u_2 &= 0.395125 + 0.148537u_1 + 0.149828u_2 + 0.151389u_3 \\ &+0.155121u_4 \end{aligned}$$

$$\begin{aligned} u_3 &= 0.559511 + 0.101827u_1 + 0.105848u_2 + 0.110684u_3 \\ &\quad +0.122130u_4 \\ u_4 &= 0.781663 + 0.0266197u_1 + 0.040406u_2 + 0.056767u_3 \\ &\quad +0.094545u_4. \end{aligned} \tag{E.6}$$

These equations are then written in a matrix form, as we did in (E.3), (E.4) in part (a), where the determinant of the coefficient matrix A is computed, and Cramer's rule of Section 1.5.4 is used to find the solution (u_1, u_2, u_3, u_4) of the system of equations in (E.6).

Since all these computations were done in detail in part (a), it is sufficient here to report in the following Table 7.10 such numerical solution $\tilde{u}_i$, $i = 1, 2, 3, 4$, which are almost equal to the exact solution $u_i = 1$ (within an error $\varepsilon_i = \tilde{u}_i - \tilde{u}_i$ of about order 10^{-7}.)

Table 7.10 Numerical (Gauss-Tchebychev) and Exact Solutions of a Fredholm Equation (Example 7)

i	x_i	$\tilde{u}_i$(num.)	$u_i = 1$ (exact)	error $\varepsilon_i = \tilde{u}_i - u_i$
1	0.102673	$1+\sim 10^{-7}$	1.0	$\sim -10^{-7}$
2	0.406204	$1+\sim 10^{-7}$	1.0	$\sim -10^{-7}$
3	0.593796	$1+\sim 10^{-7}$	1.0	$\sim -10^{-7}$
4	0.897327	$1+\sim 10^{-7}$	1.0	$\sim -10^{-7}$

We note that the accuracy of the computations is reasonable compared to that of the Gauss-Legendre rule in part (a). However, it is known that the latter can do very well as we increase the number of points N. We shall leave such observations for the exercises.

Accuracy of the Numerical Methods

For the family of the quadrature rules suggested for the numerical solution of Fredholm integral equation,

$$u(x) = f(x) + \lambda \int_a^b K(x,t)u(t)dt \tag{7.39}$$

we, of course, must first make sure that the equation has a *unique* solution $u(x)$ before embarking on using one or more quadrature rules to find its approximate samples $\{u(x_i)\}_{i=1}^N$. Next we must also have an estimate of the error for the quadrature rule (or rules). It turns out that the bound on the error for such numerical methods depends on two factors. The first is independent of the method used, and it depends on the integral equation itself as characterized by the integral operator with its kernel $K(x,t)$ in (7.39), and its nonhomogeneous term $f(x)$. As was discussed briefly in Section 5.4 for the Fredholm integral of the first kind, the first factor is described in terms of the (possibly high) sensitivity of the solution $u(x)$ to small changes in the above two characteristic parameters of the problem, namely, $f(x)$ and $K(x,t)$.

The second factor of the error bound is more related to what we are doing here, where it is a measure of the error of the rule used in approximating the integral of the equation (7.39). With the high degree quadrature rules, such error can be minimized easily if $K(x,t)u(t)$ is *well behaved*. However there is a catch here, since $u(x)$ is still the unknown to be found. A very helpful result, for directing our efforts in using the above two groups of Newton-Cotes and the Gauss quadrature rules, is that, "If the kernel $K(x,t)$ is continuous in the square; $\{a \le x \le b; a \le t \le b\}$, $f(x)$ is continuous on $a \le x \le b$, and if we also know that the equation (7.39) has a unique continuous solution, then there is a family of quadrature rules for which the error, in approximating the Fredholm integral equation (7.39), tends to zero as $N \to \infty$". Such family of rules includes, (i) the M-panel Newton-Cotes rules of degree P for any fixed P and increasing M; and, (ii) the Gauss-Legendre, and open and closed Tchebychev N-point rules for increasing N. It does not, however include the Newton-Cotes method for fixed panel M and increasing degree P.

For example, in the problem of the above Example 7, and also Example 20 of Section 5.5.1

$$u(x) = \sin x + \int_0^1 (1 - x\cos xt)u(t)dt \tag{7.42}$$

we know that the solution of this equation is $u(x) = 1$, and that the kernel $K(x,t) = (1 - x\cos t)$ is also continuous (and differentiable) on the square; $\{0 \le x \le 1; 0 \le t \le\}$. Hence, according to the above result, there is no surprise when simple quadrature rules would work as illustrated in Exercise 1(a) of Section 5.5 (with the very low value of $N = 3$) for the trapezoidal rule and with better results for Simpson's rule. However, this is not very much the case for the integral equations with kernels such as where the solutions are smooth, but the kernel $K(x,t)$

$$K(x,t) = \begin{cases} x(1-t), & 0 \le x \le t \\ t(1-x), & 0 \le t \le x \end{cases} \tag{7.43}$$

which, although it is continuous in both x and t, it does have the problem of a jump discontinuity of size 1 in its first derivative $\frac{\partial K(x,t)}{\partial x}$ as shown in (4.26) for the more general case of Green's function $G(x,t)$ in (4.25). An indication of the convergence of the numerical method for the problem (7.42) may be seen in the result of Example 20 in Section 5.5 where the trapezoidal rule was used with $N = 3$ (see also Exercise 5 in Section 5.5). For an illustration with kernel $K(x,t)$ that has a jump discontinuity in its derivative, see Exercises 8 and 6 in Section 5.5 for a nonhomogeneous and homogeneous Fredholm equations, respectively. Of course, these examples with limited small N are not enough, and we shall have a chance to illustrate such error analysis for large N in the exercises.

7.3.1 Comments on Higher Quadrature Rules for Some Singular Fredholm Equations

Some Analytical Methods

As we mentioned in the preface, the presentation of integral equations in this book assumes only basic differential equations and calculus. Thus it is not possible to touch upon the desired theory or methods of solutions of singular integral equations, which requires some preparation of at least a course in complex variables and advanced calculus. As such, our treatment in this section will be mainly illustrative in nature, where we depend on the tools that have already been developed in this book with the necessary details. This includes the use of integral transforms in Section 1.4 and the numerical methods with its basic support in Section 5.5.1 and this Section 7.3. The first *operational calculus method* covers the use of Fourier and other transforms for solving a special class of singular Fredholm integral equations. Such class includes the following singular Fredholm equation, where its integration part is in the form of a *Fourier convolution product*

$$u(x) = f(x) + \lambda \int_{-\infty}^{\infty} K(x-\xi)u(\xi)d\xi \tag{7.44}$$

as we presented it in Section 1.4.2. This method was then illustrated in Example 15 (of the same Section 1.4.2) for solving the following singular Fredholm integral equation of the second kind,

$$u(x) = e^{-|x|} + \mu \int_{-\infty}^{\infty} e^{-|x-\xi|}u(\xi)d\xi, \tag{7.45}$$

and in Example 16 of the same Section 1.4.2 for the singular *homogeneous* Fredholm integral equation,

$$u(x) = \mu \int_{-\infty}^{\infty} e^{-|x-t|}u(t)dt. \tag{7.46}$$

In Section 1.2 we classified integral equations, where we defined an integral equation as singular, when

(i) Either one of the two limits of the integral in the equation is infinite,

or

(ii) The kernel of the integral equation becomes unbounded, i.e., equations with singular kernels.

The first type of singularity is seen in the above two examples of (7.45) and (7.46). For lack of space we shall limit our general brief discussion and the numerical method, of using higher quadrature rules, to such singular Fredholm equations whose singularity is only due to the limit (or limits) of integration being infinite.[6]

[6]The interested reader may consult Baker and Miller [1977] or Delves and Mohammed [1988].

Singularity Due to Infinite Limits of the Integral

In the examples of the Fourier transforms (1.87), (1.95), (1.98) in Section 1.4.2, and the Laplace transform (1.63) in Section 1.4.1 we looked at them as Fredholm integral equations of the first kind that are singular because of the infinite limit (or limits in the case of (1.87)) of integration, while the kernels are extremely *well behaved (very smooth)* exponential functions. In the case of the Hilbert transform (1.131) we have both (strong) singularity at $x = \lambda$ for the kernel $K(x,\lambda) = \frac{1}{x-\lambda}$ and infinite limits of integration. The method of solving such (strongly) singular integral equations needs complex analysis, and falls at the center of the treatment of singular integral equations, which is termed the "*Hilbert problem*", and which we are not going to pursue in this book.

Transforming to Finite Limits

Integral equations with singularity due to the infinite limits of integration, but with well behaved kernels may be reduced to ones that are much easier to handle. For example,

$$u(x) = f(x) + \int_{-\infty}^{\infty} K(x,t)u(t)dt, \qquad -\infty < x < \infty \tag{7.47}$$

may, formally, be reduced to that of finite limits via the *change of variables*

$$t = \tan\tau, \qquad x = \tan\xi \tag{7.48}$$

where the limits of integration in both of the new variables τ and ξ are from $-\frac{\pi}{2}$ to $\frac{\pi}{2}$. Such change of variables (or *mappings*) is often used in the *numerical approximation* of this type of singular equations, where we use a quadrature rule on $(-\frac{\pi}{2}, \frac{\pi}{2})$ for the following transformed equation (7.49) instead of dealing with the infinite limits of (7.47)) (see Exercise 9 for an illustration in this general direction),

$$U(\xi) = F(\xi) + \int_{-\frac{\pi}{2}}^{\frac{\pi}{2}} H(\xi,\tau)U(\tau)d\tau \tag{7.49}$$

where $U(\xi) = u(x) = u(\tan\xi)$, $F(\xi) = f(x) = f(\tan\xi)$, and $H(\xi,\tau) = K(x,t)$ $=K(\tan\xi, \tan\tau)\cdot\sec^2\tau$. Note that the new kernel in (7.49) may have singularities at $\tau = \mp\frac{\pi}{2}$.

Another mapping that is used for the same purpose is

$$\xi = \frac{2\alpha}{x+\alpha} - 1, \qquad \tau = \frac{2\alpha}{t+\alpha} - 1 \tag{7.50}$$

which reduces the equation

$$u(x) = f(x) + \int_0^{\infty} K(x,t)u(t)dt, \quad 0 \le x \le \infty \tag{7.51}$$

to one with the finite limits $\tau = -1$ to 1,

$$U(\xi) = F(\xi) + \int_{-1}^{1} G(\xi, \tau) U(\tau) d\tau, -1 \leq \xi \leq 1 \tag{7.52}$$

where $U(\xi) = u(x)$, $F(\xi) = f(x)$, and $G(\xi, \tau) = \frac{2\alpha K(x,t)}{(\tau+1)^2} = \frac{2\alpha}{(\tau+1)^2} K(\frac{2\alpha}{\xi+1} - \alpha, \frac{2\alpha}{\tau+1} - \alpha)$.

Of course, here we notice that the new kernel has a singularity at the lower limit of integration $\tau = -1$. For the numerical methods, such singularity at an end point may be avoided by the use of an "open" quadrature rule for approximating the integral, i.e., a rule that does not use the end points for samples such as the Newton-Cotes rules of the open type (Table 7.2(b)), Maclaurin rule (Table 7.3) or the Tchebychev (open) rule (Table 7.4), as we discussed in Section 7.1.

As to the analytical treatment of singular equations of this infinite limits type (7.47), the simplest class that we can have a good hold on here is the special case where the integral is in a *Fourier convolution product form*,

$$u(x) = f(x) + \mu \int_{-\infty}^{\infty} k(x - t) u(t) dt. \tag{7.53}$$

If we let $U(\lambda)$, $F(\lambda)$ and $K(\lambda)$ be the Fourier transforms of $u(x)$, $f(x)$ and $k(x)$, respectively, then the Fourier transformation of the singular Fredholm integral equation (7.53) reduces it to an algebraic equation in $U(\lambda)$,

$$U(\lambda) = F(\lambda) + \mu K(\lambda) U(\lambda) \tag{7.54}$$

after using the Fourier convolution theorem (1.101) on the above integral as a convolution product integral in (7.53).

Now we solve for $U(\lambda)$,

$$U(\lambda) = \frac{F(\lambda)}{1 - \mu K(\lambda)}, \quad 1 - \mu K(\lambda) \neq 0. \tag{7.55}$$

So, provided that $1 - \mu K(\lambda) \neq 0$, we can use the inverse Fourier transform (1.88) on this $U(\lambda)$ of (7.55) to find the solution $u(x)$ of the singular equation (7.53),

$$u(x) = \frac{1}{2\pi} \int_{-\infty}^{\infty} e^{i\lambda x} \frac{F(\lambda)}{1 - \mu K(\lambda)} d\lambda. \tag{7.56}$$

As we mentioned at the beginning of this section, this method was illustrated for the singular nonhomogeneous Fredholm equation (7.45), and its associated homogeneous one (7.46), respectively, in Examples 15 and 16 of Section 1.4.2, respectively. Another illustration with instructive hints is done in Exercise 17 of the same section.

The Numerical Method

As we mentioned earlier, our main discussion and illustration will center around Fredholm integral equations that are singular due to (only) infinite limit (or limits) of the integral.

The main issue here, compared to the numerical methods of nonsingular Fredholm integral equations in the previous section, is how to deal with the infinite limit or limits of integration? This, we have prepared for in (7.18)–(7.24), (7.26) and (7.27) of Section 7.1, and illustrated in Examples 3 and 4, and Exercises 2 to 5 in Section 7.1.

The basic idea of the numerical methods starts with approximating the integral in an integral equation, a subject that we covered in Section 7.1 concerning quadrature rules for approximating integrals including those with infinite limits. All those preparations can be summed in the following three basic attempts:

(1) To truncate the infinite limit of integration to a finite limit L, then use a high order quadrature rule, which is, often, not good enough.

(2) To use the *Gauss-Laguerre rule* for integrals on $(0, \infty)$.

(3) To use the *Gauss-rational rule* as in (7.24), which, effectively, maps the infinite interval (a, ∞) into a finite interval (-1,1) in a new variable $U(\xi)$, whence a Gauss-Legendre rule can be easily applied for this finite interval. This was used in reducing the singular Fredholm equation (7.51) with infinite domain $(0, \infty)$ for $u(x)$ to that of (7.52) with finite domain (-1,1) for $U(\xi)$.

We shall illustrate an outline of these (simple) schemes in the following example.

Example 8

Consider the following singular integral equation.

$$u(x) = f(x) + \int_0^\infty K(x,t)u(t)dt \tag{E.1}$$

with the nonhomogeneous term

$$f(x) = (4 + x^2)^{-\frac{1}{2}} \left(1 - \frac{\pi}{4} e^{-2x}\right) \tag{E.2}$$

and a smooth kernel

$$K(x,t) = \frac{\cos xt}{(4 + x^2)^{\frac{1}{2}}(4 + t^2)^{\frac{1}{2}}} \tag{E.3}$$

where, clearly, the form of $f(x)$ is picked to have an exact solution $u(x) = \frac{1}{(4+x^2)^{\frac{1}{2}}}$ for comparing our numerical results.

The exact solution for the above equation, with the $\cos xt$ factor in the kernel, is obtained with the help of the Fourier *cosine* transform tables. From such tables we have

$$\int_0^\infty \frac{\cos xt}{4 + t^2} dt = \frac{\pi}{4} e^{-2x}. \tag{E.4}$$

Since we know the solution $u(x) = \dfrac{1}{(4 + x^2)^{\frac{1}{2}}}$ as a slowly varying function (something that may only happen for such "made up" illustration!) and combined with the kernel as another slowly varying function, we have an extra (artificial!) advantage that the user of a method does not have when looking for the solution! Again, we (purposely) made up the exact solution for comparing the results of the above different methods.

We will present here only a brief discussion of the different methods, and leave the discussion of some analysis and the numerical illustrations for a very similar problem in Exercises 8–10.

(1) The first method is straight-forward and may start with truncating the infinite limit of integration to L, then using N-point Gauss-Legendre rule on $(0, L)$. The computations may be limited to $L = 4, 8, 64$, and $N = 4, 8$.

(2) For a second method we can use N-point Gauss-Laguerre rule for $N = 4, 8$, and with the parameter β in (7.20) varying towards smaller values of $\beta = 1, \frac{1}{4}, \frac{1}{16}$. The choice for small values of β in this example is the result of our experience of the numerical integration in Example 4(a) of Section 7.1 for slowly varying function $\frac{1}{1+x^2}$, and where very small β value gave fair results. Here, knowing our solution in advance as $u(t) = (4 + t^2)^{-\frac{1}{2}}$ along with the $(4 + t^2)^{-\frac{1}{2}}$ in the kernel of (E.4) makes the integrand, aside from the oscillating $\cos xt$ factor, a slowly varying one. This is only an illustration, which suggests the importance of having some idea about the asymptotic behavior of the solution in steering our efforts towards using the most suitable (or efficient) quadrature rule.

(3) In a third method we may use the Gauss-rational rule for $N = 4, 8$, and the parameter β ranging towards larger values $\beta = 1, 4, 16$, as was suggested by our experience in Example 4(a) of Section 7.1 for approximating integrals on $(0, \infty)$ with slowly varying integrands.

The numerical results in this example may be fine for the three methods. Now we must inquire about the possible analytic reason from the theory that will help guide us for other examples. The general "regularity" or "well-behaving" conditions fall along the lines of assuming that, for the (singular) Fredholm equation (E.1) above, its kernel $K(x, t)$ is square integrable on the first quadrant $(0 < x < \infty, 0 < t < \infty)$, and that its nonhomogeneous term $f(x)$ as well as the solution $u(x)$ (if we can estimate its behavior!) are also square integrable on $(0, \infty)$. In our present example the (made up in advance) solution $u(x) = (4 + x^2)^{-\frac{1}{2}}$, and the nonhomogeneous term $f(x) = (4 + x^2)^{-\frac{1}{2}}[1 - \frac{\pi}{4}e^{-2x}]$, are clearly square integrable on $(0, \infty)$. The kernel $K(x, t) = (4 + x^2)^{-\frac{1}{2}}(4 + t^2)^{-\frac{1}{2}} \cos xt$ is also square integrable in both x and t on the first quadrant. Hence, it should not be very surprising if we see convergence for the approximate numerical methods as they are lead by the comfort in satisfying the essential conditions of the underlying theorem.

To give an illustration where we don't have the protection of the above underlying theory, we give the following example

$$u(x) = \frac{1}{4 + x^2} - \frac{\pi}{4}e^{-2x} + \int_0^\infty \cos xtu(t)dt \tag{7.57}$$

with a known solution $u(x) = \frac{1}{4+x^2}$ and a nonhomogeneous term $f(x) = \frac{1}{4+x^2} - \frac{\pi}{4}e^{-2x}$, which are both square integrable on $(0, \infty)$. However, the square integrability of $K(x,t) = \cos xt$ on the first quadrant is clearly in doubt,

$$\int_0^\infty \int_0^\infty |\cos xt|^2 dx dt = \int_0^\infty \int_0^\infty \left(\frac{1}{2} + \frac{1}{2}\cos 2xt\right) dx dt = \infty$$

which we shall leave for a simple exercise.

We leave the attempt of using the Gauss-Laguerre rule (with small β values for the slowly varying (almost constant) integrand!) and the Gauss-rational rule (with large β values) for an exercise to show that even with large N there is no sign of convergence! (see Exercise 8).

Exercises 7.3

1. Consider the Fredholm integral equation of the second kind,

$$u(x) = x^2 - 2\int_0^1 (1 + xt)u(t)dt. \tag{E.1}$$

(a) Use a four-point Tchebychev rule (7.17) for approximating the integral, and set up the system of 4×4 equations in the four unknowns u_1, u_2, u_3, and u_4. For the locations of the samples x_i, consult (7.16), and for the weights see Table 7.4.

(b) Use a four-point Gauss-Legendre rule to write the 4×4 linear equations (see Table 7.3 in Section 7.1 for the locations of the samples x_i and the weight w_i).

2. Consider the *homogeneous* Fredholm integral equation

$$u(x) = \lambda \int_0^1 (x + t)u(t)dt.$$

Use a four-point Gauss-Legendre rule to set up the 4×4 homogeneous linear equations in u_1, u_2, u_3, and u_4.

Hint: See Example 7 and Table 7.3.

3. Consider the Fredholm integral equation of the first kind in $u(x)$,

$$3x + 1 = \int_0^1 (1 + xt)u(x)dx$$

Use a four-point Tchebychev rule (7.17) for approximating the integral, and set up the resulting four linear equations in the four unknowns u_1, u_2, u_3, and u_4 at $x_1 = 0.103$, $x_2 = 0.406$, $x_3 = 0.594$ and $x_4 = 0.897$ (see (7.16) and Table 7.4). Note that we have already adjusted the locations x_i of Table 7.4 to suit the interval (0,1) of the above integral.

4. Consider the Fredholm integral equation with *degenerate* kernel

$$u(x) = \frac{x}{2} - \frac{1}{3} + \int_0^1 (x+t)u(t)dt. \qquad (E.1)$$

(a) Attempt a Gauss-Legendre rule with $N = 2, 3$, and 4 and make your conclusion concerning reaching an exact answer.

(b) Solve the integral equation by the method of Section 5.1 for the exact answer $u(x) = x$, compare with the answer in part **(a)**, and show why part **(a)** gives an exact answer.

(c) Use the exact answer of part **(b)** to verify the integral equation.

(d) Try part **(a)** with Simpson's rule (with the knowledge of the exact solution $u(x) = x$).

5. Repeat the computations and error analysis of Exercise 4 for the following Fredholm integral equation of the second kind (with *degenerate* kernel of one term),

$$u(x) = \sin x - x + \int_0^{\frac{\pi}{2}} xtu(t)dt \qquad (E.1)$$

with the following steps.

(a) Use the method of Section 5.1 for degenerate kernels to show that the exact solution is $u(x) = \sin x$.

(b) Why is that a finite degree (polynomial) quadrature rule cannot result in an exact numerical solution for this equation (E.1)?

(c) Repeat the steps of part **(a)** in Exercise 4 by attempting a Gauss-Legendre rule with $N = 2, 3$, and 4 points, and compare your results with the above exact answer.

(d) Noting part **(b)**, where the integral cannot be approximated by any finite degree (polynomial) quadrature rule, explain the reason for the good results in part **(c)**.

6. Consider the Fredholm integral equation of the *first kind*,

$$3x + 1 = \int_0^1 (1 + xt)u(t)dt$$

and its numerical setting in Exercise 2 in u_1, u_2, u_3, and u_4, where a four-point Gauss-Tchebychev rule was used. Find the determinant of the coefficient matrix, to see if Cramer's rule can be applied to find a solution.

7. Consider the *homogeneous* Fredholm integral equation,

$$u(x) = \lambda \int_0^1 (x+t)u(t)dt \qquad (E.1)$$

and its resulting 4×4 system of homogeneous equations in u_1, u_2, u_3, and u_4 after using the Gauss-Legendre rule in Exercise 2.

(a) Write the system in the matrix form $U = \lambda AU$, $(I - \lambda A)U = 0$, and set $\det(I - \lambda A) = |I - \lambda A| = 0$ to find the approximate (two) eigenvalues $\tilde{\lambda}_1$ and $\tilde{\lambda}_2$ to be compared with their exact values $\lambda_1 = -6 - \sqrt{48}$ and $\lambda_2 = -6 + \sqrt{48}$. (see part **(b)**).

(b) Find the four samples u_1, u_2, u_3, and u_4 of the two approximate eigenfunctions of (E.1) corresponding to the two (approximate) eigenvalues found in part (a). *Hint:* Note that the four sample values of the approximate eigenfunctions are very sensitive to the accuracy of the approximate eigenvalues. Indeed we had to go to ten places accuracy to get the good agreement with the exact eigenfunctions as shown in part (d).

(c) The homogeneous Fredholm integral equation (E.1) is with degenerate kernel of two terms. Use the method discussed in Section 5.1 and illustrated as in Example 2 of that section, to find the above two exact eigenvalues λ_1 and λ_2. Continue the same method to find the corresponding eigenfunctions $U_1(x)$ and $U_2(x)$. Compare these exact results with the approximate ones in part **(a)**.

(d) Use the Lagrange interpolation formula (1.153) and (1.154) to interpolate the four approximate sample values of the eigenfunctions in part **(a)** to find their *continuous* approximations $\tilde{U}_1(x)$ and $\tilde{U}_2(x)$. Graph these two functions and compare them with the graph of the exact one in part **(b)**.

Hint: See Example 7 and Table 7.3 (with its nonequidistant locations $x_1, x_2, \cdots, x_N$.

8. Consider the singular Fredholm equation

$$u(x) = \frac{1}{1+x} + \int_1^\infty e^{-2(x+t)} u(t)dt.$$

(a) Use the *rational* transformation $\xi = \frac{1}{x}$ and $\tau = \frac{1}{t}$ to reduce this equation to one with finite limits of integration $\xi = 0$ to 1 in the new solution $U(\xi) \equiv u(\frac{1}{\xi}) = u(x)$.

(b) Use a four-point Tchebychev rule on the interval $0 < \xi < 1$ for the approximate numerical set up, as 4×4 system of linear equations in the four approximate samples of the new solution $U(\xi)$, $0 < \xi < 1$.

(c) Solve the system of equations in part **(b)** to find $U(\xi_i)$, $i = 1, 2, 3, 4$, then use $x = \frac{1}{\xi}$ to report the approximate values, of the actual solution $u(x_i)$, $i = 1, 2, 3, 4$.

Appendix A
The Hankel Transforms

To further support our presentation in Sections 1.4.2 and 1.4.3 of the Fourier and other integral transforms, we will present the Hankel transforms.[1]

A.1 THE HANKEL TRANSFORM FOR THE ELECTRIFIED DISC

As was presented in (1.124)–(1.128) in Section 1.4.3, the Hankel transform $F_n(\lambda)$ of $f(x)$, $0 < r < \infty$ is defined as

$$F_n(\lambda) = \mathcal{H}_n\{f\} = \int_0^\infty rJ_n(\lambda r)f(r)dr \tag{A.1}$$

where $J_n(\lambda r)$ is the Bessel function of the first kind of order n.

As we mentioned in Section 1.4.3, we can show using (a rather lengthy!) integration by parts that the Hankel transform algebraizes the following variable coefficient part of the Bessel differential equation as in (1.127), that is,

$$\mathcal{H}_n\left\{\frac{d^2 f}{dr^2} + \frac{1}{r}\frac{df}{dr} - \frac{n^2}{r^2}f\right\} = -\lambda^2 F_n(\lambda). \tag{A.2}$$

[1]For more detailed references, see Jerri [1992] and Sneddon [1972].

The inverse Hankel transform is

$$f(r) = \mathcal{H}_n^{-1}\{F_n\} = \int_0^\infty \lambda J_n(\lambda r) F_n(\lambda) d\lambda. \tag{A.3}$$

As we indicated in (2.49)–(2.51) and Fig. 2.6 of Section 2.6.2, we will use the Hankel transform in the following example to solve for the potential distribution in three dimensions due to an electrified disk.

Example 1 The Electrified Disc: Dual Integral Equations

To illustrate the application of the Hankel transform to a boundary value problem and its resulting dual integral equations, we choose to solve for the potential distribution $u(r, z)$ due to a constant potential u_0 on a unit disk in the xy plane where it is symmetric with respect to this plane outside the disk. Hence we use the Laplace equation for the circular symmetric potential $u(r, z)$ [see Figure 2.6 and (2.49)–(2.51)],

$$\frac{\partial^2 u}{\partial r^2} + \frac{1}{r}\frac{\partial u}{\partial r} + \frac{\partial^2 u}{\partial z^2} = 0, \qquad 0 < r < \infty, \qquad 0 < z < \infty \tag{E.1}$$

and express the mixed boundary conditions at the xy plane as

$$u(r, 0) = u_0, \qquad 0 \le r < 1 \tag{E.2}$$

$$\frac{\partial u}{\partial z}(r, 0) = 0, \qquad 1 < r < \infty, \tag{E.3}$$

where the second part (E.3) of the (mixed) condition represents the symmetry of the potential with respect to the xy-plane. We note here that the radial part of the Laplacian in (E.1) is a special case of what is in (A.2) with $n = 0$; hence we use the $\mathcal{H}_0$ Hankel transform of $u(r, z)$ which is associated with J_0, the Bessel function of order zero,

$$U(\lambda, z) = \int_0^\infty r J_0(\lambda r) u(r, z) dr. \tag{E.4}$$

So if we Hankel-transform (E.1) and use (A.2), we obtain

$$\lambda^2 U(\lambda, z) = \frac{d^2 U(\lambda, z)}{dz^2} \tag{E.5}$$

whose bounded solution is

$$U(\lambda, z) = A(\lambda) e^{-\lambda z}, \ \ \lambda \ge 0. \tag{E.6}$$

To find the arbitrary function $A(\lambda)$ we need to Hankel-transform a condition on the original function $u(r, z)$ at $z = 0$, but unfortunately, this condition is given partly as $u(r, 0) = u_0$ in (E.2) and partly as $\partial u(r, 0)/\partial z = 0$ as in (E.3), which is not suitable for the Hankel transformation. So instead of finding $A(\lambda)$ for $U(\lambda, z)$, we will now find the inverse Hankel transform of $U(\lambda, z)$ in (E.6) to obtain the original function

$$u(r, z) = \int_0^\infty \lambda J_0(\lambda r) A(\lambda) e^{-\lambda z} d\lambda \tag{E.7}$$

then apply the mixed condition (E.2) and (E.3) on $u(r, z)$ in (E.7) to obtain

$$u(r,0) = \int_0^\infty \lambda J_0(\lambda r)A(\lambda)d\lambda = u_0, \qquad 0 \le r < 1, \tag{E.8}$$

$$\int_0^\infty \lambda J_0(\lambda r)A(\lambda)d\lambda = u_0, \qquad 0 \le r < 1$$

$$\frac{\partial u}{\partial z}(r,0) = \int_0^\infty -\lambda^2 J_0(\lambda r)A(\lambda)d\lambda = 0, \quad 1 < r < \infty, \tag{E.9}$$

$$\int_0^\infty -\lambda^2 J_0(\lambda r)A(\lambda)d\lambda = 0, \quad 1 < r < \infty.$$

Here we notice that the arbitrary function $A(\lambda)$ is now involved in dual integral equations (E.8) and (E.9) whose solution can be obtained, with the aid of integrals of Bessel functions (see exercise 1, Section 2.6), as

$$A(\lambda) = \frac{2u_0}{\pi}\frac{\sin\lambda}{\lambda^2}. \tag{E.10}$$

So if we substitute this in (E.7), we obtain the final solution to the electrified disk problem,

$$u(r,z) = \frac{2u_0}{\pi}\int_0^\infty J_0(\lambda r)e^{-\lambda z}\frac{\sin\lambda}{\lambda}d\lambda. \tag{E.11}$$

We leave it as an exercise to show that (E.11) satisfies (E.1), (E.2), and (E.3) (see Exercise 1).

A.2 THE FINITE HANKEL TRANSFORM

In Section 1.4 we presented the finite Fourier sine and cosine transforms, and here we present the finite Hankel transform of a function $f(r)$, defined on the finite interval $(0, a)$, as

$$F_n(\lambda_k) = \int_0^a rJ_n(\lambda_k r)f(r)dr \tag{A.4}$$

where $\{\lambda_k a\}$ are the zeros of J_n,

$$J_n(\lambda_k a) = 0, \qquad k = 1, 2, \cdots. \tag{A.5}$$

With the aid of Fourier-Bessel series we can easily see (see Exercise 3) that the inverse finite Hankel transform is

$$f(r) = \frac{2}{a^2}\sum_{k=1}^\infty F_n(\lambda_k)\frac{J_n(\lambda_k r)}{J_{n+1}^2(\lambda_k a)} \tag{A.6}$$

where the sum is over the index k of the zeros $\lambda_k a = j_{n,k}$ of J_n in (A.6).

We note that to find the inverse Finite Hankel transform $f(r)$ in (A.4), we are asking for solving (A.4) as an integral equation in $f(r)$. This solution $f(r)$ is found in terms of the Fourier-Bessel series (A.6).

Exercises: Appendix A

1. Verify that $u(r, z)$ in (E.11) of Example 1 is the solution to the problem (E.1)–(E.3).

2. The orthogonal expansion (Fourier-Bessel series) of a function $f(r)$ defined on $(0, a)$ in terms of the Bessel functions $J_n(\lambda_k r)$ is

$$f(r) = \sum_{k=1}^{\infty} c_k J_n(\lambda_k r) \tag{E.1}$$

where $\lambda_k a$ is the kth zero of the Bessel function J_n (usually written as $j_{n,k}$) and the sum is over the index k of these zeros. The Fourier-Bessel coefficients are given by

$$c_k = \frac{\int_0^a rJ_n(\lambda_k r) f(r) dr}{\int_0^a rJ_n^2(\lambda_k r) dr}. \tag{E.2}$$

   (a) Relate the Fourier coefficient c_k to the finite Hankel transform $F_n(\lambda_k)$ in (A.4) (of Section A.2).

   (b) Use (E.1) and the result in part **(a)** to verify the Fourier-Bessel series in (A.6) (the inverse finite Hankel transform) of $f(r) = 1, 0 < r < 1$. *Hint:* Use (A.6) and (A.4).

3. Fluid flow through circular aperture in a wall. It is known that the velocity potential v of the flow of a jet of perfect fluid satisfies Laplace equation (2.48) (see also (E.1) of Example 1 here).

   (a) Formulate the problem of this steady jet flow through a circular aperture of unit radius 1 in a plane rigid wall where the velocity distribution in the hole (at the wall $z = 0$ when we take z as the direction of the flow perpendicular to the wall) is given by $v(r, 0) = f(r)$ and that the slope of the velocity potential is zero on the wall (outside the hole).

   (b) Use Hankel transform to solve the problem and reduce the mixed boundary conditions to dual integral equations.

   (c) Solve the dual integral equations for the case of constant velocity potential $f(r) = 1$ at the entrance of the aperture. *Hint:* See Exercise 1, Section 2.6.

4. Find the Laplace transform of the following initial and boundary value problem in $u(r, \theta, t)$, the displacement of a membrane with zero initial displacement $u(r, \theta, 0) = 0$, $0 < r < \infty$, $0 < \theta < 2\pi$ and a given initial velocity of $\frac{\partial u}{\partial t}(r, \theta, 0) = g(r, \theta), 0 < r < \infty, 0 < \theta < 2\pi,$

$$\frac{\partial^2 u}{\partial t^2} = \nabla^2 u \equiv \frac{1}{r}\frac{\partial}{\partial r}\left(r\frac{\partial u}{\partial r}\right) + \frac{1}{r^2}\frac{\partial^2 u}{\partial \theta^2}, \quad 0 < r < \infty, \quad 0 < \theta < 2\pi, t > 0 \tag{E.1}$$

$$u(r, 0, t) = u(r, 2\pi, t) \tag{E.2}$$

$$u(r, \theta, 0) = 0 \tag{E.3}$$

$$\frac{\partial u}{\partial t}(r, \theta, 0) = g(r, \theta). \tag{E.4}$$

Hint: From (1.69), (E.3), and (E.4) it is clear that only $\frac{\partial^2 u}{\partial t^2}$ in (E.1) is suitable for Laplace transformation. So let $U(r, \theta, s) = \mathcal{L}\{u(r, \theta, t)\}$ and Laplace-transform both sides of (E.1), realizing that the differentiation with respect to r and θ on the right side of (E.1) can be exchanged with the integration with respect to t of the Laplace transformation.

Appendix B Green's Function for Various Boundary Value Problems

The following is only a collection of very useful results that center around familiar boundary value problems, associated with differential equations, along with their Green's functions that facilitate giving the Fredholm integral equation representation of the boundary value problem. In addition, and when possible, we supply the eigenvalues and eigenfunctions, which are of great value for the construction of the Green's function (Section 4.1) and the analysis and construction of the solutions to Fredholm integral equations of Chapter 5. The theory behind the derivation of some of these results may not be found detailed in this book.

B.1 GREEN'S FUNCTIONS IN TERMS OF SIMPLE FUNCTIONS

(a) $u(x) = \lambda \int_0^a K(x,t)u(t)dt$ (B.1)

$$K(x,t) = G(x,t) = \begin{cases} \dfrac{x(a-t)}{a}, & x \le t \\ \dfrac{t(a-x)}{a}, & t \le x \end{cases} \tag{B.2}$$

$$\frac{d^2u}{dx^2} + \lambda u = 0 \tag{B.3}$$

$$u(0) = 0 \tag{B.4}$$

$$u(a) = 0 \tag{B.5}$$

$$u_n = \sin\frac{n\pi x}{a}, \quad \lambda_n = \left(\frac{n\pi}{a}\right)^2 \tag{B.6}$$

(See Example 6, Section 2.5 and Example 7, Section 5.2.)

(b) $$u(x) = \lambda \int_0^1 K(x,t)u(t)dt \tag{B.1}$$

$$K(x,t) = G(x,t) = \begin{cases} \dfrac{\sinh x \sinh(t-1)}{\sinh 1}, & 0 \le x \le t \\ \dfrac{\sinh t \sinh(x-1)}{\sinh 1}, & t \le x \le 1 \end{cases} \tag{B.2}$$

$$\frac{d^2u}{dx^2} - (\lambda+1)u = 0 \tag{B.3}$$

$$u(0) = 0 \tag{B.4}$$

$$u(1) = 0 \tag{B.5}$$

$$u_n(x) = \sin n\pi x, \qquad \lambda_n = n^2\pi^2 - 1 \tag{B.6}$$

(See Exercise 3, Section 2.5.)

(c) $$u(x) = -\lambda \int_0^1 K(x,t)u(t)dt \tag{B.1}$$

$$K(x,t) = G(x,t) = \begin{cases} (1+x)t, & 0 \le x \le t \\ (1+t)x, & t \le x \le 1 \end{cases} \tag{B.2}$$

$$\frac{d^2u}{dx^2} + \lambda u = 0 \tag{B.3}$$

$$u(0) = u'(0) \tag{B.4}$$

$$u(1) = u'(1) \tag{B.5}$$

(d) $$u(x) = -\lambda \int_0^1 K(x,t)u(t)dt \tag{B.1}$$

$$K(x,t) = G(x,t) = \begin{cases} x(t-2)+t-1, & 0 \le x \le t \\ x(t-1)-1, & t \le x \le 1 \end{cases} \tag{B.2}$$

$$\frac{d^2u}{dx^2} + \lambda u = 0 \tag{B.3}$$

$$u(0) = u'(1) \tag{B.4}$$

$$u'(0) = u(1) \tag{B.5}$$

(e) $$u(x) = \lambda \int_{-1}^{1} K(x,t)u(t)dt \tag{B.1}$$

$$K(x,t) = G(x,t) = \begin{cases} \frac{1}{\pi}\sin\frac{\pi}{2}(t-x), & -1 \le x \le t \\ \frac{1}{\pi}\sin\frac{\pi}{2}(x-t), & t \le x \le 1 \end{cases} \tag{B.2}$$

$$\frac{d^2u}{dx^2} + \frac{\pi^2}{4}u - \lambda u = 0 \tag{B.3}$$

$$u(-1) = u(1) \tag{B.4}$$

$$u'(-1) = u'(1) \tag{B.5}$$

(f) $$u(x) = -\lambda \int_{0}^{1} K(x,t)u(t)dt \tag{B.1}$$

$$K(x,t) = G(x,t) = \begin{cases} \frac{1}{2}x(x-t)(t-1), & 0 \le x \le t \\ -\frac{1}{2}t(t-x)(x-1), & t \le x \le 1 \end{cases} \tag{B.2}$$

$$\frac{d^3u}{dx^3} + \lambda u = 0 \tag{B.3}$$

$$u(0) = 0 \tag{B.4}$$

$$u(1) = 0 \tag{B.5}$$

$$u'(0) = u'(1) \tag{B.6}$$

(g) $$u(x) = \lambda \int_{0}^{1} K(x,t)u(t)dt \tag{B.1}$$

$$K(x,t) = G(x,t) = \begin{cases} \frac{x^2}{6}(3t-x), & 0 \le x \le t \\ \frac{t^2}{6}(3x-t), & t \le x \le 1 \end{cases} \tag{B.2}$$

$$\frac{d^4u}{dx^4} - \lambda u = 0 \tag{B.3}$$

$$u(0) = 0 \tag{B.4}$$

$$u'(0) = 0 \tag{B.5}$$

$$u''(1) = 0 \tag{B.6}$$

$$u'''(1) = 0 \tag{B.7}$$

B.2 GREEN'S FUNCTION IN TERMS OF SPECIAL FUNCTIONS

(h) $u(x) = \lambda \int_0^\infty K(x,t)u(t)dt$ (B.1)

$$K(x,t) = G(x,t) = \begin{cases} \frac{1}{2\nu}\left(\frac{x}{t}\right)^\nu (1-t^{2\nu}), & 0 \le x \le t \\ & \nu \ne 0 \\ \frac{1}{2\nu}\left(\frac{t}{x}\right)^\nu (1-x^{2\nu}), & t \le x \le 1 \end{cases} \tag{B.2}$$

$$x^2\frac{d^2u}{dx^2} + x\frac{du}{dx} + (\lambda^2x^2 - \nu^2)u = 0, \qquad \nu \ne 0 \tag{B.3}$$

$$u(1) = 0, \quad |u(x)| < \infty, \quad 0 \le x < 1 \tag{B.4}$$

$$u_n(x) = J_\nu(\alpha_n x), \quad \lambda_n^2 = \alpha_n^2 \text{ where } J_\nu(\alpha_n) = 0 \tag{B.5}$$

Here $J_\nu(x)$ is the Bessel function of the first kind defined by the series

$$J_\nu(x) = \left(\frac{x}{2}\right)^\nu \sum_{k=0}^\infty \frac{(-1)^k}{k!\Gamma(\nu+k+1)}\left(\frac{x}{2}\right)^{2k} \tag{$B.6$}$$

$\lambda_n^2 = \alpha_n^2$, where $\{\alpha_n\}$ are the zeros of the Bessel function, $J_\nu(\alpha_n) = 0$.

For the special case of $\nu = 0$ for $J_0(x)$,

$$G(x,t) = \begin{cases} -\ln t, & 0 \le x \le t \\ -\ln x, & 0 \le x \le 1 \end{cases}$$

(i) $u(x) = \lambda \int_{-1}^1 K(x,t)u(t)dt$ (B.1)

$$K(x,t) = G(x,t) = \ln 2 - \frac{1}{2} - \begin{cases} \frac{1}{2}\ln(1-x)(1+t), & x < t \\ \frac{1}{2}\ln(1-t)(1+x), & t < x \end{cases} \tag{B.2}$$

$$(1-x^2)\frac{d^2u}{dx^2} - 2x\frac{du}{dx} + \lambda u = 0 \tag{B.3}$$

$$|u(x)| < \infty, -1 \le x \le 1 \tag{B.4}$$

$$u_n(x) = P_n(x), \lambda_n = n(n+1), n \text{ an integer} \tag{B.5}$$

Here $P_n(x)$ is the Legendre polynomial of degree n (n an integer) defined by the Rodrigues formula,

$$P_n(x) = \frac{1}{2^n n!}\frac{d^n}{dx^n}(x^2-1)^n. \tag{$B.6$}$$

Answers to Exercises

Chapter 1

Exercises 1.1, p. 20

2. (a) $f(x) = k\left[f(x) + x\dfrac{df}{dx}\right]$

   (b) $f(x) = Cx^{\frac{1-k}{k}}$, C an arbitrary constant.

5. No

7. Yes

8. Yes

12.
$$\sqrt{h(x)}u(x) = \frac{f(x)}{\sqrt{h(x)}} + \int_a^{b(x)} \frac{K(x,t)\sqrt{h(t)}u(t)dt}{\sqrt{h(x)}\sqrt{h(t)}}, \qquad (E.2)$$

   where
$$\sqrt{h(x)}u(x) \equiv \phi(x), \quad g(x) = \frac{f(x)}{\sqrt{h(x)}} \qquad (E.3)$$

   and
$$k(x,t) = \frac{K(x,t)}{\sqrt{h(x)h(t)}}. \qquad (E.4)$$

13. $\dfrac{dn_1}{dt} = k_1 n_1(t), \; \dfrac{dn_2}{dt} = -k_2 n_2(t)$

14. The resulting differential equation $\dfrac{dp}{dx} = -\sigma p$ with the initial condition $p(0) = 1$, give $p(x) = e^{-\sigma x}$

15. a) The possibly needed differentiation of the known measured data $g(x)$ is very sensitive to the inaccuracy of the measured data $g(x)$

c) $\sigma(E_{min}) = 0, \sigma(E_{max}) = \infty$.

16. $u(x) = 1 + \int_0^x (t - x)u(t)dt$

17. $u(x) = -\lambda \int_0^{\pi/2} K(x,\xi)u(\xi)d\xi, \quad K(x,\xi) = \begin{cases} \left(\dfrac{2}{\pi}\xi - 1\right)x, & 0 \le x \le \xi \\ \left(\dfrac{2}{\pi}x - 1\right)\xi, & \xi \le x \le \dfrac{\pi}{2} \end{cases}$

Exercises 1.2, p. 28

1. (a) Volterra integral equation of the second kind, nonhomogeneous.

(b) Fredholm integral equation of the second kind, nonhomogeneous.

(c) Fredholm integral equation of the first kind, singular (infinite range of integration and the kernel becomes infinite at $x = \lambda$), with difference kernel $1/(x - \lambda)$ [Fourier convolution type, see (1.72)].

(d) Volterra integral equation of the first kind, singular (the kernel becomes infinite in the range of integration at $\lambda = x$). (e) Volterra integral equation of the first kind, singular, with difference kernel $1/\sqrt{x - t}$ [Laplace convolution type, see (1.67)]. (f) Fredholm integral equation of the second kind, homogeneous, singular, with difference kernel (Fourier convolution type). (g) Fredholm integral equation of the second kind, nonhomogeneous.

2. For Exercises 1.1, #9, 10, 11, and 12.

9. Fredholm integral equation of the first kind, singular [infinite range of integration $(-\infty, \infty)$].

10. Fredholm integral equation of the first kind, singular.

11. Fredholm integral equation of the first kind, singular.

12. Volterra with $b(x)$, Fredholm when $b(x) = b$, a constant.

3. (a) Linear (e) Nonlinear

(b) Nonlinear (f) Nonlinear

(c) Linear (g) Linear

(d) Nonlinear (h) Nonlinear

4. A linear nonhomogeneous Fredholm integral equation (in three dimensions).

5. (a) Weak singularity, $\alpha = \frac{1}{2}$, $k(x,t) = 1$.

   (b) Weak singularity, $0 < \alpha < 1$, $k(x,t) = 1$.

   (c) Strong singularity or Cauchy singular for 1(c), weak singularity for 1(d), (e).

6. A Cauchy singular Fredholm integral equation of the first kind.

7. (b) Weak singularity, see part (a).

   (c) $g(x) = \frac{df}{dx} = \frac{1}{\pi}\int_0^1 \phi(t)\left[\frac{1}{x-t} + \frac{1}{x+t}\right] dt$

Exercises 1.3, p. 40

1. (b) $y(x) = -b^2 \int_0^x (x-\xi)y(\xi)d\xi + g(x)$, where

   $g(x) = \int_0^x (x-\xi)f(\xi)d\xi.$

3. (a) $u(x) = \lambda \int_0^x (x-\xi)u(\xi)d\xi + \int_0^x (x-\xi)g(\xi)d\xi + c_1 x + c_2$

4. $\frac{d^2u}{dx^2} + (2\lambda - 1)u = 0$

5. (a) $\frac{d^2u}{dx^2} + \lambda u(x) = 0$

   (b) $u(0) = 0,\ u(1) = 0$

   (c) $u_n(x) = \sin n\pi x,\ \lambda_n = n^2\pi^2;\ n = 1, 2, 3, \cdots$

Exercises 1.4, p. 73

1. (a) $\frac{\sqrt{\pi}}{2s^{3/2}}$

   (b) $\frac{\sqrt{\pi}}{s^{1/2}}$

   (c) $\frac{1}{s^2(s-2)}$

   (d) $\frac{1}{s^2} - \frac{2U(s)}{s^3}$, $U(s) = \mathcal{L}\{u(t)\}$

   (e) $\frac{1}{s-2}[sU(s) - u(0)] = \frac{sU(s)}{s-2}$

(f) $\dfrac{2s}{(s^2+1)^2}$

2. (a) $e^{(\lambda+1)x}$

(b) $\displaystyle\int_0^x e^{(\lambda+1)(x-t)}g(t)dt$

(c) $\dfrac{e^{3x}\sin\sqrt{5}x}{\sqrt{5}}$

(d) $xe^{3x}+e^{2x}+1$

(e) $\displaystyle\frac{1}{\sqrt{\pi}}\int_0^x \frac{f(t)}{\sqrt{x-t}}dt$

(f) $\displaystyle\frac{d}{dx}\left[\frac{1}{\sqrt{\pi}}\int_0^x \frac{f(t)}{\sqrt{x-t}}dt\right]$

3. (a) $f_1(t) \neq f_2(t)$ when $t=3$ as in (i) and (ii) of part (a). Indeed $f_1(t)$ and $f_2(t)$ could differ on a set of points $t_1, t_2, \cdots, t_n$ on $(0,\infty)$.

4. (a) $f(x) = \text{erf}(\sqrt{x})$

5. (a) $\dfrac{1}{s} = \sqrt{\dfrac{\pi}{s}}U(s),\ U(s) = \mathcal{L}\{u(x)\}$

(b) $u(\lambda) = \mathcal{L}^{-1}\left\{\dfrac{1}{\sqrt{\pi}}\dfrac{1}{\sqrt{s}}\right\} = \dfrac{1}{\pi\sqrt{\lambda}}$

6. (b) If $f(x)$ is such that $\dfrac{df}{dx}$ is continuous on $(0,\infty)$, and

(i) $\displaystyle\lim_{x\to\infty} f(x)e^{-sx} = 0,$ (E.1)

(ii) $\displaystyle\lim_{x\to\infty} f'(x)e^{-sx} = 0$ (E.2)

then, (1.69) follows.

7. $f(x) = 1-\cos x$

10. Let $x=0$ in (E.1) to have

$$\frac{1}{2\pi}\int_{-\infty}^{\infty} F_1(\lambda)F_2(\lambda)d\lambda = \int_{-\infty}^{\infty} f_1(-t)f_2(t)dt. \qquad (E.2)$$

Then, we consider the special case with $f_1(-t) = \overline{f_2(t)}$, which can be written as $\overline{f_2(t)} = \overline{\overline{f_2(t)}} = \overline{f_1(-t)}$. Also we have from Exercise 9(b) that $\mathcal{F}\{\overline{f_1(-t)}\} = \overline{F_1(\lambda)}$, and for our case this becomes $\mathcal{F}\{\overline{f_1(-t)}\} = \mathcal{F}\{f_2(t)\} = \overline{F_1(\lambda)}$. If we use this $f_2(t) = \overline{f_1(-t)}$ in (E.2), we obtain the *Parseval equality*,

$$\frac{1}{2\pi}\int_{-\infty}^{\infty} F_1(\lambda)\overline{F_1(\lambda)}d\lambda = \int_{-\infty}^{\infty} f_1(-t)\overline{f_1(-t)}dt$$

$$= \int_{-\infty}^{\infty} f_1(u)\overline{f_1(u)}du,$$

$$\frac{1}{2\pi}\int_{-\infty}^{\infty} |F_1(\lambda)|^2 d\lambda = \int_{-\infty}^{\infty} |f_1(t)|^2 dt \qquad (E.3)$$

after a simple change of variable $u = -t$ in the right integral, and the use of $f\bar{f} = |f|^2$.

15. (a) $F(\lambda) = G(\lambda) + K(\lambda)F(\lambda)$, $F(\lambda) = \mathcal{F}\{f(x)\}$, $G(\lambda) = \mathcal{F}\{f(x)\}$, $K(\lambda) = \mathcal{F}\{k(x)\}$

    (b) $F(\lambda) = \dfrac{G(\lambda)}{1-K(\lambda)}$, $\quad f(x) = \dfrac{1}{2\pi}\displaystyle\int_{-\infty}^{\infty} e^{ix\lambda}\frac{G(\lambda)}{1-K(\lambda)}d\lambda$

16. $\phi(x) = \dfrac{1}{2\pi}\displaystyle\int_{-\infty}^{\infty} \frac{F(t)}{1-\lambda K(t)}e^{ixt}dt \qquad$ (E.2)

    where $F(t)$ and $K(t)$ are the Fourier transforms of $f(x)$ and $k(x)$, respectively.

17.

$$\phi(x) = \frac{a^2-1}{b^2-1}e^{-|x|} + \frac{1}{b}\frac{b^2-a^2}{b^2-1}e^{-b|x|}.$$

18.

$$\phi(x) = \frac{1}{\pi}\int_{-\infty}^{\infty} \frac{e^{i\lambda x}}{1-\mu\pi e^{-|\lambda|}}\frac{\sin\lambda}{\lambda}d\lambda, \quad \mu \in \left(\frac{1}{\pi},\infty\right).$$

19. (a) $F(\lambda) = \dfrac{\Gamma(\lambda)}{a^\lambda}$, $\quad a > 0$

    (b) $U(\lambda) = \dfrac{\Gamma(\lambda)}{a^\lambda} + \Gamma(\lambda)U(\lambda)$, $\quad U(\lambda) = \dfrac{1}{a^\lambda}\cdot\dfrac{\Gamma(\lambda)}{1-\Gamma(\lambda)}$

Exercises 1.5, p. 94

1. (a) $\displaystyle\int_0^1 \frac{dx}{1+x^2} = \frac{\pi}{4} = 0.7854$

   (b) 0.7828

   (c) $|E_T(f)| \le \dfrac{(1-0)^3\cdot 2}{(12)(16)} = \dfrac{1}{96} = 0.0104$

   (d) The absolute value of the actual error from the exact value of the integral in part (a) and the numerical approximation in part (b) is

$$|0.7854 - 0.7828| = 0.0026$$

   as compared to the (larger!) estimate of its upper bound in part (c) as 0.0104.

2. (a) $\ln 2 = 0.69315$

   (b) 0.69325

(c) $|E_S(f)| < \dfrac{24(1-0)^5}{(180)(64)} = 0.00052$

(d) The absolute value of the actual error is $|0.69315 - 6.69325| = 0.0001$, while the estimate of an upper bound of such error is 0.00052 as in part (c), the latter, of course, is conservative.

3. (a) Trapezoidal rule: 0.562167

   (b) Simpson's rule: 0.561956. *Exact:* 0.561965

4. See the answer to Exercise 5 of Section 7.2.

5. (a) $\tilde{u}(x) = 2(x-0.5)(x-1) - 1.469986x(x-1) - 0.220378x(x-0.5)$.

   (b) You may consult (the rather long formula) as the answer in "The Student's Solution Manual"[1] to accompany this book (see the end of the preface for more information).

Chapter 2

Exercises 2.1, p. 102

1. (a) $N(s) = n_0 F(s) + kF(s)N(s), \quad N(s) = \dfrac{n_0 F(s)}{1-kF(s)}$

   (b) $N(s) = \dfrac{n_0}{s+c-k}, \quad n(t) = n_0 e^{-(c-k)t}, \ c > k > 0$

2. (a) $\dfrac{dn}{dt} = -kn(t), n(0) = n_0$

   (b) $n(t) = n_0 - k\displaystyle\int_0^t n(\tau)d\tau$

   (c) $n(t) = n_0 e^{-kt}$

3. (a) $n(t) = n_0 e^{-t/T} + k\displaystyle\int_0^t e^{-(1/T)(t-\tau)} n(\tau)dT, \quad N(s) = \dfrac{n_0}{s+[(1/T)-k]},$

   $N(s) = \mathcal{L}\{n(t)\}$

   (b) $n(t) = n_0 e^{-[(1/T)-k]t}$

4. (a) $g(t) = b_0 h(t)$

   (b) $b(t-\tau)\rho(\tau)m(\tau)\Delta\tau$

   (c) $\displaystyle\int_\alpha^\beta b(t-\tau)\rho(\tau)m(\tau)d\tau$

   (d) $b(t) = b_0 h(t) + \displaystyle\int_\alpha^\beta b(t-\tau)\rho(\tau)m(\tau)d\tau$

[1] Jerri, 1999

Exercises 2.2, p. 105

1. (a) $\frac{dr}{dt} = \text{Ac}$

2. (a) $I\frac{d^2\theta_s}{dt^2} = -a\phi(t) - b\frac{d\phi}{dt}$

(b) $Is^2\Theta_s(s) = -a\Phi(s) - bs\Phi(s)$, where $\Theta_s(s) = \mathcal{L}\{\theta_s(t)\}$, $\Phi(s) = \mathcal{L}\{\phi(t)\}$.

(c) $\phi(t) = (t-1)e^{-t}, \theta_s(t) = 1 + (t-1)e^{-t}$

(d) $\phi(t) = e^{-t}\left(-\frac{t^2}{2} + 2t - 1\right), \theta_s(t) = 1 + e^{-t}\left(-\frac{t^2}{2} + 2t - 1\right)$

3. (a) What should be the potential distribution on the boundary of the disk such that the potential $u(r,\theta)$ at a given point (r,θ) takes a predetermined value $u(r,\theta) = g(r,\theta)$?

Exercises 2.3, p. 112

1. (a) $D_{1,2}(x) = K\left(x, \frac{l}{2}\right) L_1\left(\frac{l}{2}\right) + K\left(x, \frac{2l}{3}\right) L_2\left(\frac{2l}{3}\right)$

(b) $D(x) = \int_0^l K(x,\xi)\rho(\xi)d\xi$, where $K(x,\xi)$ is as given in (2.26).

2. Here we consider two cases:[2]

(i) a close to an exact one, where we look at the density distribution to be zero outside the beads on $(0, \frac{l}{5})$, $(\frac{l}{4}, \frac{2l}{3})$, $(\frac{3l}{4}, l)$, and one for the two intervals of the lengths of the beads on $(\frac{l}{5}, \frac{l}{4})$ and $(\frac{2l}{3}, \frac{3l}{4})$. The answer here is

$$y(x) = g\int_{\frac{l}{5}}^{\frac{l}{4}} G(x,\xi)d\xi + g\int_{\frac{2l}{3}}^{\frac{3l}{4}} G(x,\xi)d\xi,$$

[2]The detailed solution of this problem (in over five pages), with figures is found in "The Student's Solution Manual" to accompany this book [Jerri, 1999]. See the end of the preface for more information.

$$y(x) = g\begin{cases} \dfrac{31lx}{800T_0} + \dfrac{7lx}{288T_0} = \dfrac{227}{3600T_0}x, & 0 \le x \le \dfrac{l}{5} \\ -\dfrac{1}{800}\dfrac{400x^2 + 16l^2 - 191xl}{T_0} + \dfrac{7lx}{288T_0} \\ = -\dfrac{1}{3600}\dfrac{1800x^2 + 72l^2 - 947xl}{T_0}, & \dfrac{l}{5} \le x \le \dfrac{l}{4} \\ \dfrac{9l(l-x)}{800T_0} + \dfrac{7l}{800T_0}x = \dfrac{1}{7200}\dfrac{l(81l + 94x)}{T_0}, & \dfrac{l}{4} \le x \le \dfrac{2l}{3} \\ \dfrac{9l(l-x)}{800T_0} - \dfrac{1}{288}\dfrac{144x^2 + 64l^2 - 199xl}{T_0} \\ = -\dfrac{1}{7200}\dfrac{1519l^2 - 4894xl + 3600x^2}{T_0}, & \dfrac{2l}{3} \le x \le \dfrac{3l}{4} \\ \dfrac{9l(l-x)}{800T_0} + \dfrac{17l(l-x)}{288T_0} = \dfrac{253l(l-x)}{3600T_0}, & \dfrac{3l}{4} \le x \le l \end{cases}$$

with its above five branches.

(ii) An approximation to the problem, where we take the weights of the two beads as two point forces located at their respective centers of masses $\xi_1 = \frac{1}{2}(\frac{l}{5} + \frac{l}{4}) = \frac{9}{40}l$ and $\xi_2 = \frac{1}{2}(\frac{2l}{3} + \frac{3l}{4}) = \frac{17l}{24}$.

$$y(x) = \frac{1}{T_0 l}\left(\frac{lg}{20}K\left(x, \frac{9l}{40}\right) + \frac{lg}{12}K\left(x, \frac{17l}{24}\right)\right)$$

$$y(x) = \begin{cases} \dfrac{1}{T_0 l}\left[\dfrac{lg}{20}\cdot\dfrac{31lx}{40} + \dfrac{lg}{12}\cdot\dfrac{7lx}{24}\right] = \dfrac{16gx}{15T_0}, \\ \qquad 0 \le x \le \frac{9l}{40} \\ \dfrac{1}{T_0 l}\left[\dfrac{lg}{20}\cdot\dfrac{9l}{40}(l-x) + \dfrac{lg}{12}\cdot\dfrac{7lx}{24}\right] = g\dfrac{27l + 8x}{120T_0}, \\ \qquad \frac{9l}{40} < x < \frac{17l}{24} \\ \dfrac{1}{T_0 l}\left[\dfrac{lg}{20}\cdot\dfrac{9l}{40}(l-x) + \dfrac{lg}{12}\cdot\dfrac{17l}{24}(l-x)\right] = g\dfrac{14l - 14x}{15T_0}, \\ \qquad \frac{17l}{24} \le x < l. \end{cases}$$

3. $y(x) = \dfrac{gcx}{2T_0}(x^3 - 2x^2 l + l^3)$

5. $-f(y_0)\sqrt{2g} = -\sqrt{2g}T = \displaystyle\int_0^{y_0} \frac{\phi(y)}{\sqrt{y_0 - y}}dy$

Exercises 2.4, p. 116

2. (a) $u(x) = -x + \displaystyle\int_0^x (t - x)u(t)dt$

(b) $y(x) = x + \int_0^x (t - x)y(t)dt$

4. $u(x) = -1 - \int_0^x (2x - t)u(t)dt$

Exercises 2.5, p. 122

1. $y(x) = \lambda \int_0^b K(x,t)y(t)dt, K(x,t) = \begin{cases} \frac{x}{b}(t-b), & x < t \\ \frac{t}{b}(x-b), & x > t \end{cases}$

2. (b) $\left.\frac{\partial K}{\partial x}\right|_{x>t} - \left.\frac{\partial K}{\partial x}\right|_{x<t} = \frac{\partial}{\partial x}[(x-b)(t-a)/(b-a)]$

$-\frac{\partial}{\partial x}[(x-a)(t-b)/(b-a)] = \frac{(t-a)}{(b-a)} - \frac{(t-b)}{(b-a)} = \frac{b-a}{b-a} = 1.$

(c) Away from $x = t$, the kernel $K(x,t)$ is a linear function in x, hence it satisfies $\frac{\partial^2 K}{\partial x^2} = 0, x \neq t.$

3. (a) $\frac{d^2u}{dx^2} = (\lambda + 1)u(x), u(0) = 0, u(1) = 0$

(b) $u_n(x) = \sin n\pi x, \lambda_n = -n^2\pi^2 - 1$

Exercises 2.6, p. 127

1. $A(\lambda) = \frac{2u_0}{\pi}\frac{\sin\lambda}{\lambda^2}$

2. For a rectangular aperture of width 2a, the boundary value problem in $v(x,y)$ is

$\frac{\partial^2 v}{\partial x^2} + \frac{\partial^2 v}{\partial y^2} = 0, -\infty < x < \infty, -\infty < y < \infty,$

$v(0,y) = f(y), -a < y < a$

$\frac{\partial v}{\partial x}(0,y) = 0, |y| > a$

For a circular aperture of radius a, the boundary value problem in $v(r,z)$ is

$\frac{\partial^2 v}{\partial r^2} + \frac{1}{r}\frac{\partial v}{\partial r} + \frac{\partial^2 v}{\partial z^2} = 0, 0 \le r < \infty, -\infty < z < \infty$

$v(r,0) = f(r), 0 \le r < a$

$\frac{\partial v(r,0)}{\partial z} = 0, r > a$

3. $F(x) = \int_0^\infty K(x,\lambda)B(\lambda)d\lambda, x > 0$

where $K(x,\lambda) = \begin{cases} \lambda \sin \lambda x, & x > 1, \lambda > 0 \\ \sin \lambda x, & 0 < x < 1, \lambda > 0 \end{cases}$

$F(x) = \begin{cases} 0, & x > 1 \\ f(x), & 0 < x < 1 \end{cases}$

Chapter 3

Exercises 3.1, p. 146

1. $\Gamma(x,t;\lambda) = 1 + \lambda(x-t) + \dfrac{\lambda^2(x-t)^2}{2!} + \cdots = \sum_{n=0}^{\infty} \dfrac{\lambda^n (x-t)^n}{n!} \equiv e^{\lambda(x-t)},$

   $u(x) = f(x) + \lambda \int_0^x e^{\lambda(x-t)} f(t)dt$

2. $\Gamma(x,t;\lambda) = \sinh \lambda(x-t), \quad u(x) = g(x) + \lambda \displaystyle\int_0^x \sinh \lambda(x-t) g(t)dt$

3. See answers for Exercises 1 and 2.

4. (a) $u(x) = 1 + x + \dfrac{x^2}{2!} + \cdots = \sum_{n=0}^{\infty} \dfrac{x^n}{n!} \equiv e^x$

   (b) $u(x) = e^x - 1$

5. $u(x) = \cosh x$

6. $u(x) = \dfrac{x}{2} \sin x$

9. $F(x) = -1 + Ce^{\frac{x^3}{3}},$

   $u(x) = Cxe^{\frac{x^3}{3}} = xe^{\frac{x^3}{3}}, C = 1$

10. (b) The inequalities generated for the bounds (E.1) of the iterated kernels, in particular the $n!$ in the denominator of the bound.

11. $u(x,y) = f(x,y) + \lambda \displaystyle\int_a^x \int_a^y K(x,y;\xi,\eta)u(\xi,\eta)d\xi d\eta$

    where $f(x,y)$ is a given function.

Exercises 3.2, p. 154

1. (a) $u(x) = 1 + \displaystyle\int_0^x \sin(x-t)u(t)dt$

   (b) $u(x) = 1 + \dfrac{x^2}{2}$

   (e) The answers are given with the questions of Exercises 3 and 4 of Section 1.1.

2. $u(x) = 1 - x \ln 3$

3. $u(x) = \dfrac{\sin \alpha\pi}{\pi} \dfrac{d}{dx} \displaystyle\int_0^x (x-t)^{\alpha-1} f(t)dt$ [This is equation (3.40).]

4. (a) $u(x) = xe^{x^2}$

   (b) $u(x) = e^{x^2/2}(x^2+2) - 1$

5. $f(x)$ in the Fredholm integral equation (E.2) is restricted to only the linear combination $f(x) = A \sin x + B \cos x$ of $\sin x$ and $\cos x$, where $A = \displaystyle\int_0^1 u(\xi) \cos \xi d\xi$, and $B = \int_0^1 u(\xi) \sin \xi d\xi$ are constants. On the other hand $f(x)$ of the Volterra integral equation of the first kind (E.1) can be the more general function $f(x) = g(x) \sin x + h(x) \cos x$ where $g(x) = \displaystyle\int_0^x u(\xi) \cos \xi d\xi$ and $h(x) = \displaystyle\int_0^x u(\xi) \sin \xi d\xi$.

Exercises 3.3, p. 162

1. (a), (b)

x	Exact $u(x) = \sin x$	Numerical (approximate) $u(x)$, $n = 8$
0	0.0	0.0
0.5	0.4794	0.50
1.0	0.8415	0.875
1.5	0.9975	1.03125
2.0	0.9093	0.92969
2.5	0.5985	0.5957
3.0	0.1411	0.1128
3.5	-0.3508	-0.3983
4.0	-0.7568	-0.8098

2. (a) $u(x) = x + \dfrac{2}{9} e^{3x}[1 - e^{-3x}(3x+1)]$

   (b), (c)

x	Exact $u(x)$	Numerical (approximate) $u(x)$, $n = 8$
0.0	0.0	0.0
0.625	1.4352	1.6667
1.25	9.6435	13.7124
1.875	62.0188	109.784
2.5	402.398	886.095
3.125	2,620.801	7,169.5

Table continued.

x	Exact $u(x)$	Numerical (approximate) $u(x), n = 8$
3.75	17,085.455	58,036.9
4.375	111,405.7	469,843.97
5.0	726,449.75	3,803,722.4

3. (a), (b)

x	Exact $u(x) = \cos x - \sin x$	Numerical $u(x), n = 8$
0	1.0	1.0
$\frac{\pi}{4}$	0.0	-0.1107
$\frac{\pi}{2}$	-1.0	-1.2195
$\frac{3\pi}{4}$	-1.414	-1.6740
π	-1.0	-1.2055
$\frac{5\pi}{4}$	0.0	-0.0859
$\frac{3\pi}{2}$	1.0	1.0314
$\frac{7\pi}{4}$	1.414	1.4942
2π	1.0	1.0335

(c) u_0 is arbitrary, find u_1 and u_2 in terms of u_0.

Chapter 4

Exercises 4.1, p. 193

1. (a) A set of solutions $y_n(x) = A\sin(n\pi x/l)$, $n = 1, 2, 3, \cdots$, where A is an arbitrary constant.

   (b) $y_n(x) = A\cos(n\pi x/l), n = 0, 1, 2, 3, \cdots$

   (c) $y_n(x) = A\cos(n + \frac{1}{2})\pi x/l, n = 0, 1, 2, 3, \cdots$

   (d)$y(x) = A\sin\lambda x$, $\lambda > 0$

   (e) $y(x) = A\cos\lambda x + B\sin\lambda x$, λ real

   (f) $y(x) = e^{-\lambda x}$, $\lambda > 0$,

   (g) $y(x) = \dfrac{\sinh\lambda(l - x)}{\sinh\lambda l}$

2. (b) $G(x,\xi) = \begin{cases} \dfrac{\sin b(1-\xi)\sin bx}{b\sin b}, & 0 \le x \le \xi \\ \dfrac{\sin b\xi \sin b(1-x)}{b\sin b}, & \xi \le x \le 1 \end{cases}$

3. $G(x,\xi) = \begin{cases} \dfrac{2}{\pi}x\left(\dfrac{\pi}{2}-\xi\right), & 0 \le x \le \xi \\ \dfrac{2}{\pi}\xi\left(\dfrac{\pi}{2}-x\right), & \xi \le x \le \frac{\pi}{2} \end{cases}$

4. (b) $G(x,\xi) = \begin{cases} \dfrac{\sinh b(1-\xi)\sinh bx}{b\sinh b}, & 0 \le x \le \xi \\ \dfrac{\sinh b\xi \sinh b(1-x)}{b\sinh b}, & \xi \le x \le 1 \end{cases}$

5. (i) $y(x) = \dfrac{\sinh x}{\sinh 1} - x,$

(ii) $y(x) = \dfrac{2}{\pi}\displaystyle\sum_{k=1}^{\infty}\frac{(-1)^k \sin k\pi x}{k[1+(k\pi)^2]}$

6. $y(x) = 2[\sinh x - \sinh(x-1) - \sinh 1]$

7. $y(x) = -\displaystyle\int_0^1 G(x,\xi)f(\xi,y(\xi))d\xi,\ G(x,\xi) = \begin{cases} x(1-\xi), & 0 \le x \le \xi \\ \xi(1-x), & \xi \le x \le 1 \end{cases}$

8. $y(x) = \dfrac{x\sin\lambda x}{2\lambda}$

10. $y(x) = \dfrac{1}{6}(2x^3+3x^2-17x-5) - \lambda\displaystyle\int_0^1 [G(x,\xi)y(\xi)]d\xi,$

$$G(x,\xi) = \begin{cases} (\xi-2)x+\xi-1, & 0 \le x \le \xi \\ (\xi-1)x-1, & \xi \le x \le 1 \end{cases}$$

11. $y(x) = e^x - \lambda\displaystyle\int_0^1 G(x,\xi)y(\xi)d\xi,$

$$G(x,\xi) = \begin{cases} (1+x)\xi, & 0 \le x \le \xi \\ (1+\xi)x, & \xi \le x \le 1 \end{cases}$$

12. $y(x) = \dfrac{x}{\pi}\sin\dfrac{\pi x}{2} + \dfrac{2}{\pi^2}\cos\dfrac{\pi x}{2} + \lambda\displaystyle\int_{-1}^1 G(x,\xi)y(\xi)d\xi,$

$$G(x,\xi) = \begin{cases} \dfrac{1}{\pi}\sin\dfrac{\pi}{2}(\xi - x), & -1 \le x \le \xi \\ \dfrac{1}{\pi}\sin\dfrac{\pi}{2}(x - \xi), & \xi \le x \le 1 \end{cases}$$

13. $y(x) = \sinh x + (1 - x)e^x$

14. $G(x,\xi) = \begin{cases} (\xi - 2)x + \xi - 1, & 0 \le x \le \xi \\ (\xi - 1)x - 1, & \xi \le x \le 1 \end{cases}$

15. $G(x,\xi) = \begin{cases} (1 + x)\xi, & 0 \le x \le \xi \\ (1 + \xi)x, & \xi \le x \le 1 \end{cases}$

16. $G(x,\xi) = \begin{cases} \dfrac{1}{\pi}\sin\dfrac{\pi}{2}(\xi - x), & -1 \le x \le \xi \\ \dfrac{1}{\pi}\sin\dfrac{\pi}{2}(x - \xi), & \xi \le x \le 1 \end{cases}$

17. (b) (i) A unique Green's function does exist, since the homogeneous boundary value problems has only the trivial solution $y(x) = 0$,

$$G(x,\xi) = \begin{cases} \xi - 1 + (\xi - 2)x, & 0 \le x \le \xi \\ (\xi - 1)x - 1, & \xi \le x \le 1 \end{cases}$$

(ii) The homogeneous boundary value has an infinite number of solutions $u(x) = C$ as C is arbitrary, thus the Green's function does not exist.

(iii) The Green's function does not exist, since this problem has infinity of solutions $y(x) = A\sin x$. See the answer to part (ii).

19. $G(x,\xi) = \begin{cases} \dfrac{\cos(x - \xi + \frac{1}{2})}{2\sin(1/2)}, & 0 \le x \le \xi \\ \dfrac{\cos(\xi - x + 1/2)}{2\sin(1/2)}, & \xi \le x \le 1 \end{cases}$

20. $y(x) = \dfrac{1}{4\pi}(2x - 1)\sin \pi x$

21. (b) $u_n(x) = \sin n\pi x$, $n = 1, 2, \cdots$

(d) $x = \dfrac{2}{\pi}\displaystyle\sum_{n=1}^{\infty} \dfrac{(-1)^{n+1}}{n}\sin n\pi x$, $\displaystyle\int_0^1 \sin^2 n\pi x dx = \dfrac{1}{2}$, $c_n = \dfrac{2(-1)^{n+1}}{n}$

23. (c) $f(x) = e^{2x} \simeq c_0 + c_1 x + \frac{c_2}{2}(3x^2 - 1), e^{2x} \simeq 2.639x^2 + 2.923x + 0.934$

(d)

x	e^{2x} (exact)	Three-term Legendre series approximation of e^{2x}
-0.75	0.22	0.226
-0.50	0.37	0.132
-0.25	0.61	0.368
0.00	1.00	0.934
0.25	1.65	1.830
0.50	2.72	3.055
0.75	4.48	4.610

24. (c) $g_n(y,\eta) = \frac{1}{n \sinh n\pi} \begin{cases} \sinh n(\pi - y) \sin hn\eta, & 0 < \eta < y \\ \sin hny \sin hn(\pi - \eta), & y < \eta < \pi \end{cases}$

(d) $$\begin{aligned} U(n,y) &= \int_0^\pi g_n(y,\eta) F(n,\eta) d\eta \\ u(x,y) &= \frac{2}{\pi} \sum_{n=1}^{\infty} U(n,y) \sin nx \\ &= \frac{2}{\pi} \sum_{n=1}^{\infty} \sin nx \int_0^\pi g_n(y,\eta) F(n,\eta) d\eta \end{aligned}$$

(e,f) $$\begin{aligned} &= \frac{2}{\pi} \sum_{n=1}^{\infty} \sin nx \int_0^\pi g_n(y,\eta) \left[\int_0^\pi f(\xi,\eta) \sin n\xi d\xi \right] d\eta \\ &= \int_0^\pi \int_0^\pi \left[\frac{2}{\pi} \sum_{n=1}^{\infty} g_n(y,\eta) \sin nx \sin n\xi \right] f(\xi,\eta) d\xi d\eta \\ &= \int_0^\pi \int_0^\pi G(x,y;\xi;\eta) f(\xi,\eta) d\xi d\eta \end{aligned}$$

(f) $G(x,y;\xi,\eta) = \frac{2}{\pi} \sum_{n=1}^{\infty} g_n(y,\eta) \sin nx \sin n\xi$

Exercises 4.2, p. 206

1. $y(x) = -\int_0^1 G(x,\xi) y(\xi) d\xi - 1 - x(e - 1) + e^x$

$$G(x,\xi) = \begin{cases} x(1-\xi), & 0 \le x \le \xi \\ \xi(1-x), & \xi \le x \le 1 \end{cases}$$

2. (a) $y(x) = -\lambda \int_0^{\pi/2} G(x,\xi)y(\xi)d\xi + \frac{x^4}{12} - \frac{\pi^3}{96}x,$

$$G(x,\xi) = \begin{cases} x\left(1 - \frac{2}{\pi}\xi\right), & 0 \le x \le \xi \\ \xi\left(1 - \frac{2}{\pi}x\right), & \xi \le x \le \frac{\pi}{2} \end{cases}$$

3. $y(x) = \lambda \int_0^1 G(x,\xi)y(\xi)d\xi + \frac{1}{12}x(x-1)(x^2+x-1)$

$$G(x,\xi) = \begin{cases} \frac{1}{2}x(x-\xi)(1-\xi), & 0 \le x \le \xi \\ -\frac{1}{2}\xi(\xi-x)(1-x), & \xi \le x \le 1 \end{cases}$$

4. $y(x) = \frac{1}{6}(2x^3 + 3x^2 - 17x - 5) - \lambda \int_0^1 [G(x,\xi)y(\xi)]d\xi,$

$$G(x,\xi) = \begin{cases} (\xi-2)x + \xi - 1, & 0 \le x \le \xi \\ (\xi-1)x - 1, & \xi \le x \le 1 \end{cases}$$

5. $y(x) = e^x - \lambda \int_0^1 G(x,\xi)y(\xi)d\xi,$

$$G(x,\xi) = \begin{cases} (1+x)\xi, & 0 \le x \le \xi \\ (1+\xi)x, & \xi \le x \le 1 \end{cases}$$

6. $y(x) = \frac{x}{\pi}\sin\frac{\pi x}{2} + \frac{2}{\pi^2}\cos\frac{\pi x}{2} + \lambda \int_{-1}^1 G(x,\xi)y(\xi)d\xi,$

$$G(x,\xi) = \begin{cases} \frac{1}{\pi}\sin\frac{\pi}{2}(\xi - x), & -1 \le x \le \xi \\ \frac{1}{\pi}\sin\frac{\pi}{2}(x-\xi), & \xi \le x \le 1 \end{cases}$$

7. See equation (4.65). Given $u(r,\theta)$ find $f(\rho,\phi)$ in (4.65).

Chapter 5

Exercises 5.1, p. 232

1. (a) $u(x) = \frac{2}{2-\lambda}\sin x, \lambda \neq 2$

   (b) $u(x) = x + \frac{2\lambda\pi}{1+2\lambda^2\pi^2}(\lambda\pi x - 4\lambda\pi\sin x + \cos x)$

   (c) $u(x) = 2x - \pi + \frac{\pi^2}{\pi - 1}\sin^2 x$

   (d) $u(x) = \frac{2}{4+\pi^2\lambda^2}(2\cos x + \pi\lambda\sin x)$

   (e) $u(x) = e^x + \lambda(e-2)(5x^2 - 3)$

   (f) $u(x) = 2x - \frac{8}{3}$

   (g) $u(x) = 60x^2 + 60x + 24$

2. (a) $\lambda_1 = 1/\pi, u_1(x) = \sin x, u(x) = A\sin x$, where A is an arbitrary constant.

   (b) $\lambda_1 = 2, u_1(x) = \sin x, u(x) = A\sin x$, where A is an arbitrary constant.

   (c) $\lambda_1 = 2/\pi$, $\lambda_2 = -2/\pi$, $u_1(x) = A\cos x$, $u_2(x) = B\sin x$, where A and B are arbitrary constants.

   (d) $\lambda_1 = \lambda_2 = -3, u_1(x) = u_2(x) = x - 2x^2$

   (e) $u(x) \equiv 0$

3. (a) The eigenvalues of the kernel are $\lambda_1 = 1/\pi$ and $\lambda_2 = -(1/\pi)$ with their corresponding (normalized) eigenfunctions $\phi_1(x) = 1/\sqrt{2\pi}(\sin x + \cos x)$ and $\phi_2(x) = 1/\sqrt{2\pi}(\sin x - \cos x)$, respectively.

   (i) $\lambda = 1/\sqrt{2\pi}$ is not an eigenvalue, i.e., $\lambda = 1/\sqrt{2\pi} \neq \mp 1/\pi$, hence the problem in (E.1) has a unique solution for arbitrary $f(x)$, which, of course includes the given function $f(x) = x^2$.

   (ii) $\lambda = 1/\pi$ is one of the eigenvalues $\lambda_1 = 1/\pi$, which corresponds to the eigenfunction $\phi_1(x) = 1/\sqrt{2\pi}(\sin x + \cos x)$.

   Since the kernel is symmetric, Theorem 2 requires that in order for (E.1) to have an infinity of solutions, the given nonhomogeneous term $f(x) = \sin 3x$ must be orthogonal to the eigenfunction $\phi_1(x)$ corresponding to $\lambda_1 = 1/\pi$, which happened to be the case since $\int_0^{2\pi}(\sin x + \cos x)\sin 3x dx = 0$.

   (b) (i) $\lambda = 1/\sqrt{\pi}$, $f(x) = x^2$

   $$u(x) = x^2 + \frac{4\pi^{\frac{1}{2}}}{1-\pi}\left[\left(1 - \pi^{\frac{1}{2}}\right)\sin x + \pi^{\frac{1}{2}}\left(1 - \pi^{\frac{1}{2}}\right)\cos x\right]$$

   (b) (ii) $u(x) = \sin 3x + \frac{c}{\pi}[\sin x + \cos x]$, c is an arbitrary constant (an infinity of solutions!).

5. The eigenvalue in problem 2(b) is $\lambda_1 = 2$, and the answer in 1(a) requires $\lambda \neq 2$. This is consistent with Fredholm alternative (Theorem 1), which insists that $\lambda \neq \lambda_1 = 2$ for the nonhomogeneous equation in 2(a) to have a

unique solution. When $\lambda = \lambda_1 = 2$ in the equation of problem 1(a), then according to the Fredholm alternative of Theorem 2, this equation will have a solution only if its nonhomogeneous term (here $f(x) = \sin x$) is orthogonal to the eigenfunction $\phi_1(x) = \sin x$ (found in problem 2(b)) corresponding to its eigenvalue $\lambda_1 = 2$, which is not the case since $\int_0^{2\pi} \sin x \sin x dx = \int_0^{2\pi} \sin^2 x dx = \pi$. Hence for $\lambda = 2$, the equaiton in problem 1(a) **does not** have a solution.

6. (a) $u(x) = C(x - x^2)$, C arbitrary

   (b) $u(x) = C = 0$,

   (c) $u(x) = C|x|$, C arbitrary

7. The kernel $K(x,t) = \sin(x - t)$ of 1(d) must not have real eigenvalues, as it is clear form the answer of problem 2(d) with its denominator of $4 + \pi^2\lambda^2$ not vanishing for all real values of λ! (Indeed, following the steps around (E.6) of Example 4 we find the two eigenvalues as pure imaginary ones $\lambda = \pm\frac{2i}{\pi}$.)

8. $u(x) \equiv 0$ in problem 2(a).

9. (a) $\lambda_1 = 2, \lambda_2 = -2$

   (b) $\lambda \neq \pm 2$

   (c) The two systems corresponding to λ_1 and λ_2 become either incompatible or redundant. So there exists either no solution or not a unique solution depending on $f(x)$.

   (d) $\lambda = \lambda_1 = 2, \phi_1(x) = A(1 - x)$; $\lambda = \lambda_2 = -2, \phi_2(x) = B(1 - 3x)$

   (e) $u(x) = f(x) + A(1 - x)$ (an infinity of solutions!)

10. (a) $u(x) = \dfrac{1 \pm \sqrt{1 - 4b\lambda}}{2\lambda}$

    (b) $u(x) = \dfrac{1}{\lambda}$

11. (a), (b) The error for both cases is of about the order 10^{-2}, so it appears that we have to include many more terms of the Maclaurin series of $\cos xt$ than the above two or three!

12. (a) $v(x) = e^x - x - 0.50102x^2 - 0.167x^3 - 0.0418x^4$

    (c)

x	0.0	0.1	0.25	0.5	0.75	1.0
Approximate Solution	1.0	0.999989	0.999937	0.999962	1.00144	1.00833
Exact solution, $u(x) = 1$	1.0	1.0	1.0	1.0	1.0	1.0

13. (a) $v(x) = x + \int_{-1}^{1} \sum_{n=0}^{N} \frac{x^n t^n}{n!} v(t)dt$

    (b) $v(x) = 3x$

14.

$$\begin{aligned}\overline{K_1(x,y)} &= \int \overline{K(s,x)\overline{K(s,y)}}ds \\ &= \int \overline{K(s,x)}\ \overline{\overline{K(s,y)}}ds \\ &= \int K(s,y)\overline{K(s,x)}ds = K_1(y,x)\end{aligned}$$

after using $\overline{\overline{K}} = K$. The same is done for $K_2(x,y)$.

15. The eigenvalue of the kernel is $\lambda_1 = -2$ corresponding to the eigenfunction $\phi_1(x) = \sin(\ln x)$.

    In case (i), $\lambda = 3 \neq -\lambda_1$, hence according to Theorem 1 of the Fredholm alternative, the integral equation (E.1) has a unique solution

$$u(x) = x^3 + \frac{2\lambda}{\lambda+2}\sin(\ln x), \quad \lambda = 3 \neq \lambda_1 = -2$$

    (see steps (E.3)–(E.6) in Example 5).

    For case (ii) we have the eigenvalue $\lambda = -2$ in (E.1), which is equal to $\lambda_1 = -2$ the eigenvalue of the kernel. So we appeal to Theorem 4 besides Theorem 1, which requires the eigenfunction $\phi_1(x) = c$ (found in Example 6) to be orthogonal to $f(x)$ (here $f(x) = \cos \pi x$), which is the case. Hence we follow the steps (E.3)–(E.6) in Example 5 to find a solution or (solutions) to (E.1) with $f(x) = \cos \pi x$. The final answer is an infinity of solutions $u(x) = \cos \pi x - 2c_1 \sin(\ln x)$ for (E.1) with $\lambda = \lambda_1 = -2$ and $f(x) = \cos \pi x$.

Exercises 5.2, p. 251

1. (b) $\frac{d^2u}{dx^2} + u = 0$, $u(x) = A\cos x + B\sin x$

   (c) $\phi_1(x) = \sqrt{\frac{2}{\pi}}\cos x$, $\phi_2(x) = \sqrt{\frac{2}{\pi}}\sin x$

(e) $\displaystyle\int_0^{\pi}\int_0^{\pi}\cos^2(x+t)dxdt = \frac{\pi^2}{2}$

(f) $K(x,t) = \cos x \cos t - \sin x \sin t = \cos(x+t)$

2. $\displaystyle u(x) = x - \lambda\left(\frac{-4}{\pi}\frac{\cos x}{\lambda - 2/\pi} + 2\frac{\sin x}{\lambda + 2/\pi}\right)$

3. (a) The kernel is square integrable since $\displaystyle\int_0^{\frac{\pi}{2}}\int_0^{\frac{\pi}{2}} K^2(x,t)dxdt < \infty$, due to $|\sin x \cos t|^2 \leq 1$ and $|\cos x \sin t|^2 \leq 1$.

(b) $\displaystyle\frac{d^2u}{dx^2} + (1+\lambda)u = 0, u(0) = 0, u\left(\frac{\pi}{2}\right) = 0$

(c) $\displaystyle u(x) = \cos 2x + \lambda\sum_{k=1}^{\infty}\frac{2a_k}{\sqrt{\pi}}\frac{\sin\sqrt{1+\lambda_k}x}{\lambda_k - \lambda}$, where $\sqrt{1+\lambda_k}\frac{\pi}{2} = \pi k$, and

$$a_k = -\frac{2}{\sqrt{\pi}}\left[\frac{(-1)^{k+1}}{2(2k-2)} + \frac{(-1)^{k+1}}{2(2k+2)} - \frac{1}{2(2k-2)} - \frac{1}{2(2k+2)}\right]$$
$$= \begin{cases} \dfrac{2}{\sqrt{\pi}}\dfrac{k}{k^2-1}, & k \text{ even} \\ 0 \quad k & \text{odd,} \end{cases}$$

(d) $\lambda = 2 \neq \lambda_n = 4n^2 - 1, n = 1, 2, \cdots$. Hence, according to the Fredholm alternative (as stated in Theorem 1), the integral equation (E.1) has a unique solution.

4. (b) See (E.3) and (E.4) of Example 8 for $u(x) = 1$ and $f(x) = \frac{1}{2}(x - x^2)$, respectively.

Exercises 5.3, p. 269

1. (a) $C_1 = -\frac{1}{2}, B_1(x,t) = -x - t + 2xt + \frac{2}{3}$

$C_2 = \frac{1}{3}, B_2 = B_3 = B_4 = \cdots = 0$

$C_3 = C_4 = \cdots = 0$

$$D(x,t;\lambda) = (x - 2t) + \lambda\left(x + t - 2xt - \frac{2}{3}\right)$$

$$D(\lambda) = 1 + \frac{\lambda}{2} + \frac{\lambda^2}{6}$$

Resolvent kernel: $\displaystyle\Gamma(x,t;\lambda) = \frac{x - 2t + \lambda(x + t - 2xt - 2/3)}{1 + \lambda/2 + \lambda^2/6}$

$$u(x) = x^2 + \frac{\lambda}{1 + \lambda/2 + \lambda^2/6}\left[\frac{x}{3} - \frac{1}{2} + \frac{\lambda}{6}\left(\frac{1}{6} - x\right)\right]$$

(b) $\Gamma(x,t;-1) = \dfrac{e^{x-t}}{2}, \quad \Gamma(x,t;\lambda) = \dfrac{e^{x-t}}{1-\lambda}, \quad u(x) = \dfrac{1}{2}e^x$

(c) $\Gamma(x,t;\lambda) = \dfrac{4xt - x^2 - \lambda[2x^2t - (4/3)x^2 + x - (4/3)xt]}{1 - \lambda + \lambda^2/18},$

$u(x) = \dfrac{3x(2\lambda - 3\lambda x + 6)}{\lambda^2 - 18\lambda + 18}$

(d) $\Gamma(x,t;\lambda) = \dfrac{\sin(x+t) + (\pi\lambda/2)\cos(x-t)}{1 - \pi^2\lambda^2/4}, \quad \lambda^2 \neq \dfrac{4}{\pi^2}$

$u(x) = 1 + \dfrac{1}{1 - \pi^2\lambda^2/4}(\lambda^2\pi \sin x + 2\lambda \cos x), \quad \lambda^2 \neq \dfrac{4}{\pi^2}$

2. (b) $u(x) = f(x) + \lambda \int_0^1 \frac{xe^t}{1-\lambda} f(t)dt$

3. $\Gamma(x,t;\lambda) = \sin(x - 2t), u(x) = 2$

4. See answers to problems 1(d) and 5.

$$u(x) = 2 + \frac{3}{1 - \pi^2\lambda^2/4}(\lambda^2\pi \sin x + 2\lambda \cos x), \quad \lambda^2 \neq \frac{4}{\pi^2}$$

5. $\Gamma(x,t;\lambda) = \sum_{I=0}^{\infty} \lambda^i K_i(x,t),$

$K_{2i}(x,t) = \left(\dfrac{\pi}{2}\right)^{2i-1} \cos(x-t), i = 1, 2, \cdots$

$K_{2i+1}(x,t) = \left(\dfrac{\pi}{2}\right)^{2i} \sin(x+t), i = 0, 1, 2, \cdots$

$u(x) = 3 + \dfrac{3}{1 - \pi^2\lambda^2/4}(\lambda^2\pi \sin x + 2\lambda \cos x), \quad \lambda^2 \neq \dfrac{4}{\pi^2}.$

6. (a) $\lambda_1 \sim 5$

   (b) $\lambda_1 \sim 2.58$

7. (a) (i) $S_3(x) = 4.1818 \sin x$

   (ii) $S_8(x) = 4.91689 \sin x - 1.32080 \sin 2x + 0.28337 \sin 3x - 0.03132 \sin 4x$

   (b)

x	-1	-0.5	0.0	0.5	1.0
Exact, $u(x) = 3x$	-3.0	-1.5	0.0	1.5	3.0
$S_3(x)$	-3.519	-2.005	0.0	2.005	3.519
$S_8(x)$	-3.0001	-1.50005	0.0	1.50005	3.0001

8. (a) $S_4(x) = 1 - 1.40383x + 0.249985x^2 + 0.0218413x^3$. For the collocation points $0, \frac{1}{3}, \frac{2}{3}, 1$; $S_4(x) = 1 - 1.493788x + 0.464096x^2 - 0.1025009x^3$.

(b) $S_2(x) = -0.8861 \sin x + \cos x$

(c) $S_2(x) = e^{-x} - \frac{x}{2}$

9. (a) (i) $S_3(x) = 3.2997 \sin x$

10. (a) $S_4(x) = 1 - 1.5x + 0.5x^2 - 0.1x^3$ [very close to the collocation method of exercise 8(a)]

(b) $S_2(x) = -0.8162 \sin x + 0.8997 \cos x$

(c) $S_2(x) = e^{-x} - \dfrac{x}{2}$ (exact solution)

Exercises 5.4, p. 282

2. (b) For $f(x) = \dfrac{1}{2}(x - x^2)$ and $u(x) = 1$ on (0,1), see their Fourier coefficients in (E.3) and (E.4) of Example 8, respectively.

3. (a)

$$\begin{aligned} f(x) &= \int_a^b \left[\sum_{k=1}^{n} a_k(x) b_k(t) \right] u(t)dt \\ &= \sum_{k=1}^{n} a_k(x) \int_a^b b_k(t)u(t)dt \\ &= \sum_{k=1}^{n} c_k a_k(x), \quad c_k = \int_a^b b_k(t)u(t)dt \end{aligned}$$

where c_k is defined by the above integral as given in (5.8).

(b) No, since a piecewise continuous function $u(t)$ in the integral of (E.1) can still give a continuous output $f(x)$ to (E.1), i.e., there is the smoothing of the integration operation on $u(t)$!

(c) $f(x) = \sum_{n=1}^{\infty} c_n \phi_n(x)$,

$u(x) = \sum c_n \lambda_n \phi_n(x)$.

For this solution $u(x)$ to exist, its series must be convergent, but knowing that the eigenvalues λ_n increase with n (see Example 7 with its $\lambda_n = n^2\pi^2$), then c_n must die out (relatively) very fast which puts a lot of strain (or restriction) on the class of functions $f(x)$ that allow its associated Fredholm integral equation of the first kind (E.1) to have a solution.

4. $f(x)$ is restricted to a linear combination of the eigenfunctions of $K(x,t)$, i.e.,

$$f(x) = \sum_{i=1}^{n} b_i \phi_i(x),$$

where

$$b_i = \int_a^b u(t)c_i(t)dt$$

5. (a) (i) $\lambda_1 = \frac{1}{\pi}$ corresponding to $\phi_1(x) = \sin x + \cos x$, $\lambda_2 = -\frac{1}{\pi}$ corresponding to $\phi_2(x) = \sin x - \cos x$.

(ii) $\int_0^{2\pi} \sin(x+t)u(t)dt =$

$= \sin x \int_0^{2\pi} u(t)\cos t dt + \cos x \int_0^{2\pi} u(t)\sin t dt$

$= b_1 \sin x + b_2 \cos x$

(c) If there are infinite eigenfunctions for the kernel, and they are complete, then there can be no nontrivial function $g(x)$ that is orthogonal to all of the eigenfunctions (see a version of Picard's theorem in Theorem 7).

7. (b) No, we cannot limit ourselves to search for only continuous input $u(x)$ (see part (a).)

(c) Here $u(t)$ can have a large jump discontinuity, for example, which corresponds to practically no change in the continuous $f(x)$, i.e., the input and output are not related in a continuous way. Thus the problem is called ill-posed, i.e., not stable, since a very small change in the input may cause a very large change in the output.

8. If $f(t) \neq 0$, this would mean that such a static load of density distribution $\rho(x) = f(x)$ in (2.28) gives no deflection, i.e., $y(x) \equiv 0$ at any point, but we expect some deflection $y(x)$ as in (2.28) and Fig. 2.2.

Exercises 5.5, p. 295

1. (a)

(i) 1.005133, 1.00380, 1.004589, 1.00888

(ii) 1.00042, 1.00038, 1.00034, 1.00032, 1.00031, 1.00032, 1.00035, 1.00042, 1.00051, 1.00063, 1.00078

(b) Example 20: See the answer for 1(a)i, ii.
Example 6: 1.003, 1.0027, 1.0024, 1.0021, 1.002, 1.002, 1.002, 1.0025, 1.0037, 1.0056, 1.009

2. (a), (b)

x	(a) Simpson's rule	(b) Exact	Example 20 (Trapezoidal rule)
0	0.99987	1	1.013
0.5	0.99992	1	1.009
1.0	0.99967	1	1.021

3. (a)

(i) $u(0) = 1.0, u(0.5) = 0.367, u(1.0) = -0.11019$ [see part (b)]

(ii) $u_0 = 1.0$, $u_1 = 0.85493$, $u_2 = 0.7189$, $u_3 = 0.59109$, $u_4 = 0.47069$, $u_5 = 0.35699$, $u_6 = 0.249$, $u_7 = 0.147213$, $u_8 = 0.05007$, $u_9 = -0.04260$, $u_{10} = -0.1320$ [see part (b)]

(b) (a,i)

x	0.0	0.5	1.0
Collocation (Example 16), $S_3(x) = 1 - 1.441x + 0.310x^2$	1	0.3568	-0.1310
Numerical, trapezoidal rule	1	0.367	-0.1102
Exact solution, $u(x) = e^{-x} - x/2$	1	0.3565	-0.1321

(a,ii)

x	0	.1	.2	.3	.4	.5
Approximate						
Part (a,i)	1	–	–	–	–	0.367
Part (a,ii)	1	0.855	0.719	0.60	0.47	0.357
Exact, ($u(x) = e^{-x} - x/2$)	1	0.855	0.719	0.591	0.47	0.3565
Approximate $u(x) = S_3(x) = 1 - 1.441x + 0.310x^2$	1	0.859	0.7242	0.5956	0.4732	0.3568

.6	.7	.8	.9	1.0
–	–	–	–	-0.11
0.249	0.147	0.05	-0.043	-0.132
0.249	0.147	0.05	-0.043	-0.132
0.247	0.143	0.046	-0.046	-0.131

4. (a) $u(-1) = 1.5$, $u(0) = 1.34$, $u(1) = 1.5$

(c) $u_0 = 1.51134$, $u_1 = 1.47727$, $u_2 = 1.44503$, $u_3 = 1.41543$, $u_4 = 1.38912$, $u_5 = 1.36656$, $u_6 = 1.34798$, $u_7 = 1.33355$, $u_8 = 1.32326$, $u_9 = 1.31709$, $u_{10} = 1.31504$, $u_{11} = 1.31504$, $u_{12} = 1.51135$, $u_{13} = 1.51137$, $u_{14} = 1.51142$, $u_{15} = 1.521$, $u_{16} = 1.511$, $u_{17} = 1.5998$, $u_{18} = 1.49989$, $u_{19} = 1.50131$, $u_{20} = 1.51134$

5. (a) $\Delta t = \dfrac{b-a}{n}$,

$$\left(1-\frac{\Delta t}{3}K_{0,0}\right)u_0 - 4\frac{\Delta t}{3}K_{01}u_1 - \cdots - 4\frac{\Delta t}{3}K_{0,2n-1}u_{2n-1} - \frac{\Delta t}{3}K_{0,2n}u_n = f_0$$

$$-\frac{\Delta t}{3}K_{10} + \left(1 - 4\frac{\Delta t}{3}K_{10}\right)u_1 - 2\frac{\Delta t}{3}K_{12}u_2 \cdots - 4\frac{\Delta t}{3}K_{1,2n-1}u_{2n-1} - \frac{\Delta t}{3}$$
$$\cdot K_{1,2n}u_n = f_1$$
$$\vdots$$
$$-\frac{\Delta t}{3}K_{n0} - \cdots - 4\frac{\Delta t}{3}K_{1,2n-1}u_{2n-1} + \left(1-\frac{\Delta t}{3}K_{n,n}\right)u_n = f_n$$

(b)

x	$u(x)$
0	0.9987
0.5	0.9992
1.0	0.99967

(c) Comparison of results:

x	Exact	Example 6 (Exercise 1(b))	Simpson rule
0	1.0	1.003	0.99987
0.5	1.0	1.002	0.99992
1.0	1.0	1.009	0.99967

6. (a) $u(0) = 0.0$, $u(1/2) = 1.73$ (corresponds to $\lambda = 16$), $u(1) = 0.0$

7. (a) $u_0 = 0, u_1 = 1.73, u_2 = -1.73, u_3 = 0$, largest finite eigenvalue is $\lambda = 27$.

(b) The closest exact eigenvalue $\lambda_2 = (2\pi)^2 \approx 39$, which corresponds to $n = 2$ for $u_2(x) = \sqrt{2}\sin 2\pi x$.

x	0	1/3	2/3	1
Numerical, $n = 3$	0.0	1.73	-1.73	0.0
Exact, $u_2(x)$ $= \sqrt{2}\sin 2\pi x$	0.0	1.23	-1.23	0.0

8. Numerical method: $u(0) = 0.0$, $u(1/4) = 0.46$, $u(1/2) = 0.90$, $u(3/4) = 1.28$, $u(1) = 1.58$

9. $u(x) \approx \tilde{u}(x) = (1)2\left(x-\frac{1}{2}\right)(x-1) - (0.367)4x(x-1) + (-0.1102)(2)(x)\left(x-\frac{1}{2}\right)$

$$= 2(x - \frac{1}{2})(x - 1) - 1.468x(x - 1) - 0.2204x\left(x - \frac{1}{2}\right)$$

See the following table for comparison of the results.

x_i	$u(x_i)$ exact	$\tilde{u}(x)$ (interp.)	$u(x_i)$ (num.exerc.3a(ii))
0.000	1.00000	1.00000	1.00000
0.100	0.85484	0.86094	0.85493
0.200	0.71873	0.72810	0.7189
0.300	0.59082	0.60150	0.59109
0.400	0.47032	0.48114	0.47069
0.500	0.35653	0.36700	0.35699
0.600	0.24881	0.25910	0.249
0.700	0.14659	0.15742	0.147213
0.800	0.04933	0.06198	0.05007
0.900	−0.04343	−0.02722	−0.0426
1.000	−0.13212	−0.11020	−0.1320

Chapter 7

Exercises 7.1, p. 348

1. $\displaystyle\int_0^\infty \frac{dx}{100 + x^2} = \int_0^\infty e^{-\beta x} \cdot \frac{e^{\beta x}}{100 + x^2} dx,$
let $u = \beta x$, hence $du = \beta dx$ and the above equation becomes

$$\int_0^\infty \frac{dx}{100 + x^2} = \frac{1}{\beta} \int_0^\infty e^{-u} \cdot \left[\frac{e^u}{(\frac{u}{\beta})^2 + 100}\right] du$$

$$\approx \frac{1}{\beta} \sum_{i=1}^{N} w_i \left[\frac{e^{u_i}}{(\frac{u_i}{\beta})^2 + 100}\right].$$

(i) If we use the Gauss-Laguerre rule with $N = 8$, $\beta = 0.2$, we have an approximate answer of 0.14999. The exact answer is $\frac{\pi}{20} = 0.1570796$.

(ii) For the Gauss-rational rule, we have $x = v(\xi) = \dfrac{2\beta}{\xi + 1} - \beta$,

$$\int_0^\infty \frac{dx}{100 + x^2} = 2\beta \int_{-1}^{1} \frac{f(v(\xi))}{(1 + \xi)^2} d\xi = 2\beta \int_{-1}^{1} \frac{F(\xi)}{(1 + \xi)^2} d\xi.$$

For $N = 8$, $\beta = 10$, we prepare the data for $w_i \dfrac{F(\xi_i)}{(1 + \xi_i)^2}$, and use the Gauss-Laguerre quadrature to have the approximate value of the integral 0.15707944. The exact answer is $\frac{\pi}{20} = 0.15707963$.

2. $\int_0^\infty e^{-x}\sin x dx = \int_0^\infty e^{-\beta x}e^{\beta x}e^{-x}\sin x dx$
let $u = \beta x$, the above equation becomes,

$$\int_0^\infty e^{-x}\sin x dx = \frac{1}{\beta}\int_0^\infty e^{-u}e^{u}e^{-(\frac{u}{\beta})}\sin\left(\frac{u}{\beta}\right)du.$$

(i) If we use $\beta = 1$, we have

$$\frac{1}{\beta}\int_0^\infty e^{-u}e^{u}e^{-(\frac{u}{\beta})}\sin\left(\frac{u}{\beta}\right)du = \int_0^\infty e^{-u}\sin u\ du$$

and an eight-point Gauss-Laguerre rule gives a value of 0.499988. The exact answer is 0.5.

(ii) For the Gauss-rational rule, let $x = \dfrac{2\beta}{\xi+1} - \beta$, then

$$\int_0^\infty e^{-x}\sin x dx = 2\beta\int_{-1}^1 \frac{F(\xi)}{(1+\xi)^2}d\xi$$

where

$$F(\xi) = e^{-\frac{2\beta}{\xi+1}+\beta}\sin\left(\frac{2\beta}{\xi+1} - \beta\right).$$

For $N = 8$, $\beta = 10$ we prepare the data in the same way as in part (a) to have $2\beta\sum_{i=1}^{8} w_i\dfrac{F(\xi_i)}{(1+\xi_i)^2} = 0.4955$. The exact answer is 0.5.

3. Since the integrated function in (E.2), with respect to $\rho(y) = e^{-y}$ on $(0,\infty)$, is $f(y) = (y+1)^3$, a polynomial of degree $m = 3$, then a two-point Laguerre rule will give the exact result because $m = 3 \leq 2(N) - 1 = 2(2) - 1 = 3$. With $x_1 = 2+\sqrt{2}, x_2 = 2-\sqrt{2}, w_1 = \frac{1}{4}(2-\sqrt{2})$, and $w_2 = \frac{1}{4}(2+\sqrt{2})$, the two-point shifted Laguerre rule in (E.2) gives

$$\int_1^\infty e^{-x}x^3 dx \approx \frac{16}{e}.$$

5. (a) The result is not exact for two-point Gauss-Hermite rule because the function $f(x) = x^4$ is of degree $4 > 2N - 1 = 2(2) - 1 = 3$.

Exercises 7.2, p. 357

1. (a) $u_0 = u(0) = 1 - 0 - 0 = 1$

Using the *trapezoidal rule* (for u_1) with $\Delta t = 0.05, t_0 = 0, t_1 = 0.05$,

$$u_1 = 0.89 + \frac{0.05}{2}[3.29u_0 + 3u_1], \quad u_1 = 1.051081$$

Simpson's Rule $, \Delta t = 0.05, t_2 = 0.10, h = \dfrac{0.1}{2}$

$$u_2 = 0.76 + \frac{0.1}{6}[3.56u_0 + 13.16u_1 + 3u_2], \quad u_2 = 1.105127$$

Simpson's-$\frac{3}{8}$ Rule $, \Delta t = 0.05, t_3 = 0.15,$

$$\begin{aligned} u_3 &= 0.61 + \frac{3}{8}(0.05)[3.81u_0 + 10.68u_1 + 9.87u_2 + 3u_3], \\ u_3 &= 1.161784385 \end{aligned}$$

Extended Simpson's Rule $, \Delta t = 0.05, t_4 = 0.20,$

$$\begin{aligned} u_4 &= 0.44 + \frac{0.05}{3}[4.04u_0 + 15.24u_1 + 7.12u_2 + 13.16u_3 + 3u_4], \\ u_4 &= 1.221334 \end{aligned}$$

(b) The comparison of these results with the exact solution $u(x) = e^x$ are presented in the following table.

i	x_i	$\bar{u}_i$ (num.)	$u_i = e^{x_i}$ (exact)	error $\varepsilon_i = \tilde{u}_i - u_i$
0	0.00	1.000000	1.000000	0.000000
1	0.05	1.051081	1.051271	-1.9001 $\times 10^{-4}$
2	0.10	1.105127	1.105171	-4.4128 $\times 10^{-4}$
3	0.15	1.161784	1.161834	-4.9858 $\times 10^{-5}$
4	0.20	1.221334	1.221403	-6.8356 $\times 10^{-5}$

See also the answer to Exercise 4(e) for a bit more accurate results.

2. (a) $u_1 = 1.05263$

$u_3 = 1.16344$

$u_5 = 1.28201$

$u_7 = 1.41863$

$u_9 = 1.56959$

(b) The comparison of the numerical results with the exact solution $u(x) = e^x$ is given in the following table.

i	x_i	$\tilde{u}_i$ (num.)	$u_i = e^{x_i}$ (exact)	error $\varepsilon_i = \tilde{u}_i - u_i$
1	0.05	1.052632	1.051271	1.361 $\times 10^{-3}$
3	0.15	1.163435	1.161834	1.601 $\times 10^{-3}$
5	0.25	1.282014	1.284025	-2.011 $\times 10^{-3}$
7	0.35	1.418631	1.419068	-4.364 $\times 10^{-4}$
9	0.45	1.569590	1.568312	1.278 $\times 10^{-3}$

3. For $\Delta t = 0.02, u_0 = u(0) = 1,$

$$\begin{aligned} 0.97u_1 &= 0.9584 + \frac{0.02}{2}[3.1189u_0], \\ u_1 &= 1.020190 \end{aligned}$$

and if we proceed parallel to Exercise 1,

$$u_2 = 1.040810.$$

In the same way we obtain

$$\begin{aligned} u_3 &= 1.06183, \\ u_4 &= 1.083286. \end{aligned}$$

The comparison of these numerical results with the exact solution $u(x) = e^x$ is given in the following table.

i	x_i	$\tilde{u}_i$ (num.)	$u_i = e^{x_i}$ (exact)	error $\varepsilon_i = \tilde{u}_i - u_i$
0	0.00	1.000000	1.000000	0.000000
1	0.02	1.020190	1.020201	-1.165×10^{-5}
2	0.04	1.040809	1.040811	-9.911×10^{-7}
3	0.06	1.061836	1.068365	-9.465×10^{-7}
4	0.08	1.083286	1.083287	-1.1910×10^{-6}

4. (a) Using (the coarse!) $\Delta t = 0.5$, with the trapezoidal rule-Simpson's rule combination as in Exercise 1, we have $u_0 = 1.0$

$$\begin{aligned} u_1 &= 1 - 2(0.5) - 4(0.5)^2 + \frac{1}{4}\left[3 + 6\left(\frac{1}{2}\right) - 4\left(\frac{1}{4}\right)\right] u_0 \\ &\qquad + 3\left(\frac{1}{4}\right) u_1 \\ 0.25u_1 &= -1 + \frac{5}{4}u_0, \quad u_1 = 1.0 \end{aligned}$$

In the same way we obtain

$$u_2 = -1.66667$$

(b) With the exact answer $u(x) = e^x > 0$, this negative value for $u_2 = -1.66667$ suggests a break down of the numerical method due to the inaccuracy inherent in $u_1 = 1.0$, which is very far from its exact value of $e^{0.5} = 1.6487$. In Exercise 1 with $\Delta t = 0.05$, we had $u_1 = 1.051081$ compared to its exact value of $e^{0.05} = 1.05127$, i.e., a good enough resolution of $\Delta t = 0.05$ that preserves the characteristics of the solution $u(x) = e^x$ inside the integral.

5. $\tilde{u}(x) = -1333.333333(x-0.05)(x-0.01)(x-0.15) + 4204.324x(x-0.1)(x-0.15) - 4420.508x(x-0.05)(x-0.15) + 1549.045333x\,(x-0.05)(x-0.1)$. The comparison of this approxiamte interpolated result $\tilde{u}(x)$ of the numerical values and the exact answer $u(x) = e^x$, is given in the following table.

Approximate (interpolated) $\tilde{u}(x)$ and exact $u(x) = e^x$ solutions of (E.1)

x	$\tilde{u}(x)$	$u(x)$
0.00	1.00000	1.00000
0.04	1.04062	1.04081
0.10	1.10513	1.10517
0.14	1.15026	1.15027
0.20	1.22070	1.22140
0.24	1.26922	1.27125
0.30	1.34388	1.34986
0.40	1.47184	1.49182
0.50	1.60175	1.64872
0.60	1.73078	1.82212
0.70	1.85609	2.01375
0.80	1.97486	2.22554
0.90	2.08424	2.45960
1.00	2.18141	2.71828

6. (a) $\phi(\xi) = \xi + \int_0^{\xi} (\xi - \tau)\phi(\tau)d\tau$

   (b) $\phi(\xi) = \sinh\xi$,

   $$u(x) = \phi\left(\frac{1}{x}\right) = \sinh\left(\frac{1}{x}\right).$$

Exercises 7.3, p. 370

1. (a) First we use the transformation $\xi = 2t - 1$, $d\xi = 2dt$ to have an integral on the symmetric interval $-1 < \xi < 1$ ready for using the Tchebychev rule (7.17),

$$u(x) = x^2 - \int_{-1}^{1} \left(1 + \frac{(\xi+1)}{2}x\right) u\left(\frac{\xi+1}{2}\right) d\xi. \qquad (E.1)$$

$$1.505271u_1 \;+0.52085309u_2 + 0.5304834u_3 + 0.54606562u_4 = 0.01054174 \qquad (E.2)$$

$$0.52085309u_1 \;+1.5825008u_2 + 0.62060115u_3 + 0.68224891u_4 = 0.16500169 \qquad (E.3)$$

$$\begin{aligned} 0.5304834u_1 \quad &+0.62060115u_2 + 1.6762968u_3 + 0.76641459u_4 \\ &= 0.35259369 \end{aligned} \tag{E.4}$$

$$\begin{aligned} 0.54606562u_1 \quad &+0.68224891u_2 + 0.76641459u_3 + 1.90259787u_4 \\ &= 0.805195744 \end{aligned} \tag{E.5}$$

(b) For the use of the Gauss-Legendre rule, we employ the same transformation $\xi = 2t - 1, d\xi = 2dt$ to have an integral on the symmetric interval $-1 < \xi < 1$ to be ready for Table 7.3.

$$\begin{aligned} u_1 \quad &= (0.069432)^2 - [(0.347855)(1 + (0.069432)^2)u_1 \\ &+(0.652145)(1 + (0.069432)(0.3300095))u_2 \\ &+(0.652145)(1 + (0.069432)(0.6699905))u_3 \\ &+(0.347855)(1 + (0.069432)(0.930568))u_4, \end{aligned}$$

$$\begin{aligned} 1.34953194u_1 \quad &+0.66708774u_2 + 0.682482u_3 + 0.37033033u_4 \\ &= 4.8208 \times 10^{-3} \end{aligned} \tag{E.6}$$

and in the same way we obtain the following results for x_2, x_3, and x_4,

$$\begin{aligned} 0.35582548u_1 \quad &+1.72316768u_2 + 0.79633636u_3 + 0.45467998u_4 \\ &= 0.108906272 \end{aligned} \tag{E.7}$$

$$\begin{aligned} 0.3640368u_1 \quad &+0.79633636u_2 + 1.94488459u_3 + 0.56473275u_4 \\ &= 0.448887272 \end{aligned} \tag{E.8}$$

$$\begin{aligned} 0.37033033u_1 \quad &+0.8524163u_2 + 1.05873896u_3 + 1.6490824u_4 \\ &= 0.86595682 \end{aligned} \tag{E.9}$$

2. $u(x) = \lambda \int_0^1 (x + t)u(t)dt.$

For the four-point Gauss-Legendre rule, we let $\xi = 2t - 1$, $d\xi = 2dt$, to have the above integral defined on the symmetric interval (-1,1) to be ready for Table 7.3.

$$u(x) = \frac{\lambda}{2} \int_{-1}^{1} \left[x + \frac{\xi + 1}{2}\right] u\left(\frac{\xi + 1}{2}\right) d\xi,$$

then we substitute for $x_1 = 0.069432$, $x_2 = 0.330010$, $x_3 = 0.593796$ and $x_4 = 0.897327$ to have the first linear equation (E.1). The four resulting linear homogeneous equations in u_1, u_2, u_3, and u_4 are:

$$\begin{aligned} u_1 \quad &= \lambda[0.024152u_1 + 0.130247u_2 + 0.241105u_3 \\ &+0.173928u_4] \end{aligned} \tag{E.1}$$

$$\begin{aligned} u_2 \quad &= \lambda[0.069474u_1 + 0.215214u_2 + 0.326072u_3 \\ &+0.219249u_4] \end{aligned} \tag{E.2}$$

$$u_3 = \lambda[0.128606u_1 + 0.326072u_2 + 0.436931u_3 + 0.278381u_4] \quad (E.3)$$

$$u_4 = \lambda[0.173928u_1 + 0.41104u_2 + 0.521898u_3 + 0.323703u_4]. \quad (E.4)$$

3. Here we use the same four points x_i, $i = 1, 2, 3, 4$ of the Tchebychev rule (7.17) after making the change of variable $\xi = 2t - 1$ to have the following integral of (E.1)

$$3x + 1 = \int_0^1 (1 + xt)u(t)dt \quad (E.1)$$

ready for Table 7.4 that needs symmetric limits of integration,

$$2(3x + 1) = \int_{-1}^{1} \left(1 + x\frac{\xi + 1}{2}\right) u\left(\frac{\xi + 1}{2}\right) d\xi.$$

The resulting four linear homogeneous equations in x_1, x_2, x_3 and x_4 are:

$$5.232 = 1.0105u_1 + 1.041706u_2 + 1.060967u_3 + 1.092131u_4 \quad (E.1)$$

$$8.874 = 1.041706u_1 + 1.1650u_2 + 1.24120u_3 + 1.3645u_4 \quad (E.2)$$

$$11.125 = 1.060967u_1 + 1.24120u_2 + 1.352594u_3 + 1.532829u_4 \quad (E.3)$$

$$14.7679 = 1.092131u_1 + 1.3645u_2 + 1.532829u_3 + 1.805196u_4 \quad (E.4)$$

4. (a) $N = 2$ yields the exact answer, see part (b).

$$u(x) = \frac{x}{2} - \frac{1}{3} + \int_0^1 (x + t)u(t)dt$$

Let $\xi = 2t - 1$, $d\xi = 2dt$,

$$u(x) = \frac{x}{2} - \frac{1}{3} + \frac{1}{2}\int_{-1}^{1} \left(x + \left(\frac{\xi + 1}{2}\right)\right) u\left(\frac{\xi + 1}{2}\right) dt$$

$$N = 2$$
$$u_1 = 0.211325, \quad u_2 = 0.788675$$

$$N = 3$$
$$u_1 = 0.112701, \quad u_2 = 0.50, \quad u_3 = 0.887298$$

$$N = 4$$
$$u_1 = 0.069432, \quad u_2 = 0.330010, \quad u_3 = 0.669991, \quad u_4 = 0.930568$$

Also, see part (b).

(b) The computations, using the method in Section 5.1 for the present equation with degenerate kernel, show that the exact solution is $u(x) = x$. Part (a) gives an exact answer with $N = 2$, since with the exact solution $u(x) = x$, the integrand in the integral above is of degree $2 \leq 2N - 1 = 2(2) - 2 = 3$, hence it is approximated exactly by the two-point Gauss-Legendre rule. Of course, what concerns the numerical rule is that the error at the samples locations, in this case x_1 and x_2, vanishes.

(d) Simpson's rule will have an infinite accuracy as a rule of degree 2, which is the degree of the polynomial $(x + t)u(t) =(x + t)t = xt + t^2$ integrated with respect to t in (E.1).

Simpson's rule with:

(i) $N = 2, h = \frac{1}{2}, \quad u_1 = 0.5, \quad u_2 = 1.0$

(ii) $N = 3, h = \frac{1}{3}, \quad u_1 = \frac{1}{3}, \quad u_2 = \frac{2}{3}, \quad u_3 = 1.0$

(iii) $N = 4, h = \frac{1}{4}, \quad u_1 = \frac{1}{4}, \quad u_2 = \frac{1}{2}, \quad u_3 = \frac{3}{4}, \quad u_4 = 1.0$

5. (a) We have a degenerate kernel where the method of Section 5.1 results in the exact solution as $u(x) = \sin x$.

(b) Since the exact solution is $u(x) = \sin x$, the integrand in (E.1) is $(xt)\sin t$, which is not a polynomial in t. Hence no *finite* degree (polynomial) quadrature rule can approximate it exactly.

(c) Let $\xi = \frac{4}{\pi}t - 1, d\xi = \frac{4}{\pi}dt$,

Gauss-Legendre rule,

(i) $N = 2$
$u_1 = 0.320388, \quad u_2 = 0.924892$

(ii) $N = 3$
$u_1 = 0.176134, \quad u_2 = 0.707221, \quad u_3 = 0.984574$

(iii) $N = 4$
$u_1 = 0.108847, \quad u_2 = 0.495471, \quad u_3 = 0.868624, \quad u_4 = 0.994058$

See the following table for $N = 2, 3$, and 4,

Gauss-Legendre, $N = 2, 3$ and 4

i	x_i	$\tilde{u}_i$ (num.)	u_i (exact)	$\varepsilon_i = \tilde{u}_i - u_i$
		$N = 2$		
1	0.331948	0.320388	0.325886	-0.00549755
2	1.238848	0.924892	0.945409	-0.02051700
		$N = 3$		
1	0.177031	0.176134	0.176108	2.58×10^{-5}
2	0.785398	0.707221	0.707107	1.1×10^{-4}
3	1.393765	0.984574	0.984371	2.0×10^{-4}
		$N = 4$		
1	0.109064	0.108847	0.108847	-4.8×10^{-8}
2	0.518378	0.495471	0.495472	-2.3×10^{-7}
3	1.052419	0.868624	0.868624	-4.7×10^{-7}
4	1.461733	0.994058	0.994058	-6.5×10^{-7}

(d) The reason for the good results is the smoothness of the integrand $t \sin t$ of $K(x,t) = xtu(t) = xt \sin t$ in (E.1).

6. The determinant is zero, since the kernel is symmetric, and with using the Tchebychev rule this symmetry is preserved for the coefficients matrix, as seen in the answer to Exercise 2. With this symmetry, the determinant vanishes, which excludes the use of Cramer's rule. So we must first check the theory of Fredholm integral equations of the first kind in Section 5.4 for ensuring the existence of the solution to the above integral equation before we embark on a numerical solution. See Section 5.4, where the condition for the existence of a unique solution of Fredholm integral equations of the first kind is, in general, (much) more restrictive than that of Fredholm integral equations of the second kind (Sections 5.1-5.3).

7. (a)

$$\begin{bmatrix} u_1 \\ u_2 \\ u_3 \\ u_4 \end{bmatrix} = \begin{bmatrix} 0.024152\lambda & 0.130247\lambda & 0.241105\lambda & 0.173927\lambda \\ 0.069474\lambda & 0.215214\lambda & 0.326073\lambda & 0.219249\lambda \\ 0.128606\lambda & 0.326073\lambda & 0.436931\lambda & 0.278381\lambda \\ 0.173927\lambda & 0.411040\lambda & 0.521898\lambda & 0.323703\lambda \end{bmatrix} \begin{bmatrix} u_1 \\ u_2 \\ u_3 \\ u_4 \end{bmatrix} \tag{E.1}$$

$|I - \lambda A| = 0$ results in $\tilde{\lambda}_1 = -12.92820323$, $\tilde{\lambda}_2 = 0.9282$, to be compared with the given exact eigenvalues of $\lambda_1 = -12.928$ and $\lambda_2 = 0.92820$, which shows no difference (see part (b)).

(b) (i) $\tilde{\lambda}_1 = -12.92920323$, $u_1 = \tilde{U}_1(0.0694320) = U(0.0694320) = 0.8797402479$ for normalizing to the exact eigenfunction $U_1(x)$ at $x = 0.0694320$. $u_2 = -0.4284098556$, $u_3 = 0.1604561724$, $u_4 = 0.61179665745$.

(ii) $\tilde{\lambda}_2 = 0.928203232$, $u_1 = \tilde{U}_2(0.06932) = U_2(0.06932) = 1.1202560$ for normalizing to the exact eigenfunction $U_2(x)$ at $x = 0.0694320$. $u_2 = 1.5715944$, $u_3 = 2.160457$, $u_4 = 2.611795$.

(c) $u(x) = \lambda \int_0^1 (x+t)u(t)dt$

The two eigenvalues are

$$\lambda_1 = -6 - \sqrt{48}, \quad \lambda_2 = -6 + \sqrt{48}.$$

The first eigenfunction corresponding to the $\lambda_1 = -6 - \sqrt{48}$ is

$$u_1(x) = \frac{6 + \sqrt{48}}{(4 + \frac{1}{2}\sqrt{48})} x - 1$$

The second eigenfunction corresponding to $\lambda_2 = -6 + \sqrt{48}$, is

$$u_2(x) = \frac{(-6 + \sqrt{48})}{(4 - \frac{1}{2}\sqrt{48})} x + 1$$

(d) The interpolations of the four sample values of the two approximate eigenfunctions of (E.1) are

(i)

$$\begin{aligned}
\tilde{U}_1(x) &= 6.528178475(t - 0.330010)(t - 0.669990)(t - 0.930570) \\
&\quad -8.052147011(t - 0.0694320)(t - 0.669990)(t - 0.930570) \\
&\quad -3.015830589(t - 0.0694320)(t - 0.330010)(t - 0.930570) \\
&\quad +4.539799149(t - 0.0694320)(t - 0.330010)(t - 0.669990)
\end{aligned}$$

and

(ii)

$$\begin{aligned}
\tilde{U}_2(x) &= -8.313307847(t - 0.330010)(t - 0.669990)(t - 0.930570) \\
&\quad +29.54016446(t - 0.0694320)(t - 0.669990)(t - 0.930570) \\
&\quad -40.60839619(t - 0.0694320)(t - 0.330010)(t - 0.930570) \\
&\quad +19.38153958(t - 0.0694320)(t - 0.330010)(t - 0.669990)
\end{aligned}$$

corresponding to the two approximate eigenvalues $\tilde{\lambda}_1 = -12.9292$ and $\tilde{\lambda}_2 = 0.928203232$, respectively.

The comparison of these two continuous approximations $\tilde{U}_1(x)$ and $\tilde{U}_2(x)$ of the eigenfunctions $U_1(x)$ and $U_2(x)$ of (E.1) are given in the following two tables, respectively.

The Interpolated Approximation $\tilde{U}_1(x)$ and the exact first eigenfunction $U_1(x)$ of (E.1), $\tilde{\lambda}_1 = -12.92820323$, $\lambda_1 = -12.928$

x	Approx $\tilde{U}_1(x)$ $\tilde{\lambda}_1 = -12.92820323$	Exact $U_1(x)$ $\lambda_1 = -12.928$
0.0	−1.000007	−1.000000
0.1	−0.826801	−0.826795
0.2	−0.653595	−0.653590
0.3	−0.480389	−0.480385
0.4	−0.307183	−0.307180
0.5	−0.133977	−0.133975
0.6	0.039229	0.039230
0.7	0.212435	0.212436
0.8	0.385641	0.385641
0.9	0.558847	0.558846
1.0	0.732053	0.732051

The Interpolated Approximate $\tilde{U}_2(x)$ and the exact second eigenfunction $U_2(x)$ of (E.1). $\tilde{\lambda}_2 = 0.928203232$, $\lambda_2 = 0.92820$

x	Approx $\tilde{U}_2(x)$ $\tilde{\lambda}_2 = 0.928203232$	Exact $U_2(x)$ $\lambda_2 = 0.92820$
0.0	1.000048	1.000000
0.1	1.173260	1.173205
0.2	1.346473	1.346410
0.3	1.519686	1.519615
0.4	1.692898	1.692820
0.5	1.866111	1.866025
0.6	2.039323	2.039230
0.7	2.212536	2.212436
0.8	2.385748	2.385641
0.9	2.558961	2.558846
1.0	2.732173	2.732051

8. (a) $U(\xi) = \dfrac{\xi}{1+\xi} + \displaystyle\int_0^1 e^{-2(\frac{1}{\xi}+\frac{1}{\tau})} \cdot \frac{1}{\tau^2} U(\tau)d\tau$

(b)
$$\begin{aligned}
4.7u_1 + 3.4u_2 + 1.5u_3 + 1.2u_4 &= 0.9 \times 10^{10},\\
11.1u_1 + 7.1u_2 + 3.2u_3 + 2.4u_4 &= 0.3 \times 10^{4},\\
4.6u_1 + 3.4u_2 + 1.5u_3 + 1.1u_4 &= 0.4 \times 10^{3},\\
14.4u_1 + 10.5u_2 + 5.4u_3 + 3.6u_4 &= 0.5 \times 10^{2},
\end{aligned}$$

(c)

ξ	$x = \frac{1}{\xi}$	$u(x)$
0.8973	1.1145	-1.79 $\times 10^{-10}$
0.5938	1.6841	-7.87 $\times 10^{-9}$
0.4062	2.4615	-4.13 $\times 10^{-10}$
0.1029	9.7371	1.04 $\times 10^{-10}$

Appendix A

Exercises Appendix A, p. 376

2. (a) $c_k = \dfrac{F_n(\lambda_k)}{\int_0^a rJ_n^2(\lambda_k r)dr} = \dfrac{2F_n(\lambda_k)}{a^2 J_{n+1}^2(a\lambda_k)}$

See Bessel functions integrals for the integral in the denominator;

$$\int_0^a rJ_n^2(\lambda_k r)dr = \frac{a^2}{2} J_{n+1}^2(a\lambda_k).$$

(b) $f(r) = 2\sum_{k=1}^{\infty} F_n(\lambda_k) \dfrac{J_0(\lambda_k r)}{a^2 J_{n+1}^2(a\lambda_k)}, F_n(\lambda_k) = \int_0^a rJ_0(\lambda_k r) f(r)dr$

3. (a) $\dfrac{\partial^2 v}{\partial r^2} + \dfrac{1}{r}\dfrac{\partial v}{\partial r} + \dfrac{\partial^2 v}{\partial z^2} = 0,$

$v(r, 0) = f(r), 0 \le r < a$
$\dfrac{\partial v}{\partial z}(r, 0) = 0, r > a$

(b) $\dfrac{d^2 V(\lambda, z)}{dz^2} = \lambda^2 V(\lambda, z);\;\; V(\lambda, z) = A(\lambda)e^{-\lambda z}$

$v(r, z) = \int_0^{\infty} \lambda J_0(\lambda r) A(\lambda) e^{-\lambda z} d\lambda$

where $A(\lambda)$ is the solution of the dual integral equations

$$f(r) = \int_0^{\infty} \lambda J_0(\lambda r) A(\lambda) d\lambda, 0 < r < a$$

$$0 = \int_0^{\infty} \lambda^2 J_0(\lambda r) A(\lambda) d\lambda, r > a$$

(c) See the answer to Exercise 1, Section 2.6.

4. $\nabla^2 U(r, \theta, s) - s^2 U(r, \theta, s) = -g(r, \theta).$

References

Abramowitz, M., and I.E. Stegun, *Handbook of Mathematical Functions*, Dover, New York, 1965.

Atkinson, K.E., *The Numerical Solution of Integral Equations of the Second Kind*, Cambridge University Press, New York, 1997.

Anderson, R.S., and F.R. DeHoog, *Application and Numerical Solutions of Integral Equations*, Sijthoff en Noordhoff, Alphen aan den Rijn, The Netherlands, 1980.

Anton, H., *Calculus with Analytic Geometry*, 5th ed., John Wiley and Sons, Inc., New York, 1995.

Arfken, G., *Mathematical Methods for Physicists*, Academic Press, New York, 1970.

Baker, C.T., and G.F. Miller, *Treatment of Integral Equations by Numerical Methods*, Oxford University Press, London, 1977.

Bell, E.T., *The Development of Mathematics*, McGraw-Hill, New York, 1945, pp. 524–30.

Bôcher, M., *Integral Equations*, Cambridge University Press, Cambridge, 1914.

Briggs, W.L., and V.E. Henson, *The DFT-An Owner's Manual for the Discrete Fourier Transforms*, SIAM, 1995.

Brigham, E.O., *Fast Fourier Transform and its Applications*, Prentice-Hall, Englewood Cliffs, N.J., 1988.

Churchill, R.V., *Operational Mathematics*, McGraw-Hill, New York, 1972.

Cochran, J.A., *The Analysis of Linear Integral Equations*, McGraw-Hill, New York, 1972.

Coddington, E.A., and N. Levinson, *Theory of Ordinary Differential Equations*, McGraw-Hill, New York, 1964.

Collatz, L., *Functional Analysis and Numerical Methods*, Academic Press, New York, 1966.

Corduneanu, C., *Principles of Differential and Integral Equations*, Allyn and Bacon, Boston, 1971.

Courant, R., and D. Hilbert, *Methods of Mathematical Physics*, Vol. 1, Chapter 3, Interscience, New York, 1953.

Davis, P.J., *Interpolation and Approximation*, 2nd ed. Blaisdell, New York, 1965.

deHoog, F.R., Review of Fredholm Integral Equations of the First Kind, in *The Application and Numerical Solution of Integral Equations*, Andersen, R.S., et. al., eds. Sijthoff and Noordhoff, the Netherlands, 1980.

Delves, L.M., and J.L., Mohammed, *Computational Methods for Integral Equations*, Cambridge University Press, London, 1988.

Ditkin, V.A., and A.P. Prudnikov, *Integral Transforms and Operational Calculus*, Pergamon Press, Elmsford, N.Y., 1965.

Erdelyi, A., W. Magnus, F. Oberhettinger, and F.W. Tricomi, *Tables of Integral Transforms*, Vol. 1, McGraw-Hill, New York, 1954.

Golberg, A.M., *Solution Methods for Integral Equations: Theory and Applications*, Plenum Press, New York, 1979.

Green, C.D., *Integral Equation Methods*, Barnes & Noble, New York, 1969.

Hildebrand, F.B., Integral Equations, Chapter 3 in *Methods of Applied Mathematics*, 2nd ed., Prentice-Hall, Englewood Cliffs, N.J., 1965.

Hochstadt, H., *Integral Equations*, Wiley, New York, 1973.

Jerri, A.J., *Introduction to Integral Equations with Applications*, Marcel-Dekker Inc., New York, 1985. (The first edition of the present book.)

Jerri, A.J., Elements and Applications of Integral Equations, UMAP J., Vol. 7, pp. 45–80, 1986 (Also UMAP Module #609, 1982.)

Jerri, A.J., *Student's Solution Manual to Accompany "Introduction to Integral Equations with Applications – Second Edition, Wiley & Sons", with Additional Solved Problems*, Sampling Publishing, 1999 (see the end of the preface for more information).

Jerri, A.J., *Integral and Discrete Transforms with Applications and Error Analysis*, Marcel Dekker Inc., New York, 1992.

Jerri, A.J., *The Gibbs Phenomenon in Fourier Analysis, Splines and Wavelet Approximations*, Kluwer Academic Publishers, Boston, 1998.

Jerri, A.J., A Recent Modification of Iterative Methods for Solving Nonlinear Problems. In *the Mathematical Heritage of C.F. Gauss* (G.M. Rassias, ed.), an invited paper, pp. 379–404. World Scientific Publishing, Singapore, 1991.

Jerri, A.J. and R.L. Herman, The solution of Poisson-Boltzmann equation between two spheres — Modified iterative methods, J. Sci. Comp., 11, pp. 127–153, 1996.

Jerri, A.J., R.L. Herman, and R.H. Weiland, The modified iterative method for nonlinear concentration in cylindrical and spherical pellets, J. Chem. Eng. Commun., 52, pp. 173–193, 1987.

Kanwal, R.P., *Linear Integral Equations, Theory and Technique*, Academic Press, New York, 1971.

Kanwal, R.P., *Linear Integral Equations*, Second Edition, Birkhäuser, Boston, 1997.

Keyfitz, N., Population Waves, Chapter 1 in *Population Dynamics* (T.N.E. Grenville, ed.), Academic Press, New York, 1972.

Kondo, J., *Integral Equations*, Clarendon Press, Oxford, 1991.

Kress, R., *Linear Integral Equations*, Springer-Verlag, New York, 1989.

Lonseth, A., *Sources and Applications of Integral Equations*, SIAM Review, **19**, pp. 241–278, 1977.

Miller, R.K., *Nonlinear Volterra Integral Equations*, W.A. Benjamin, Menlo Park, CA, 1971.

Noble, B., *Applications of Undergraduate Mathematics in Engineering*, Macmillan, New York, 1967.

Papoulis, A., *Systems and Transforms with Applications in Optics*, McGraw-Hill, New York, 1968.

Petrovskii, I.G., *Integral Equations and Their Applications*, Vol. 1, Pergamon Press, Oxford, 1960.

Pogorzelski, W., *Integral Equations and Their Applications*, Macmillan (Pergamon), New York, 1966.

Porter, D. and D.G. Stirling, *Integral Equations - A Practical Treatment, from Spectral Theory to Applications*, Cambridge University Press, Cambridge, 1991.

Rashevsky, N., *Mathematical Biophysics: Physico-Mathematical Foundations of Biology*, Vol. 1, Dover, New York, 1960.

Roberts, G.E., and H. Kaufman, *Tables of Laplace Transform*, W.B. Saunders, Philadelphia, 1966.

Sneddon, I.N., *The Use of Integral Transforms*, McGraw-Hill, New York, 1972.

Srivastava, H.M., and R.G. Buschman, *Convolution Integral Equations*, Wiley, New York, 1977.

Stakgold, I., *Green's Functions and Boundary Value Problems*, Wiley-Interscience, New York, 1979.

Tricomi, F.G., *Integral Equations*, Dover, New York, 1985.

Volterra, V., *Theory of Functionals and of Integral and Integro-Differential Equations*, Dover, New York, 1959.

Weinberger, H.E., *A First Course in Partial Differential Equations*, Blaisdell, New York, 1965.

Widder, D.V., *The Laplace Transform*, Princeton University Press, Princeton, 1946.

Wing, G.M., *A Primer on Integral Equations of the First Kind - The Problem of Deconvolution and Unfolding*, SIAM, Philadelphia, 1991.

Wolf, K.B., *Integral Transforms in Science and Engineering*, Plenum, New York, 1978.

Index